Illustrated Encyclopedia of
Applied and
Engineering Physics

Volume I
A–G

Illustrated Encyclopedia of
Applied and Engineering Physics

Volume I
A–G

Robert Splinter, PhD

CRC Press
Taylor & Francis Group
Boca Raton London New York

CRC Press is an imprint of the
Taylor & Francis Group, an **informa** business

CRC Press
Taylor & Francis Group
6000 Broken Sound Parkway NW, Suite 300
Boca Raton, FL 33487-2742

© 2017 by Taylor & Francis Group, LLC
CRC Press is an imprint of Taylor & Francis Group, an Informa business

No claim to original U.S. Government works

Printed and bound in India by Replika Press Pvt. Ltd.

Printed on acid-free paper
Version Date: 20160831

International Standard Book Number-13: 978-1-4987-4078-4 (Hardback)

Library of Congress Cataloging-in-Publication Data

Names: Splinter, Robert, author.
Title: Illustrated encyclopedia of applied and engineering physics / Robert Splinter.
Description: Boca Raton, FL : CRC Press, Taylor & Francis Group, [2016] | "2016 | Includes bibliographical references and index.
Identifiers: LCCN 2015040711| ISBN 9781498740784 (alk. paper : v. 1) | ISBN 1498740782 (alk. paper : v. 1) | ISBN 9781498740821 (alk. paper : v. 2) | ISBN 9781498740838 (alk. paper : v. 3) | ISBN 1498740839 (alk. paper : v. 3)
Subjects: LCSH: Physics--Dictionaries.
Classification: LCC QC5 .S65 2016 | DDC 530.03--dc23
LC record available at http://lccn.loc.gov/2015040711

Visit the Taylor & Francis Web site at
http://www.taylorandfrancis.com

and the CRC Press Web site at
http://www.crcpress.com

Contents

Preface

The purpose of this *encyclopedia* is to provide a single, concise reference that contains terms and expressions used in the study, practice, and applications of physical sciences. The reader will be able to quickly identify critical information about professional jargon, important people, and events. This encyclopedia gives self-contained definitions with essentials regarding the technical terms and their usages and information about important people in the following areas of physics:

- Acoustics
- Astronomy/astrophysics
- Atomic physics
- Biomedical physics
- Chemical physics
- Computational physics
- Condensed matter
- Dynamics
- Electromagnetism
- Electronics
- Energy
- Engineering
- Fluid dynamics
- General
- Geophysics
- High-energy physics
- Imaging
- Instrumentation
- Materials sciences
- Mechanics
- Meteorology
- Nanotechnology
- Nuclear physics
- Optics
- Quantum physics
- Relativistic physics
- Rheology
- Sensing
- Signal processing
- Solid-state physics
- Theoretical physics
- Thermodynamics
- Ultrafast phenomena

This reference differs from the standard dictionaries in its inclusion of numerous illustrations, including photographs, micrographs, diagrams, graphs, and tables, which support the textual definitions and draw the reader into the explanation to enhance didactic value. Together, these over 2500 entries will educate the reader about the current practice of physics and its applications in biomedicine, materials sciences, chemical engineering, electrical engineering, mechanical engineering, geology, astronomy, meteorology, and energy.

It is envisioned that novices and trainees, in addition to seasoned professionals, will find this resource useful, both for sustained reading and for taking a dip into the topics periodically. The contents are also designed to help the professionals who are new to a work environment and recently enrolled students who need to become more familiar with terminology and nuances relevant to certain research and applications. Moreover, it will assist in understanding the primary literature as well as technical reports and proposals. Finally, any student from the high school to graduate levels should be able to benefit from the broad and applied emphasis of this concise encyclopedia, which may support undergraduate courses in applied sciences for nonscience majors.

AUTHOR

Robert Splinter MSc PhD—University of North Carolina at Charlotte, North Carolina

MATLAB® is a registered trademark of The MathWorks, Inc. For product information, please contact:

The MathWorks, Inc.
3 Apple Hill Drive
Natick, MA 01760-2098 USA
Tel: 508-647-7000
Fax: 508-647-7001
E-mail: info@mathworks.com
Web: www.mathworks.com

Author

Robert Splinter, PhD, obtained his master of science degree in applied physics from the Eindhoven University of Technology, Eindhoven, the Netherlands, and his PhD from the VU University of Amsterdam. Dr. Splinter built his career as a scientist and technology manager in biomedical engineering. His work is dedicated to resolving issues in device development with a particular focus on medicine and biology through the development of novel diagnostic techniques and treatment methods using all multidisciplinary aspects of engineering and applied physics.

He cofounded several companies in biomedical engineering and worked for several established metrology and medical device companies. In addition, Dr. Splinter worked in clinical settings, prototyping, and validating devices using the full practical and theoretical knowledge of physics, engineering, electrical engineering, chemistry, and biology. He is an associate professor (Adj.) in the Department of Physics at the University of North Carolina at Charlotte.

Abbe, Ernst Karl (1840–1905)

[general, imaging, optics] A physicist from Germany (Prussian Empire) who worked for CARL ZEISS (1816–1888) in microscopy development, providing systematic advancement in LENS design and quality of the IMAGE formation based on his mathematical formulation of light REFRACTION. Additional efforts in telescopic and PHOTOGRAPHY equipment made considerable qualitative improvements. Ernst Abbe, as the heir of the will of Carl Zeiss, was placed in charge of the factory upon Zeiss' death in 1888. Abbe was also the cofounder of the GLASS factory Schott in Jena, Germany. Abbe formally introduced the concept of NUMERICAL APERTURE (see Figure A.1).

Figure A.1 Ernst Karl Abbe (1840–1905), a scientist from Germany. (Courtesy of Emil Tesch, taken around 1905.)

Abbe condition

[general, instrumentation, optics] (syn.: resolution) {use: imaging} The ability to distinguish two neighboring points, the RESOLVING POWER between these two points based on the angular resolution of the instrument (EYE, MICROSCOPE). This concept was independently described by both the German physicist ERNST KARL ABBE (1840–1905), while working for CARL ZEISS (1816–1888) in 1878, and the English physicist JOHN WILLIAM STRUT, THE 3RD BARON RAYLEIGH (1842–1919). When two objects are projecting their images through the same APERTURE, each object will generate its own diffraction pattern: AIRY DISK. The Airy disk of object 1: image A will overlap with the Airy disk of object 2: image B. Under the condition that the first maximum from the center of image A is adjacent to the first maximum of image B, and not overlapping, the two IMAGES can be separated or resolved. The RESOLUTION is a direct function of the WAVELENGTH λ of the

ELECTROMAGNETIC RADIATION used for imaging. The RAYLEIGH CRITERION for a circular aperture (e.g., LENS) will require corrections for the separation between the edges, and results in the fact that the ANGLE ω that can separate two objects in respect to the aperture of the viewing device (pupil, lens, and aperture, as in a CAMERA OBSCURA) needs to satisfy the following angular definition that defines the two points that can be resolved: $\omega = \lambda/n\sin(\omega) \equiv 1.22\lambda/D$, in AIR at 550 nm for HUMAN EYE, where D is the diameter of the opening or the diameter of the aperture and ω is the half-angle of the cone of rays emitted from the opening and $n\sin(\omega) = \text{NA}$ is the NUMERICAL APERTURE (NA). A bird of prey has a very high resolution due to the fact that the eye only has a relatively small angle of VISION resulting from a densely packed RETINAL RODS configuration and a large pupil diameter, approximately $\omega = 0.18$ arcmin compared to the FOVEA of the human eye with both CONES and RODS: $\omega = 0.30$ arcmin. In addition, some birds of prey have a double foveae that allows for simultaneous high-resolution (this is in particular called the *convexiclivate fovea*) and low-resolution vision (*see* REFRACTION; *also see* RAYLEIGH CRITERION) (see Figure A.2).

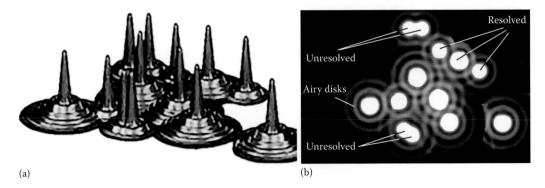

(a)   (b)

Figure A.2 Abbe criterion: (a) Airy disks created by diffraction originating from the edges of an aperture for a number of sources, allowing only sources that are separated by the Abbe criterion to be distinguished separately and (b) diffraction pattern with minima and maxima.

Abel, John Jacob (1857–1938)

[biomedical, chemical] A professor of pharmacology from the United States who invented and prototyped the first DIALYSIS machine in 1912; the process was then called vividiffusion and was intended to prepare the treatment of patients with kidney disease. The first human applications were by GEORG HAAS (1886–1971) in 1924 in Germany. Abel also discovered the amino acids as constituents of BLOOD (see Figure A.3).

Figure A.3 John Jacob Abel (1857–1938) from the United States. (Courtesy of Doris Ulmann.)

Aberration, chromatic

[optics] A LENS (i.e., TRANSPARENT shape or curved MIRROR) may have different focal points for different WAVELENGTHS. The INDEX OF REFRACTION of most transparent solids is a function of the wavelength; hence the REFRACTION will change across the SPECTRUM. A PRISM produces a RAINBOW due to the fact that the optical material (crown glass, flint glass, plastic, etc.) refracts at each surface with an ANGLE specific for the incident wavelength. Transparent materials under stress also generate density variations resulting in a specific localized index of refraction, which in turn results in stress-induced COLOR patterns under WHITE LIGHT illumination, allowing this phenomenon to be used for quality control purposes (see Figure A.4).

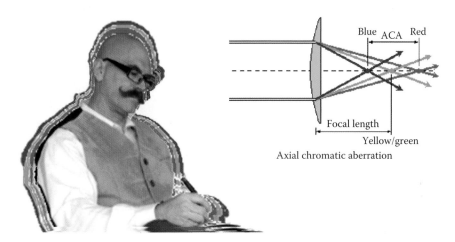

Figure A.4 Graphic illustration of chromatic aberration.

Aberration, spherical

[optics] Linear LIGHT ray distortion resulting from imperfections in the LENS or MIRROR design or imperfections in the lens material (mirror coating) with an inhomogeneous INDEX OF REFRACTION across the volume of the lens. The edges of a lens (i.e., curved mirror or transparent shape) may focus at a different location than the rays passing close to the OPTICAL AXIS (see Figure A.5).

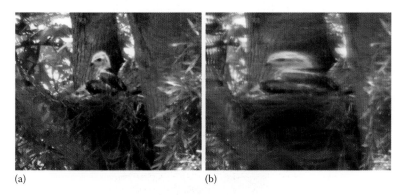

(a)   (b)

Figure A.5 Graphic illustration of spherical aberration: (a) undistorted picture of a buzzard chick and (b) distorted view obtained with badly shaped and uncorrected lenses illustrating the spherical aberration resulting from inferior optical instruments.

A-bomb

[atomic, nuclear] *See* ATOMIC BOMB.

Absolute energy

[thermodynamics] Internal ENERGY under conditions of SPECIAL RELATIVITY, such as nuclear mechanics and the STATE OF A SYSTEM. When the state of a system A changes from one configuration ("1") to another ("2"), the energy provided as WORK as a function of position z is initially written as: $W_{12}^A = Mg(z_2 - z_1)$. By including the special relativity conditions when considering the nuclear interactions, this becomes: $W_{12}^A = Mc^2 \left(e^{gz_2/c^2} - e^{gz_1/c^2} \right)$, where c is the speed of light in VACUUM and M is the MASS at $z = 0$. The energy E of the system A with movement at velocity v and MOMENTUM "$\vec{p}$" is compliant with $M\sqrt{(E^2 - |\vec{p}|^2 c^2)} = mc^2$, where m is the mass of the system. This translates into an energy equation for the two-state system, assuming one GROUND STATE, with momentum $\vec{p} = 0$ by definition, and both states are otherwise equal in energy. The energy for the two-state system can now be written as $E_1 = E_2 = m_2 c^2 \sqrt{1 + (|\boldsymbol{p}_2|^2/m_2^2 c^2)}$. For a system (PARTICLE, MOLECULE, ATOM, or [electric/magnetic/EM] FIELD without external force), the ground-state mass can be derived from MASS-SPECTROSCOPY measurements and is defined as m_g^{free}. Expanding the process for a multitude of states all in ground state ($?_g$), this denotes the total energy of the free system F, with n states and r types of ELEMENTS or constituents as: $E_g^F = \sum_{i=1}^r n_i \left(m_g^{\text{free}} \right)_i c^2$, where $\left(m_g^{\text{free}} \right)_i$ is the ground-state mass for the ith free constituent under special relativistic conditions. Now the absolute energy can be derived with the help of the following equation: $E_0^A = \sum_{i=1}^r n_i \left(m_g^{\text{free}} \right)_i c^2 + V_0 \left(n, \beta, \text{internal forces} \right)$, describing the absolute energy values of each of the respective states, and $V_0 \left(n, \beta, \text{internal forces} \right)$ is the configuration of external ($\beta_1, \beta_2, \beta_3, \ldots, \beta_y,$) and internal forces. These forces can be turned on or off based on the relativistic conditions.

Absolute entropy

[thermodynamics] $S(A) = \int_0^A dQ/T$ (*see* ENTROPY), which is undefined for the ABSOLUTE ZERO temperature. The inclusion of a constant (ENTROPY CONSTANT OF A GAS) compensates to define the entropy of every system at absolute zero to equate to zero. Stated in its original form by WALTHER HERMANN NERNST, Nernst's "heat theorem" formulates that the entropy equation is applied only to condensed systems; however, it was later proven to hold true for gaseous systems as well.

Absolute gravitational acceleration (g_0)

[astrophysics, general] GRAVITATIONAL ACCELERATION as the result of the attraction between two bodies resulting from the mass of one primary body (M) calculated at a point with respect to the square of radius from the core (R), with no other forces acting (e.g., no rotation: CENTRIPETAL FORCE): $g_0 = GM/R^2$ (see Figure A.6).

Figure A.6 Illustration how gravitational acceleration will result in increasing velocity on a downhill run for a roller-coaster ride.

Absolute pressure (*P*)

[fluid dynamics, general] PRESSURE measured against absolute VACUUM, in contrast to ATMOSPHERIC PRESSURE, units (TORR) (see Figure A.7).

Figure A.7 A pressure gauge providing absolute pressure value. The mirror annulus allows for accurate determination when viewing the position of the needle covering its own image.

Absolute simultaneity

[computational, general] A virtual concept proposed under special relativistic conditions describing the apparent correlation between multiple inertial systems outlining a frame of reference. Absolute simultaneity entails that the observations made by observers of two simultaneous event observed in one reference frame midway in space between the two events would also be experienced by observers in another inertial system as simultaneous. This, in fact, is not necessarily true, primarily because of the specific definition of the boundary conditions of each reference frame in space and time. The discrepancy results from the relativistic concept of TIME DILATION, where time is relative to the observer's inertial frame of reference and consequently DISTANCE is subjected to the boundary conditions of the reference frame of the observer (see Figure A.8).

Figure A.8 The principle of simultaneity. Two events that occur simultaneously in two separate locations may not be observed as equivalent between two observers in separate frame of reference, one reference frame traveling close to the speed of light.

Absolute streamline density ($\rho(u)$)

[imaging] Concept in imaging derived from seed point analysis revealing "fiber orientation" in a POLYMER or biological medium composed of a seemingly fibrous structure.

Absolute streamlines

[fluid dynamics] Concept used in turbines and generators indicating the direction and FLOW velocity of a FLUID in MOTION, the vector length or number of lines indicates the velocity. The absolute streamline points in the direction of LIFT for a turbine blade. The BERNOULLI'S EQUATION for the absolute streamline in a TURBINE is given by $\left(P_2 - P_1\right)/\gamma = \left[\left(V_1^2 - V_2^2\right)/2g\right] - \left[u\left(V_{\theta 1} - V_{\theta 2}\right)/g\right]$, where $u\left(V_{\theta 1} - V_{\theta 2}\right)/g$ is Euler's turbine equation (see Figure A.9).

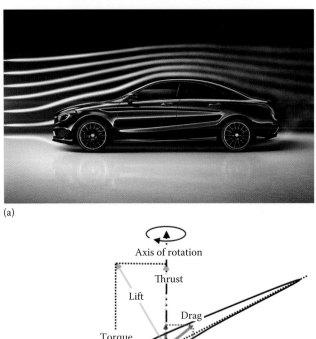

(a)

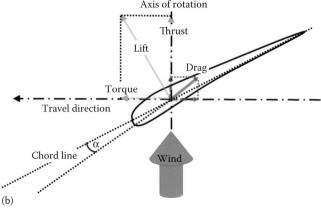

(b)

Figure A.9 Streamlines illustrate the local direction and magnitude of fluid flow in either a tube or around an object. The density of the streamlines provides a visual reference to the magnitude (as vector indication): (a) visual display of streamlines generated by water-vapor jets at fixed locations at a distance from the object subjected to flow. The flow around the Mercedes CLA automobile illustrates the lowest coefficient of flow friction for a production automobile to this date ($C_d = 0.23$). (Courtesy of Daimler AG.) The stream lines of flow provide the information that can be used to calculate lift, thrust drag, and torque resulting from the motion of the aerofoil of a wind generator blade, respectively, a boat propeller, rotating in a viscous fluid. The reported first use of wind-powered water well pumps dates back to the Persian Empire, 900 AD. (b) Diagram of forces acting on a horizontal axis wind turbine (HAWT); torque and thrust on a wind-generator blade section under the influence of streamlines. The streamlines are the vector resultants from the wind direction and magnitude and the respective travel direction of the propeller. The streamlines for a rotating propeller are continually varying and are also modified by the flow-induced vortices. *(Continued)*

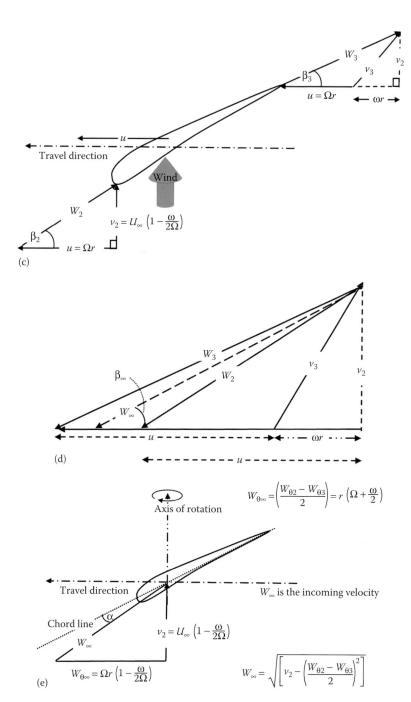

Figure A.9 (Continued) Streamlines illustrate the local direction and magnitude of fluid flow in either a tube or around an object. The density of the streamlines provides a visual reference to the magnitude (as vector indication): (c) diagram of velocity vectors with respect to the streamlines on the inlet and outlet side of a propeller blade. The angular velocity of the propeller is indicated by Ω with respect to the radial distance from the axis r, with the relative flow velocity as a function of angle ω. (d) Trigonometry of the velocity vectors outlined for the streamlines in (b) and (e) derived effective incoming fluid stream velocity on the rotating blade.

Absolute temperature

[thermodynamics] Temperature based on the KELVIN SCALE with the lower limit set as the point that cannot exchange any heat. Because the CARNOT CYCLE is not ideal and violates the SECOND LAW OF THERMODYNAMICS, not all heat dissipation is converted into work. The theoretical implication is that the ABSOLUTE ZERO can never be reached for a functional volume of medium.

Absolute temperature (*T*)

[general, thermodynamics] The absolute temperature is defined as the inverse of the partial derivative of the ENTROPY (*S*) of a system with respect to the ENERGY (*E*) of the system as a function of state defined by the amount of the constituent *n*—the system may consist of *r* constituents (n_i, $i = 1, 2, \ldots, r$), and environmental parameters: β^* common to the many assorted states of the system, which is composed of, for instance, volume, pressure, and more, and is formulated as $T = \left[\partial S(E) / \partial E \right]_{n, \beta^*}^{-1} = \left(\partial E / \partial S \right)_{n, \beta^*}$. The temperature is the result of kinetic energy from the MOLECULAR MOTION, which can be split into four categories: vibrational, scissor action, rotation, and potential energy. The temperature (*T*) results from solving the equation $(1/2)mv_{av} = (3/2)kT$, where v_{av} is the average VELOCITY of the atoms and molecules in the medium with mass *m*, and *k* is the BOLTZMANN CONSTANT. Absolute temperature is expressed in KELVIN (K).

Absolute temperature scale

[general, thermodynamics] Measure of temperature proposed by WILLIAM THOMSON (1824–1907), *also* LORD RAYLEIGH, in 1848: the scale is based on the thermodynamic concepts of the CARNOT ENGINE, where the high temperature (T_H) ENERGY intake (Q_H) compared to the lower temperature (H_L) exhaust energy (Q_L) is correlated as $Q_L / Q_H = T_L / T_H$. The Carnot engine is assumed to operate from the upper temperature of boiling water and the lower temperature at FREEZING water, the scale is based on subdividing the CARNOT CYCLE in 100 subunits, each performing an equal amount of work (i.e., equal area) with the isotherms separating the cycles defined as 1 K apart. The scale is in perfect agreement with that of the CONSTANT-VOLUME GAS THERMOMETER, primarily because a PERFECT GAS will obey the Carnot equation. The only flaw in the reasoning is that not all the energy difference in the Carnot cycle is converted into work (ΔW) (see Figure A.10).

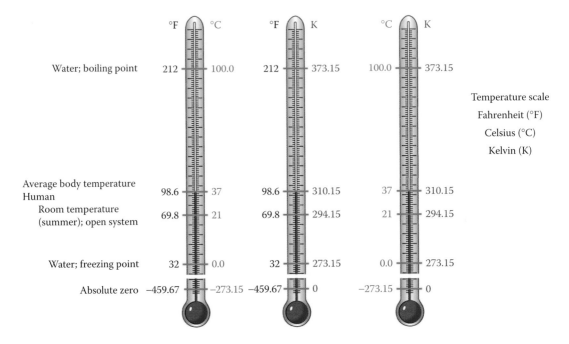

Figure A.10 Thermometers representing the different scales and the reference to absolute zero.

Absolute viscosity (μ)

[biomedical, fluid dynamics] The DYNAMIC VISCOSITY of a LIQUID, in contrast to the KINEMATIC VISCOSITY. It is a measure of internal RESISTANCE of a liquid. The absolute VISCOSITY is determined from the tangential force per unit area (SHEAR STRESS: τ) required to maintain a constant VELOCITY in the x-direction when applied on a plane that is at a fixed DISTANCE (in the y-direction) from a stationary parallel surface separated by the liquid of interest. Using a NEWTONIAN LIQUID, the shear stress in the liquid as a function of distance to the plane under LAMINAR FLOW conditions is given as $\tau = \mu\left(dx/dy\right)$ (Poise: p or Ns/m^2 or Pa*s or kg/ms) (see Figure A.11).

Figure A.11 Viscous friction from a dense fluid. The kinematic viscosity endured by this bullet fired into water generates a shear stress at the interface creating a turbulent boundary layer. The shear stress additionally slows down the bullet, so that it may be captured with moderate high-speed photography.

Absolute zero

[general] The energetic point where all MOLECULAR MOTION has ceased and the temperature as defined has reduced to zero (i.e., entropy = 0). This temperature point equates to −273.15°C or −459.67°F. The principle uses the gas thermometer (based on GALILEO GALILEI's (1564–1642) THERMOSCOPE concept, c. 1592) and a statement by JOHANN HEINRICH LAMBERT (1728–1777) (in 1779) that the lowest temperature should equate to the absence of the exchange of heat; as the temperature approaches zero theoretically, the volume of the GAS in the THERMOMETER will also approach zero. This hypothetical limit can never be reached universally due to conflict with the SECOND LAW OF THERMODYNAMICS; however, this condition could occur in the center of a larger medium while excluding direct contact with the containment walls.

Absorbed dose (D_c)

[biomedical, energy] In RADIATION THERAPY, this indicates the potential for the biological effect caused by radiation delivered and absorbed at locations $\vec{r}$ in the direction $\vec{s}$ with respect to the surface of an absorbing volume. The absorbed dose (D_c) is a direct function of the fluence $\vec{L}\left(\vec{r},\vec{s}\right)$ of photons with ENERGY $\left(E\right)$ with respect to the absorbed energy per unit MASS (m) expressed by the MASS ENERGY absorption coefficient (μ_{en}/ρ), also referred to as *mass–energy attenuation*: μ_{en}/ρ, $D_c = \left(\Delta E_{absorbed}/m\right) = \left[\vec{L}\left(r,\theta,\phi\right)\cdot\vec{r}\right]E\left(\mu_{en}/\rho\right)$. The absorbed dose is different from the KERMA (also "collision Kerma," K_c) because the latter describes the energy transferred between the radiation applied and the medium the radiation is migrating through, not the absorbed radiation. Absorbed dose is expressed in Gray: $[Gy] = 1$ J/kg $= 100$ rad. Absorbed dose is different from "radiation exposure" (see Figure A.12).

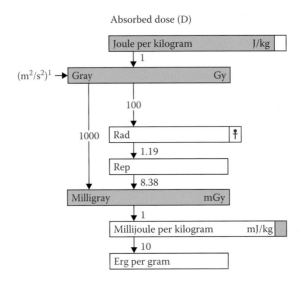

Absorbed dose (D)

Figure A.12 Absorbed radiation dose equivalent table. Does not provide indication of magnitude for biological or chemical effect.

Absorptance (μ*)

[atomic, chemical] Ratio of the "absorbed" RADIANT ENERGY ($\Psi_0 - \Psi(z)$), after traversing a thickness (d) in an arbitrary direction within a medium, to the incident radiant energy (Ψ_0) expressed as $[\Psi_0 - \Psi(z)]/\Psi_0 = \mu^*$, in contrast to ABSORPTION, which yields the ABSORPTION COEFFICIENT (μ_a). For a BLACK BODY, the ABSORPTANCE equals unity ($\mu^* = 1$).

Absorption

[fluid dynamics, general, nuclear, optics] The conversion of ENERGY from an external source during interaction on an atomic level. The electronic interaction can be described by the LORENTZIAN ELECTRON OSCILLATOR MODEL. Pertaining to the loss of one or more modalities (e.g., mass), attributes (e.g., gaseous absorption = chemical), or phenomena (e.g., ELECTROMAGNETIC RADIATION). (1) Conversion of one form of energy (primarily ELECTROMAGNETIC energy) into another form of energy when interacting with a medium primarily THERMAL ENERGY (i.e., DISSIPATIVE ABSORPTION); (2) incorporation of one substance/constituent into another source of MATTER; and (3) diminished AMPLITUDE as a result of RESISTANCE, for example, oscillating string (GUITAR, PIANO, etc.).

Absorption, by solids

[acoustics, atomic, mechanics, optics, solid-state] When waves or chemicals pass through a solid they are reduced in MAGNITUDE, either by formation of bonds (chemicals) or by dissipation into THERMAL ENERGY or other forms of energy, such as ELECTRICITY. An acoustic WAVE is made audible when absorbed by solid, where ELECTROMAGNETIC RADIATION can produce electricity (i.e., PHOTOELECTRIC EFFECT) or temperature (i.e., VIBRATION).

Absorption, of gamma rays

[atomic] High-energy PHOTON absorption in the nuclear core of an ATOM.

Absorption, of radiation, heat

[atomic] Conversion of ELECTROMAGNETIC ENERGY to THERMAL ENERGY. The dispersed absorption (e.g., BEER–LAMBERT LAW) creates thermal gradient resulting in the FLOW of energy from higher to lower temperature (i.e., heat).

Absorption, of X-rays

[atomic] High-energy PHOTON absorption in the atomic electron transitions (as described in the BOHR ATOM model).

Absorption, optical by gases

[atomic] Molecular or atomic absorption of ELECTROMAGNETIC RADIATION resulting in element-specific black lines in the transmitted spectrum that can be used for identification of the constituents of the GAS. Generally, the pressure of the gas has to be relatively low to be able to accurately identify the characteristic lines (see Figure A.13).

Figure A.13 Spectral profile of the absorption of sunlight in transmission through atmospheric nitrogen (~70% of total gas volume), in comparison to the spectral emission of the heated nitrogen gas (bottom right).

Absorption coefficient ($\propto$)

[general, nuclear, optics, solid-state] The proportional DECAY equivalence of a scalar quantity (ΔN) with respect to the original quantity (N_0) and the path length traveled (z) in a medium that reduces the original quantity due to conversion of ENERGY or state $\Delta N = -\propto N_0 z$, with SOLUTION $N = N_0 e^{-\alpha z}$, which is known as the BOUGUER LAW or Beer–Lambert law (*see* EINSTEIN COEFFICIENT *and* ATTENUATION COEFFICIENT).

Absorption coefficient, linear

[atomic] *See* ABSORPTION COEFFICIENT. In one direction, as expressed by α in: $-dI/I = \alpha * dz$.

Absorption coefficient, mass (μ/ρ)

[atomic, nuclear, solid-state] Attenuation (μ) of ELECTROMAGNETIC RADIATION due to COMPTON EFFECT, PHOTOELECTRIC EFFECT, and PAIR PRODUCTION per unit DENSITY (ρ) of atoms in an absorbing volume of material. This attenuation process in the z-direction still obeys the BEER–LAMBERT LAW, and is written as $I_z = I_o e^{-(\mu/\rho)\rho z} = I_o e^{-\mu_m m_a}$.

Absorption coefficient, X-ray

[atomic, nuclear] The element-specific absorption is directly related to the orbital ENERGY levels of electrons and nucleons. In general, lower energy X-RAY photons will have a higher PROBABILITY of interaction on an atomic level. The values for virtually all ELEMENTS and various chemical combination are tabulated in dedicated reference books. Because the absorption of X-ray RADIATION is an energy phenomenon, the energy of the radiation needs to be considered when calculating the radiant intensity: $I = \int_{I_{v0}}^{I_{v\infty}} I_o(E) e^{-\mu(E)e} dE$. Composite materials require that the respective contributions to the absorption of the individual chemical elements needs to be taken into consideration with the respective impact, expressed as $I = \int_{I_{v0}}^{I_{v\infty}} I_o(E) e^{-\sum_i \mu_i(E)e_i} dE$.

Absorption cross sections for neutrons

[atomic, nuclear] σ, expressed as the PROBABILITY of a PARTICLE hitting a nuclear target with various light-reaction products as a result of the interaction, expressed as $(I/A) = n_a v_a$; $\sigma = \sum_j \left(N_j / n_a v_a n A \Delta z \right)$ (see Figure A.14).

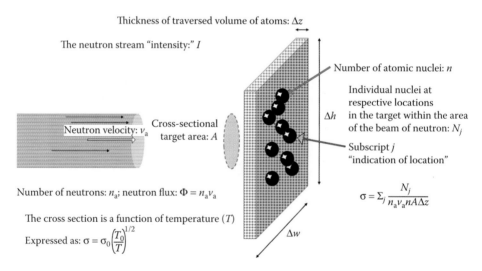

Figure A.14 Graphic representation of collision interaction, representative of absorption cross section.

Absorption cross sections for photons

[optics, solid-state] σ_a, radiant ENERGY is incident on a PARTICLE (molecule, ATOM) and is in its entirety converted into heat by definition of absorption (neglecting the release of lower energy photons as a contribution in absorption). The ratio of the total incident radiant power (P_a) on the volume containing the particle to the incident radiant energy on the particle within a specific SOLID ANGLE $\left(\Psi(\Omega') \right)$, written as $\sigma_a(\Omega') = P_a / |\Psi(\Omega')|$. Generally the CROSS SECTION will be the average over a grouping within a volume (expressed in units [m^2]).

Absorption curve

[nuclear] Because of ENERGY interaction of specific mechanisms with a medium, the DECAY of the source energy or source material relies on the characteristic source energy or particulate size. For instance He^{2+}, monoenergetic electrons, beta rays, gamma rays, and visible light will have distinctly different decay patterns as a function of DISTANCE (see Figure A.15).

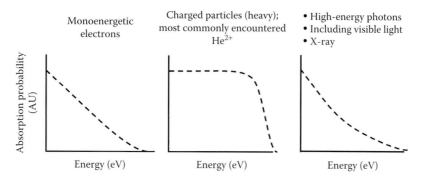

Figure A.15 Representations of the absorption curve energy spectrum for various interactions, based on particle and wave interaction.

Absorption edge

[nuclear] A ramp (not a discontinuity) in spectral performance of the ABSORPTION as a function of WAVELENGTH, primarily for semiconductor materials. Transition between the strong absorption of short wavelengths and the much weaker long-wavelength absorption, *also* "BAND-EDGE," because there is a significant step in the absorption, respectively, the REFLECTION pattern is a function of wavelength. The onset wavelength of the ramp is a direct function of the transition between VALENCE and CONDUCTION ENERGY bands. The absorption edge is partially dependent on the PHOTOELECTRIC EFFECT (see Figure A.16).

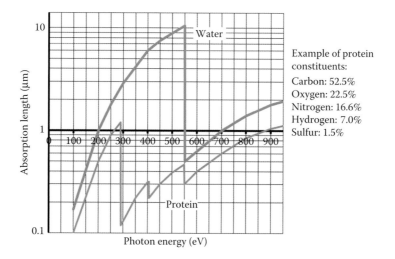

Figure A.16 Representation of the absorption edges at specific energy levels for photon interaction based on the available energy configuration of the molecules in the target specimen. The water absorption edge is representative of a specific vibrational resonance, whereas the protein is much more complex in its energy structure and associated translational, vibrational, and rotational resonance interactions.

Absorption edge for X-rays

[atomic, nuclear] Discontinuity in the ABSORPTION CURVE as a function of WAVELENGTH corresponding to a RESONANT absorption associated with a nuclear transition (as described energetically by the BOHR ATOMIC model and, for instance, the LYMAN SERIES) (see Figure A.17).

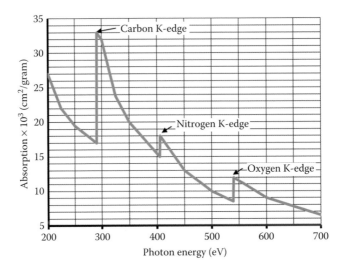

Figure A.17 Representative X-ray absorption spectrum for K radiation (electron orbit $n = 1$, the K-shell of the atom) for carbon, nitrogen, and oxygen atoms, respectively.

Absorption spectrum

[atomic, nuclear] Selective annihilation of specific WAVELENGTHS (i.e., SPECTRAL LINES) of ELECTROMAGNETIC RADIATION emitted from a BROADBAND source when passing through a medium (e.g., VAPOR, GAS, or transparent solid). The transmitted spectral profile displays a decreased MAGNITUDE compared to the incident profile of light at wavelengths that are conducive to absorption as a result of, for instance, the PHOTOELECTRIC EFFECT, conversion into HEAT, or PHOTON energy conversion to lower energy. The spectrum is typically a considered characteristic of the illuminated material (see Figure A.18).

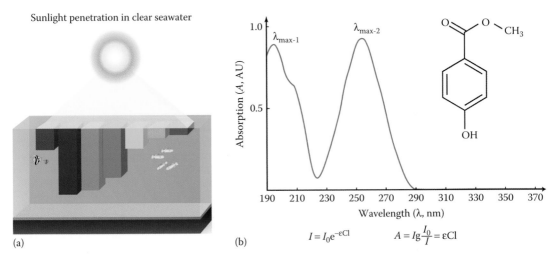

$$I = I_0 e^{-\varepsilon Cl} \qquad A = \lg\frac{I_0}{I} = \varepsilon Cl$$

Figure A.18 Artist impression of absorption spectrum for water (a), the transmission of long wavelength ultraviolet for water yields a design requirement for fiber-optics operating with excimer laser (e.g., 308 nm) to have high OH contents (b), compared to low OH requirement for optical instruments operating in the red and infrared.

Abundance of stable isotones

[nuclear] The collection of ELEMENTS has a much larger number of stable isotones than unstable ones. This is primarily based on the high NEUTRON SEPARATION ENERGY for nuclei with a MAGIC NUMBER neutron quantity in the NUCLEUS.

AC

[electronics, general] Alternating current (abr.). In contrast to a fixed POLARITY ELECTRICAL POTENTIAL driving a constant CURRENT (i.e., DC). Generally pertaining to a single rhythm OSCILLATION in the AMPLITUDE of the ELECTRIC CURRENT as well as the associated driving VOLTAGE (i.e., AC voltage). The ELECTRONS will FLOW alternatively forward and backward in a CLOSED CIRCUIT. The principle applies both to a CURRENT SOURCE and a VOLTAGE SOURCE. The delivery of alternating current is usually provided in "SINGLE-PHASE" two-wire mechanism. The delivery of multiple single-phase currents that are OUT-OF-PHASE with each other (originally devised by NIKOLA TESLA [1856–1943]) overcomes the periodic lag of power delivery. The ideal configuration of maximum efficacy for power consumption is three wires that are 120° out of PHASE (i.e., 3-PHASE AC) (see Figure A.19).

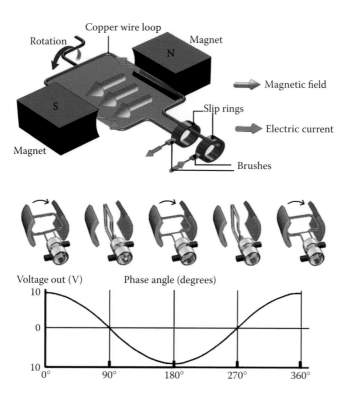

Figure A.19 Representation of the generation of alternating current resulting from a periodically increasing and decreasing captured magnetic field by an electric loop. Flux captured: dB/dt; B=MAGNETIC FIELD passing through an area outlined by wire-loop, i.e. MAGNETIC FLUX.

AC capacitive circuit

[electronics, general] Electric circuit containing at least one CAPACITOR, which is charged and discharged under the following conditions: $q = CV_m \sin \omega t$, where q is the charge on the capacitor, C is the capacitance, V_m is the driving voltage, t is the elapsed time, and ω is the position in the oscillatory cycle or angular frequency. (*Note*: Current through the capacitor $I = dq/dt$.) The delay factor produced by the charging/discharging of the capacitor brings the voltage across the capacitor out of PHASE (delayed) from the driving alternating current/voltage by one quarter cycle (i.e., $\pi/2$ rad or 90°) (see Figure A.20).

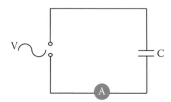

Figure A.20 Circuit diagram of a capacitor that is charged and discharged with the period of the alternating current applied by the sinusoidal fluctuating power supply.

AC circuits

[electronics, general] Electric circuits can contain any or all of the following components in PARALLEL and/or SERIES configurations: CAPACITOR (C), CURRENT SUPPLY, GENERATOR, INDUCTOR (L), RESISTOR (R), SUPERCONDUCTOR, VOLTAGE SUPPLY (BATTERY and device), and wire. Each component has its own typical frequency-dependent characteristics that influence the TIME-dependent ELECTRICAL CURRENT and VOLTAGE at any location in the circuit.

AC generator

[electronics, energy, general] Mechanism using a mechanical force to move an area (A), respectively, a MAGNETIC FIELD ($\vec{B}$) (which encompasses a magnetic FLUX: Φ_m through the normal projection of an area A) and induce a VOLTAGE (electromotive force voltage: ε_{emf}) that drives a CURRENT that alternates over TIME (t) with the direction of the magnetic field as described by the induced voltage or electromotive force $\varepsilon_{emf} = \Delta\Phi_m/\Delta t = \Delta\left(\vec{B} \odot \vec{A}\right)\big/\Delta t = l*\left(\vec{v} \otimes \vec{B}\right)$, whereas the magnetic field changes with velocity v and the field works on a circuit with length l, which can be the length of continuous loops in a coil. The current follows from OHM'S LAW: $I = V/R = \varepsilon_{emf}/R$, where R is the compound electric RESISTANCE of the closed circuit (see Figure A.21).

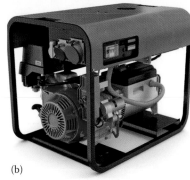

Figure A.21 Example of (a) an alternating generator, (b) a gasoline-powered generator, and (c) an electrical power generator.

AC induction motor

[electronics, general] *See* INDUCTION MOTOR.

Acceleration (*a*)

[general, geophysics, mechanics, nuclear] The incremental rate of change in velocity $\vec{v}$ of an object in direction (angular acceleration) or MAGNITUDE (linear acceleration) $(\vec{a} = d\overrightarrow{v(t)}/dt)$, under the influence of the sum of all external forces $(\Sigma F = ma)$. The direction of MOTION is taken as positive (m/s^2). For averaging the calculation of acceleration, average acceleration (a_{av}), greater time steps are used: $\vec{a} = \Delta\vec{v}/\Delta t$ (m/s^2) (see Figure A.22).

Figure A.22 Increase or decrease in linear acceleration. A change of direction and resulting decrease in velocity in the direction of prior flight pattern as well as new direction with increasing velocity in that direction as a result of acceleration for the Thunderbirds flight team.

Acceleration, angular (α)

[general, mechanics] The incremental rate of change of ORBITAL VELOCITY or ANGULAR SPEED (ω) expressed as $\alpha = d\omega/dt$ (rad/s^2).

Acceleration, centripetal (*a*$_c$)

[general, mechanics] (i.e., center-seeking) A rotating object is subjected to a continuous force that applies a change in direction irrespective of time. The rate of change in directional VELOCITY (vector velocity) provides the angular acceleration, which is directly linked to the CENTRIPETAL FORCE changing the direction. The acceleration tied to the angular MOTION is defined by the ANGULAR VELOCITY (ω) as $a = v^2/r = \omega^2 r$, where v is the tangential component of the motion of an object in a circular path. This acceleration is directed toward the center of the circular motion (m/s^2) (see Figure A.23).

Figure A.23 Force balance in centrifuge (a,b) ensuring that the center of mass of the baskets is in equilibrium, causing the baskets to tip (c). The centripetal acceleration provides the force applied by the hangers to the baskets from flying off into a tangential path. Because of the centripetal force, the mass distribution in the test tube mounted in the basket is redistributed with heavy media toward the bottom of the test-tube (farthest outward point, due to tangential force) and lighter media closer to the axis of rotation. Similarly, in (d) the skateboarder is held normal to the curved surface due to the centripetal force, which in this situation exceeds the gravitational force resulting from the angular acceleration exceeding the threshold value for balance. In the rollercoaster (e), the centripetal acceleration exerted by the seats of the passengers will ensure a "comfortable" seat while rushing through the loop upside-down. Meanwhile, the back-rest will provide the (normal) force on the passengers to make them move forward. A force diagram provides the resultant force on the passenger.

Acceleration, due to gravity (g)

[general, geophysics, mechanics] *See* GRAVITATIONAL ACCELERATION. The acceleration of a freely falling body in a VACUUM; 9.80665 m/s^2 at $45°$ LATITUDE and at sea level.

Acceleration, in harmonic motion

[general, mechanics] SIMPLE HARMONIC OSCILLATION is described by the sinusoidal MOTION pattern of an object, operating at a single FREQUENCY. This means that the motion periodically changes direction as well as the VELOCITY. Using the fact that the unidirectional acceleration is the derivative of the motion velocity the acceleration is defined as $a = -A\omega^2 \cos \omega t$ (m/s^2).

Acceleration, radial (a$_r$)

[general, mechanics] *See* CENTRIPETAL ACCELERATION.

Acceleration, relationship to force and mass

[general] *See* NEWTON'S SECOND LAW.

Acceleration, tangential (a$_t$)

[general, mechanics] The rate of change in the tangential VELOCITY (v_t) between the final velocity (v_{tf}) and initial velocity (v_{ti}) measured spanning a time frame Δt, of a PARTICLE performing a circular MOTION defined as $a_t = (v_f - v_i)/\Delta t$, under the uniform motion $a_t = 0$ (m/s^2).

Acceleration in cyclotron

[atomic] A system made of two short-length, large-radius half-tubes (split along the longitudinal direction of the tube; apposing "D" shapes), which are placed under the influence of a MAGNET, creating a MAGNETIC FIELD along the length of the axis of the tubular geometry (*see* CYCLOTRON). ACCELERATION OF CHARGED PARTICLES (=ION) relies on NEWTON'S THIRD LAW, AMPÉRE'S LAW, and influenced by HANS CHRISTIAN OERSTED (1777–1851) to yield the MAGNETIC FORCE (F_m) on a PARTICLE with CHARGE q as $F_m = q * (\vec{v} \otimes \vec{B})$, with the magnetic field $\vec{B}$ and the injected particle velocity "$\vec{v}$," providing the acceleration $a = Z_e * (\vec{v} \otimes \vec{B})/Au$, where Z_e is the VALENCE charge of the particle, A is the ATOMIC NUMBER of the ion, and u is the unified ATOMIC MASS, considering that effectively only one particle at a time is manipulated (see Figure A.24).

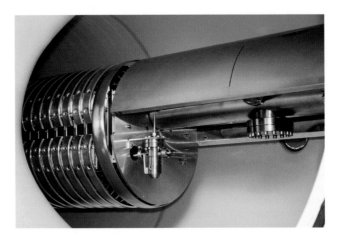

Figure A.24 Detail of particle accelerator, manipulating the path and velocity of ions by means of orchestrated magnetic field vectors in a confined environments.

Accelerator, particle

[atomic, nuclear] *See* PARTICLE ACCELERATOR.

Accelerator, Van de Graaff generator

[atomic] *See* VAN DER GRAAFF GENERATOR.

Acceptor impurity

[atomic] Atoms and molecules deviating from the norm or the GROUND STATE configuration such as charge carriers in SEMICONDUCTORS. For instance, in a crystalline configuration, an impurity ATOM may have one VALENCE electron that has the possibility of residing in CONTINUUM with the free electrons or getting bound to the atom in a preferential state.

Accommodation, of the human eye

[general] The ability to focus due to ADAPTIVE OPTICS in the LENS of the EYE. Muscular (i.e., CILIARY MUSCLE) ACTIVITY controls changes in the radius of curvature of the lens (CRYSTALLINE LENS) that determine the focal length based on the "LENS EQUATION," and is referred to as ACCOMMODATION. With the ciliary muscle totally relaxed, the eye is focused at infinity or defined as the "far point," which in practice means anything beyond 5 m DISTANCE. The total contracted ciliary muscle focuses the eye in the "NEAR-POINT," which allows correct and undeformed sharp focused VISION of an object placed at approximately 25 cm distance for the average human being. Aberrations in the vision achieved by the accommodation of the crystalline lens are primarily defined by two conditions: presbyopic (far-sighted), which is also classified as hypermetropia, and MYOPIA (near-sighted), respectively (see Figure A.25).

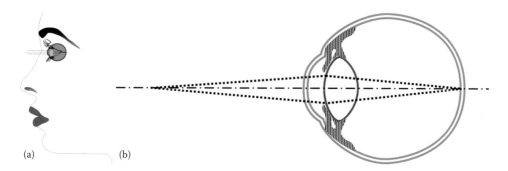

Figure A.25 (a) The manipulation of the focal length of the eye by means of the ciliary muscle (i.e., accommodation) and (b) relaxing for focusing an object at close proximity on the same retinal distance as a far-away object, allowing the lens to bulge-out. The focal length is a function of the radius of curvature of the curved surfaces enclosing the gel lens calculated with the "lens equation."

Achromatic lens

[optics] A composite LENS with two or more layers, each respective material of uniform index of REFRACTION that compensate each other (eliminating DISPERSION) for the inherent refractive CHROMATIC ABERRATION of the individual material properties.

Achromatic prism

[optics] A composite PRISM with two or more prisms, each of uniform index of REFRACTION that compensate each other (eliminating DISPERSION) for the inherent refractive CHROMATIC ABERRATION of the individual material properties, allowing the emitted light to maintain the incident COLOR temperature, only offset in optical path.

Acid

[biomedical, chemical] LIQUID with pH less than 7 (pH < 7), that is, a high concentration of hydrogen ions. Acids are corrosive media used to dissolve solids, such as metals and cement; for instance, hydrochloric acid is used to remove excess cement in construction (see Figure A.26).

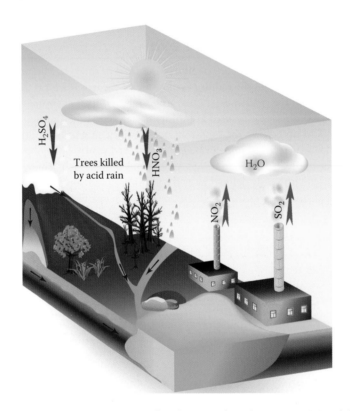

Figure A.26 Diagram of the formation and impact of acid rain. High acidity is corrosive and disintegrates biological life forms as well as inorganic media. One example of acid rain impact on the urban environment is the diminished structural integrity of concrete, visible in buildings and statues.

Acoustic impedance (Z)

[biomedical, computational, mechanics] It is a measure of how reactive the medium is to pressure wave disturbance. Acoustic impedance implies that the medium has both elastic and viscous properties. The acoustic impedance relates to ULTRASONIC IMAGING. The mismatch of acoustic impedance at an interface between two media will result in a wave propagation that obeys SNELL'S LAW in both REFRACTION and REFLECTION. The acoustic impedance is the ratio of the pressure wave MAGNITUDE (P) to velocity of a PARTICLE (v) in the medium and is written as $Z = P/v = \rho v$, where ρ is the local transporting medium density. For low frequencies, the acoustic impedance reverts to acoustic resistance, sometimes also referred to as ACOUSTIC REACTANCE.

Acoustic impedance tomography

[acoustics] Imaging modality that uses acoustic transmission and REFLECTION signals to form a density IMAGE of an inhomogeneous medium. This method of imaging is used in industrial quality control (nondestructive) and defect detection, medical diagnostics, and guidance (SONOGRAPHY or sonar). A special case is ULTRASONIC IMAGING (*see* ULTRASOUND IMAGING) (see Figure A.27).

Figure A.27 Artist impression of acoustic impedance imaging, locating the size of fish and the size of the school of fish. Alternatively, sonographically outlining the contour and composition of the sea floor.

Acoustic intensity ($I_{acoustic}$)

[acoustics] The acoustic intensity, also called "sound level," is directly proportional to both the square of the AMPLITUDE of the compression wave (A) and the square of the FREQUENCY (ν) as $I_{acoustic} \propto \nu^2 A^2 = \nu^2 P_{acoustic}^2$, because $P_{acoustic} \propto A$ (see Figure A.28).

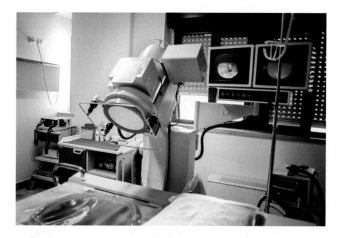

Figure A.28 Acoustic lithotripsy used for the breakdown of kidney stones or gall stones by means of focused high-energy mechanical vibration.

Acoustic intensity method (AIM)

[fluid dynamics] Mechanism of action to "visualize" the acoustic energy flow and determine the acoustic power. The visualization process takes places through NUMERICAL reconstruction. Experimental applications may use, for instance, OIL-FILM METHODS. More advanced experimental methods include hot-wire ANEMOMETER, PITOT TUBE, LASER SPECKLE velocimetry, SCHLIEREN IMAGING, and injection-tracer methods (e.g., smoke and dye). One specific application may use the STEREOSCOPIC SOUND intensity MEASUREMENT and calculated reconstruction of the SONIC PRESSURE vectors distribution (e.g., finite ELEMENT computation) in the respective plane defined by the placement of MICROPHONES. One specific application is in background noise-reduction architectural design validation.

Acoustic power ($P_{acoustic}$)

[general] The concepts of power and intensity in ACOUSTICS are directly related to the rate of MECHANICAL ENERGY transfer and the rate of energy transfer across a unit surface area. The average acoustic power ($P_{acoustic}$) and intensity ($I_{acoustic}$) are linked by the following equation: $I_{acoustic} = P_{acoustic}/A$.

Acoustic reactance

[biomedical, computational, mechanics] *See* ACOUSTIC IMPEDANCE.

Acoustic resistance (R_a)

[biomedical, computational, mechanics] It is a measure of how resistant the medium is to pressure wave disturbance. Acoustic resistance implies attenuation only, elastic and viscous properties. RESISTANCE is expressed as $R_a = \left(\rho G_\infty\right)^{1/2}$, where ρ is the density of the medium and G_∞ the stiffness or SHEAR MODULUS.

Acoustic waves

[general] *Also* SOUND WAVES are periodic COMPRESSION and EXPANSION patterns propagating through a MEDIUM. Acoustic waves are LONGITUDINAL waves, in contrast to ELECTROMAGNETIC waves (i.e., TRANSVERSE). The AMPLITUDE of acoustic waves is derived from the MOLECULAR MOTION or the local PRESSURE. Acoustic waves require a medium, unlike electromagnetic waves (see Figure A.29).

Figure A.29 Illustration of a mechanical surface wave resulting from the impact of a drop on the interface between air and water.

Acoustic waves, intensity-level (β)

[acoustics, general, mechanics] Sound intensity is frequently expressed as a relative MAGNITUDE, referenced against the intensity (i.e., pressure) threshold for audibility at 1000 Hz. The HEARING threshold is standardized as $I_{\text{acoustic}_0} = 1.0 \times 10^{-12}$ W/m^2. The concept is derived from the nonlinear (actually logarithmic, such as many human sensory sensations) response of the human EAR; a sound level that is ten times larger is perceived as only twice as loud. The intensity level is expressed in DECIBEL (dB), which is derived as $\beta = 10 \log_{10}(I/I_0)$.

Acoustical engineering

[biomedical, fluid dynamics, general] The science and ENGINEERING aspect of generating, propagating, transmitting and receiving, and processing of longitudinal pressure waves, primarily identified as SOUND; theoretical and practical implications of moving solid bodies, liquids, and gaseous mixtures. Because of the mechanism of action, the field of ACOUSTICS includes elastic theory in mechanical engineering and FLUID- and AERODYNAMICS. In principle, the field of acoustics was established in the last part of the seventeenth century with the publication of works by the French physicist and mathematician JOSEPH SAUVEUR (1653–1716), who incidentally was deaf himself (see Figure A.30).

Figure A.30 (a) Example of mechanical design to provide a unique sound generated by the metal straight flute played by Ian Anderson of the rock group Jethro Tull and (b) drawing of the pioneer in acoustical analysis: Joseph Sauveur (1653–1716), a physicist from France.

Acoustical holography

[acoustics] Using both AMPLITUDE (pressure) and PHASE of the mechanical waves collected by specific sensors (*see* ACOUSTIC INTENSITY METHOD), an IMAGE can be formed from dispersed and reflected SOUND waves similar to optical holography. Generally, ACOUSTICAL HOLOGRAPHY is a near-field technique with RESOLUTION and accuracy directly dependent on the number of sensors. A reconstruction algorithm uses finite ELEMENT methods and computational PHYSICS as required for the complexity of the diagnostic design. The three-dimensional representation of the ENERGY distribution yields the geometric and material information of the polymorphic construction under investigation. Time-domain analysis can additionally provide

MOTION artifacts such as FRICTION (and other density and contact surface interaction related characteristics) (see Figure A.31).

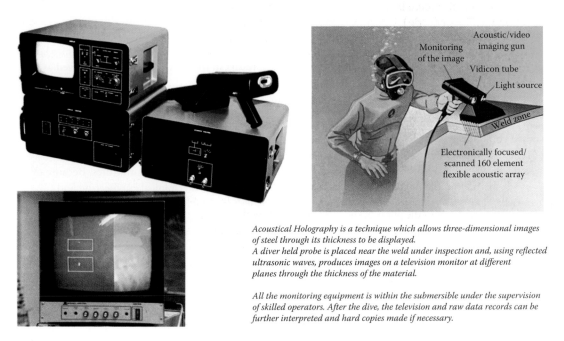

Monitoring of the image

Acoustic/video imaging gun

Vidicon tube

Light source

Weld zone

Electronically focused/ scanned 160 element flexible acoustic array

Acoustical Holography is a technique which allows three-dimensional images of steel through its thickness to be displayed.
A diver held probe is placed near the weld under inspection and, using reflected ultrasonic waves, produces images on a television monitor at different planes through the thickness of the material.

All the monitoring equipment is within the submersible under the supervision of skilled operators. After the dive, the television and raw data records can be further interpreted and hard copies made if necessary.

Figure A.31 Artist impression of nondestructive testing using acoustic imaging. Based on the interference pattern between the incident and backscattered pressure wave an image can be reconstructed resulting from amplitude as a function of time of flight. The interference principle provides accuracy within a half-wavelength depth of field. The lateral resolution is primarily a function of the focusing ability of the device.

Acoustics

[acoustics, biomedical, fluid dynamics, general, mechanics] The experience of SOUND and mechanical WAVES (primarily in the short WAVELENGTH range below 40,000 Hz), these waves are COMPRESSION WAVES or TRANSFERS WAVES. In architecture and structural ENGINEERING, this relates to the propagation and remission REVERBERATION of sound in a semi- or fully enclosed environment. The word "ACOUSTICS" is derived from the Greek word ακουστικοσ, meaning "able to be heard." Acoustics is also the science of (1) creating sound and either (2) maintaining the FREQUENCY SPECTRUM most accurately or using the information incorporated in the transmission through a medium or architecture, and subsequent (3) detection and analysis in order to derive information from the obstacles in the path, BATS providing the most obvious example. Propagation of density waves in solids or liquids is called an ELASTIC WAVE. The propagation of a homogeneous plane density WAVE traveling in the z-direction can be described as the changes in a slab of undisturbed compressible medium with the cross-sectional area $\hat{A}$ and thickness dz, contained between planes at location z and $z + dz$ and mass $\rho_0 \hat{A} dz$, where ρ_0 is the density of the undisturbed medium at location z, with pressure P_0. Upon compression/EXPANSION resulting from the acoustic wave, the plane at z will be displaced by ζ, whereas the plane at location $z + dz$ will be displaced by $\zeta + (d\zeta/dz)dz$ with $(d\zeta/dz) \ll 1$, with a local pressure gradient resulting in accelerated MOTION of the FLUID plane over the DISTANCE ζ. This situation may best describe a PIPE organ or a flute. Under this condition, the process can be assumed to be ISOTHERMAL, which will not be the case for a large change in motion or significant pressure changes. The net force acting on the surface at location z that is directly proportional to the local transient excess pressure ($P_{AC} = P_t - P_0$), with P_t the instantaneous pressure at any point in the wave, expressed as: $dF_z = \{P_{AC} - [P_{AC} + (dP_{AC}/dz)dz]\}\hat{A} = -(dP_{AC}/dz)dz\,\hat{A}$ as derived from NEWTON'S EQUATION OF MOTION. Under NEWTON'S SECOND LAW, the force per unit volume is now defined as $-(dP_{AC}/dz) = \rho_0(d^2\zeta/dt^2)$. Because the medium is considered compressible, the cross-sectional area will be considered constant for now, yielding for the CONSERVATION OF MASS of the elastically deformed medium: $\rho \hat{A} dz \left[1 + (d\zeta/dz) \right] = \rho_0 \hat{A} dz$, and the condensation/EVAPORATION at point z can be defined as $\Xi = (\rho - \rho_0)/\rho_0$. Assuming only small amplitude, with respective small condensation and displacement, yields the EQUATION OF CONTINUITY as $\Xi = -(\partial \zeta/\partial z)$. Under continuously varying pressure (i.e., sound wave), the condensation will continuously change sign, condensation/evaporation (compression/expansion, respectively). Because the movement in the medium is so small and of relatively short duration, the process can be considered adiabatic. Using the BOYLE–GAY–LUSSAC LAW applied to an isothermal and ADIABATIC process gives $P_{AC}/P_0 = (\rho/\rho_0)^\gamma$, where γ is the adiabatic constant ($\gamma = 1.4$ for AIR). After substitution of Newton's second law of motion can be written as $\left(d^2 \zeta / dt^2 \right) = \Phi'^2 \left[\left(d^2 \zeta / dz^2 \right) - (\gamma + 1)(d\zeta/dz)(d^2 \zeta/dz^2) \right]$, where $\Phi' = \gamma P_0 / \rho_0$ is the velocity of propagation of the acoustic wave front in the z-direction. In order to solve for displacement, a zero order approximation only considers the first term: $d^2 \zeta / dt^2 = \Phi'^2 \left(d^2 \zeta / dz^2 \right)$, with SOLUTION $\zeta = f \left[t - (z/\Phi') \right]$. Under a driving compression with simple HARMONIC OSCILLATION $f(t) = C_1 \cos(\omega t)$, where C_1 is the AMPLITUDE, or stroke of the compression (e.g., sound-board and piston), and ω is the ANGULAR VELOCITY with associated frequency $\nu = \omega/2\pi$, the displacement will be described by $\zeta(t) = C_1 \cos\omega\left[t - (z/\Phi') \right] + \left[C_1^2 (\gamma+1)\omega^2 / 8\Phi'^2 \right] z \left\{ 1 - \cos\omega\left[t - (z/\Phi') \right] \right\}$. The solution now no longer describes a SIMPLE HARMONIC MOTION of the particles in the plane of the medium; the second term captures the distortion imposed by the condensation/vaporization process. On a larger scale, this process can be used to describe the destructive impact (e.g., ultrasonic cleaner) or to facilitate drug-delivery in biomedical application. The processes of condensation and vaporization in these two examples involve farther deviation from harmonic oscillation to include BUBBLE expansion and CAVITATION. The science of architectural acoustics can be attributed to WALLACE CLEMENT SABINE (1868–1919), a Harvard professor of PHYSICS. The design of a concert hall is one example involving acoustics; other applications are in SPEAKER boxes and the musical instruments themselves. The TRANSVERSE PRESSURE WAVE generated by a specific source will be encountering obstacles that will reflect and change the spectral profile before the sound is detected. One detection device is the EAR (see Figure A.32).

(a)   (b)

Figure A.32 (a) Pioneer of acoustical design Wallace Sabine (1868–1919) from the United States. (b) Modern acoustically accurate concert hall design.

Rhythmic mechanical COMPRESSION and EXPANSION WAVE at FREQUENCIES ranging from 10 to 30,000 Hz. This phenomenon also generates SOUND when exposed to the human EAR. Frequencies in excess of 30,000 Hz are generally categorized as ULTRASOUND. In addition to human speech, communication, and music, acoustics is also used for sensing and diagnostics. In quality control, acoustics is used for nondestructive testing, whereas in biological applications, ultrasound can IMAGE inside live tissue up to great depths without the risk for side-effects from exposure to the ENERGY source. In comparison, other medical imaging techniques such as X-RAY, MRI/FMRI and PET have health concerns associated with the respective probing energy sources. In other modalities, animals use acoustics for the equivalence of VISION, where frequencies up to 300,000 Hz are used, for instance, by BATS for guidance. Acoustics is also used for guidance by dolphins, submarines, and to assist the blind through obstacles and surface irregularities. The human ear is generally sensitive in the frequency range of 20 Hz–20 kHz, with peak sensitivity around 1000 Hz (*see* HEARING) (see Figure A.33).

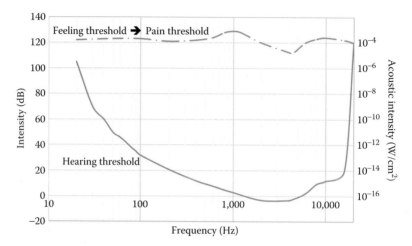

Figure A.33 The human acoustic perception threshold spectrum as well as pain level in the ear when exposed to sound of various magnitudes.

Acoustics, superposition of wave

[general] *See* WAVE SUPERPOSITION.

Acoustoelectric effect

[acoustics] Generation of ELECTRICITY in a semiconductor material (PIEZOELECTRIC SEMICONDUCTORS) and certain metals when an acoustic wave passes through a medium block. The medium will need to have access to mobile charges for the phenomenon to become apparent. The mechanical strain created by the acoustic deformation locally generates electric field gradients, which travel with the WAVE, generating a mechanism to induce the FLOW of charges. In nonpiezoelectric media, the mechanism of action is the induction of a shift in CARRIER ENERGY of the molecular composition in response to the strain. The resulting formation of electric current is actually correlated to the SOUND intensity but not to frequency. The apparent correlation between moving charges and acoustic energy alludes to the fact that an external electric field may be used to amplify the mechanical deformation. This phenomenon is supported in several piezoelectric materials, general under an applied field in access of 700 V/cm, at specific material-dependent RESONANCE frequencies (relevant parameters are constituents and dimensions). The theoretical explanation of the acoustoelectric gain/acoustoelectric amplifier mechanism is the fact that a condition much be reached where the charge carrier DRIFT VELOCITY must match the SHEAR WAVE VELOCITY (see Figure A.34).

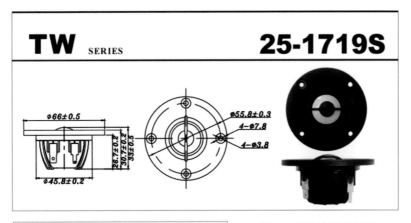

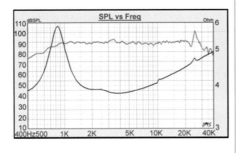

Figure A.34 Ceramic speaker using the acoustoelectric effect to generate a sound wave by applying an electronic frequency spectrum to a piezoelectric transducer. (Courtesy of TB Speakers.) The same principle applies to the transducers used in ultrasonic imaging, only operating in the megahertz frequency range instead of the kilohertz spectrum.

Acoustoelectric imaging

[acoustics] Using the ACOUSTOELECTRIC EFFECT in biological media lends itself for in vivo functional imaging because of the localized changing cellular DEPOLARIZATION in the tissue resulting from compression and EXPANSION. The subdermal electrical ACTIVITY can be collected cutaneous (in contact) by a configuration of ELECTRODES in various array configurations. Sources of electrical depolarization are MUSCLE CONTRACTION and NERVE CONDUCTION. FINITE ELEMENT computational three-dimensional reconstruction may be possible based on the respective MONOPHASIC, BIPHASIC, or TRIPHASIC characteristics of the depolarization influences. *Also see* ELECTROMYOGRAM (EMG; muscular activity IMAGE) *and* VOLUME CONDUCTION (see Figure A.35).

Figure A.35 The collection of the changes in electrical muscle activity because of the changes in density associated with muscle contraction, next to the physical volume conduction aspects of the acute depolarization itself. Once the muscle is starting to contract the depolarization wavefront is altered because of the changes in electrical conductivity due to density changes (i.e., the "acoustic effect"). In this manner, the muscle contraction with respect to frowning or smiling can be quantified. Note that one uses more muscles to smile than to frown.

Actin

[biomedical] Protein that is an active part of the CONTRACTION mechanism of STRIATED MUSCLE. ACTIN in filament shape is a globular POLYPEPTIDE chain made out of helical POLYMER consisting primarily of amino acids. Actin interacts with MYOSIN for contraction (see Figure A.36).

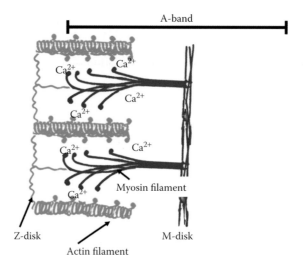

Figure A.36 Graphical representation of the microscopic filament in the muscle paired with myosin that provides a means for contraction by incremental shortening of the muscle file, allowing the actin to be attracted by sequential myosin molecular chains, gradually moving the actin toward the M-band, one molecular "feather" at a time.

Actinium

[nuclear] Symbol (Ac), radioactive ELEMENT in "METAL"-group IIIb, atomic number $Z = 89$ with a half-life of 21.6 years, electron configuration: $[\text{Rn}]7s^2 6d^1$, meaning the inner core has the electron configuration as radon (Rn). Commonly found in uranium ores and is a segment of the uranium DECAY chain, which has inspired the decay: $A = 4n + 3$ chain of uranium-235 to be named the ACTINIUM series.

Actinium, radioactive series

[atomic, nuclear] DECAY chain within the uranium decay commonly used to determine geologic age using the ACTINIUM-D/radium-G ratio to ascertain age beyond a billion years.

Action at a distance

[general, nuclear] Objects exerting forces without apparent physical contact across an intervening void. A few examples include GRAVITATIONAL FORCE, VAN DER WAALS FORCE, STRONG NUCLEAR FORCE, and WEAK NUCLEAR FORCE.

Action potential

[biophysics, chemical, electromagnetism, energy] Specific cells have the ability to exchange ions that will generate an electrical potential across the cellular MEMBRANE, the NERNST POTENTIAL. When the TRANSMEMBRANE NERNST POTENTIAL exceeds a set threshold (cell-specific) the membrane opens the ION gates and the ELECTRIC POTENTIAL (V) changes. This process becomes self-sustaining and supports the migration along the length of the CELL MEMBRANE. Specific cells capable of generating an ACTION POTENTIAL are nerve cells and cardiac MUSCLE cells. This electrical potential is part of both an active and passive transmembrane ion transportation system and is short-lived (order of milliseconds). The CELL membrane can be part of either the axon or dendrite from a nerve cell, or a cardiac muscle cell grouping. The minimally required ion current to initiate membrane DEPOLARIZATION is referred to as RHEOBASE. The stimulus to reach an action potential can be of high intensity, short duration or low intensity and long duration. The excitation mechanism of action is cell specific, such as PRESSURE (e.g., touch), TEMPERATURE, or MOTION (e.g., HEARING), chemical (e.g., taste), or LIGHT (e.g., EYE). The cell communicates with neighboring cells as well as with the autonomic and PARASYMPATHETIC NERVE SYSTEM to regulate their own chemical ACTIVITY as well as balance with the organ or tissue they are an integral part of. Nerve cells communicate observations and commands respectively to and from the brain as well as autonomic and parasympathetic nerve system to achieve a predefined goal. Some of the specific functions of action-potential transmission are heartbeat, VISION, hearing, flight from harm, or danger, and for general sensing purposes such as carbolic acidity related to carbon dioxide SATURATION of the BLOOD (RESPIRATION), the acidity of the blood and lymph, core temperature, and proprioception, to name a few. Before describing how electrical impulses travel through a MOTOR unit, the concept of MEMBRANE POTENTIAL (V_m) needs to be reviewed (*see* MEMBRANE POTENTIAL). In addition to the capacitive charge under equilibrium, the electronic model for an action potential involves the FLOW of ionic current (displacement current) that needs to be characterized by resistive ELEMENTS. The ion pumps (e.g., Na^+/K^+-pump) are voltage sources in this electronic modeling. The capacitive discharge of one segment of the cell (i.e., nerve) is the initiation of an action potential, causing the neighboring elements to follow suit by chemical and electronic COHESION of the membrane. The DEPOLARIZATION will travel over the membrane along the length of the cell as if it were a series circuit, and hence the electric discharge wave-propagation will obey the TELEGRAPH EQUATION. SIR ALAN LLOYD HODGKIN (1914–1998) and SIR ANDREW FIELDING HUXLEY (1917–2012) built on these previously devised equations and developed an equation that factors in not only relative ion concentrations and permeability but also mean conductance through individual channels and inherent membrane properties such as capacitance and OPEN-CHANNEL and closed-channel probabilities. $i_m = i_{mI} + C_m \left(\partial V_m / \partial t \right) = \left(1/R_i + R_e \right) \left(\partial^2 V_m / \partial z^2 \right)$, or when considering the propagating action potential described by the respective ion migrations: $i_m = \left(V_m - V_{Na} \right) g_{Na} + \left(V_m - V_K \right) g_K + \left(V_m - V_{ion-leak} \right) g_{ion-leak} + C_m \left(\partial V_m / \partial t \right)$, where z is the DISTANCE along the length of the depolarizing cell (e.g., nerve axon [nonmyelinated/unmyelinated] or muscle), i_m

the membrane current per unit length, i_{mI} the ionic constituent of the transmembrane current per unit length, C_m the MEMBRANE CAPACITANCE per unit length, t time, R_i and R_e the intracellular and extracellular RESISTANCE, respectively. With g_s, the ionic conductance for ion S, this coefficient become the ruling parameter in the depolarization process. The ionic conductance is defined with a time dependence that relates to a variety of factors, including the transfer rate for a fixed number of particles (n_i) from closed to open state of the pores in the membrane expressed by the transfer rate coefficient: α_{n_i}, similarly from open to closed: β_{n_i}, with n_i the fraction of particles engaged in the "open" state and the compliment $1 - n_i$. The PROBABILITY of open/closed state is linked as: $dn_i/dt = \alpha_{n_i}(1 - n_i) - \beta_{n_i}n_i$, also known as the "transition probability." The conductance for each of the ions is time dependent when the transition probability is included, specifically for POTASSIUM (K): $g_K = \bar{g}_K n_K^4$ (4 open units) and for SODIUM (Na): $g_{Na} = \bar{g}_{Na} m_{Na}^3 h$, where m_{Na} is the sodium equivalent open units (3 open) and h the closed number of units (1 this case). There is a time constant associated with the migration process, which is defined as $\tau_n = 1/(\alpha_n + \beta_n)$, with this parameter the probability of "open" unit is defined by the following equality: $\tau_{n_K}(V_m)dn_K/dt = n_{K\infty}(V) - n_K$, same for "$m_{na}$" and the steady-state "probability" is delineated as $n_\infty = \alpha_n/(\alpha_n + \beta_n)$, which allows the "port open" to be defined as: $n_{n'} = n_\infty - (n_\infty - n_0)^{-t/\tau_n}$, with t representing time and n_0 the initial "open" probability. The propagation of the membrane voltage depolarization is governed by the "telegraph equation," including a WAVE characteristic as well as a "DIFFUSION" (i.e., transmembrane ionic diffusion, both active and passive) part: $(d^2V_m/dt^2) - (c_1 + c_2)(dV_m/dt) = c_3^2(d^2V_m/dz^2) + c_1c_2V_m$, where $c_3^2 = 1/L_pC_m$, $c_1 = (1/C_m)(1/R_mdz)$, and $c_2 = R_p/L_p$ with z being the direction of propagation along the length of the cell (axon), and subscripts p and m with respect to the direction of propagation and membrane, respectively. Generally, the inductance (L), capacitance (C), and resistance (R) are all defined per unit length of the membrane. The action potential migration reduces to $d^2V_m/dt^2 = c_3^2(d^2V_m/dz^2)$ in the absence of RESISTANCE and transmembrane ionic current, which is a plain HARMONIC OSCILLATION, which will never apply. In case of a myelinated axon, the action potential "jumps" from the "NODE OF RANVIER" to subsequent "node of Ranvier," which has a distinctly different propagation pattern than for the unmyelinated axon or muscle as described first. At each node of Ranvier, the initiation process is the arrival of an electrical current, which changes the local transmembrane potential due to the localized intercellular POLARIZATION of the membrane, raising the transmembrane potential above the threshold for the ion gates to open and the cell membrane is depolarized only at the node of Ranvier in a circumferential manner. The localized depolarization creates a potential difference with the next node of Ranvier, while the prior "node" is still in restoration model (ionic efflux/influx) to reestablish reobase. The potential gradient in this case will induce an ordinary current along the intercellular surface of the membrane. The current is much faster than the migration of the membrane depolarization train, virtually instantaneous (i.e., "speed of light"), which makes the MYELINATED NERVE a much faster CONDUCTOR.

Depolarization patterns

The MEMBRANE current balance in this case is written as $2\pi r_1 \Delta z c_m (\partial \varepsilon_m/\partial t) + 2\pi r_1 \Delta z(i_m - i_e) = (\pi r_1^2/r_L)(\partial \varepsilon_m/\partial z)_{left} + (\pi r_1^2/r_L)(\partial \varepsilon_m/\partial z)_{right} = (\partial/\partial z)[(\pi r_1^2/r_L)(\partial \varepsilon_m/\partial z)]\Delta z$, where i_* is the current density over the surface of the cylinder of the axon (with radius [r_1]), ε_m is the electromotive force transmembrane potential generated by the ION migration such as the Na/K-pump, the variable z can now no longer be treated as continuous. The ability to support a current the entire length of the axon would require an excessive amount of ENERGY and would be difficult to control. The electrochemical communication between clusters of nerve cells at the synaptic connections between sequential cells will create a mechanism of amplification as well as redundancy. The myelinated impulse transmission can be described as a discrete or step-wise. The internal resistance (R_i) through compartments enveloped by the Schwann CELL are defined by the intrinsic RESISTANCE of the cell and its ionic migration (r_i), where the cross-sectional area is defined by the radius of the axon (r_1) as follows: $R_i = r_i\ell/\pi r_1^2$ where ℓ is the DISTANCE between the nodes of Ranvier. The capacitance across the membrane of, for instance, an axon (C_m) is the localized capacity of the ion exchange (c_m) multiplied by the surface area (A) of the axon shell enclosed by the myelin sheath ($A = 2\pi r_1\ell$) as $C_m = 2\pi r_1\ell c_m(2r_1/r_2 - r_1)$. The length over which there is no ACTION POTENTIAL, because there is no transmembrane ion exchange (the distance between the nodes of Ranvier), is defined as the ELECTROTONIC LENGTH (λ_ℓ): $\lambda_\ell = \sqrt{(\pi r_1^2 R_m/2\pi r_1 R_\ell)} = \sqrt{(r_1 R_m/2R_\ell)}$. The DEPOLARIZATION process, by default, has a resistor–capacitor (RC) time constant: $\tau_\ell = r_m c_m$ because that the membrane is a CAPACITOR with a resistive transmembrane ion flow. For a large electrotonic length (i.e., the Schwann cell covered section of the myelinated axon), the depolarization potential needs to be larger than for a short electrotonic length (unmyelinated axon; here, the electrotonic length is from the next ion port over). The depolarization process in a myelinated axon becomes especially important when considering multiple sclerosis, which destroys the myelin. In order to demonstrate the complexity of the segmented depolarization, the modified WAVE EQUATION for the myelinated axon is given without providing the full SOLUTION. The full SOLUTION requires advance mathematics:

$$\lambda_\ell^2\left[\partial^2\varepsilon_m\left(z,t\right)/\partial z^2\right]-\varepsilon_m\left(z,t\right)-\tau_m\left[\partial\varepsilon_m\left(z,t\right)/\partial t\right]=I_{\mathrm{diff}}\left(z,t\right)=-\lambda_{\ell n}^2\left[\partial^2 V_s\left(z,t\right)/\partial z^2\right],$$ where V_s defines

the potential difference between two neighboring nodes of Ranvier, and I_{diff} the ion DIFFUSION displacement current (see Figure A.37).

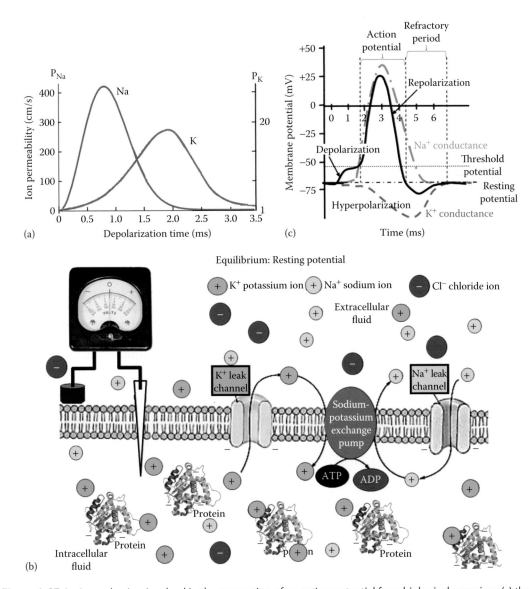

Figure A.37 Ionic mechanism involved in the generation of an action potential for a biological organism: (a) the transmembrane ionic diffusion rate of two of the primary ions: potassium (K^+) and sodium (Na^+) from inter- and extracellular locations involved in the formation of the action potential, (b) the mechanism of action with respect to the transport of sodium and potassium by means of active pump mechanism as well as gates that open or close, providing the opportunity for passive ion diffusion, and (c) outline of the typical depolarization and repolarization process and time frame for the generalized action potential.

Activation energy, fission

[atomic] A CHEMICAL REACTION requires a minimum amount of KINETIC ENERGY that colliding MOL-ECULES allow to initiate a reaction. Once the reaction is in progress, it will release enough ENERGY to maintain the reaction.

Active transport

[biomedical] The transportation of MOLECULES with a mechanism of action that requires ENERGY to move the molecules against ELECTRICAL and/or CONCENTRATION GRADIENTS. The primary ENERGY source is ADENOSINE TRIPHOSPHATE (ATP), while LIGHT and ELECTRONS can also act as driving mechanism (see Figure A.38).

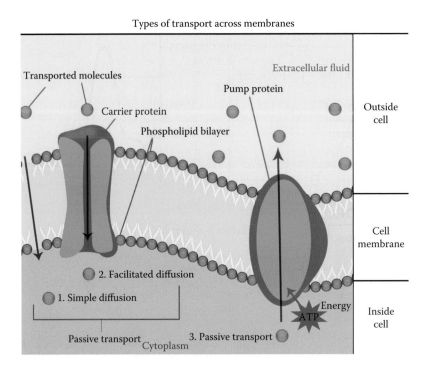

Figure A.38 The biological active transport of ions, nutrients and oxygen, and waste through the phospholipid bilayer cell membrane.

Active transport, primary

[biomedical] Transmembrane transportation by means of moving molecules, nutrition, and/or waste products by formation of "pockets" (vacuole) in the CELL MEMBRANE that under external ENERGY supply migrate from intra- to extracellular and vice versa. The transport processes are, respectively, called EXOCYTOSIS and ENDOCYTOSIS. Endocytosis can be separated in LIQUID "consumption" as PINOCYTOSIS and in particle-intake as PHAGOCYTOSIS.

Active transport, secondary

[biomedical] Assisted transmembrane transportation using the ENERGY released by one SOLUTE moving with the CONCENTRATION GRADIENT to move another solute against the concentration gradient. The transport can be parallel and matching to the driving gradient (i.e., cotransport or symport) or antiparallel (i.e., countertransport or anti-transport).

Activity

[thermodynamics] Coefficient of a constituent i in a mixture (γ_i) indicator of deviation of the characteristics of the components of a condensed AGGREGATION of a nonideal GAS MIXTURE operating as an IDEAL GAS from ideal behavior. The activity coefficient of the constituent i is defined as $\gamma_i = a_i/x_i$, where a_i is the activity of constituent i, and x_i is the molar fraction of the ith component of the mixtures expressed by Dalton's law: $x_i = P_i/P$ with P the total pressure of the mixture. The activity ties together with the chemical potential μ_i as follows: $\mu_i = \mu_{i0} + RT \ln x_i + RT \ln \gamma_i$, where R is the Boltzmann constant and T is the temperature.

Activity ($\alpha(T, P, P_0)$)

[thermodynamics] Derived from the interpretation of the ESCAPE TENDENCY within the CHEMICAL POTENTIAL: $\mu(T,P)$ of a MEDIUM operating under the IDEAL GAS LAW at certain TEMPERATURE (T) and PRESSURE (P) relative to a reference state at identical temperature but equilibrium pressure P_0. The excess chemical potential with respect to the reference state is the activity, defined as $\alpha(T,P,P_0) = \exp\left\{\left[\mu(T,P) - \mu(T,P_0)\right]/RT\right\} = \pi(T,P)/\pi(T,P_0)$, where $\pi(T,P)$ is the FUGACITY, the chemical potential and R the Boltzmann constant.

Activity, defined (R_a)

[general, nuclear] Rate of generation of DECAY particles (i.e., RADIATION: dN_r) over a time span dt from a grouping of RADIOACTIVE NUCLEI: N during ISOTOPE degradation: $R_a = dN_r/dt = \lambda N = \lambda N_0 e^{-\lambda t}$, where λ is the decay constant. Commonly used units for ACTIVITY are BECQUEREL (Bq) (SI), Rutherford (R) and Curie (Ci) (non-SI) $\left(1 Bq = 1\,\text{disintegration/s}; 1R = 1 \times 10^6\,\text{disintegrations/s}; 1Ci = 3.70 \times 10^{10}\,\text{disintegrations/s}\right)$.

Activity, metabolic

[biomedical] *See* METABOLIC ACTIVITY.

Activity, molecular

[thermodynamics] The respective sum of the kinetic and potential ENERGY of the molecules.

Acute angle

[general] An ANGLE in a triangle on either ends opposite the right angle used to determine sine and cosine functions. The choice of the appropriate acute angle depends on the prevalent reference frame; generally, the sine provides the vertical segment, and the triangle can be flipped 180° for either acute angles.

Acute respiratory distress syndrome (ARDS)

[biomedical] Increase of the permeability of the endothelial MEMBRANE in the PULMONARY microvasculature. The transcapillary filtration pressure in the lungs, as subjected to STARLING FORCES based on respective OSMOTIC PRESSURES, reduces the oxygen exchange from alveoli to blood in addition to creating an imbalance in fluid exchange, potentially leading to chronic pulmonary hypertension.

Adaptive optics

[astronomy/astrophysics, biomedical, optics] A means or mechanism that provides compensation for transmission of artifacts resulting from distortions created by the transmitting medium, for example, compensation with regard to the Abbe diffraction limit. Generally, an electronic and mechanical feedback system using optical sensors (e.g., digital CAMERA) measuring PHASE aberrations drives an iterative adjustment to deformable OPTICS, either MIRROR or LENS, or both, in the imaging device configuration, eliminating the detected aberrations. Corrections can be for atmospheric disturbances that generate discontinuities and gradients in the index of REFRACTION along the path of the observed light. Galactic distortions can, for instance, originate from dust clouds, magnetic and electric field lines, and thermal emissions, for example, stars (*also see* EYE) (see Figure A.39).

Figure A.39 Autofocusing targeting scope mounted on a rifle.

Adaptive time step

[computational, fluid dynamics] Various DIFFUSION ALGORITHMS and other DIFFERENTIAL EQUATIONS can exhibit multiple time scales. This requires the use of robust variable step-time integrators to efficiently solve for these conditions computationally. One computational approach is the use of the second-order TRAPEZOID RULE using an explicit method such as the Adams–Bashforth method. This allows for well-maintained error control. The trapezoid rule is nondispersive and stable, to commensurate with a second-order spatial discretization. The simplest Adams–Bashforth–Moulton pair is, for instance, readily available as a MATLAB® function. This approximation is particularly well suited for long-time integration. In the following approach, the incremental time steps increase to infinity at the end of the series. According to the trapezoid rule, the following vector approximation can be made: $\zeta_n \cong \zeta(t_n)$, implementing a time step Δt_n. The increment can be computed as $\zeta_{n+1} \cong \zeta(t_n + \Delta t_n)$ by solving the following implicit system: $\zeta_{n+1} = \zeta_n + (1/2)\Delta t_n \left(\dot{\zeta}_{n+1} + \dot{\zeta}_n\right) = \zeta_n + M^{-1}\left[f - (1/2)A\left(\zeta_{n+1} + \zeta_n\right)\right]$, where A is the sum of a skew symmetric convection matrix and a symmetric positive-definite diffusion matrix, and A the mass matrix tied together by a system of coupled nth order ORDINARY DIFFERENTIAL EQUATIONS as $M\dot{\zeta} + A\dot{\zeta} = f; \zeta(0) = \zeta_0$. The adaptive time step function will have a truncation error defined by $\zeta_{n+1}{}^{*} = \zeta_n + \Delta t_n \dot{\zeta}_n + (1/2)\Delta t_n^2 \left[\left(\dot{\zeta}_n + \dot{\zeta}_{n-1}\right)/\Delta t_{n-1}\right]$, which follows from the second-order Adams–Bashforth method.

Addition theorem for velocities

[atomic, mechanics, nuclear] In classical MECHANICS, velocities of the moving object and the moving reference frame with respect to a fixed point of reference can be added as vectors, adding the x- and y-components of the respective aspects of MOTION to be recombined as a resultant vector. This principle can readily be explained with the Galilei transform in which the coordinates moves as $x' = v'_x t'$ and $y' = v'_y t'$, because the time t in the reference frame may not be standard. This addition in the experiment of Fizeau requires the theoretical derivation to include the Lorentz transformations as well to account for velocity changes because of the medium (i.e., the local index of REFRACTION).

Additive coloration

[general, optics] Combination of colors creating new colors, such as that based on the red–green–blue (RGB) base three-primary COLOR set (trichromatic). The RGB system is used in television and PHOTOGRAPHY. In this configuration, all three colors combined yield WHITE LIGHT. Another color definition is the International Commission on Illumination (CIE) color space CHROMATICITY diagram, which was designed in 1931 (*see* **CIE 1931 COLOR SPACE**) (see Figure A.40).

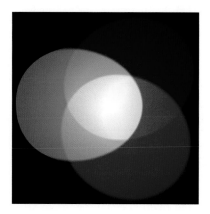

Figure A.40 Color map used to illustrate the concept of additive color in illumination. Mind that in solid coloring, the additive concept works differently, because the perceived color is based on absorption, meaning white has no pigments in the surface of the observed object.

Additive process

[biomedical] In regenerative MEDICINE, the formation of SCAFFOLD structures can rely on additive and subtractive processes. The additive processes involve the assembly from fundamental structural building blocks (e.g., protein chains, including soy, and polymers) or monolayers, including mineralization that form during a "growth" process.

Additive property

[thermodynamics] Property that is subject to the superposition principle in a linear system. This applies to system properties describing the convolution of a mixture of two or more states of a single system. Examples are ENERGY and entropy. Additionally, physical parameters of constituents are additive, such as pressure, volume, and mass.

Additivity, of energy

[thermodynamics] *See* **ADDITIVE PROPERTY**. This principle is governed by the FIRST LAW OF THERMODYNAMICS.

Additivity, of energy differences

[thermodynamics] The total difference in ENERGY between two IDENTICAL STATES of two respective systems is equal to the sum of the differences of the individual systems.

Additivity, of entropy

[thermodynamics] With any change of state of a system the entropy (S_i) will remain constant (spontaneous—reversible process) or will increase (irreversible process), also called the "principle of non-decrease of entropy": $S_1 - S_2 \geq 0$ (*also see* **ADDITIVE PROPERTY**).

Additivity, of entropy differences

[thermodynamics] The total entropy difference between two states of the individual components of a composite system is equal to the sum of the entropy differences of the respective constituents. A composite system C from constituents B and A with states 1 and 2 is expressed as follows: $S_{22}^C - S_{11}^C = \left(S_2^A - S_1^A \right) + \left(S_2^B - S_1^B \right)$.

Adenosine diphosphate

[biomedical] ADP is a chemical chain that is an integral part of the cellular renewable ENERGY supply through the chemical linking of one phosphate ION in the process of forming ATP (*see* ADENOSINE TRIPHOSPHATE) with the assistance of approximately 6.7 kcal/mol of energy. This energy is, for instance, supplied by the aerobic breakdown of GLUCOSE and fatty acids.

Adenosine triphosphate

[biomedical] ATP is a chemical chain that is an integral part of the cellular renewable ENERGY supply through the chemical release of one phosphate ion in the process of forming ADP (*see* ADENOSINE DIPHOSPHATE), producing approximately 6.7 kcal/mol of energy that is used in the cell's NUCLEUS for its METABOLISM as well as for MEMBRANE ACTIVITY such as "ACTIVE TRANSPORT." The exact amount of energy released depends on various cellular and environmental conditions. The chemical reaction transpires as follows: $ATP + H_2O \leftrightarrow ADP + P_-^+ \, 30543.2 \, J/mol$ (see Figure A.41).

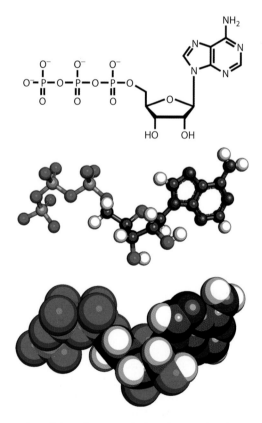

Figure A.41 Graphical representation of the adenosine triphosphate molecule.

Adhesion

[general, solid-state, thermodynamics] Chemical, electrostatic, or mechanical bond between two surfaces. The effect plays a significant role in producing FRICTION as a binding force between molecules of separate substances, or individual constituents of different nature (e.g., different materials). Contrast to COHESION (see Figure A.42).

Figure A.42 Adhesion has many commercial varieties, slow or fast curing. Illustrated is the adhesive power of chewing-gum "rubber" on the sole of a shoe.

Adhesion strength

[biomedical, chemical] Bonding strength (force per bond: f_a) between CELL and free surface based on LIGAND interaction. This concept was introduced in 1978 by GEORGE IRVING BELL (1926–2000) based on ligand "concentration" $[L]$ and a length parameter (γ_ℓ) derived from the association rate with rate constant $k_r = k_{r0} \exp\left(\gamma_\ell F / N_C k_b T\right)$ (under the influence of external force F, where N_C represents the receptor–ligand complexes and the Boltzmann coefficient $k_b = 1.3806488 \times 10^{-23}\, \mathrm{m^2 kg/s^2 K}$ at temperature T) defined as $f_a = F/n_B \approx 0.7\left(k_b T/\gamma_\ell\right)\ln\left([L]/K_D^0\right)$, where K_D^0 is the dissociation rate constant without stress and n_B is the number of formed bonds. This principle can also apply to the mechanism of action supporting muscular contraction.

Adhesive work

[biomedical, thermodynamics] W_A is the strength of ADHESION to a SOLIDS (s) surface, primarily used for LIQUIDS (l) as expressed by SURFACE TENSION in $W_A = W_{sl} = \gamma_{sv} + \gamma_{lv} - \gamma_{sl} + \pi_e$, where the surface tension between the LIQUID and the liquid–vapor and solid and its VAPOR are represented by γ_{lv} and γ_{sv}, respectively, and the tension between the surface of the solid with the liquid is γ_{sl}. The VAPOR PRESSURE (π_e) of the liquid in contact with the surface in equilibrium as it flows out is usually negligible. This value can also be measured with a STRAIN-GAUGE or can be derived from the GIBBS ENERGY of ADHESION: $W_A = -G_A$.

Adiabat

[general, thermodynamics] The curve in the P–V diagram (pressure [P] *vertically* against volume [V] *horizontally* diagram) representing the changing conditions under ADIABATIC COMPRESSION of a gas. While compressing the gas, the work ($W < 0$) performed will increase the internal ENERGY ($\Delta U = -W$) (*Note*: CONSERVATION OF ENERGY principle), thus increasing the temperature (T). Because the Boyle–Guy–Lussac law holds true for constant temperature when relating pressure to volume, the adiabatic compression deviates from the IDEAL GAS LAW represented by the "ADIABAT" curve (see Figure A.43).

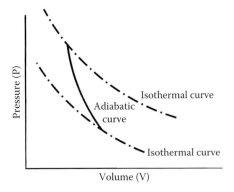

Figure A.43 Graphical representation of the adiabatic process transfer between two isothermal processes.

Adiabatic approximation

[computational, thermodynamics] It is also called "BORN–OPPENHEIMER APPROXIMATION." The basic strategy is to first hold the external parameters fixed and solve for the behavior of the system. Next, allow the parameters to change at the end of the initial calculations. This may involve an iteration process to solve for the full initial to final transition. On an atomic level, the situation refers to solving for the WAVE functions of the nuclei and the electrons. In this case, an assumption is made that the electrons are much lighter than the nuclear constituents, in the range of 3 to 4 orders of MAGNITUDE. The lighter electrons that are bound in the POTENTIAL WELL of the nuclei move significantly faster than the heavy nuclei. The adiabatic approximation assumes that the electrons follow the nuclear constituents in their MOTION, and hence the electron wave function is coupled (*also see* **BORN–OPPENHEIMER APPROXIMATION**). Returning to the initial approach, the SCHRÖDINGER EQUATION and the resulting wave function and the associated VIBRATION frequency of the nuclei of a molecule can be obtained by assuming a nuclear dimension R. The value of R yielding the lowest total ENERGY can be derived while solving the electron eigenstate problem for fixed nuclei and subsequently calculating for this condition the net nucleus–nucleus force.

Adiabatic availability

[thermodynamics] Because the entropy affects the usefulness of the ENERGY of a system, the adiabatic availability is inversely proportional to the entropy. The adiabatic availability is the energy difference between the lowest EQUILIBRIUM STATE above the GROUND STATE of a system and the highest excited state.

Adiabatic change

[fluid dynamics, general, mechanics, thermodynamics] An IDEAL GAS can be subjected to a change in pressure or volume while preventing the exchange of heat with the surroundings, for instance, the rapid down stroke of a hand PUMP or the EXPANSION of a CO_2 or N_2O_2 container (e.g., fire extinguisher or whipped cream canister). The first situation results in a temperature rise whereas the second one yields a temperature drop, primarily because of the work applied on the system or by the system, respectively. An example in MECHANICS: a PENDULUM moving without FRICTION at constant and lossless MOTION has the length of the pendulum gradually lengthened, thus gradually changing the motion, which in this case in done adiabatically. The length is altered throughout the swinging process, not abruptly at, for instance, the highest point of the swing. A mechanical equivalent can be found in the elastic deformation of the length of a pendulum under tension, presumably because of MATERIAL FATIGUE.

Adiabatic compression

[thermodynamics] The compression of an IDEAL GAS under adiabatic conditions.

Adiabatic demagnetization

[general] Mechanism used in the attempt to attain the ABSOLUTE ZERO, on both a MACROSCOPIC scale and a nuclear scale. The temperature manipulation uses the magnetic orientation of the atomic and nuclear components. The temperature follows the mechanism $\Delta T = W_M / \rho c$, where W_M is the magnetic HYSTERESIS work per unit volume, ρ is the density, and c is the SPECIFIC HEAT. On a nuclear level, demagnetization is applied to the nuclear SPIN phenomena and to associated magnetic moments. The MICROSCOPIC magnetic moments result from the following mechanisms: ELECTRON SPIN and electron ORBITAL ANGULAR MOMENTUM, next to the elusive PROTON and NEUTRON spins. An externally applied alternating strong MAGNETIC FIELD applied under adiabatic conditions disorients the nuclear magnetic moments, which will fall back to their GROUND STATE by taking away from the internal ENERGY, thus reducing the temperature on a nuclear level. The nuclear adiabatic magnetization reaches the lowest temperatures so far (approximately $T = 2 \times 10^{-8}$ K).

Adiabatic expression

[thermodynamics] The EXPANSION of an IDEAL GAS under adiabatic conditions.

Adiabatic flame temperature

[thermodynamics] Temperature of the flame products after the ignition of a combustible GAS that produces no work, nor is there an exchange of heat with the ambient environment.

Adiabatic process

[biomedical, general, thermodynamics] A process in which there is no exchange of HEAT. It is derived from the Greek word "adiabatos," which translates as "not to be passed." This condition is satisfied under one of two circumstances: INSULATION or THERMAL EQUILIBRIUM with the surroundings. This does not equate to ISOTHERMAL, because the INTERNAL ENERGY may still change due to WORK. An adiabatic process falls under the FIRST LAW OF THERMODYNAMICS. In an adiabatic process, however, no exchange or conversion of heat takes place, that is, no change of heat for acquisition or loss of MECHANICAL ENERGY.

ADP

[biomedical] *See* ADENOSINE DIPHOSPHATE.

Adsorption

[fluid dynamics, general, nuclear] It is similar to ABSORPTION but limited to SURFACE interaction only. This phenomenon is primarily the result of surface FORCES (chemical attraction) and frequently results in the MOLECULAR binding of a medium to an engulfing substance either in GAS form or in the form of SOLUTION. There is a direct correlation between the VAPOR PRESSURE of a LIQUID medium and the gas adsorption at the surface of the liquid. The basis of the attraction is the fluctuating ELECTRIC FIELDS of molecular DIPOLES and POLARIZED atoms under the influence of the surrounding bulk volume of constituent,

A

particularly the direct neighbors closest to the surface; these forces between neutral ATOMIC and molecular substances are grouped under VAN DER WAALS FORCES (see Figure A.44).

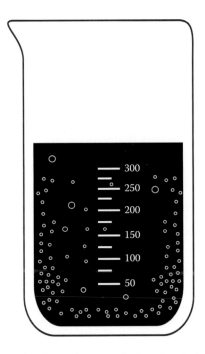

Figure A.44 Artist impression of the molecular exchange at the interface between two media.

Aeolian tones

[acoustics, fluid dynamics, mechanics] SOUND made by wires in the wind resulting from KÁRMÁN EDDY CURRENT VORTEX STREET formation—very familiar sounds on a sailboat under strong wind. Placing a taught wire between the thumb area of two hands pressed together only at the thumbs and ball of the hand (as "in prayer") will leave a small gap that when blowing on the wire will create AEOLIAN TONES (practice and skills required). More specific and everyday examples are the wind rushing past cables under tension or obstructions (*also see* TACOMA NARROWS BRIDGE) (see Figure A.45).

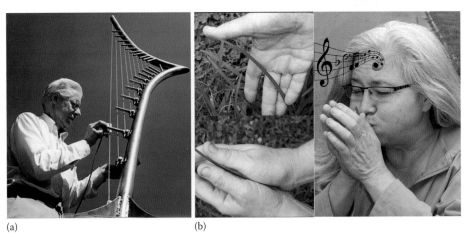

(a)   (b)

Figure A.45 (a) Aeolian harp. (b) It is also possible to make sounds with a blade of grass under tension between the thumbs.

Aerodynamics

[fluid dynamics, general, mechanics] The study of the relative MOTION of fluids, particularly gaseous substances around objects with the inherent forces. This field of ENGINEERING evaluates the forces acting on the object because of the gaseous FLOW or the respective motion of the object as well as the motion artifacts with the flow pattern, such as TURBULENCE, DRAG, LIFT, and velocity-layering effects (i.e., VISCOSITY), as a function of the flow in proximity to the object. The shape and size of the object and relative motion are also part of the theoretical analysis in regard to the following fluid-dynamic dimensionless parameters: Brinkman number, Bond number, DEAN NUMBER, Eckert number, GRASHOF NUMBER, MASS TRANSFER Biot number, Prandtl number, REYNOLDS NUMBER, Strouhal number, and Weber number. Other aspects include sound, MACH NUMBER, and sonic/supersonic classification (*also see* COMPUTATIONAL FLUID DYNAMICS) (see Figure A.46).

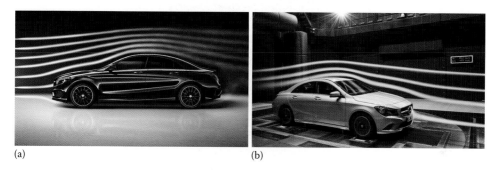

(a)   (b)

Figure A.46 (a) Streamlines of the air-flow pattern for a Mercedes CLA (Courtesy of Daimler AG.), the production automobile with the lowest coefficient of air friction at the time of this publication ($C_d = 0.23$). Note that Formula 1® race cars are unique, single production vehicles and have a much lower coefficient of friction and (b) different perspective of flow dynamics for Mercedes CLA automobile. (Courtesy of Daimler AG.)

Aerofoil

[fluid dynamics] *See* AIRFOIL.

Aerofoil section

[fluid dynamics] WING cross-sectional designs. A wing may be constructed with several AEROFOIL sections, blending into each other to maximize the AERODYNAMICS. The structural behavior of the aerofoil can be analyzed based on a model that allows for linear bending as well as torsional springs with inherent dampeners that can be assumed to be attached to the elastic axis of the aerofoil. The mechanical parameters are captured as the mass (m), the static moment about the elastic axis (S_α), a DAMPING coefficient for the TORSION (plunge) (c_b), and stiffness coefficient in plunge (K_b), next to the moment of INERTIA with respect to the total wing-mass (I_α) in reference to the elastic axis, with a torsional damping constant (c_α), and torsional rigidity (stiffness coefficient K_α), where the plunge deflection with respect to the elastic axis is h, and α is the PITCH

ANGLE in the direction of nose-up tilting around the elastic axis. The LIFT (F_L) for FLUID with density ρ can be defined as $-F_L = m\left(d^2h/dt^2\right) + S_\alpha\left(d^2\alpha/dt^2\right) + c_b\left(dh/dt\right) + K_b h = -(1/2)\rho U^2 c C_{F_L}\left(\alpha, \dot{\alpha}, \ddot{\alpha}, \dot{h}, \ddot{h}\right)$, respectively, where $\dot{\alpha} = d\alpha/dt$, $\ddot{\alpha} = d^2\alpha/dt^2$, $\dot{h} = dh/dt$, $\ddot{h} = d^2h/dt^2$. The static moment with respect to the elastic axis (I_T) is described by $I_T = S_\alpha\left(d^2h/dt^2\right) + I_\alpha\left(d^2\alpha/dt^2\right) + c_\alpha\left(d\alpha/dt\right) + K_\alpha\alpha$, for an aerofoil with chord-length c, with midpoint $b = c/2$. The laminar horizontal FLOW velocity is U (*also see* **KUTTA–JOUKOZSKY LAW**) (see Figure A.47).

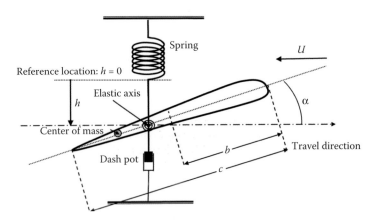

Figure A.47 Graphical representation of the mechanisms involved in the operation of an aerofoil, such as the wing of a plane.

Aerosol

[biomedical, fluid dynamics] It is the suspension of submicroscopic particles in a gaseous volume. Aerosols can be natural, for example, volcanic dust, histamines, and sea SALT, or man-made (anthropogenic), for example, drug (inhaler) and soot. Aerosols can act as condensation nuclei by providing a catalyst for RAIN generation and fog formation (see Figure A.48).

Figure A.48 The use of aerosols in paint, hair spray, cooking oil, and a large variety of other applications.

Aerostatics

[fluid dynamics, general, mechanics] Study of the equilibrium in fluids, coinciding with hydrostatics, with the main difference that aerostatic gasses are compressible fluids whereas liquids are not (see Figure A.49).

Figure A.49 Representation of static volume of enclosed gas used in an advertising "blimp," according to the earlier concept of the Zeppelin, designed by Graf Ferdinand von Zeppelin (1838–1917) from the Prussian Empire/Germany.

Aether

[general, geophysics, optics] (*also* ETHER) The medium initially assumed to be the means for transmittance of ELECTROMAGNETIC RADIATION (ARISTOTLE [384–322 BC]), later reiterated by Christiaan Huygens. Initially, ARMAND HIPPOLYTE LOUIS FIZEAU (1819–1896) concluded that the aether concept does not hold true for light passing through a wavy water surface; there is a lag in the propagation pattern, primarily because of REFRACTION. The aether concept was formally disproved with the MAXWELL EQUATIONS, describing the transport of ENERGY without the need for a CARRIER substance and allowing for transmittance through VACUUM. This is in contrast to ACOUSTIC WAVES mandatorily traveling in a mechanical medium composed of solids, liquids, gases, and/or PLASMA.

Affinity of a chemical reaction mechanism (Y_j)

[thermodynamics] Indication of PROBABILITY to initiate a chemical reaction, for example, binding, ELECTRON RELEASE, and establishing chemical matrix configuration (e.g., diamond pattern). The probability in turn is directly linked to the ENTROPY of the STATE of the SYSTEM. Under CHEMICAL EQUILIBRIUM, the "affinity" is zero for each respective REACTION MECHANISM. The affinity expression for the respective reaction mechanisms (indicated by the order number j is defined as the partial derivative of the entropy (S_ϵ) as a function of the respective (chemical-) REACTION COORDINATES (ϵ_j):
$Y_j = \left(\partial S_\epsilon / \partial \epsilon_j\right)_{U,V,n_a,\nu,\epsilon} = \sum_{i=1}^r \left(\partial S_{off} / \partial n_i\right)_{U,V,n} \left(\partial n_i / \partial \epsilon_j\right)_{n_a,\nu,\epsilon} = -\sum_{i=1}^r \left(\mu_i / T\right) \nu_i^{(j)}$, where S_{off} indicates the entropy of a stable equilibrium condition where all reactions and interactions are turned off, n the respective amounts of the contributing r constituents of the medium, with internal ENERGY U and volume V, and the respective constituents each have a chemical potential μ_j, at temperature T, the coefficient $\nu_i^{(j)}$ indicates the STOICHIOMETRIC parameters at each state.

AFM

[biomedical, general, nuclear, solid-state] *See* ATOMIC FORCE MICROSCOPE.

Age

[nuclear] The lifetime, from inception to age, can be determined from ISOTOPE DECAY using the HALF-LIFE of specific ELEMENTS. There is, however, a distinct difference between the RADIOACTIVE years and the calendar years. This distinction comes from the fact that all dating assumptions need to be satisfied and ideally backed-up with historical data (recent human history, ancient Greek and Arabic records). The choice of the ELEMENT by definitions limits the time range and level of accuracy to determine the age of an object containing a specific NUCLEAR reactive group of ATOMS (*see* RADIOACTIVE DATING). For instance, the rate of DECAY or carbon-14 has a HALF-LIFE of 5,730 years, and in naturally occurring CARBON, there is one ^{14}C ATOM for every trillion ^{12}C atoms. However, at a certain point in time, the chance of decay has decreased to one event per hundreds (-thousands?) of years, so you either measure this decay event or you do not; hence the time limit for accurate CARBON DATING is approximately 70,000 years, depending on the measuring

technique and quality of the devices. At 40,000 years, the background RADIATION starts to play a significant role when not properly accounted for or shielded against. Hence, taking the proper precautions in determining the age of an object is critical.

Age of the earth's crust

[atomic, nuclear] The EARTH'S CRUST has solidified over time in location and ranges in thickness from 20 km to close to 70 km. Typically, the age of the earth's crust is split on geological phenomena as in younger than 30 million years and older than 30 million years ($\sim 3.0 \times 10^7$ years). In comparison, the EARTH is believed to have formed 4.54 billion years ago (4.54×10^9 years $\pm$ 1%) based on Uranium dating. In reference, the AGE of the MILKY WAY is currently estimated at 11 to 13 billion years (1.2×10^{10} years $\pm$ 10%) (see Figure A.50).

Figure A.50 Continental drift was first hypothesized in 1596 by Abraham Ortelius (1527–1598) from Belgium/the Netherlands (painting by the Belgian master Peter Paul Rubens). The concept was verified and validated by Alfred Wegener (1880–1930) from Germany in 1912.

Aggregation

[general, thermodynamics, solid-state] State of aggregation, *see* **PHASE** (of medium): the condition in which a medium or substance presents itself: solid, LIQUID, or GAS. This means "the conglomeration to one whole." However, the aggregation generally refers to a broader range of material characteristics. In general, a phase (e.g., gas, liquid, solid, PLASMA, superfluids, supersolids, and Bose–Einstein condensates) consists of a single form of aggregation, whereas a form of aggregation may consist of a mixture of several phases, or can be the NONEQUILIBRIUM STATE of a medium. One such example of aggregation states is the hydrophobic or hydrophilic aggregation of solid particles in a SOLUTION. For instance, in a non-wetting medium during hydrophobic aggregation, gas cavities have been observed between the solid and the liquid (see Figure A.51).

Figure A.51 Quartz aggregate.

Agonist

[biomedical] Molecule that facilitates a chemical bond formation while attached to a soluble factor consisting of proteins pertaining to mediation of CELL actions and cell growth.

Air

[general] Atmospheric GAS MIXTURE composed of the following constituents on a MOLE basis, excluding water VAPOR: nitrogen (N_2, 78.08%), oxygen (O_2, 20.95%), argon (Ar, 0.94%), carbon dioxide (CO_2, 0.03%), Hydrogen (H_2, 0.01%), neon (Ne, 0.0015%), helium (He, 0.00015%), and other TRACE gasses depending on location, altitude, and weather conditions such as sulfur products, methane, krypton (Kr), nitrous oxide (N_2O), xenon (Xe), and ozone (O_3). The average MOLECULAR WEIGHT of atmospheric air is $M_{air} = 28.964 \times 10^{-3}$ kg/mol. Dry air may be considered an IDEAL GAS, with specific heat ration $\gamma_{air} = 1.4$ and SPECIFIC HEAT under constant pressure $c_{p\_air} = \gamma_{air}\left(R/M_{air}\right)\left(\gamma_{air} - 1\right) = 1.4\left[8.314/\left(28.964 \times 10^{-3}\right)\right]0.4 = 1\text{kJ/kg K}$, with the universal gas constant $R = 8.314$ J/mol K.

Air bubble method

[fluid dynamics] The release of air bubbles under upward FLOW conditions used to determine the flow velocity of the LIQUID. The preferred mechanism is the hydrogen bubble. The DISTANCE per unit time of the air bubble can provide a noninvasive mechanism for flow control measurements. Other applications are in velocity profiling. Competitive techniques include the smoke wire method and the spark tracing method.

Air capacitor

[general] CAPACITOR that uses an air-gap as an INSULATOR between the two POLES of the capacitor. This type of capacitor was widely used in AM and FM RADIO devices in the late twentieth century (see Figure A.52).

Figure A.52 Capacitor plates with air gap that can change the opposing surface area by means of turning a dial. This type of air-capacitor was used in the early design radio for tuning purposes, changing the resonant frequency by changing the capacitor value in the electronic resonator.

Air columns

[general] Wind instruments use standing waves in air columns as a means to amplify the generation of SOUND due to RESONANCE effects. The creation of a standing WAVE in an air column has to satisfy a multiple of quarter or half wavelengths depending on whether the column is closed on one end or open on both ends

(e.g., pan flute). The length of the column is frequently controlled by vents in the side of the column that are closed off by fingers (e.g., straight flute) or mechanical keys and levers (e.g., saxophone) (see Figure A.53).

Figure A.53 The saxophone is one of many instruments that generate musical tones that can be modified by changing the length of the air-column, as illustrated by President William (Bill) Jefferson Clinton (1946–) playing the saxophone in the dacha of the president of the Russian Federation Boris Yeltsin (1931–2007) in 1994. (Courtesy of Bob McNeely, William J. Clinton Presidential Library, Little Rock, Arkansas.)

Air drag

[computational, fluid dynamics] *See* AIR FRICTION.

Air friction

[general] Resistive FORCE associated with movement through AIR. The resistive force experience by the object is proportional to the relative VELOCITY (v), the density of the medium, the exposed area normal to the direction of MOTION as well as the surface shape and the associated REYNOLDS NUMBER next to the surface finish/treatment. All of the latter conditions can be captured as a constant, assuming no change in density (b): $F_r = -bv$. Using the sum of forces $\sum F = F_{\text{driving}} - bv = ma = m(dv/dt)$, the velocity approached the limit of the TERMINAL VELOCITY when the resistive force approach the driving force for motion, during the downward free fall, this force would be the WEIGHT (F_g) of the object with mass m under GRAVITATIONAL ACCELERATION: g ($F_g = mg$) which yields $dv/dt = g - (b/m)v$. Under high-speed movement through air, the approximation for the resistive force becomes proportional to the square of the velocity, such as for airplanes (frequently well in excess of 600 km/h), sky divers (200–300 km/h), baseballs (>150 km/h), and, in some instances, cars: $F_r = -(1/2)\rho DAv^2$, where A is the cross-sectional normal area of the moving object measured in a plane perpendicular to its velocity, ρ is the DENSITY of air, and D is the DRAG COEFFICIENT (coupled to the Reynolds number), a dimensionless empirical quantity. Spherical objects for instance will have a drag coefficient of approximately 0.5 for but can have a value as large as 2 for objects that are flat or irregular. The theoretical approach predominantly relies on the assumption that the object motion and the FLOW of air (i.e., relative velocity) is in the same one-dimensional orientation; however, in practice, this may not always be satisfied. Because of differential dependencies, solving this type of practical problem will require the use of the EULER METHOD.

Airfoil

[general] An object with a specific shape that is designed to create a LIFT action when exposed to FLUID FLOW (GAS or LIQUID). The lift is the result of the Bernoulli principle, for example WING, sail, and ski-jumper in free flight (see Figure A.54).

(a)   (b)   (c)

Figure A.54 (a) Typical airfoil: airplane wing. The streamlines with respect to the wing surface can be modified to influence the lift by (b) opening and (c) closing the "flaps" at the wing tip.

Air–fuel ratio

[energy, thermodynamics] An EXOTHERMIC REACTION is the OXIDATION of a reactant such as hydrogen and fossil fuels, that is, gasoline and coal, combined in a COMBUSTION mechanism of action. The smallest amount of dry AIR (STOICHIOMETRIC amount) needed for a self-sustained combustion is calculated theoretically based on the chemical binding between the constituents of the hydrocarbon as $C_kH_l + kO_2 + (l/4)O_2 = kCO_2 + l(1/2)H_2O$, which brings the oxygen ($[n_{ox}]$) to fuel ($[n_{fuel}]$) molar ratio to $n_{ox}/n_{fuel} = k + (l/4)$, after considering the molar fraction of 20.95% oxygen in air, the molar fraction of oxygen in dry air yields $1\,mol\,[O_2] = 4.77\,mol\,[air]$, and hence the ideal theoretical air-fuel mixture for fossil fuel is $(n_{air}/n_{fuel}) = 4.77[k + (l/4)]$, or based on mass $(m_{air}/m_{fuel}) = 4.77[k + (l/4)](28.96/M_{fuel})$, where M_{fuel} is the MOLECULAR WEIGHT of the hydrocarbon fuel.

Airy, George Biddell (1801–1892)

[general] He is an English astronomer and mathematician. Airy helped establish GREENWICH as the prime meridian. Other work included the mathematical description and validation of PLANETARY ORBITS and the determination of the mean density of EARTH using a PENDULUM method. Airy's work involved the planetary forces and mathematical description of the PLANETARY MOTION in the SOLAR SYSTEM, providing detailed NUMERICAL values of unprecedented accuracy in his time (see Figure A.55).

Figure A.55 George Biddell Airy (1801–1892), picture taken in 1891.

Airy disk

[general, optics] DIFFRACTION LIMITED IMAGE of a LIGHT SOURCE through a circular APERTURE. ELECTROMAGNETIC RADIATION passing through an aperture will experience IMAGE formation influenced by PHASE differences in the secondary sources of the ORIFICE area creating a FRAUNHOFER DIFFRACTION pattern. The central bright round projection is referred to as the "AIRY DISK," the edge of the disk, or first minimum in the INTERFERENCE pattern is at an ANGLE with the normal to the orifice expressed as $\theta \approx \sin\theta = 1.22(\lambda/D)$, where λ is the wavelength and D is the diameter of the aperture. This diffraction pattern is the main limitation in RESOLUTION for most image formation devices as described by the RAYLEIGH CRITERION or ABBE CONDITION (see Figure A.56).

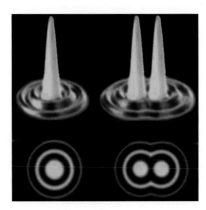

Figure A.56 Radiance pattern for an diffraction pattern Airy disks and the mechanism to distinguish adjacent points when the maxim of the Airy pattern of one object falls at least at a distance of the first minimum of the Airy disks of the other object viewed by the device limited in resolution based on its design.

AIS (abbreviated injury scale)

[biomedical] Injury-scale severity developed by the American Medical Association (AMA) and the American Association for Automotive Medicine (AAAM) to rank the physical damage to bone and mechanical performance as well as soft-tissue and related internal injuries. The ranking will guide the treatment procedure and the assessment of tolerance and assigning a tolerance curve to quantitative data. For physical trauma, the tolerance curve groups the level of injury with respect to the applied FORCE and the duration of the force. This guide is also used in the evaluation of the effectiveness of personal safety features and devices such as seat belts, AIR bags, and helmets.

Al-Bīrūnī (Abū al-Rayḥān Muḥammad ibn Aḥmad al-Bīrūnī) (973–1048)

[general] Philosopher and scientist in the Persian Empire, from what is now known as a geographic section split between Uzbekistan and Turkmenistan. His contribution includes the principles of experimental investigation in MECHANICS (empirical/heuristic approach), and he introduced an EXPANSION on the ARCHIMEDES PRINCIPLE to introduce the RELATIVE DENSITY, a dimensionless parameter to describe MASS.

Albumin

[biomedical] Protein (MOLECULAR WEIGHT of 75,000) with average size of 700 nm. Albumin can contribute to the intra- and extracellular OSMOTIC PRESSURE, based on the VAN'T HOFF EQUATION (Starling equation). The CAPILLARY REFLECTION coefficient (Staverman reflection coefficient) for albumin is 0.8, meaning low permeability. However, a low protein diet may damage the vascular permeability, causing the WALL to leak out and filling the extra-cellular space with LIQUID because of the distortion of the hemodynamic and osmotic balance in the body.

Alcohol

[chemical] Carbohydrate chain with hydroxyl group (–OH).

Alfvén, Hannes Olof Gösta (1908–1995)

[astronomy/astrophysics, computational, electromagnetism, fluid dynamics, mechanics, plasma] An engineer and physicist from Sweden who started out in electrical engineering. Alfvén received the Nobel Prize in Physics in 1970 for his work on magneto-hydrodynamics. The waves in a fluidic magnetic system known as magneto-hydrodynamic waves, also referred to as "Alfvén waves." Hannes Alfvén is also known as the father of the concept of plasma physics. His work contributed substantially to the current understanding of the earth's magnetosphere, revealing many new aspects. Other efforts were in the description of lightning bolts as part of the general concepts of the mobility of charge particles in free space (see Figure A.57).

Figure A.57 Hannes Olof Gösta Alfvén (1908–1995) from Sweden.

Alfvén number ($A_L = (v_{fluid}/v_a)$)

[astronomy/astrophysics, computational, fluid dynamics, geophysics, mechanics] A dimensionless number used in flowing molten substances, where $v_a = \left(B_0 / \sqrt{\mu_{mag}\rho} \right)$ is the Alfvén velocity and v_{fluid} is the fluid flow velocity (*also see* magnetic mach number).

Alfvén theorem

[astronomy/astrophysics, computational, fluid dynamics, mechanics] Perfectly conductive fluids confined by a cylindrical configuration of a magnetic field will remain for eternity within the magnetic field lines. This applies to plasma eruptions from a star, resulting in a steady-state flux being emitted from the star.

Alfvén velocity $\left(v_a = \left(B_0 / \sqrt{\mu_{mag}\rho} \right) \right)$

[fluid dynamics, geophysics] The velocity of flow of a magnetic fluid (e.g., magma and interstellar gas) under the influence of an external magnetic field, where B_0 is the magnetic field strength, μ_{mag} is the magnetic permeability, and ρ is the density.

Algebra

[computational] Elementary mathematics based on the four concepts of addition, subtraction, multiplication, and division. Because of the limited principles, the concept of "limit" in range of numbers is not in the algebraic vocabulary. In "modern algebra," the concept of letters representing numeric variables was introduced as an EXPANSION.

Alhazen (Abū 'Alī al-Ḥasan ibn al-Ḥasan ibn al-Haytham) (965–1040)

[general, optics] An Iranian philosopher who introduced most of the formal scientific methods (i.e., validation of experimental observation by testing a hypothesis), especially in OPTICS. He is the founder of analytic geometry, applying ALGEBRA to geometry. He established the rudimentary workings of the EYE and disputed the long-standing theory that light is emanating from the eye to facilitate VISION as postulated by PTOLEMY (90–168 AD) and EUCLID (325–270 BC). In his book *Kitab al-Manazir* (*Book of Optics*), he reportedly described the observed injuries to the RETINA of the eye resulting from exposure to direct sunlight (see Figure A.58).

Figure A.58 Graphical interpretation of the likeness of Alhazen (965–1040) from Persia/Iran.

Aliasing

[computational] SIGNAL distortion resulting from signal-processing techniques, generally referring to the loss of acuity and discrimination between two independent signals that are acquired simultaneously.

Alkali metal

[general] First column of the PERIODIC TABLE OF ELEMENTS. ELEMENT with one unpaired electron; this electron is easily shared or released making the alkali metals a very chemically active group and are rarely found as a single element in nature. Because alkali metals are highly electropositive, they have the lowest IONIZATION enthalpies in the periodic table; ionization instantaneously provides them with the inert-gas electron configuration (most stable). The alkali-metal elements in this first column are LITHIUM (Li, 3 electrons; $[He]2s^1$, where [He] is the basis for the configuration of the inner electrons), SODIUM (Na, 11 electrons; $[Ne]3s^1$), POTASSIUM (K, 19 electrons; $[Ar]4s^1$), rubidium (Rb, 37 electrons; $[Kr]5s^1$), caesium (Cs, 55 electrons; $[Xe]6s^1$), and Francium (Fr, 87 electrons; $[Rn]7s^1$). Hydrogen (H; $1s^1$) has only one electron but is always paired with hydrogen and is not considered an alkali metal. All alkali metals react strongly with halogens and form salts, and

interaction with WATER (H_2O) gives alkaline hydroxides. Other potential but unconfirmed alkali metals are ununennium (element $Z = 238$) and unhexennium (element $Z = 338$).

Alkaline earth metals

[general] Second column of the PERIODIC TABLE OF ELEMENTS. They are elements with two unpaired electrons: beryllium (Be, 4 electrons; [He]$2s^2$, where [He] is the basis for the configuration of the inner electrons), magnesium (Mg, 12 electrons; [Ne]$3s^2$), calcium (Ca, 20 electrons; [Ar]$4s^2$), strontium (Sr, 38 electrons; [Kr]$5s^2$), barium (Ba, 56 electrons; [Xe]$6s^2$), and radium (Ra, 88 electrons; [Rn]$7s^2$—unstable and radioactive). All alkaline earth metals react strongly with halogens and form salts.

Alkaloid

[chemical] Base, nonacid.

Allievi, Lorenzo (1856–1941)

[fluid dynamics] He is an electrical engineer and HYDRAULICS expert from Italy. Allievi provided a detailed hydrodynamic analysis of the WATER HAMMER effect, which caused catastrophic damage to the hydroelectric power plant at his work location in Papigno (see Figure A.59).

Figure A.59 Lorenzo Allievi (1856–1941) from Italy.

Allievi's equation, valve closure

[fluid dynamics] This equation applies to the pressure ratio between the closed-valve position and the open tube, however, only when the valve-closing time t_c is considered slow or $t_c > 2l/v_s$, where l is the length of the PIPE from the reservoir (or section with virtually unlimited free flowing LIQUID) to the VALVE, and $v_s = \sqrt{(K/\rho)/[1+(D/b)(K/E)\text{const}]}$ the speed of WAVE propagation (e.g., SOUND) in the liquid with the liquid FLOW density ρ, K the bulk modulus of the liquid with regard to compressibility, D the inside diameter of tube with wall-thickness b, E the Young's modulus of the flexible pipe, and const is the correction factor for dimensional and material factor affecting the speed of sound. Under this condition, the pressure ratio can be derived as $P_{max}/P_0 = 1+(1/2)\left(n^2 + n\sqrt{n^2+4}\right)$, where P_{max} is the highest pressure generated when the valve is closed, P_0 is the pressure in the pipe when the valve is open, $n = \rho l v_f/P_0 t_c$ and v_f is the flow VELOCITY (open valve). The equation does not take FRICTION into consideration and is ideal under all other circumstances.

Allotropy

[mechanics, solid-state] The phenomenon describing the changes in crystalline structure of specific materials under the influence of changing pressure or temperature. Another factor that can attribute to the final configuration of the crystalline structure is material preparation. The ELEMENTS most known for allotropy are carbon, IRON, and sulfur, with carbon being most known for the difference between graphite and diamond. The allotropy of the crystalline configuration can be identified by either a thermodynamically stable or a metastable structure.

Allowed beta transition

[nuclear, solid-state] RADIOACTIVE DECAY with BETA PARTICLE release where zero ORBITAL ANGULAR MOMENTUM is ejected. The DECAY constant decreases approximately proportional to the fifth power of the decay ENERGY under this condition. This phenomenon was described by Enrico Fermi in 1934. The allowed beta decay is captured under the condition: $L_\beta = 0$, where L_β is the nuclear orbital angular momentum for beta decay: $L_\beta = M_\beta v_t r$, with v_t the tangential component of the velocity, and M_β the mass of the beta PARTICLE at orbit radius r.

Allowed transition

[atomic, energy, mechanics, nuclear, quantum, solid-state] Transition between two ENERGY states of the ATOM with emission of potentially a PARTICLE or otherwise ELECTROMAGNETIC RADIATION. RADIOACTIVE DECAY transitions of atomic nuclei are exempt from exclusion rules and have a high PROBABILITY of occurring. For emission of electromagnetic RADIATION, two sets of rules apply: conservation of ELECTRON SPIN quantum number $\Delta m_s = 0$, or the fact that the TOTAL ANGULAR MOMENTUM change must be one $\Delta j = \pm 1$, because the PHOTON has an intrinsic angular momentum of 1. The atomic transitions supporting the photon emission and angular momentum exchange are supported by the DIPOLE MODEL for the electrons in orbit around the NUCLEUS. For particle emission, one additional rule is added: $\Delta j = 0$ with the exception of a transition from $j = 0$ to $j = 0$. The transition also satisfies the CONSERVATION OF ENERGY law which entails that the wavelength of the emitted or absorbed light matches the energy difference between the two states of the atom. Also, under certain conditions (*Note*: FLUORESCENCE) there must be a change in PARITY, which translates in that the following WAVE EQUATION condition must be satisfied for the density of states (LAPORTE RULE, transition momentum integral): $\iiint \Psi_f^* \Delta V \Psi_0 dx dy dz \neq 0$, where Ψ_f^* is the complex conjugate WAVE FUNCTION of the final state, Ψ_0 is the initial state, and ΔV is the change in potential energy or the interaction potential. The wave function can be found by solving the SCHRÖDINGER EQUATION. For instance, during beta decay, the parity change is $\Delta \pi_P = (-1)^{L_\beta}$ and the PARITY CONSERVATION RULE accounts for this as $\pi_P = \pi_D (-1)^{L_\beta}$.

Alpha decay

[atomic, general, nuclear, quantum, solid-state] (syn.: ionizing radiation) {use: biomedical, fundamental} Release of ALPHA PARTICLE (α) emitted from the NUCLEUS of primarily heavy atoms (isotopes). The accompanying change in atomic number of the parent has a release of ENERGY involved difference in REST MASS (nonrelativistic: $E = mc^2$) between the parent and the "offspring," *also* DISINTEGRATION ENERGY or ALPHA DECAY ENERGY (Q_d, expressed in MeV). The relationship between the DECAY time, or HALF-LIFE ($\tau_{1/2}$), and the decay energy was derived by JOHN MITCHELL NUTTALL (1890–1958) and JOHANNES WILHELM GEIGER (1822–1945) in 1911 is $\log(\tau_{1/2}) \cong \left(aZ/\sqrt{Q_d} \right) + b$, with for heavy nuclei $a = 1.454$ and $b = -46.83$ constants, and the decay time depends on the atomic number (Z). The alpha decay process can be described energetically as if the alpha particle is TUNNELING through a POTENTIAL BARRIER of Coulomb forces, both attractive and repulsive (*also see* QUANTUM MECHANICS OF ALPHA DECAY). An alpha particle released from a large ATOM by overcoming a significant BINDING ENERGY, which can be represented as a barrier. The barrier tunneling effect is illustrated in the figure below. Uranium-238 has a decay series that contains several steps where alpha particles are emitted. An alpha particle released from a large atom by overcoming a significant binding energy, which can be represented as a barrier. The barrier tunneling effect is illustrated in the figure.

Alpha decay barrier penetration

[atomic, general, nuclear, quantum, solid-state] Because the ALPHA PARTICLE is released from the NUCLEUS, it will need to penetrate a POTENTIAL BARRIER, that is $V = 2eZ_D(e/r)$, as free particle because the ENERGY of the alpha particle is less than the BINDING ENERGY. For instance, in an ATOM with atomic number $Z = 90$, the NUCLEAR BINDING ENERGY is greater than 8.6 MeV, whereas the alpha particle carries approximately 4.87 MeV. The attractive forces are both Coulomb and nuclear, which forms a POTENTIAL WELL. This TUNNELING can be accounted for with the QUANTUM THEORY. The PROBABILITY of alpha escape (λ_α) is proportional to the number of collisions with the WALL of the potential well, multiplied by the probability of the PARTICLE escaping the well (classical MECHANICS): $\lambda_\alpha \approx (v_{in}/R)P$, where v_{in} is the velocity of the alpha particle in the nucleus, the potential well "radius" is $R = R_0 A^3$, where R_0, radius constant, is 1.4×10^{-15} m and A is the atomic number. The probability is $P = \exp\left[-(4\pi Z_D e^2/\hbar v) + (8/\hbar)\sqrt{(Z_D e^2 M_0 R)}\right]$, with eZ_D the NUCLEAR CHARGE, e the electron charge, and $\hbar$ (h-bar) Plank's constant divided by 2π and the REDUCED MASS of the alpha particle (accounted for due to the nuclear recoil): $M_0 = (M_\infty M_D)/(M_\infty + M_D)$, where M_∞ is the alpha particle mass and M_D the mass of the nucleus (see Figure A.60).

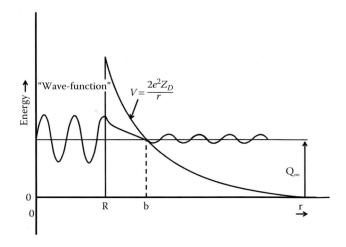

Figure A.60 Energy plot representing the potential well for alpha decay for an atomic nucleus.

Alpha decay energy

[atomic, general, nuclear, quantum, solid-state] (Q: ${}^A_Z P \rightarrow {}^{A-4}_{Z-2}O + {}^4_2 He + Q$, with ${}^A_Z P$ the parent ATOM, ${}^4_2 He$ the ALPHA PARTICLE, and ${}^{A-4}_{Z-2}O$ the offspring and where the rest MASS ENERGY is $Q = (m_P - m_O - m_\alpha)c^2$; this goes primarily to the kinetic energy of the alpha particle, with residual energy in nuclear recoil. Sometimes, two alpha particle are emitted in the RADIOACTIVE DECAY process, however, with different respective energies. The difference in energy is covered in CONSERVATION OF ENERGY by emission of a GAMMA RAY with the remaining energy. This process is seen in the radium decay process to the unstable radon ISOTOPE (radon NUCLEUS) with $E_\alpha^1 = 4.78$ MeV and $E_\alpha^2 = 4.59$ MeV next to 0.19 MeV gamma ray emission.

Alpha particle

[atomic, biomedical, nuclear, quantum, solid-state] α PARTICLE with the following constituents: two protons and two neutrons, equivalent to a helium NUCLEUS, with associated double POSITIVE CHARGE. The mass is 4.002764 amu (atomic mass units). The alpha particle has little kinetic energy and will be stopped by a sheet of paper compared to several CENTIMETERS of aluminum for γ-rays. The alpha particle can be detected by means of a SCINTILLATION COUNTER (e.g., GEIGER COUNTER), a BUBBLE CHAMBER (CLOUD CHAMBER) or specialized SOLID-STATE devices (e.g., crystal or plastic, with embedded phosphor or anthracene that fluoresces).

Alternating current (AC)

[general] A FLOW of electrons periodically changing to reverse direction. (*Note*: The current direction is in the opposite direction of the flow of electrons).

al-Zarqali (Abū Isḥāq Ibrāhīm ibn Yaḥyā al-Naqqāsh al-Zarqālī) (1029–1087)

[general] Scientist from Spain, also "Arzachel." He is the first to document the statement before JOHANNES KEPLER (1571–1630) that PLANETARY ORBITS may be elliptical. Al-Zarqali an instrument maker, astronomer, and philosopher from what we now consider to be Spain, then momentarily under Arab–Islam rule, made his statement on the "oval" orbit of Mercury in 1081. He also made several other corrections to the established Ptolemaic model (see Figure A.61).

Figure A.61 Postage stamp with graphical interpretation of the likeness of al-Zarqali (1029–1087) from Spain.

AM radio

[general] AMPLITUDE modulated transmission of electromagnetic waves used for wireless SIGNAL transmission. The mechanism of operation is distinctly different from FM radio. The RADIO signal is generated by an alternating DIPOLE in oscillatory fashion over the length of an ANTENNA generating a CARRIER WAVE with a fixed frequency, which is modulated in amplitude by the superimposed signal WAVE. The accompanying alternating electric and MAGNETIC FIELD is described for the dipole (*see* DIPOLE, ALTERNATING). The superposition principle applies to electromagnetic waves. The broadcast frequency range is split in the following wavelength bands. The carrier wave is denoted by three denominations. The first transmission designation is long wave radio: 148.5–283.5 kHz. Long-wave (LW-) radio generally uses 9 kHz channel spacing. Next is medium-wave (MW-) radio: 520–1,710 kHz. Channel spacing varies between 9 and 10 kHz. Medium-wave radio is associated with the generally available information channels and music for commercial use. The last band is short wave broadcasting in the 1.711–30.0 MHz range. Short-wave (SW-) radio is subdivided into 15 bands; channels are generally separated by 5 kHz (see Figure A.62).

Figure A.62 An old-fashioned AM radio.

Amagat, Emile Hilare (1841–1915)

[thermodynamics] A scientist and physicist from France who had experimentally validated the superposition of the IDEAL GAS LAW for a mixture of gases.

Amagat's law of additive volumes

[thermodynamics] The assembly of the volumes of a number r gasses combined under same pressure P and temperature T of the amounts n_i representing the individual constituents yields an IDEAL GAS MIXTURE volume under equivalent conditions, that is, the sum of the constituents: $V(T,P,n) = \sum_{i=1}^{r} n_i(RT/P) = \sum_{i=1}^{r} n_i v_{ii}(T,P)$.

Amber

[general] Fossilized pine tree resin that is electrically charged when rubbed with fur or cloth. The first recorded observation of this ELECTRIC CHARGE phenomenon was by the Greek scientist THALES OF MILETUS (c. 624–546 BC) in around 590 BC. Later, PLATO (c. 427–347 BC) also described the effects of Amber as having attractive powers. The Greek name for amber is "Electron" (ἤλεκτρον), establishing the initial nomenclature for modern electrical ENGINEERING.

Amorphous media

[general] Solids that behave in a fashion primarily know to liquids. GLASS is in an amorphous state, technically a LIQUID but with fixed and rigid shape. Other familiar concepts include shaving cream, certain make-ups, chocolate mousse, and butter. Additionally, volcanic magma could be considered in this group.

Ampère, André Marie (1775–1836)

[electromagnetism, general] A French mathematician and physicist during the time of Napoleon Bonaparte (1769–1821). He developed mathematical relationships between the MAGNETIC FIELD and electric current, recognizing certain innate characteristics of current before the development of the atomic model by NIELS BOHR (1885–1962). The unit for current was named after him. His mathematical verve laid the foundation for many electromagnetic and electrodynamic laws. Ampère's work was taken on by the Danish scientist HANS CHRISTIAN ØRSTED (1777–1851) resulting in several significant experimental validations of Ampère's theories. Their combined work led to the discovery of a link between MAGNETISM and ELECTRICITY in 1819. His mathematical contributions were in the recognition of the values imbedded in the second-order derivative in partial differential equations (see Figure A.63).

Figure A.63 André Marie Ampère (1775–1836).

Ampère

[biomedical, electronics] The MAGNITUDE of a current in two wires that are lined up parallel to each other at separation DISTANCE $r = 1\mathrm{m}$ that produces a MAGNETIC FORCE (F_B) per unit length (ℓ) between the two wires of $2 \times 10^{-7}\,\mathrm{Nm}^{-1}$, based on the Lorentz force: $F_B/\ell = \mu_o^{\mathrm{magn}}I_1I_2/2\pi r$, where I_i is the respective electrical current in either wires and μ_o^{magn} the permeability of the medium the wire is suspended in. The force is the direct result of the MAGNETIC FIELD (**B**) generated by the current in each wire as $\mathbf{B} = \mu_o^{\mathrm{magn}}I_1/2\pi r$. The ampère is also equivalent to the FLOW of charge of $1A = 1C/s$, one coulomb per SECOND.

Ampère (A)

[biomedical, general] Unit of electric current expressed as charge per unit time $[C/s]$, also 3×10^{19} electrons/s. It has been originally defined as the amount of current that will deposit 1.118 milligram of silver per SECOND on the CATHODE of a CONDUCTOR placed in a silver SOLUTION along with the ANODE. It is often described as measured by an ammeter.

Ampère's law

[general] Electrical equivalence described by MARIE AMPÈRE (1775–1846) between the current through a CONDUCTOR and the MAGNETIC FIELD produced by this current. Because magnetic field lines form a closed loop, the summation of the magnetic field along individual segments of an arbitrary loop that encloses one or more current carrying conductors with respective currents is directly proportional to the sum of the enclosed currents. Only the components of the magnetic field tangential to the loop are added for the loop, expressed as $\sum \vec{B}_{//}/\Delta l = \mu_0 \sum I$, or in integral form: $\oint \vec{B} \cdot d\ell = \mu_0 I$ or $\oint \vec{H} \cdot d\ell = I$ for any arbitrary closed loop, where $\vec{B}$ is the magnetic field strength, $\vec{H}$ is the magnetic FLUX density, which has a HYSTERESIS with respect to the magnetic flux as described by JAMES CLERK MAXWELL (1831–1879), ℓ the path of the loop of integration around the current I, and μ_0 the DIELECTRIC permittivity. An alternative way to define Ampère's law is as follows: $\nabla \times \vec{H} = \boldsymbol{j}$, where $\boldsymbol{j}$ is the current density. An equivalent mathematical description was derived by PIERRE-SIMON DE LAPLACE (1749–1827) based on work by JEAN-BAPTISTE BIOT (1774–1862) and FÉLIX SAVART (1791–1841). James Maxwell later introduced a revision incorporating the FLOW of charges (i.e., the electric flux: $\phi_e = EA = q/\epsilon_0$, with E the electric field, A the area, q the ELECTRIC CHARGE, and ϵ_0 the dielectric permeability) as displacement current $\left(\epsilon_0\left(d\phi_e/dt\right)\right)$ defined by $\oint \vec{B} \cdot d\ell = \mu_0 I + \mu_0\epsilon_0\left(d\phi_e/dt\right)$.

Amplification (M_A)

[electronics, general, optics] The process of increasing the AMPLITUDE of a SIGNAL, field-strength or MOTION by means of a mechanism that requires the supply of external ENERGY. Generally amplification is expressed in DECIBEL (dB). Electronic amplification of an electrical signal increases the voltage amplitude and hence the current from input to output with the use of an operational amplifier. Alternatively the use of current amplifiers is also practiced, as are transconductance amplifiers; supplying an output current proportional to the input voltage. Respectively, transresistance amplifier which generates an output voltage that is proportionate to the input current. Typically, filters are added to reduce the impact of NOISE in the amplification process. Filters are only beneficial when the signal frequency spectrum is reasonably well defined, hence rejecting portions of the FREQUENCY SPECTRUM that is considered to supply noise. Mechanical amplification can be obtained by means of a resonant cavity (e.g., casing of a string instrument), or continuous excitation at the resonant frequency (e.g., TACOMA NARROWS BRIDGE, Puget Sound–Washington, got destroyed because of the wind blowing on the suspension cables, forcing the cables in VIBRATION that matched the resonant frequency of the concrete and METAL main structure of the bridge; construction of the bridge started in 1938, opened in July 1940, and the bridge was destroyed in November 1940). The amplification of electric or magnetic fields can be achieved through induction under Faraday's law and LENZ'S LAW.

Amplifier, electronic

[general] *See* OPERATIONAL AMPLIFIER (OPAMP).

Amplitude

[general] The maximum value of the displacement in a mechanical oscillatory MOTION, or the MAGNITUDE of the maximum field strength of an oscillating electric or MAGNETIC FIELD.

Amplitude modulation (AM)

[general] Superposition of two waves: a CARRIER WAVE and a modulating SIGNAL wave. The information stored in the AMPLITUDE is retrieved by tuning to one specific carrier wavelength and deconvolving or demodulating the superimposed information stream by a tuning circuit composed of capacitors, resistors, and inductors (potentially imbedded in solid-state ELECTRONICS; *see* INTEGRATED CIRCUIT). The signal can be electromagnetic (e.g., *see* AM RADIO, TELEVISION) or mechanical (e.g., ULTRASOUND) (see Figure A.64).

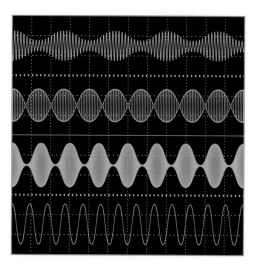

Figure A.64 Signal pattern with imbedded amplitude modulation.

Anaerobic respiration

[biomedical] Cellular METABOLISM operating without OXYGEN. The process changes the chemical structure of CARBOHYDRATES to generate ATP; however no REDUCTION takes place with a relatively low conversion efficiency in comparison to AEROBIC respiration. The short-term ENERGY that is stored is rapidly depleted causing LACTIC ACID concentrations to accumulate to the point where MECHANICAL EFFORT by the MUSCLE becomes impossible. The point at which the muscle becomes subject to this lactic poisoning is referred to as the "anaerobic threshold."

Analysis of the motion of a fluid element

[fluid dynamics] Using STOKES ANALYSIS, the respective components (x, y, z) in a three-dimensional space can be presented as having three orthogonally directed VELOCITIES (u, v, w).

Anatomy

[biomedical, general] A study of the construction and composition of biological entities. This science is mainly static; it does not concern itself with the way body part move. The study of the biological structure started with the first official documented proof in 1543 from ANDREAS VESALIUS (1515–1564): Humani Corporis Fabrica. Additionally, LEONARDO DA VINCI (1452–1519) contributed to this new line of thought with extensive drawings and theoretical analysis (see Figure A.65).

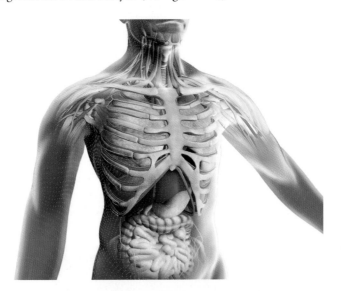

Figure A.65 Example representation of human anatomy.

Anderson, Carl David (1905–1991)

[atomic, energy, general, nuclear] An American physicist who is known for the discovery of several ELEMENTARY PARTICLES such as the POSITRON in 1932 and the MUON in 1938. He started out under ROBERT A. MILLIKAN (1868–1953) in CLOUD-CHAMBER analysis for COSMIC RAYS. His positron discovery was a direct validation of the theoretical description of this phenomenon by PAUL ADRIEN MAURICE DIRAC (1902–1984) (see Figure A.66).

Figure A.66 Carl David Anderson (1905–1991) from the Nobel Prize files.

Andromeda Galaxy

[astronomy, astrophysics, general] It has a spiral configuration. Andromeda is part of the Andromeda CONSTELLATION, with a name derived from Greek mythology, namely the daughter of CASSIOPEIA. It is located directly beneath the Cassiopeia constellation in the northern sky, visually in the direction of the NORTH POLE. The GALAXY is cataloged as both M31 and NGC 224 in the Messier astronomical catalog (denoted by the letter M), as defined by Charles Messier in 1764, and the "New General Catalogue" (NGC), respectively. The NGC was introduced by JOHN LOUIS EMIL DREYER (1852–1926) in 1888. Andromeda is the closest galaxy with respect to our MILKY WAY (see Figure A.67).

Figure A.67 Reconstructed image forming an impression of how Andromeda is outlined from our point of view.

Anechoic

[acoustics, electromagnetism, imaging] Structure designed to eliminate and/or prevent reflections (see Figure A.68).

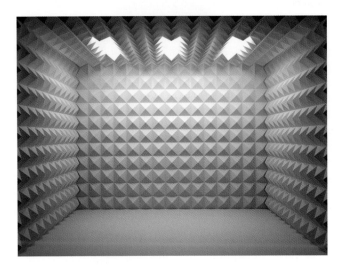

Figure A.68 An anechoic chamber.

Anelasticity

[mechanics, solid-state] A time-dependent variant of HOOK'S LAW OF ELASTICITY for the relationship between STRAIN (ϵ) and STRESS (σ): $\epsilon = J\sigma$. An anelastic solid will display a lag between FORCE (F) and stress, while the relationship between stress and strain is preserved permitted there is an allotment for time to allow the Hook's equilibrium to be reached. Stress and strain will following the behavior described as $J_R\sigma + \tau J_u(d\sigma/dt) = \epsilon + \tau(d\epsilon/dt)$, which solves for a constant stress as displaying what is known as "CREEP." The equation defines the following concepts: J_u is the unrelaxed COMPLIANCE that describes the response to an instantaneous force: $J_u = \left[\epsilon(0)/\sigma_0\right]_{t=0}$, $J_R = \lim_{t\to\infty}\left(\epsilon/\sigma_0\right)$ is the relaxed compliance, and τ is the RELAXATION TIME of the transformation under a force applied over a time t at constant stress (see Figure A.69).

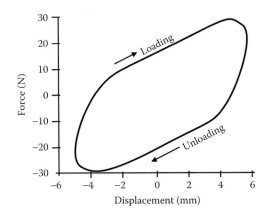

Figure A.69 Hypothetical mechanical behavior for an anelastic medium under applied force.

Anemometer

[fluid dynamics] A device for measuring wind VELOCITY (v) and airflow. The first wind anemometer was invented by Leon Battista Alberti (1404–1472) in 1450. Several mechanisms are available that can provide FLUID velocity based on interaction of AIR flow with a sensing mechanism. The cup anemometer rotates under the applied force by the wind and the angular velocity directly correlates to the wind velocity taking the arm length with respect to the cup catching the wind into account. The hot-wire anemometer has a glow wire that will have a HEAT TRANSFER with its direct environment proportional to the wind velocity. Optical device used for FLOW measurement use either light SCATTERING by particles in the fluid (air, LIQUID) or variations in density expressed as variations in refractive index. Examples of optical anemometers include interference-fringe anemometer and laser-Doppler anemometer. The interference-fringe method depends on the refractive index gradients resulting from flow velocity gradients and a split laser beam that is converged and recombined after passing through a stream creating INTERFERENCE patterns on transmission similar to Moiré interferometry, whereas the DOPPLER technique measures the velocity from particles based on frequency shift. The fringe INTERFEROMETER detects a "beat" frequency (v_d) that is proportional to the local velocity perpendicular to the intersection plane of the two laser beams with wavelength λ_0 at a relative ANGLE with each other (θ) expressed by $v_d = \left(2v/\lambda_0\right)\sin\left(\theta/2\right)$. The sonic or acoustic anemometer is based on the speed of SOUND combined with the velocity of the air-flow to provide a time-delayed detection of the sound source. Pitot tubes (also called "Prandtl tubes"), on the other hand, are used to determine fluid flow velocity by determining the difference between the static pressure (P_s) and dynamic pressure ($P_d = (1/2)\rho v^2$) in the flow pattern of fluids. The Pitot tubes rely on the Bernoulli equation as $P_{tot} = P_s + (1/2)\rho v^2$, where P_{tot} is the total pressure and ρ the local fluid density (in this case assuming an INCOMPRESSIBLE FLUID). The Pitot mechanism is a variety of the VENTURE EFFECT. The anemometer is a crucial tool in determining surface layer effects in the lower atmospheric BOUNDARY LAYER. Specifically, the recognition of TURBULENCE is caused by relief patterns as well as the confluence of airstreams. One of the newer mechanisms used to measure airflow is by means of ULTRASOUND, for example, A-Probe ring anemometer from ATI. Other alternatives use a wire heated by electrical current, which will lose heat due to convection (see Figure A.70).

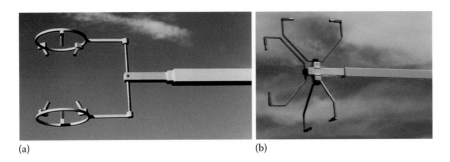

(a)   (b)

Figure A.70 Example of an anemometer used to measure wind velocity and direction. (a) One of a series of five ultrasonic (nonmechanical and can be used under virtually any weather conditions) anemometers from Applied Technologies, Inc., Longmont, CO. According to 2014 calibration tests, this series of anemometers has been verified as one of the most accurate ultrasonic anemometers available for scientific research. (b) Different ultrasonic anemometer design from Applied Technologies, Inc.

Aneurysm

[biomedical, fluid dynamics] Localized widening of an artery, usually combined with a thinning of the vessel WALL. The incident is attributed to both intra-arterial pressure and FLOW phenomena. Generally, the segment of a vessel that is involved in the development of an aneurysm has TURBULENCE or low-flow due to the vessel geometry and infrastructure. Aneurysms often develop in curved vessel segments or near bifurcations. The arterial flow creates shear stress on the surface of the inner lining of the wall, which induces change in the chemical interaction between specific BLOOD components and the lining of the vessel wall. In addition, the flow associated with the vascular geometry has a complex turbulent structure that indices a frictional BOUNDARY LAYER, which applies a tangential force on the material structure. As a result of all the physical characteristics, both material and chemical changes are initiated, which in turn worsen the initial flow conditions. An aneurysm has a weakened integrity of the vascular wall, which may rupture under certain conditions, specifically under increased local pressure. Aneurysms in the brain and in the AORTA frequently have lethal consequences. Because of the turbulent conditions in the aneurysm, the blood itself is also affected, potentially leading to thrombus formation (blood clot) as well as the formation of EMBOLISM (BUBBLE formation, sealing of the vessel for flow; fat globule) (see Figure A.71).

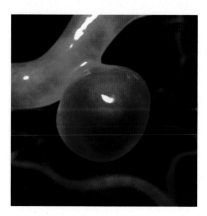

Figure A.71 Representation of a vascular aneurysm. These types of thin-walled vessel protrusions can burst, resulting in massive blood loss and, depending on location, instantaneous death.

Angle

[general] For example θ and α, the geometric span of an arc (i.e., curved length of line with constant curvature) between two lines, planes, or spaces that intersect in one point or shapes that do not intersect but have respective axes that when extended intersect. The ANGLE can be measured/defined between two points on a surface with no constant curvature such as the contour of a WING of an airplane as measured between two points that are separated close enough that by first- or second-order approximation the DISTANCE to a center-point of curvature (i.e., origin) can be considered identical. Also, the use of tangents to respective surfaces can be defined by an angle at the point of extended intersect. *Note*: The straight line connecting the two ends of an arc is called a "chord." Angles can be in two or three dimensions, three-dimensional angles are called "SOLID ANGLE." Angles are defined in either degrees (°) and radian (rad), or in three-dimensional steradian, the circumference of a circle has 360 degrees (360°) or 2π radian [2π rad], whereas in three dimensions, the solid angle spans 4π steradian (4π sr). The radian is defined as the ratio of the length of the bow of an arc and its radius (length to arc from point of curvature).

Ångström, Anders Jonas (1841–1874)

[general] A Swedish astronomer and physicist who dedicated his work to the explanation of predominantly the solar spectral lines (see Figure A.72).

Figure A.72 Painting of Anders Jonas Ångström (1841–1874) by an unknown artist.

Ångström, Knut Johan (1857–1910)

[optics, thermodynamics] A Swedish physicist famous for his discoveries in the INFRARED part of light. He is the son of Anders Jonas Ångstrom.

Ångstrom unit

[atomic, nuclear] A measure of size named after ANDERS JONAS ÅNGSTRÖM (1841–1874), 10^{-10} m $= 10^{-1}$ nm.

Angular acceleration (α)

[biomedical, general, mechanics] The rate of change in ANGULAR VELOCITY ($\omega(t)$) with respect to a reference frame or fixed point for an object moving in curved trajectory, or the rate of change in direction of axis of rotation. The definition of angular velocity is $\alpha = d\omega/dt$ (rad/s^2).

Angular displacement (θ)

[biomedical, general, mechanics] In rotational KINEMATICS, the ANGLE of the curved circular path of a moving point on an object revolving around a central axis measured with respect to the axis of rotation in reference between two points in time. The angular displacement relates to the arc of the circular trajectory (s) as a function of DISTANCE (r) to the axis of rotation as $s = r\theta$.

Angular frequency (ω)

[general] Rotational velocity of an object revolving around its own axis or an object in orbit around a FOCAL POINT, generally consisting of an object with attractive force (electrical or gravitational). Additionally, the angular frequency of a periodic event is associated with the sinusoidal periodicity, which can also be plotted as a circular phenomenon with changing AMPLITUDE as a function of ANGLE (α) as a function of time (t): $\omega = \partial\alpha/\partial t$.

Angular impulse

[biomedical, general, mechanics, quantum] Change in angular momentum. A gymnast performing a (dual, triple) somersault can change the angular momentum when touching the GROUND/balance-beam, or touching the second bar on the uneven bar exercise (*also see* ANGULAR INERTIA) (see Figure A.73).

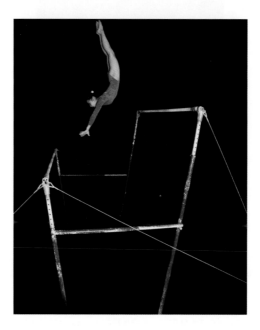

Figure A.73 A gymnast who changes angular impulse when touching an object in their path.

Angular inertia ($I\alpha$)

[biomedical, general, mechanics, quantum] Moment of INERTIA of a rotating body multiplied by the angular acceleration. The angular inertia is directly correlated to the TORQUE (τ) applied to the revolving object (revolving around its own axis or around another object while in orbit): $\tau = I\alpha$. One particular example is found in an avalanche or rock-slide (see Figure A.74).

Figure A.74 The angular inertia of a rolling object, seemly unstoppable, as experienced during an avalanche.

Angular magnification (M_α)

[general, optics] The ratio of the ANGLE of the incoming beam (acceptance angle) as observed with a LENS (aided: α_{aid}) to the angle of observation received without a lens (unaided: α_{un}): $M_\alpha = \alpha_{aid}/\alpha_u$ (see Figure A.75).

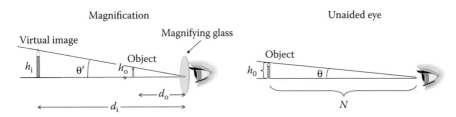

Figure A.75 Angular magnification under the influence of a lens in the path of vision, specifically pertaining to the use of eye glasses (spectacles).

Angular momentum (L)

[biomedical, general, mechanics, quantum] Representation of the propensity of a rotating body to remain in rotation at constant angular velocity around the axis of symmetry, or the axis of rotation, similarly described by the NEWTON'S FIRST LAW of MOTION or LAW OF INERTIA. There is both a classical and a QUANTUM mechanical interpretation. The classical angular momentum references the momentum (p) of a moving object(s) with mass m (respectively: center of mass) taking in the perpendicular direction to a line connecting with the origin of a reference frame (at a DISTANCE, r), or equivalently a fixed point in space as the axis of rotation. This is expressed as $L = r_\perp m \odot v = r \otimes p$. The fixed point can be the attachment of an arm/LEVER, such as the elbow. Consider the rotational counterpart to linear momentum. When applied to a rotating unit of mass the angular momentum can be related to the moment of inertia (I) as $L = I\omega$, with $\omega = v_\parallel/r$ the angular velocity around the axis of rotation, and $v_\parallel$ the tangential velocity of the moving entity. *Note*: under this definition even a PARTICLE moving in a straight line can have an angular momentum. On a planetary level, Kepler's law is a direct consequence of the CONSERVATION OF ANGULAR MOMENTUM. On an atomic or elementary particle level, each atomic or subatomic constituent has a fundamental angular momentum, and this phenomenon is described in quantum mechanical terms. For instance, the revolution on its own axis for an electron the angular momentum is quantized and is named the "intrinsic SPIN," not to be confused with the "ORBITAL ANGULAR MOMENTUM" of the NUCLEON circling a nuclear grouping. The quantum-mechanical SPIN angular momentum of a nucleon is defined as $L = s\hbar$, where s is the spin quantum-number. Neutrons, electrons, and protons have an intrinsic spin of $(1/2)\hbar$. The TOTAL ANGULAR MOMENTUM for the combined respective contributions of the constituents obeys the superposition principle; this means that for even atomic number, however, that nucleon pairs need to be broken up, whereas odd-odd nuclei are not restricted in their excitation and joining mechanisms. The TOTAL NUCLEAR angular momentum for each constituent: $j\hbar$ is governed by the quantum-mechanical coupling rules, yielding the vector addition of the intrinsic spin and the orbital angular momentum: $\ell\hbar$, with ℓ the azimuthal—or orbital-quantum number and j the total QUANTUM NUMBER, which in this case is limited to the values: $j = \ell + (1/2)$ and $j = \ell - (1/2)$. The angular momentum is now given as: $L = \sqrt{(v(v+1))}\hbar \approx v\hbar$ ($v = 0, (1/2), 1, 1(1/2), 2, 2(1/2), \ldots$), indicating the limitation to the maximum value of the vector component in any direction. For even mass-number-A nuclei only the integers of v apply and for odd-A nuclei the values are $n + (1/2)$; n = integer. In this notation, $\hbar$ is the Plank's constant h divided by 2π; $h = 6.62606957 \times 10^{-34}$ m^{2}kg/s. This forms the basis for the theoretical description of the filling order in the Bohr model and the filling constraints of the ENERGY levels in the nuclear quantum well (i.e., maximum

A

occupation numbers). The net rate of change in angular momentum equals the torque on the moving compilation of mass or respective the sum of the torque on all moving masses. One final quantum-mechanical interpretation of angular momentum is the PRECESSION of a PROTON in an external MAGNETIC FIELD as experienced under nuclear MAGNETIC RESONANCE imaging (NMR/MRI). The precession of the angular momentum of the proton is initiated by the application of torque resulting from the interaction with an external MAGNETIC FIELD causing a change in angular momentum that is perpendicular to the existing angular momentum of the proton. The inherent magnetic moment (vector) will now gyrate in a cone-shaped revolution around the applied magnetic field. The rate of precession is called the "LARMOR FREQUENCY" (i.e., LARMOR PRECESSION) (see Figure A.76).

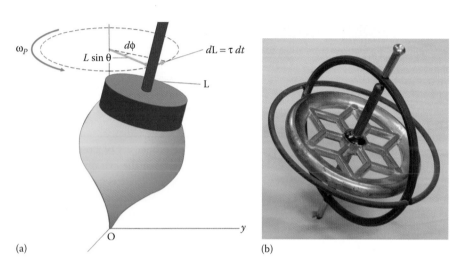

(a)   (b)

Figure A.76 (a) The angular momentum of objects performing a revolution. The best-known example is the spinning top (b); however, the same principles apply on microscopic level for electron spin as well as for galactic objects, including planets, black holes, and galaxies.

Angular momentum, conservation

[atomic, general, mechanics] When no external torque is applied to a system, the TOTAL ANGULAR MOMENTUM will be conserved because there is no change in internal ENERGY content. During RADIOACTIVE DECAY, the CONSERVATION OF ANGULAR MOMENTUM places restrictions on the PARITY change of the process as well as determining the "allowed" transitions and defining FERMI DECAY and GAMOW–TELLER DECAY constraints.

Angular quantum number (azimuthal)

[atomic, nuclear] QUANTUM NUMBER defining the angular momentum L for electrons in orbit (ℓ) as $= \hbar \sqrt{\ell(\ell+1)}$ ($\hbar$ is the Plank's constant h divided by 2π; $h = 6.62606957 \times 10^{-34}\,\mathrm{m}^2\mathrm{kg/s}$). It is also used in Azimuthal angular quantum number.

Angular speed (ω)

[general] The angular (θ) rate of change in a curved (~circular) orbit: ω = *d*θ/*dt* (see Figure A.77).

Figure A.77 The rate of angle change velocity for an object in continuously changing angle motion around one or more principal axis (an ellipse track has two principal axes).

Angular velocity (ω)

[general, mechanics, quantum] (rad/s) The rate of change in angular position with respect to a reference frame or fixed point for an object moving in curved trajectory; the definition of angular velocity is ω = *d*θ/*dt*. This angular velocity is constant for every point in uniform circular MOTION, since the rate of change in angular position is independent of the DISTANCE to the origin. The angular velocity also relates to the hypothetical circular orbit of the time-dependent sinusoidal AMPLITUDE of an alternating SIGNAL, which repeats every 2π radians or one period. Examples of waves with a sinusoidal pattern are ELECTRO-MAGNETIC RADIATION and SOUND. The tangential VELOCITY (*v*) of a point in circular motion will increase with distance (*r*) to the axis as: *v* = *r*ω.

Anion

[biomedical, chemical, energy, electronics] PARTICLE (ION) with POSITIVE CHARGE, which is attracted by the ANODE.

Anisotropic oscillator

[computational, mechanics] Mechanical OSCILLATOR with more than one degree of freedom that has a different elastic modulus or direction specific elastic spring constant for various directions. Most three-dimensional system is anisotropic, including, but not limited to, a violin or a drum next to layered AIR masses with different temperatures. Another example would be a PENDULUM with the axis not perfectly in perpendicular orientation to the normal force. For electromagnetic oscillations this also applies to materials with BIREFRINGENCE, providing DISPERSION. In QUANTUM MECHANICS, this also applies to a system with mass m, solving for the SCHRÖDINGER EQUATION with the three-dimensional Hamiltonian: $H = -\left(\hbar^2/2m\right)\nabla^2 + (m/2)\left(\omega_1^2 x_1^2 + \omega_2^2 x_2^2 + \omega_3^2 x_3^2\right)$, with three-directional angular frequencies ω_i and $\hbar = h/2\pi$, with $h = 6.62606957 \times 10^{-34} \, \text{m}^2\text{kg/s}$ Planck's constant. In the isotropic case $\omega_1 = \omega_2 = \omega_3 = \omega$.

Annealing

[computational, engineering, solid-state] Two different meanings are associated with annealing, one for the solid-state ENGINEERING application and the other in computational science. In material properties, annealing references a heat treatment. The surface may be chemically altered to make it more resistant to scratches or thermally modified to make the material resilient to applied force in order to induce geometric modifications. Rapid heating and slow cooling will reduce the mechanical strength, whereas regular heating and fast cooling can increase the mechanical strength, as used to give hardness to the blade of a sword. In computational sciences, annealing of data refers to the process of reducing the SOLUTION domain in order to limit the complexity of a system, making it unsolvable. In QUANTUM MECHANICS, the quantum annealing technique provides the perturbation mechanism of action for combinatorial optimization of a GROUND STATE problem for systems with a crystal structure. The crystalline, or glassy DYNAMICS, refers to developing process at extreme slow incremental evolution, also referred to as "relaxation." The relaxed state will have amorphous quantum states on MACROSCOPIC scale. In quantum mechanics, the quantum annealing provides the tools to solve for the time-dependent SCHRÖDINGER EQUATION in a real-time process within a range of constraints and approximations.

Annihilating pair

[atomic, biomedical, mechanics, quantum, solid-state] The combination of two complementary particles forming a new PARTICLE while losing their own identity. The joining of the two particles may also release ENERGY. This release can be in the form of ELECTROMAGNETIC ENERGY such as gamma RADIATION. One example of an annihilation pair is a positron–electron pair. A POSITRON (β^+) is a short-lived particle that has the same mass as an electron and is positively charged, and it has the same rest energy as an electron. Generally, the travel DISTANCE of a positron is limited due to the density of the free electrons in the volume surrounding the positron, primarily less than 1 millimeter, as described by the ANNIHILATION RATE. A free positron will virtually immediately annihilate with any of the many free electrons in the residing medium. This reaction is governed by CONSERVATION OF ENERGY. The energy of the electron before collision with a positron is marginally more than the rest energy. The momentum of the electron on impact can be taken as negligible. The positron–electron annihilation process releases a substantial amount of energy equivalent to the REST MASS of the initial particles, described by conservation of energy as: $m_{pos}c^2 + m_e c^2 - 2h\nu_\gamma = 0$, where ν_γ is the frequency of the emitted radiation (gamma: γ), and m_{pos} and m_e the mass of the positron and electron, respectively. The annihilation energy is liberated as two gamma QUANTA, each with energy of 511 keV, radiating perfectly perpendicular to each other.

Annihilation rate (Γ)

[atomic, biomedical, nuclear, quantum, solid-state] The inverse of the positron lifetime. The positron lifetime is correlated to the electron density, yielding for the annihilation rate (*also see* DIRAC RATE): $\Gamma = Z_{eff} \pi r_e^2 c n_m$, with Z_{eff} the effective number of free electrons per ATOM/molecule, n_m the molecular, respectively, atomic number density, c the speed of light, and r_e the electron radius. The effective number of free electrons per atom/molecule: Z_{eff} will depend on the specific ISOTOPE. Additionally, the kinetic ENERGY of the emitted positron will equivalently impact the energy exchange, proportionally raising the energy of gamma pair due to CONSERVATION OF ENERGY constraints.

Annulus, rotating liquid

[fluid dynamics] In a QUASI-STEADY STATE, the VORTEX in AIR during a TORNADO (CYCLONE) or a swirl in a coffee cup or a whirlpool in the ocean can be approximated by a rotating annulus. Additionally, a LIQUID bearing could also be included; however, this will introduce internal FLOW and stresses. In the theoretical (and practical) approach, rotating fluids become rigid object parallel to the axis of rotation (z-axis). The case of the rotating annulus was treated initially by the French mathematician and scientist PIERRE-SIMON, MARQUIS DE LAPLACE (1749–1827) in his treatise for Saturn's rings in 1787 as well as the mathematician and scientist HENDRIK ANTOON LORENZ (1853–1928) from the Netherlands developed theoretical models on this principle. The symmetrically moving body (also grouping of liquid) revolves around an axis which can be chosen in the z-direction, with an equatorial dissecting plane of symmetry. Generalized, this will be treated as an elliptical path, having two foci. However, the deviation from circular is small (DISTANCE from physical center to either FOCAL POINT divided by the smallest distance from the center to the edge <<1). The MOTION will have potential ENERGY as a function of location $\Omega = \rho\pi\left(\alpha_0 x^2 + \gamma_0 z^2\right) + \text{constant}$. The pressure as a function of location in the annulus is expressed as:

$$\left(P/\rho\right) = (1/2)\omega^2\left(D+x\right)^2 - \Omega + \left[s/\sqrt{\left(D+x\right)^2 + z^2}\right] + \text{constant}.$$

The regular round annulus (still under the assumption that deviation from circular is negligible) will have a moment of INERTIA associated with its revolution; for a flat (i.e., cylindrical) annulus all the mass will be equally distributed yielding: $I = M\left\{\left[R^2 + \left(R + \Delta R\right)^2\right]/2\right\}$, where R is the inner radius of the circular annulus, with WALL thickness ΔR and mass M. While for a tubular annulus the mass is proportional to the "thickness" of the annulus as a function of radius to the center, giving:

$$I = 4\pi\rho\sqrt{\left[\left(8r + 39r^2 R + 30rR^2 + 10R^3\right)/15\right]}$$

with r the radius of the tube itself in the loop configuration and presumably uniform density ρ (see Figure A.78).

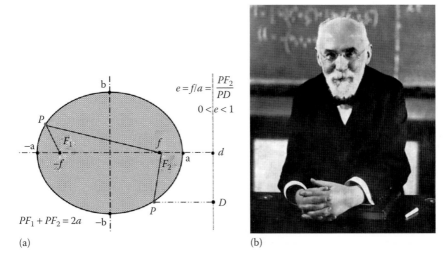

Figure A.78 (a) Geometric configuration of a rotating fluid annulus. (b) Hendrik Antoon Lorentz from the Netherlands. (Courtesy of the Boerhaave Museum, Leiden, the Netherlands.)

Anode

[biomedical, chemical, electromagnetism, general] Positively charged ELECTRODE, usually accompanied by the negatively charged CATHODE, as introduced by MICHAEL FARADAY (1791–1867). The word "anode" in Greek stands for the direction from which the SUN rises. Both a cathode and anode submerged in a SOLUTION of salt, ACID or base will divide the constituents of the respective electrolyte(s) to have the positive (CATION) and negative (ANION) charged ions to diffuse, respectively, to the cathode and anode.

Anomalous Zeeman effect

[atomic, general, nuclear] Random splitting of transition lines between electronic states under the influence of an external MAGNETIC FIELD. This phenomenon is a deviation from the theoretical explanation under the ZEEMAN EFFECT (*see* ZEEMAN EFFECT, ANOMALOUS). The counter phenomenon is the NORMAL ZEEMAN EFFECT.

Antagonist

[general, mechanics] Opposite in direction, mainly used for forces. ANATOMY: two muscles that operate the same joint but in opposing direction and with equal order of MAGNITUDE, one flexing the other under controlled relaxation (see Figure A.79).

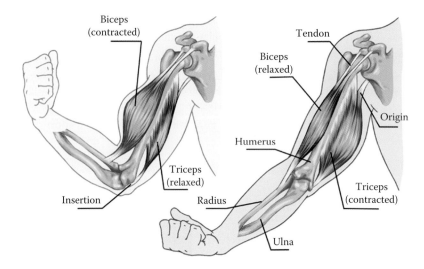

Figure A.79 Specifically in muscular action the torque on a joint cannot be reversed by the same muscle, requiring an antagonist to apply the force to return to the point of origin.

Antenna

[electromagnetism, general] System of conducting constructions that relay electric signals that are converted in the antenna to ELECTROMAGNETIC RADIATION to the outside (transmission) or collecting EM radiation for conversion to electrical signals (reception), for example, RADIO, television, and telephone. Various designations of antennas are used such as Hertz-A, Doublet-A, T-Antenna, and Parabolic (see Figure A.80).

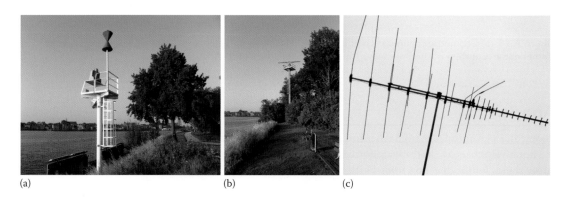

(a) (b) (c)

Figure A.80 (a–c) Variety of transmitting and receiving antennas operating over a broad electromagnetic band.

Antileptons

[general, high-energy, nuclear, quantum, solid-state] Counter particles of the family of leptons in the format of ANTIMATTER (*see* **ANTIMATTER**).

Antimatter

[atomic, general, high-energy, nuclear, quantum, solid-state] Phenomenon derived from QUANTUM mechanical theory in the form of: $E^2 = p^2c^2 + m^2c^4$. This equation has two roots, both positive and negative: $E = \pm c\sqrt{p^2 + m^2c^2}$. The negative roots would represent negative ENERGY, which is unattainable in classical terms, and even under quantum conditions this presented a problem. The negative terms are considered the physical embodiment of the dereliction from a genuine PARTICLE location and associated energy state (i.e., opposite state/particle), in inclusive mirror IMAGE from the real particle. Dirac's view on solid state has all the seemingly virtual counter states in the UNIVERSE occupied in inverted REFLECTION from the real quantum states. The antimatter particle on the other side of the looking GLASS has identical mass as its counterpart but negative energy and opposite charge. An example of an experimentally verified antimatter particle was the introduction of the positron to represent the "antielectron" concept. The creation and annihilation of antimatter obeys the CONSERVATION OF ENERGY principle, making it inherently more challenging to create an antiproton than making a positron based on the respective REST MASS energies. Some artificially produced antimatter particles are POSITRONIUM, antiprotonic hydrogen, antialpha, positron, ANTINEUTRON, antiproton, and concomitantly all subnuclear particles.

Antineutrino ($\overline{v}^*$)

[atomic, general, high-energy, nuclear, quantum, solid-state] Elementary PARTICLE with no CHARGE and negligible MASS compared to an electron emitted to preserve SPIN, momentum, and ENERGY primarily in nuclear DECAY as well as other processes (CONSERVATION OF SPIN, CONSERVATION OF MOMENTUM, CONSERVATION OF ENERGY, and CONSERVATION OF MASS). The antineutrino is produced in a similar process as the neutrino during the breakdown of a NEUTRON: $n \rightarrow p + e + \overline{v}^*$, where n is a neutron, p is a PROTON, and e an electron. ANTIMATTER particle counterpart of NEUTRINO (*see* **NEUTRINO**). The electron and anti-neutrino balance each other in such a way that when the electron ENERGY (i.e., KINETIC ENERGY) is high the antineutrino energy is low and vice versa. The antineutrino has an intrinsic ANGULAR MOMENTUM of $L = \hbar/2$, with $\hbar = h/2\pi$ the reduced PLANCK'S CONSTANT. The rotational direction of the antineutrino SPIN (clockwise/counter clockwise) is given by the LEFT-HAND RULE. The antineutrino is the counter part of the neutrino, which has a spin given by the RIGHT-HAND RULE, which is opposite from that of the antineutrino.

Antineutron

[atomic, general, high-energy, nuclear, quantum, solid-state] ANTIMATTER particle counterpart of NEUTRON (*see* **NEUTRON**).

Antinodes

[general] Crest and through at maximum AMPLITUDE (both positive and negative) in WAVE pattern in contrast to the nodes which are always in EQUILIBRIUM STATE (amplitude zero), no deflection (see Figure A.81).

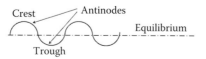

Figure A.81 Antinodes of a waveform: crest and through.

Antipodal point

[computational] A point on a closed circumference that is diametrically opposed in location. Antipodal pairs are used on polygons to define the edges, while on a circle the points are dividing the circumference in two exact halves.

Antiproton ($\bar{p}$)

[atomic, general, high-energy, nuclear, quantum, solid-state] ANTIMATTER particle counterpart of PROTON (*see* PROTON).

Antiquarks ($\bar{q}$)

[atomic, general, high-energy, nuclear, quantum, solid-state] ANTIMATTER particle counterpart of elementary PARTICLE under the grouping quarks (*see* QUARKS).

Antireflection coating

[general] DIELECTRIC coating (e.g., PLASTICS) with a thickness that is a quarter WAVELENGTH ($d = \lambda/4$; quarter-wave plate) of the peak in the optical spectrum of interest. Antireflective coatings on EYE glasses are generally in the range where the eye is most sensitive: yellow-green. The coating design is configured to provide a REFLECTION coefficient between the coating and AIR to be equal to the reflection coefficient between the transparent medium (e.g., GLASS) and the coating. The coating and the medium will have a different index of REFRACTION n. The reflection coefficient for light polarized perpendicular to the PLANE OF INCIDENCE: $R_\sigma = \left(n_1 \cos\theta_i - n_2 \cos\theta_r\right)/\left(n_1 \cos\theta_i + n_2 \cos\theta_r\right)$ (ANGLE of incidence θ_i versus the refracted angle θ_r, SNELL'S LAW, theorem of Malus) is different from the light polarized parallel to the plane of incidence: $R_\pi = \left(n_1 \cos\theta_r - n_2 \cos\theta_i\right)/\left(n_1 \cos\theta_r + n_2 \cos\theta_i\right)$. The same principle can be applied in mechanical dampening by means of a mechanical transition. For instance, for two strings with mass per unit length m_i' the reflection coefficient at the joining point is $R = \left(\sqrt{m_1'} - \sqrt{m_2'}\right)/\left(\sqrt{m_1'} + \sqrt{m_2'}\right)$ (see Figure A.82).

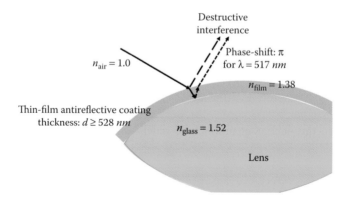

Figure A.82 Antireflective coating concept applied to corrective lenses in eye glasses.

Aorta

[biomedical, general] Compliant BLOOD vessel at the egress route of blood from the left ventricle behind the aortic VALVE. The AORTA has three sinuses that facilitate the outflow of blood from the HEART by means of allowing for TURBULENCE. The sinuses are outward bulges in the contour of the aortic WALL directly downstream from the three respective leaflets of the aortic valve. The track of the aorta starts at the heart and arcs over to drop down into the abdominal cavity where it is named "abdominal aorta," from where it splits in the lower abdomen/groin into the common iliac arteries. In the arc (approximately at a DISTANCE of 5–10 cm from the heart, respective to AGE, gender, and height) the aorta has two arteries leading off to the brain: the left- and right-subclavian arteries, respectively, further branching off into the carotid arteries (see Figure A.83).

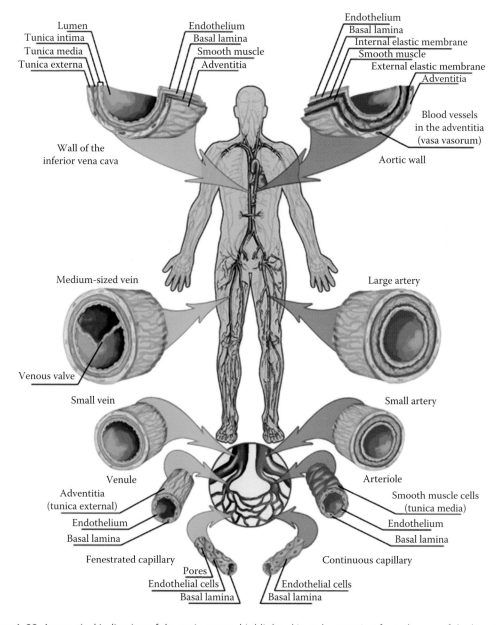

Figure A.83 Anatomical indication of the main artery, highlighted in red, emerging from the top of the heart at the left ventricle: the aorta. For humans, the aorta is several centimeters in diameter (depending on the size and stature of the person, as well as the athletic history and genetic predisposition). The aorta of a blue whale (a mammal—only several hundred remaining, on the verge of extinction) has a diameter in excess of 23 cm (attached to a heart, which is the size of a Volkswagen Beetle). The blue whale is the largest currently living mammal, with the length reaching up to 30 m.

Aperiphractic

[computational, fluid dynamics] Region in an irrotational segment of a moving FLUID, pertaining to Maxwell theory, that is defined by one or more closed boundaries. The closed spherical surface may be contracted down to a point without losing the confinement for the region.

Aperture

[acoustics, general, optics] Optical opening area, also referred to in PHOTOGRAPHY as "*f*-stop" (see Figure A.84).

Figure A.84 The aperture of a camera is used to set the light exposure and control the depth of field for a sharp image, primarily used in analog cameras, but can be found on advanced digital cameras as well; the aperture is an inverse indication, with 3.5 on this lens yielding the largest opening and 22 being the smallest opening. The aperture is the fraction of the focal length over the limiting diameter in the optical system, hence the aperture for a zoom-lens changes with the selected field of view. A large aperture will result in an extensive depth-of-field, whereas a small aperture number has a short depth of field, making only the object that is in focus sharp, blurring the background and foreground. The alternative term "*f*-stop" refers to the concept of focal ratio.

Apollo

[general] A rocket named for the Greek god "Apollo" in a long line of spacecrafts send out to investigate the UNIVERSE directly surrounding the EARTH, as well as and in particular the MOON, in addition looking out without atmospheric INTERFERENCE. The Apollo rocket system was constructed from three components, where the Earth launch tail-end (launch rocket) was discarded at elevation of approximately 68 km. The Apollo program included 11 manned spaceships next to technically three unmanned Apollo missions in the period 1961–1972 as part of the U.S. President Dwight David Eisenhower (1890–1969) and U.S. President John Fitzgerald Kennedy's (1917–1963) space program. The third human flight was with the Apollo 11,

aiming for an up-close and personal lunar observation. In 1969, the space ship Apollo 11 was the first craft to provide a manned landing on the Moon with the, after collecting the Lunar Module (moon shuttle and lunar rover) within orbit, manned by the astronauts from the United States: Neil Armstrong (1930–2012), and Edwin Eugene "Buzz" Aldrin (1930–), while Michael Collins (1930–) remained in orbit around the Moon with the remaining rocket. The use of a lunar-stationary rocket was at the time required to ensure the technical SOLUTION for a return to Earth. Later this issue was resolved with the Space-Shuttle program, a spaceship resembling a plane. Space-Shuttle program was initiated in 1969 with first flight in 1981 (see Figure A.85).

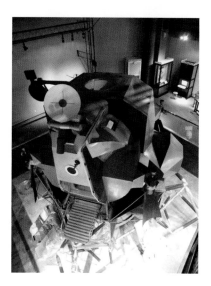

Figure A.85 A representative reconstruction of the Apollo moon landing shuttle, the Lunar Module.

Apparent mass

[general] *See* REDUCED MASS.

Apparent viscosity (η_{app})

[biomedical] Variability in VISCOSITY based on FLOW conditions, specifically CAPILLARY FLOW. This phenomenon is directly tied in with the FAHRAEUS–LINDQVIST EFFECT. The apparent viscosity is related to the standard KINEMATIC (*also* "DYNAMIC") VISCOSITY (η_K) and the radius of the CAPILLARY tube (R) in which the flow takes place as well as the thickness of the cell-free layer (δ), which is of particular importance in colloid solutions such as BLOOD: $\eta_{app} = \eta_K \left\{ 1 - \left[1 - (\delta/R) \right]^4 \left[1 - \left(\eta_K / \eta_b \right) \right] \right\}$, where η_b is the viscosity of whole blood in the center of the flow.

Aqueous humor

[biomedical, optics] LIQUID transparent media filling the space between the cornea and the front of the CRYSTALLINE LENS of the HUMAN EYE. The liquid aqueous humor can morph to remain compliant with the changing curvature of the LENS during accommodation while maintaining a continuous transition between the optical media of the cornea and the lens to avoid distortions resulting from diffraction when air-bubbles would be formed if the medium was more rigid. The aqueous humor has an index of REFRACTION of $n_{\text{aqueous humor}} = 1.337$, keeping in mind that water has $n_{\text{H}_2\text{O}} = 1.331$, and AIR $n_{\text{air}} = 1.0001$ (see Figure A.86).

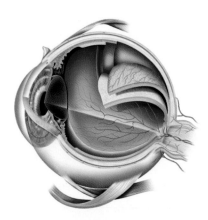

Figure A.86 Aqueous humor is the gel-fluid filling the main compartment of the eye, the volume between the lens and the retina. Increased pressure from excessive aqueous humor fluid will result in glaucoma, which can lead to blindness.

Arago, François Jean Dominique (1786–1853)

[general] A physicist, mathematician, and astronomer, as well as a politician from France. Working with AUGUSTIN-JEAN FRESNEL (1788–1827), he supported the WAVE theory of light in 1811. His theoretical work on the correlation between electrical current and voltage drew the attention of his fellow countryman ANDRÉ MARIE AMPÈRE (1775–1836) in the 1820s. Based on the wave theory, he also argued that the velocity of propagation must be a function of the medium and worked with ARMAND HIPPOLYTE LOUIS FIZEAU (1819–1896) and JEAN-BERNARD-LÉON FOUCAULT (1819–1868) to provide the proof for this in 1850. François Arago also effectively supported the development of the initial stages of PHOTOGRAPHY (DAGUERREOTYPE) by LOUIS-JACQUES-MANDÉ DAGUERRE (1787–1851). François Arago was also the twenty-fifth prime minister of France (see Figure A.87).

Figure A.87 Portrait of Francois Jean Dominique Arago (1786–1853), engraving by Alexandre Vincent Sixdeniers (1795–1846) from a painting by Henry Scheffer (1798–1862).

Arc discharge

[atomic, nuclear] In an ionizing medium (fluids, primarily gasses) the presence of a high ELECTRIC FIELD (E) can induce a displacement current. The IONIZATION materializes as an arc, with ENERGY released as photons. The formation of the arc is described by the empirical equation: $\sigma E^2 + (1/r)(d/dr)\left[rK(dT/dr)\right] - \epsilon' + a = 0$, where σ is the electrical conductivity, K the THERMAL CONDUCTIVITY, T the local temperature of the GAS with dissociated molecules, respectively, atoms and associate free electrons, r the DISTANCE from a pole with an applied electrical potential, ϵ' the volumetric EMISSION COEFFICIENT, and a the broad spectral volumetric absorption coefficient of the medium for all electric frequencies involved in the process (*also see* CORONA (-DISCHARGE), IONIZATION, LIGHTNING) (see Figure A.88).

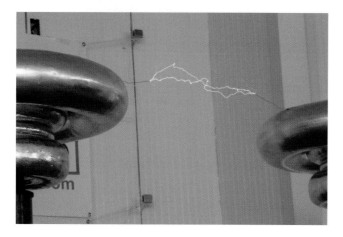

Figure A.88 Arc discharge in a Tesla coil configuration designed by Robert Beck.

Archibald approach to equilibrium method

[biomedical, chemical, mechanics] Mechanism of action used to determine MOLECULAR WEIGHT of the SOLUTE under ULTRACENTRIFUGE treatment. At the point of equilibrium between DIFFUSION and sedimentation the concentration $[C]$ will depend on the DISTANCE (r) to the axis of rotation as a function of ANGULAR VELOCITY ω. The molecular WEIGHT (M) distribution under ABSOLUTE TEMPERATURE (T) and rotational angular velocity: ω, with the assigned partial specific volume of the solute $\bar{V}$ expressed as $M(r) = RTr[C]/(1-\rho\bar{V})\omega^2(d[C]/dr)$, where R is the UNIVERSAL GAS CONSTANT, and ρ is the SOLVENT density based on the dissolved constituent.

Archimedes

[biomedical, fluid dynamics, general] A mathematician and scientist (287–212 BC) from Syracuse then Greece, Sicily, now Italy. Most known for his formulation of BUOYANCY and additionally he developed a water corkscrew used for irrigation, and he presented the mathematical description of "regular bodies," that is, Archimedes bodies. According to urban legend, Archimedes was taking a batch when he realized the concept of buoyancy, that is, Archimedes' principle, and ran naked on the street calling out eureka ("I found it") (see Figure A.89).

Figure A.89 Artistic representation of a person known as Archimedes (287–212 BC) in the form of a statue.

Archimedes number ($Ar = g\rho L^3/\eta^2$)

[astrophysics, fluid dynamics, geophysics, mechanics] The ratio of gravitational forces with respect to viscous forces, where g is the gravitational constant, ρ the density of the FLUID, L the characteristic length, and η the VISCOSITY.

Archimedes' principle

[fluid dynamics, general] A solid body in a LIQUID will experience an upward directed force that is equal but opposite to the WEIGHT of the displaced liquid, defined by Archimedes (287–212 BC) (see Figure A.90).

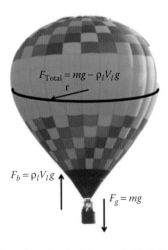

Figure A.90 Hot-air balloon levitation based on the Archimedes' principle.

A

Aristocles (427–347 BC)

[general] He is better known under the name Plato. A scientist and philosopher from Greece (*see* **PLATO (427–347 BC)**).

Aristotle (384–322 BC)

[biomedical, general] A Greek scientist from Stagira, Thracia, during the Macedonian era. He is a student and apprentice of PLATO (427–347 BC). Aristotle was the first to recognize that the EYE captured light reflected from an object, not that the eye emitted probing light itself. He postulated the concept of AETHER, the medium which allows light and SOUND to travel. He also wrote several transcripts on the biology, biological functioning, and reproduction of animals. On the mammalian experience, he posed the three states of existence: vegetative (nourish and reproduction), sensation (cognitive awareness and desire), and rationalism (decision making, only for higher mammals, i.e., humans). His work proposes that decisions are made based on emotions and are ruled by the HEART, not the brain as postulated by his preceding and contemporary scholars such as Plato and Democritus (460–370 BC). In addition to his scientific work, he was intrigued by politics and wrote a detailed description of the governmental structure and arrangements of the very differently organized 158 Greek states/countries (see Figure A.91).

Figure A.91 Statue of Aristotle, artistic interpretation of period data and descriptive documentation by his contemporaries.

Aristyllus

[astrophysics, general] A fourth century BC, Greek scientist and astrophysicist. Together with TIMOCHARIS, they documented the first list of visible fixed stars.

Arndt–Schultz law

[biomedical, energy] Biostimulation. Physiological formulation of the biological stimulus of physiological ACTIVITY expressed as "weak stimuli excite physiological activity; moderately strong stimuli favor physiological activity; strong ones retard physiological activity and very strong ones arrested physiological activity" (see Figure A.92).

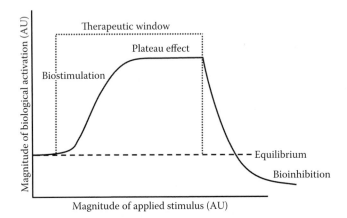

Figure A.92 The Arndt–Schultz law. In chemical reactions and biological activity, there will be an increase in chemical activity due to the administration of a stimulus (e.g., temperature and hormone) until a plateau is reach, while further increase of catalyst will result in decline and eventual inhibition.

Arrhenius, Svante August (1859–1927)

[atomic, computational, quantum] A Swedish chemist and professor of PHYSICS at the University of Stockholm. He worked with FRIEDRICH WILHELM OSTWALD (1853–1932), FRIEDRICH WILHELM KOHLRAUSCH (1840–1910), LUDWIG BOLTZMANN (1844–1906), and JACOBUS HENRICUS VAN'T HOFF (1852–1911). Combining the theoretical efforts on OSMOTIC PRESSURE by Van't Hoff with his own ELECTROLYTIC DISSOCIATION theory opened up a new era in chemistry. For this revealing contribution, he received the Nobel Prize in Chemistry in 1903. Other credits go to his work on serum therapy (see Figure A.93).

Figure A.93 Portrait of Svante August Arrhenius (1859–1927). Svante Arrhenius attending the 1922 Solvay Conference "First Chemistry Conference."

Arrhenius equation

[atomic, computational, quantum] Expression of the temperature dependence of the reaction rate k of a chemical reaction developed by SVANTE AUGUST ARRHENIUS (1859–1927), in 1889, based on the work by van't Hoff: $k = A_n e^{-(E_a/R)1/T}$, with E_a the reaction ENERGY, R the universal GAS constant, T the ABSOLUTE TEMPERATURE, and A_n a unit equalizer constant, with n the order of the reaction. The reaction rate constant depends on the order of the reaction in value and units, for a first order reaction the units are s^{-1}, which can be interpreted as the number of collisions leading to a chemical reactions per SECOND. In DIFFUSION limited reactions the unit constant A_n will play a more significant role than under standard pressure and temperature kinetic reactions. In general PHYSICS and mathematics expressions, the term "Arrhenius equation" refers to an exponential DECAY with a straightforward exponential coefficient.

Arrhenius equation for electrolytic solution

[atomic, computational, quantum] Devised the following equation for the IONIZATION (α_e) of an ELECTROLYTE: $\alpha_e^2 [c]/(1-\alpha_e) = k_i$, with $[c]$ the concentration, and k_i the ionization constant, which was shown to be independent of the concentration itself by the law of dilution of Ostwald.

Arrhenius number ($\alpha^* = E_0/RT$)

[atomic, energy, nuclear, thermodynamics] The relative aspect of activation ENERGY (E_0) with respect to THERMAL ENERGY, where R is the universal GAS constant, and T the temperature.

Asteroid

[astronomy] Galactic PROJECTILE constructed of solids (rocks and metals), ranging in size from millimeters to in the order of thousand kilometers. Asteroids orbit the SUN, mainly in what approximates circular ORBITS, primarily in a belt between the paths of MARS and JUPITER, with the largest identified asteroid named Ceres with size of approximately 1000 km. Asteroids in our SOLAR SYSTEM alone have quantities of millions. However, some asteroids also have larger trajectories around the Sun, intersecting the earth's orbit on occasion. In particular, every year in the first or second week of August, the EARTH traverses a cloud of asteroids, generating a few days of METEOR showers that light up the sky. In contrast, comets are made of GAS, ice, and rock. Comets also have a tail, made from vapors and GAS. There are only several thousand recorded and verified comets in our field of interest. Meteors are celestial objects that enter the Earth's ATMOSPHERE (see Figure A.94).

Figure A.94 Asteroid/meteor impact crater near Flagstaff, Arizona.

Asthenosphere

[acoustics, astronomy/astrophysics, chemical, electromagnetism, fluid dynamics, geophysics, mechanics, solid-state, thermodynamics] One layer of the crust of the planet's surface. The asthenosphere is primarily composed of a plastic medium. The asthenosphere is the underlying structural formation that provides the mechanism that supports SEISMIC ACTIVITY and volcanic activity (see Figure A.95).

Oceanic crust
Continental crust
Lithosphere
Asthenosphere
Upper mantle
Lower mantle
Outer core
Inner core

Figure A.95 Asthenosphere layer of the earth's crust.

Astigmatism

[optics] The formation of a focal IMAGE of a narrow bundle of light formed by a converging LENS generally results in two lines, that are both sharp images, respectively, which are orthogonal to each other in two planes carved out by the optical axis and perpendicular to each other, however, not necessarily in the same location on the optical axes, generally a finite DISTANCE apart. The images convolve into an oval shape proceeding down the optical path from the primary to the secondary focal line. The image formation yields a distance between the sharp projections of the two perpendicular lines. The distance between these two points on the optical axis is referred to as the astigmatic difference. There is, however, a circular image formed in between the two respective locations, but his image is not sharply defined but accepted as the FOCAL POINT. In the EYE, astigmatism represents a geometric contraction in one or more directions, resulting in a distorted image on the RETINA. The distorted image can be corrected by astigmatic lenses

that are defined with a cylindrical axis of correction to the normally applied symmetrical concave or convex lens surface (see Figure A.96).

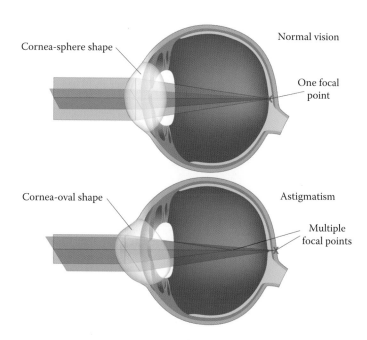

Figure A.96 Deviation of the optical axis by misshaped lens of the eye produces a distortion of the image formed on the retina known as "astigmatism."

Astronomical units (AU)

[general] The average DISTANCE from the EARTH to the SUN: $1\,AU \cong 1.5 \times 10^{11}\,m$.

Astrophysics

[astronomy, astrophysics, computational, energy, general, nuclear, quantum] Science of galactic entities, more specifically their characteristics such as ELECTROMAGNETIC emanation, MAGNETIC FIELDS, and magnetic interactions between PLANETS, SOLAR SYSTEMS, and GALAXIES; TEMPERATURE, COMPOSITION, WEIGHT, GRAVITY and GRAVITATIONAL INTERACTIONS, ORBIT, and places with respect to other galactic bodies or groupings (e.g., galaxies) are also referred to as the PHYSICS of ASTRONOMY. In addition in astrophysics, theoretical, and mathematical concepts are used to make predictions and derive mechanical properties of these galactic bodies without the ability to stage controlled experimental investigations. One of the many topics of interest is celestial MECHANICS (e.g., KEPPLER'S LAW). Studying the spectral emissions of remote stars, for instance, can provide information about the GRAVITATION field strength and field line distribution based on the observed ZEEMAN EFFECT. Certain STARS have revealed gravitation forces in the order of several thousand Gs. Other interests of astrophysics are related to forming a model of how the EARTH came to existence and how the global environment is shaped the way it is (e.g., SEASONS, CIRCADIAN RHYTHM, THERMAL HYSTERESIS, FLUID pathways [wind and rivers] and indirectly impacting our daily WEATHER). Additional information is revealed about particulate (i.e., nuclear) and ELECTROMAGNETIC RADIATION, which provided early documented data on ELEMENTARY PARTICLES and the range of the ELECTROMAGNETIC SPECTRUM.

Atmosphere

[general] The earth's atmospheric composition is made up of the following components in the lower layer: nitrogen (N_2, 78.08%), oxygen (O_2, 20.09%), argon (Ar, 0.9%), carbon dioxide (CO_2, 0.03%), neon (Ne, 0.002%), helium (He, 0.0005%), methane (Me, 0.0005%), next to water VAPOR and several chemicals resulting from environmental pollution (see Figure A.97).

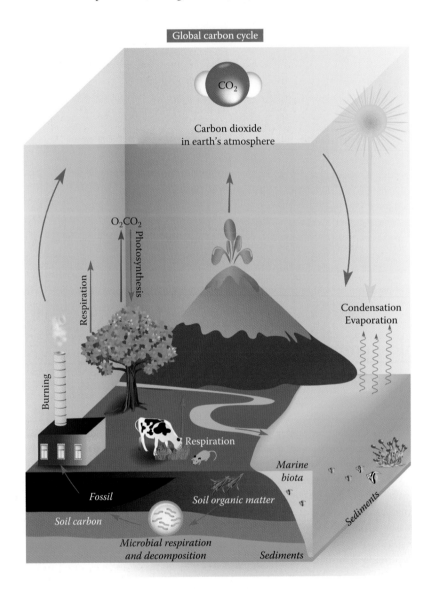

Figure A.97 Drawing of the composition of the atmosphere, specifically geared to the carbon cycle in carbon oxidation (metabolic activity in biology and fossil-fuel burning) with the reconstitution in pure oxygen for consumption and oxidation (i.e., exothermic reaction).

Atmosphere, unit (**atm**)

[general] The unit ATMOSPHERE was indirectly introduced in 1643 by the Italian scientist EVANGELISTA TORRICELLI (1608–1647). The atmosphere is still in common use next to the bar $\left(1\,bar = 10^5\,Pa\right)$ and millimeters mercury (1 atm = 760 mmHg), and in some countries pounds per square inch $\left(1\,PSI = 6.8948 \times 10^3\,Pa\right)$ but was replaced by Pascal (1 atm = 1.01325×10^5 Pa = 101.325 kPa) in based on the efforts by BLAISE PASCAL (1623–1662).

Atmospheric layers

[energy, fluid dynamics, geophysics] The ATMOSPHERE is demarcated into various layers with respect to the most significant physical parameters within a specific range of altitudes. The various atmospheric layers are (starting at sea-level outward): TROPOSPHERE, STRATOSPHERE, MESOSPHERE, THERMOSPHERE, EXOSPHERE, IONOSPHERE, and also includes the MAGNETOSPHERE (see Figure A.98).

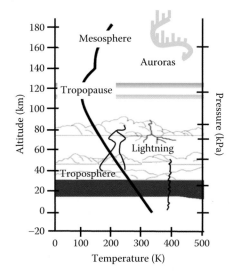

Figure A.98 Detailed diagram outlining the various established atmospheric layers with their respective nomenclature and the reigning temperature ranges.

Atmospheric physics

[astrophysics, atomic, energy, fluid dynamics, general, geophysics] Field of PHYSICS dealing with physical phenomena in the EARTH's ATMOSPHERE. Specific phenomena of interest in the LOWER ATMOSPHERE pertain to METEOROLOGY and CLIMATOLOGY. Additional interests in the UPPER ATMOSPHERE or IONOSPHERE are in relation to communication and the long-term influences on SEASONAL as well as DIURNAL variations. Examples of the latter include the environmental impact of the OZONE LAYER (see Figure A.99).

A

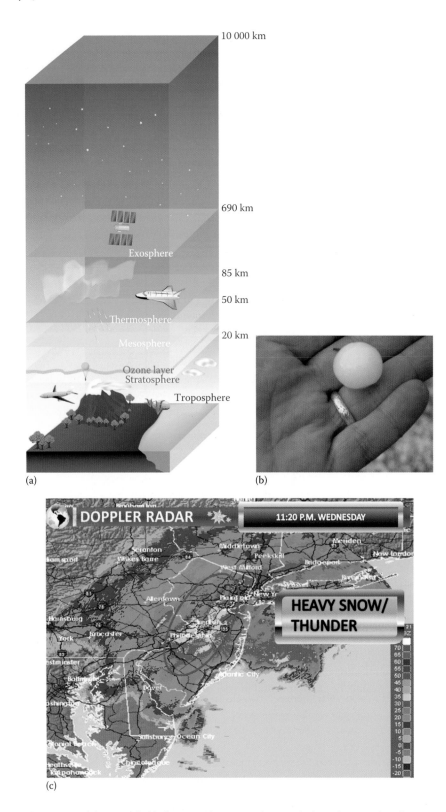

10 000 km

690 km

Exosphere

85 km

50 km

Thermosphere

20 km

Mesosphere

Ozone layer
Stratosphere

Troposphere

(a)

(b)

DOPPLER RADAR   11:20 P.M. WEDNESDAY

HEAVY SNOW/
THUNDER

(c)

Figure A.99 (a) Diagram of distinguishable layers in the atmosphere with their characteristic features. (b) Image of hail precipitation occurring under specific meteorological conditions. (c) Weather conditions obtained by means of Doppler imaging systems designed by atmospheric physicists.

Atmospheric pressure

[fluid dynamics, general] (standard), Pressure measured against atmospheric pressure, also called "GAUGE PRESSURE." VACUUM conditions are frequently expressed as negative pressure. Atmospheric pressure is generally the weight of the column of AIR per unit area $P = \int \rho(r) g dr$ at a certain location with respect to sea-level in altitude (dr), where g represents the GRAVITATIONAL ACCELERATION and $\rho(r)$ the atmospheric density as a function of the DISTANCE to sea level into outer space, which is highly variable over the distance (r) from the surface to the outer outer-layer or IONOSPHERE. The pressure will also change because of geophysical conditions such as temperature and conglomerations of air layers (*also see* WEATHER).

Atom

[atomic, general, nuclear, quantum, solid-state] Smallest self-contained matter unit that has all the primary and secondary characteristics of the MACROSCOPIC ELEMENT, which is embedded in the name that is derived from the Greek word *atomos* (ἄτομος), meaning indivisible. The concept of a discrete structure of MATTER dates back to the Greek philosophers, specifically LEUCIPPUS (fifth century BC) from ancient Greece, who produced the first documented hypothesis of an elementary PARTICLE that contains all the characteristics of a larger assembly object in approximately 450 BC. Later in 420 BC DEMOCRITUS (c. 460–370 BC) made a similar statement in this regard. A quantitative validation of this hypothesis was not obtained until approximately the year 1790 when ANTOINE-LAURENT DE LAVOISIER (1743–1794) submitted that total mass is conserved in any CHEMICAL REACTION no matter what physical shape (LIQUID, SOLID, and GAS). Based on these and other observations, JOHN DALTON (1766–1844) postulated the first chemical reaction statement in 1808, including that every pure chemical is made up of same elementary components, forming the element foundation. The ELEMENTS themselves were also considered to be composed of a unique grouping of identical components: atoms. The designation "element" however was conceived long before then, more precisely in 1661, introduced by ROBERT BOYLE (1627–1691). Additional contributions from AMEDEO AVOGADRO (1776–1856) in 1811, who concluded that not all elements consist of single atoms, some are biatomic forming MOLECULES, which provided supporting evidence with regard to the chemical mass statements made by Dalton. Avogadro's work led to the postulation of the AVOGADRO'S NUMBER quantifying the mass of elements, yielding the atomic mass or ATOMIC WEIGHT as a relative number. The atomic mass for the documented atoms ranges from 1 (HYDROGEN) to 250 as a rounded-of integer, (mass number: A). The atomic structure was further defined by the work of JOSEPH JOHN THOMSON (1856–1940) with the discovery of the electron in 1897. Refinements were made by NIELS HENRIK DAVID BOHR (1885–1962) based on his description of the QUANTUM model of the HYDROGEN ATOM in 1915. ERNEST RUTHERFORD (1871–1937) along with HANS GEIGER (1822–1945) and ERNEST MARSDEN (1889–1970) and provided ALPHA PARTICLE scattering evidence in the period 1908–1913 making way for the BOHR ATOM model by providing proof of the quantization of electronic ENERGY levels. The alpha particle SCATTERING experiments of Rutherford, Geiger, and Marsden revealed that the nucleus has to be of relatively large mass in comparison with the electrons and the nucleus would have a POSITIVE CHARGE. It was concluded that the (relative minority of) alpha particles were recoiled by ELASTIC COLLISION with the nucleus, not by the electrons (*also see* **BOHR ATOM**). In 1913, HENRY (HARRY) GWYN JEFFREYS MOSELEY (1887–1915) delivered experimental proof of quantized X-RAY emission, conforming the energy exchange (CONSERVATION OF ENERGY) between two electron levels ($E(n_i)$) in the form of photons with a specific WAVELENGTH (λ) matching an energy content: $E_{photon} = E(n_2) - E(n_1) = h\nu = hc/\lambda$, with associated PHOTON frequency $\nu = c/\lambda$, where c is the speed of light and h is PLANCK'S CONSTANT. Many of the transitions were documented as BREMSSTRAHLUNG defined, for instance, by the BALMER, LYMAN, and PASCHEN SERIES. The work of ERWIN SCHRÖDINGER on WAVE MECHANICS published in 1927 provided additional background for the development of the atomic model. Specifically, Schrödinger's work led to the introduction of three QUANTUM NUMBERS for the electron in orbit: n the PRINCIPAL QUANTUM NUMBER ($E_n = -\left(2\pi^2 k_0^2 e^4 m_e Z^2 / h^2\right)\left(1/n^2\right)$, where Z is the ATOMIC NUMBER, m_e is the electron mass, e is the ELECTRON CHARGE, h is PLANCK'S CONSTANT, and k_0 the Coulomb force constant), and two ORBITAL ANGULAR MOMENTUM QUANTUM NUMBERS: the ORBITAL ANGULAR MOMENTUM QUANTUM NUMBER ℓ for the MAGNITUDE

$\left(m_{\mathrm{angular}} = [\ell(\ell+1)]^{1/2}\hbar\right.$, where $\hbar = h/2\pi$) and the ORBITAL MAGNETIC QUANTUM NUMBER m_ℓ for the magnitude of the projection in the direction of a specific axis ($L_{z\,\mathrm{orbital\,angular:projection}} = m_\ell \hbar$). The atom has a core made up of PROTONS and NEUTRONS providing the ATOMIC NUMBER (Z) and electron charge due to the difference between the number of positive protons and the balance of electrons in orbit around the NUCLEUS. The core (nucleus) has no specific role in the physical description of chemical interaction and quantum structure apart from the fact that it describes the energy field (i.e., POTENTIAL WELL) that confines the electrons. The nucleus is binding the electrons by COULOMB FORCE, but in order for the electrons not to spiral into the opposite charge nucleus the electrons will need to be in rotational MOTION as described by Rutherford, and supported by the work of Schrödinger. The size of the nucleus was derived from the ALPHA PARTICLE collisions. In the collision the kinetic energy (KE_α) of the alpha particle with mass m_α and velocity v_α needs to convert into potential energy (PE_α) described by the Coulomb energy. This will take place at the radius (R) of the nucleus described as: $KE_\alpha = (\pi/2)\,m_\alpha v_\alpha{}^2 = PE_\alpha = 4\left(k_0 Z e^2/R\right)$, yielding a nuclear size in the order of 10^{-14} m, while the atom has dimensions in the order 10^{-10} m leaving a significant amount of empty space. The size of the average atom was derived experimentally in the late 1800s simply from volumetric calculations. When OIL is allowed to spread out on a water surface, the final result will be a single molecular layer, spread out over a surface area that can be measured. Relating this area to the initial volume of oil will give a reasonable approximation of the thickness and hence the size of the molecule. Other means of deriving atomic size (L_{specific}) applied to atomic gases rely on AVOGADRO'S NUMBER (N_a) for the VOLUME (V) of one MOLE (M_m) of GAS with mass M_m and density: $\rho = M_m/V = M_m/N_a L_{\mathrm{specific}}{}^3$ (see Figure A.100).

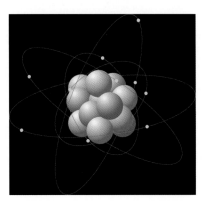

Figure A.100 Hypothetical model of the structure of an atom.

In property classification, the atom is the smallest quantity of a single element that has all the characteristics that can be attributed to any large quantity of this element and verified against scientific standards.

Atomic bomb

[general, nuclear, solid-state] Nuclear fuel device using URANIUM ($^{238}_{92}\mathrm{Ur}$) or PLUTONIUM ($^{239}_{94}\mathrm{Pu}$) (both can be derived from uranium), conceived by ROBERT J. OPPENHEIMER at the Los Alamos, NM research facility with the assistance of NIELS BOHR, JAMES CHADWICK, OTTO FRISCH, ENRICO FERMI, and RICHARD P. FEYNMAN under direction of General Groves. The atomic self-sustaining FISSION reaction for the unstable ISOTOPE U-235 initiated by a NEUTRON proceeds as follows: $^{1}_{0}\mathrm{n} + {}^{235}_{92}\mathrm{U} \rightarrow {}^{236}_{92}\mathrm{U}^* \rightarrow {}^{145}_{56}\mathrm{Ba} + {}^{92}_{36}\mathrm{Kr} + 3{}^{1}_{0}\mathrm{n} + E_{\mathrm{fission}}$, leaving three neutrons to initiate other fission reactions each. Only 0.72% of all uranium occurring as $^{235}_{92}\mathrm{U}$, whereas $^{238}_{92}\mathrm{U}$ (99.27% of all uranium in natural state) is stable and a TRACE occurrence of $^{234}_{92}\mathrm{U}$. Both barium and krypton have a BINDING ENERGY of approximately 8.5 MeV for the 145 and 92 nucleons, respectively, whereas uranium has a binding energy of 7.6 MeV for the 235 NUCLEON, leaving approximately 216 MeV for the fission ENERGY of one uranium ATOM. This reaction was postulated by Leo Szilard, a Hungarian

A

physicist, leading to the development of the atomic bomb in 1941. The nuclear reaction requires a critical mass of greater than 10 kg to be able to support sustained NUCLEAR FISSION. The critical mass is accomplished by forcefully joining two segments of subcritical mass under explosive conditions (detonation). At the moment, the critical mass is exceeded the fission process becomes a self-sustaining CHAIN REACTION fueled by emitted neutrons, while at subcritical mass to many of the neutrons that are released will exit from the uranium or plutonium and the chain reaction is not sustained (see Figure A.101).

Figure A.101 Image of the explosion of an atomic bomb in atmosphere.

Atomic clock

[general] Clock based on the transition between the lowest ENERGY state to the second-lowest state of cesium-133 triggered by the absorption of incident ELECTROMAGNETIC RADIATION. The transition generates a frequency signature that is defined and measurable to an accuracy of 10^{-13}, or defining the second as 9,192,631,770 oscillations on EM RADIATION resulting from the transition. This definition/standard hold true with precision of 1 SECOND deviation over a period of 300,000 years.

Atomic emission

[general, nuclear, optics] The emission of a PHOTON from the (negative-) ACCELERATION of an EXCITED atomic ELECTRON returning to its GROUND STATE. The ATOM can be raised to an excited state with an ENERGY equal or greater than the energy difference between the orbital states (i.e., the BOHR ATOM model). The excitation may result from COLLISION or ELECTROMAGNETIC RADIATION. This process is the primary mechanism of action for the operation of LASER.

Atomic energy

[general, nuclear, solid-state] An ATOM stores ENERGY from the electrical attraction between the positive NUCLEUS and orbiting negative ELECTRONS (the BOHR ATOM model). The electrical force described by COULOMB FORCE ($F_e = k_0(Ze)(e)/r^2$) equals the CENTRIPETAL FORCE ($F_c = m_e(v^2/r)$) of the orbit for equilibrium to be maintained. The GROUND-STATE ENERGY of each ELECTRON ORBITAL MOTION identified by the PRINCIPAL QUANTUM NUMBER: n, is expressed as: $E_0 = -\left(2\pi^2 k_0^2 e^4 m_e/h^2\right)\left(Z^2/n^2\right)$.

Atomic force microscope (AFM)

[biomedical, general, solid-state] Imaging device that maps the surface FORCE and surface elevation by means of electrical attraction between a nanoprobe and the ATOMIC and MOLECULAR structure of the MEDIUM under investigation using the COULOMB FORCE to generate a signal that is represented as a function of the two-dimensional location of the probe. The AFM can typically measure forces better than $1\,fN$. The concept of the AFM is based on the principles of the scanning tunneling microscope (STM) developed by the scientists GERD BINNIG (1947–) and HEINRICH ROHRER (1933–2013), both from Switzerland, working for IBM. The AFM provides resolution at least a thousand times greater than optical diffraction limit. The AFM moves a hyperfine needle over a surface by means of piezoelectrical OSCILLATION. The concept resembles braille reading to an extent. Generally, the mechanical MOTION of the CANTILEVER is recorded by means of a reflected laser beam; the position of the reflected ray on a photo-array correlates linearly with the probe's motion in response to the scan across the surface relief. Under these operating conditions, the use of a stylus with tip size of several nanometer the horizontal resolution is approximately 0.2 nm, and progressively improving, while the vertical resolution is better than 0.1 nm. The topography can be determined in full-contact/DRAG mode, or in tapping mode, respectively. The tapping mode relies on a DRIVEN DAMPED HARMONIC OSCILLATOR. The AFM resolution is in the order of single ATOM vertically. The resolutions differs due to the lateral use of piezo drives (more course) and springs vertically. The electronic feedback on the piezo deflection allows MEASUREMENT of electrical forces between the probe and the surface of less than $1\,fN$. After SIGNAL PROCESSING an IMAGE is created on atomic scale showing the force distribution and hence the localized (molecular-) material properties (see Figure A.102).

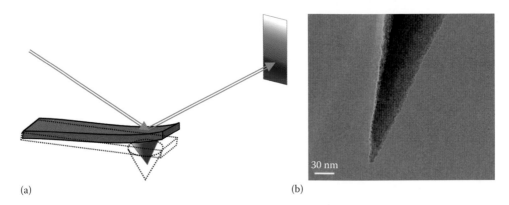

(a) (b)

Figure A.102 (a) Mechanism of operation of the atomic force microscope. (b) Electron microscopy image of the deflecting needle tip (PointProbe®) used in the AFM.

Atomic line spectra

[atomic, chemical, energy, mechanics, optics, quantum, thermodynamics] Emission lines of atomic excitation (*see* LINE SPECTRA).

Atomic mass, unit (*u*)

[general, nuclear, solid-state] One twelfth of the mass of standard CARBON ($^{12}_{6}$C); $1u = 1.660540 \times 10^{-27}\,\text{kg}$.

Atomic model, Bohr

[atomic] *See* BOHR ATOMIC MODEL.

Atomic model, Rutherford

[atomic] *See* **Rutherford model**.

Atomic model, Thomson

[atomic] *See* **Thomson atomic model**.

Atomic number (Z)

[atomic, general, nuclear] The quantity of protons in the NUCLEUS of an ATOM, also providing the total POSITIVE CHARGE. The concept was introduced in 1913 by Niels Henrik David Bohr (1885–1962) with assistance from Henry (Harry) Gwyn Jeffreys Moseley (1887–1915).

Atomic weight

[atomic, nuclear] *See* ATOMIC MASS.

ATP (adenosine triphosphate)

[biomedical, chemical, energy, thermodynamics] Biological renewable CHEMICAL ENERGY source. ATP is a chemical substance that can exchange a single phosphate molecule under hydrolysis with release of ENERGY. This process can easily be reversed by means of external energy, either released from metabolic process and, under certain conditions, by means of solar RADIATION. The average energy release is 7.3 kcal/mole: $ATP + H_2O \leftrightharpoons ADP + phosphate + energy$, producing ADENOSINE DIPHOSPHATE (ADP). The depleted ATP, that is, ADP can quickly be replenished through the phosphate release from stored creatine-phosphate (specifically in MUSCLE) next to GLYCOGEN (70% stored in muscle, 20% in the LIVER, and the remaining 10% floating freely in BLOOD to provide an emergency backup). In the biological CELL, there is a secondary and tertiary chemical energy process produced by NADH (nicotinamide adenine dinucleotide: FAD) in the form of $2NADH + 2H^+ + O_2 \leftrightharpoons 2NAD^+ + 2H_2O + energy$, respectively, $2FADH_2 + O_2 \leftrightharpoons 2FAD + 2H_2O + energy$ (flavin adenine dinucleotide: FAD) (see Figure A.103).

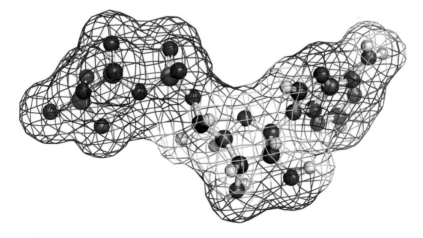

Figure A.103 Graphical interpretation of the adenosine triphosphate (ATP) molecule.

A

ATPases

[biomedical, energy, thermodynamics] Active ionic reactions or ION transfer (in biological systems) that consume the ENERGY from ATP conversion. ATPases are proteins that have been conserved throughout evolution and are grouped in three categories: P-type, V-type, and F-type. The P- and V-type ATPases are involved in processes where the ATPases consume energy, whereas the F-type ATPases provides a major source of ADENOSINE TRIPHOSPHATE (ATP, a nucleoside triphosphate renewable energy source used in cells as a coenzyme). The P-type ATPases include the cellular MEMBRANE ion gradient maintenance through the Na^+/K^+ ATPases as well as Ca^{2+} ATPases, specifically in MUSCLE and participates in the phosphorylated intermediate steps of the phosphor reactions. The V-type is involved in the regulation of the acidity (pH) of the CELL, in particular the accumulation of H^+ in the vesicle lumen. ATPase F-type drives the ATP synthase by means of the inner mitochondrial membrane. In this process, under aerobic conditions, a single GLUCOSE molecule is converted in pyruvate molecules in two-to-one ratio, yielding 2 ATPs; where subsequently the respective pyruvate molecules can produce up to 32 TP molecules from ADENOSINE DIPHOSPHATE (ADP) in a citric ACID cycle within the MITOCHONDRIA. The general role of ATPases is in the recovery process after an ACTION POTENTIAL has been generated and the CELL MEMBRANE needs to brought back to equilibrium NERNST POTENTIAL. In response to the uncontrolled K^+ efflux after the action potential that opens the ion-gates in the membrane for free migration, the cell reaches hyperpolarization. The ATPases subsequently initiate the Na^+/K^+-pump mechanism to regulate the inner- and extracellular ion concentrations, moving three (3) Na^+ ions out of the cell against two (2) K^+ ions. The Na^+/K^+-pump ACTIVITY is regulated by means of membrane receptors that respond to fluctuations in cyclic adenosine monophosphate (cAMP, or 3′-5′-cyclic adenosine monophosphate; a second membrane messenger molecule). In addition, hormones such as insulin, thyroid-hormone, and aldosterone increase the Na^+/K^+-pump mechanism, whereas dopamine in the KIDNEY in particular inhibits the activity (see Figure A.104).

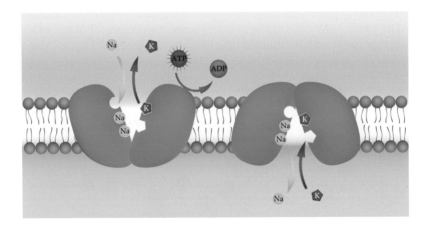

Figure A.104 The role of ATP in the membrane energy exchange supporting active processes, such as the sodium-potassium pump.

Atwood, George (1746–1807)

[general, mechanics] A British physicist known for his computational work on MECHANICS. He conceived a machine that can be used to experimentally determine accelerated MOTION (i.e., GRAVITATIONAL ACCELERATION), aptly named after him as the Atwood machine. The Atwood machine uses a PULLEY system with two weights suspended on either side from a string hanging over the pulley that remains in balance if both weights are equal, when a mass is added to either sides, the system of weights will perform linear accelerated motion and can be used for validation (see Figure A.105).

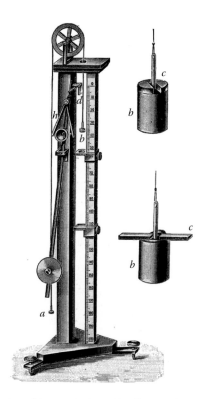

Figure A.105 Drawing of the Atwood machine, as designed by George Atwood (1746–1807) for testing of mechanical principles.

Auger, Pierre Victor (1899–1993)

[atom, energy, general, solid-state] A nuclear physicist from France, who started out working under JEAN PERRIN (1870–1942) on the PHOTOELECTRIC EFFECT. Pierre Auger provided elementary insight in the interaction of electrons on an atomic level. In electron microscopy, the AUGER ELECTRONS captured from SCATTERING events are named after him (see Figure A.106).

Figure A.106 Pierre Victor Auger (1899–1993).

Auger effect

[atomic, nuclear, solid-state] Atomic or molecular electronic transition from higher to lower energetic state in which another electron (Auger electron) is released and no RADIATION is emitted. First described by PIERRE AUGER (1899–1993). This process usually takes place in high energetic electron ORBITS close to the NUCLEUS and is primarily associated with the interaction of X-RAY with MATTER. Auger transitions can be extremely fast; one example is the COSTER–KRONIG TRANSITION. The radiationless transit PROBABILITY is described mathematically as $w_{\mathrm{fi}} = \hbar^{-2}\left|D_{\mathrm{irect}} - E_{\mathrm{xchange}}\right|^{2}$ (see Figure A.107).

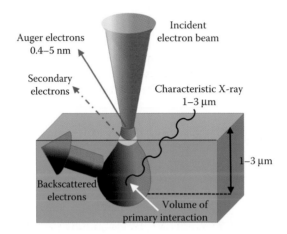

Figure A.107 Graphical representation of the electron interaction with a conductive medium, releasing Auger electrons from the surface with reduced energy.

Auger electrons

[atomic, imaging] Scattered electrons used to provide ENERGY information about the structural configuration of an atomic structure under electron microscopy imaging. Incident primary electrons interact in many different ways: IONIZATION with release of secondary electrons, REFLECTION, REFRACTION, LUMINESCENCE, and desorption (process of molecular breakdown) next to the AUGER EFFECT. During the interaction of primary electrons from the electron beam inner electrons may be ejected, creating a vacancy that requires filling with an electron from a lower energy (higher) orbit. The RADIATION emission resulting from the Auger electron is characteristic for the ATOM in question and can hence be used for atomic and molecular specification of the object under investigation.

Aurora

[general] A stream of IONS, caused by SOLAR FLARES, accelerating in the Earth's ATMOSPHERE under GRAVITATIONAL pull near the POLES. The ions travel through space and become entangled in the EARTH'S GRAVITATIONAL FIELD to be drawn into the NORTH AND SOUTH POLES. The charges (q) moving with

VELOCITY ($\vec{v}$) are subject to NEWTON'S THIRD LAW and will experience a FORCE (F_B) resulting from the MAGNETIC FIELD ($\vec{B}$). The force on the charge is, with AMPÈRE'S LAW, the magnetic equivalence of the LORENTZ FORCE, written as: $F_B = q\vec{v} \times \vec{B}$, causing the ION to move in a direction that follows from the RIGHT-HAND-RULE, migrating in a circular MOTION (more specifically: FLEMING'S RIGHT-HAND RULE). The force is perpendicular to the MAGNETIC FIELD direction while the force is also perpendicular to the velocity. The magnetic force will continuously attempt to balance against the CENTRIPETAL FORCE: $F_C = mv^2/R$, where m is the mass of the PROJECTILE and R designates the radius of curvature of the circular MOTION. The charge will however lose KINETIC ENERGY because of external forces (e.g., FRICTION and collision) causing it to slow down, which results in a spiraling motion toward to EARTH's surface. The phenomenon observed in the NORTH POLE (AURORA BOREALIS) is generally more impressive than on the SOUTH POLE (AURORA AUSTRALIS). As the ions are accelerated and collide with atmospheric molecules, they emit ELECTROMAGNETIC RADIATION that causes the polar sky to light up with a stream of colored light at approximately 95 km altitude. The phenomenon at such high altitude explains why it is visible down to the 60th LATITUDE and on occasion as far down as to the 30th latitude (see Figure A.108).

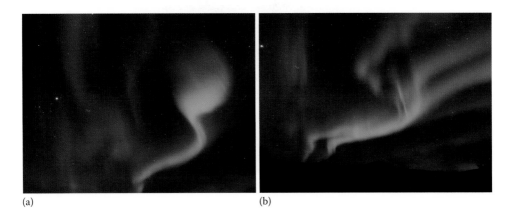

(a)   (b)

Figure A.108 (a,b) Aurora borealis. (Courtesy Joe Morris.)

Aurora Australis

[astrophysics, atomic, energy, geophysics] AURORA on the SOUTH POLE, *also* SOUTHERN LIGHT (*see* AURORA).

Aurora Borealis

[astrophysics, atomic, energy, geophysics] AURORA on the NORTH POLE, *also* NORTHERN LIGHT (*see* AURORA).

Auscultatory method for blood-pressure determination

[acoustics, biomedical, fluid dynamics] From Latin: auscult, meaning "listen." It is a method using a SPHYGMOMANOMETER in combination with a stethoscope, where the former is used to measure compression force on an artery in the arm with a mercury column indicating the applied pressure in mmHg while listening for the SOUND made by BLOOD seeping through a small slit as the artery is opened up in hemostatic balance with the applied external pressure from an inflated (RIVA-ROCCI) CUFF around the arm that is slowly emptying out. The "listening" applies to the Korotkov sounds made when the blood is forced through the narrow opening acting as letting AIR out of a balloon making a screeching sound, in this case only intermittently when the SYSTOLIC PRESSURE exceeds the externally applied pressure (see Figure A.109).

Figure A.109 Measuring blood pressure based on the audible sounds resulting from the periodic blood flow (acting like a kazoo; toy blow instrument) generating mechanical vibrations when blood is forced through a pressurized closed venturi (Korotkoff sounds). The audible aspect is referred to as the "auscultatory method," in contrast to, for instance, Doppler flow recordings.

Autocorrelation

[computational, theoretical] A process of mathematically identifying a SIGNAL with respect to itself, past and future events. In comparison, cross-correlation compares a signal to a "universal" standard. Correlation theory is concerned with quantifying the similarity between two signals. It has wide application, ranging from radar signal processing to character recognition. The procedure of cross-correlating the probing, incident signal with the measured signal in order to highlight time-lag between the two signals for feature recognition with the medium the measured signal has returned from. Specifically, the spectral analysis of the two signals is imbedded in the resolved harmonics of the autocorrelated signal. The autocorrelation process is performed as a function of time, thus providing time resolved information, specifically with respect to nonperiodic functions. The defined (in most cases "electronic") process (X) changes with time and holds the input or initial value as well as the returned value at different points in time (time-lag between input and MEASUREMENT: time t and time $t - \tau$, respectively). The signal will have a mean value at time i: μ_i and associated variance: σ_i^2 (with standard deviation σ_i). The autocorrelation between the events recorded at times t and $t - \tau$ is defined by $R(t, t - \tau) = E\left[\left(X_t - \mu_t\right)\left(X_{t-\tau} - \mu_{t-\tau}\right)\right]/\sigma_t \sigma_{t-\tau}$ (also referred to as the "autocorrelation coefficient," due to the NORMALIZATION $1/\sigma_t\sigma_{t-\tau}$, and the fact that all measurements are taken against the mean μ_t), where $E\left[(X_t - \mu_t)(X_s - \mu_s)\right]$ expresses the "expected value" operator. Alternatively, the correlation between two variables x (with mean μ_x and standard deviation σ_x) and y (with mean μ_y) expressed similarly as $R(x, y) = \sum_N (x_t - \mu_x)(y_t - \mu_y)/\sigma_x \sigma_y$. Generally, the autocorrelation process needs to be performed within a spectral range, depending on the source and the application as well as the type of information sought after. The value of $R(t, t - \tau)$ falls in the range $[1, -1]$, where $R(t, t - \tau) = 1$ (respectively $R(x, y) = 1$) indicates perfect correlation and $R(t, t - \tau) = -1$ references total anti-correlation. When considering a signal that can be described as a well-defined function $f(t)$ the [continuous] autocorrelation process becomes the convolution (*) between the function at time t with lag τ expressed as $R_{ff}(\tau) = \left(f(t) * f^*(-t)\right)(\tau) = \int_{-\infty}^{\infty} f(t + \tau) f^*(t) dt = \int_{-\infty}^{\infty} f(t) f^*(t - \tau) dt$, where f^* is the complex conjugate. The autocorrelation process also allows for tracking through time by artificially increasing the time-lag in the correlation process, hence specifically increasing the temporal resolution, which may be limited by the device specifications or based on the preferred "distance (x)" the collected signal is returning from. In this case, the DISTANCE can be physical depth, linked to TIME (t) by the speed of propagation (v) as $x = vt$, or is the delayed signal resulting from, for instance, atomic RESONANCE. The spatial information retrieval is used in OPTICAL COHERENCE TOMOGRAPHY and a variety of other tomographic imaging techniques, which employ the simultaneous identification of "resonance" characteristic recovery for identification of atomic, energetic, and anatomic details such as primarily used in ULTRASOUND, and with selective applications for PET, and MRI, for resolution enhancement. The autocorrelation process also defines the "visibility" concept, how well can two signals be separated, that is, location information as well as time information resolved. The visibility is defined as the difference in intensity between the highest AMPLITUDE and lowest amplitude.

Autocorrelation function

[computational] It represents the degree to which the field correlates to itself at an earlier time: $\gamma(\tau) = E(t)E(t-\tau)/E(t)^2$, where $E(t)$ represents the "expected value" operator for the function or SIGNAL as a function of time, meaning the weighted average of all possible values. This weighted average can be acquired from multiple recurring measurements on the repetitive signal, generally adhering to a PROBABILITY distribution, for instance, NORMAL DISTRIBUTION or Gaussian (*see* AUTOCORRELATION).

Autocrine signaling

[biomedical, chemical, energy] Cellular secretion of soluble factors that regulate the hemostasis and often also control the growth rate of the CELL.

Average binding energy ($E_{bind, avg}$)

[nuclear] The NUCLEAR BINDING ENERGY for a solid or a LIQUID is a constant, allowing the heat of vaporization to account for the binding energy. The heat of vaporization (Q) is the total amount of work required to generate n separate molecules by dissociation of m gram of liquid (or solid) substance, as provided by $E_{bind, avg} = Q/n = QM_0/m$, where $m = nM_0$, and M_0 is the mass of one molecule (see BINDING ENERGY).

Avogadro, Lorenzo Romano Amedeo Carlo (1776–1856)

[general] A physicist, engineer, and chemist from Italy; Count of Quaregna and Cerreto [Lorenzo Romano Amedeo Carlo Avogadro di Quaregna e di Cerreto]. In 1811, Amedo Avogadro made a statement introducing the word and definition "molecule." The statement contains two hypothesis that are generally accepted as true. The first hypothesis is that gaseous media can exist in molecular form. The second hypothesis

A

verbalizes that equal volumes of different gasses in molecular form under the same pressure and temperature will consist of the identical number of molecules, that is, Avogadro's law (see Figure A.110).

Figure A.110 Representation of Lorenzo Romano Amedeo Carlo Avogadro (17776–1856).

Avogadro's constant (N_A)

[atomic, biomedical, general, nuclear, thermodynamics] Standard for an arbitrary ELEMENT or chemical substance describing the quantity of atoms or molecules in a gram-molecular weight, respectively, *also* Avogadro's NUMBER. The constant defining the number of moles of GAS in $1 cm^3$ under standard temperature (0°C) and pressure (1 atm) [STP]. $N_A = 6.022140857(74) \times 10^{23} \pm 1.8 \times 10^{16}$ (2011 definition). In addition, the Avogadro's number is associated with the MOLE as the number of atoms in a quantity of pure carbon-12, where 1 mole of ^{12}C weighs exactly 12 g, and contains Avogadro's number in atoms of carbon. This was defined by Amedeo Avogadro (1776–1856) around 1811, following on the work of John Dalton (1766–1844) in 1805.

Avogadro's law

[atomic, fluid dynamics, general, nuclear] Equal VOLUMES of different GASSES under equal TEMPERATURE and PRESSURE contain an equal amount of MOLECULES. Expressed by Amedeo Avogadro (1776–1856) in 1811. This evolved in the definition of the UNIVERSAL GAS CONSTANT that states the number of any gaseous molecules in a volume of $1\,\text{cm}^3$ under standard temperature ($0\,^\circ\text{C} = 273.15\,\text{K}$) and standard pressure ($1\,\text{atm} = 1.01325 \times 10^5\,\text{Pa}$) [STP].

Avogadro's number (N_A)

[atomic, biomedical, general, nuclear, thermodynamics] Standard for an arbitrary ELEMENT or chemical substance describing the quantity of atoms or molecules in a gram-molecular weight respectively, equating to 6.02214×10^{23} atoms or molecules/mole (*see* AVOGADRO'S CONSTANT).

Azimuthal (orbital angular) momentum quantum number (ℓ)

[atomic, high-energy, nuclear, quantum, solid-state] QUANTUM NUMBER introduced to define the additional DEGREES OF FREEDOM with respect to the nuclear and electron MOTION for the ATOM next to the principle quantum number (n). The azimuthal momentum quantum number or orbital angular momentum quantum number describes the degree of rotation and associated "FINE-STRUCTURE" splitting resulting from PRECESSION. The angular momentum associated with the wavemechanics of an electron in orbit is $L = \hbar\sqrt{\ell(\ell+1)}$, $\hbar = h/2\pi$, with $h = 6.62606957 \times 10^{-34}\,\text{m}^2\text{kg/s}$ the Planck's constant, and $\ell = 0, 1, 2, \ldots, n-1$. Electron transitions are governed by the PAULI EXCLUSION PRINCIPLE (see Figure A.111).

Figure A.111 Representation of the angular (azimuthal) momentum quantum number associated with the energy state of the atomic electron spin.

Bacher, Robert Fox (1905–2004)

[nuclear] A physicist from the United States. One of the principal scientists on the MANHATTAN PROJECT. The work of Robert Bacher provided significant insight into the structure of the atomic NUCLEUS (see Figure B.1).

Figure B.1 Robert Fox Bacher (1905–2004). (Courtesy of Los Alamos National Laboratory, Los Alamos, New Mexico.)

Backscatter

[acoustics, optics] Optical and/or pressure waves that are reemitted from within a medium that thus carry information about the object the waves interacted with. In OPTICS, the angular spread of the backscattered light will provide details about the size of the SCATTERING cross-sectional area, described by JOHN WILLIAM STRUT, THE 3RD BARON RAYLEIGH (1842–1919) and GUSTAV ADOLF FEODOR WILHELM LUDWIG MIE (1869–1957). In ACOUSTICS, the resonant remittance of SOUND waves provides details about the elastic modulus of the region and the discontinuity with the surrounding volume (see Figure B.2).

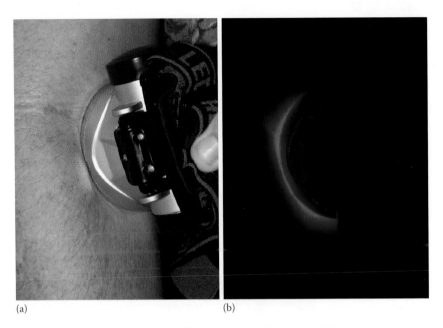

(a)   (b)

Figure B.2 (a) Image representing the concept of backscatter. (b) The arm with little pigment scatters the incident light back, primarily the red light from the white light source pressed up against the skin shown in (a).

Bacon, Francis, Sir (1561–1626)

[general] A scientist and philosopher from Great Britain (later known as Lord Verulam). The work of Sir Bacon describes the equivalence of MOTION and heat; following in the footsteps of PLATO (427–347 BC), he defined the equivalence of motion (kinetic ENERGY) and thermal phenomena. Additional work by Sir Bacon provided insight into the charges released when certain LATTICE structures are broken, and he described the sparks generated when sugar crystals are broken (see Figure B.3).

Figure B.3 Sir Francis Bacon (1561–1626), engraved by James Posselwhite.

Bacon, Roger (1214–1294)

[general] A scientist and philosopher from Great Britain, as well as Franciscan friar. The work of Roger Bacon with respect to OPTICS, corrective VISION more precise, was published in the *Opus Majus* in 1267.

Bacteriorhodopsin

[biomedical, chemical, thermodynamics] A transmembrane protein used in ENERGY conversion and ACTIVE TRANSPORT for harvesting light energy and converting it to electrostatic energy in the form of PROTON transport (PTR: $^1_1H^+$) to be transported out of the CELL for energy delivery elsewhere. Bacteriorhodopsin is part of certain rod-CELL MEMBRANE structures, pertaining to intensity VISION (not necessarily for COLOR [cone] vision). The light-induced proton transport against the proton gradient is a significant feature in bioenergetics. Bacteriorhodopsin is a CHROMOPHORE that converts light into electrochemical energy through a protonated Schiff base within a time frame in the order of 2 femtoseconds. The chemical process following the light conversion results in chemical conversion to bathorhodopsin after 2000 fs = 2 ps, initiating a change in MEMBRANE permeability leading up to cellular DEPOLARIZATION. Only several photons acquired simultaneously are required to produce a full depolarization, resulting in the electrical stimulus conveying the observation of several photons to the brain. On a local cellular membrane level, the effects of the interaction of a single photon can be verified. The chemical process involves the isomerization of carbon–carbon double bond (C-11) with a high energy GROUND STATE which is sterically strained. In the isomerization, the order/sequence of certain atomic structures with the large molecule is rearranged, forming another molecule with the same atoms in total. Additional chemical processes in the integral process of light reaction are the electron coupled and PROTON transfer in complex IV of cytochrome-*c*-oxidase (electron transport chain). The energetic process of proton transfer can be described through a SOLUTION of the SCHRÖDINGER WAVE EQUATION for the VALENCE bond RESONANCE structures, with WAVE functions representing the protonated and deprotonated sets of acceptors with valence bond resonance: Ψ_1 representing retinal (Schiff base) amino ACID Asp85 and Ψ_2 representing retinal amino acid Asp-H. The wave functions for the chemical structures in bacteriorhodopsin are defined through: $\Psi_1: R_1 \stackrel{\cdots}{=} C(H) = N^+(R_2)H - A_1$, A_1 representing the acceptor (Asp85) and R_i the rest-terms of the large molecule (MOLECULAR WEIGHT of bacteriorhodopsin 35,200 ± 1,200), N stands for nitrogen, C stands for carbon, and H stands for hydrogen; Ψ_2: $R_3 \stackrel{\cdots}{=} C(H) = N^+(R_4)H - A_2$, A_2: Asp-H. The respective diabatic SURFACE ENERGY states for the two conditions and photochemical interactions and diabatic states associated with the two wave functions (1 and 2) are $\bar{\varepsilon}_1 = \varepsilon_{QCFF/PI}(SBH^+) + \varepsilon'_{Asp85} + \varepsilon^{(1)}_{SS'} + \varepsilon^{(1)}_{Ss} + \varepsilon_{ss} + \alpha^{(1)}$ and $\bar{\varepsilon}_2 = \varepsilon_{QCFF/PI}(SB) + \varepsilon'_{AspH85} + \varepsilon^{(2)}_{SS'} + \varepsilon^{(2)}_{Ss} + \varepsilon_{ss} + \alpha^{(2)}$, respectively; the surface energy of the chromophore is given by $\varepsilon_{QCFF/PI}$, which incorporates the potential from the solvent/protein system (QUANTUM mechanical consistent force field method for Pi-electron systems [QCFF/PI]) for the Schiff base, ε' is an empirical electron valence bond (EVB) description of Asp85 or AspH85, $\varepsilon_{SS'}$ is the interaction between the classical EVB and π-electron systems, and $\varepsilon^{(i)}_{Ss}$ represents the interaction between the SOLUTE (S) and the SOLVENT (s) in the given state, while $\alpha^{(i)}$ defines the "gas-phase shift," incorporating the covalent influences of the water environment on the proton transfer. The classical

EVB and π-electron systems can be treated as regular EVB because the steric interactions between the two fragments and the classical electrostatic interactions can be considered under classical mechanical terms instead of QUANTUM MECHANICS. The Franck–Condon factor (FCF) for the process between the two states is given as $FCF = \left| \int_{-\infty}^{\infty} \Psi_n^{1*}(x) \Psi_m^2(x) dx \right|$. Incorporating the GROUND STATE (E_g) into the energy balance is accomplished through the use of the off-diagonal coupling ELEMENT tying the wave functions H^{12} in the form: $E_g = (1/2)\left\{ (\varepsilon_1 + \varepsilon_2) - \left[(\varepsilon_1 - \varepsilon_2)^2 + 4(H^{12})^2 \right]^{1/2} \right\}$. The depolarization is fueled by the proton motive force (PMH(ΔpH); compare to electromotive force), consisting of a CHEMICAL GRADIENT (ΔpH) and a pure charge gradient ($\Delta\Psi$) expressed as $PMH(\Delta p_{proton}) = \Delta pH + \Delta\Psi$ (see Figure B.4).

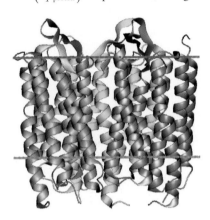

Figure B.4 Graphical representation of bacteriorhodopsin as part of the membrane proton-pump mechanism, the red line illustrates the external side of the bilipid membrane. (Courtesy of Andrei Lomize.)

Baffled piston

[acoustics, imaging] Acoustical screening for vibrating piston to steer the acoustic profile, focus as well. The baffle provides a filtering mechanism to counteract the out-of-phase waves originating from other parts in the moving system, for instance, the backside of the piston. The backside of the piston has an extended travel DISTANCE for the pressure WAVE and is hence not necessarily in PHASE with the pressure wave emerging at the front. The baffled piston provides the mechanism used to acquire SIGNAL used for velocity-based near-field acoustic holographic imaging. In the holographic imaging process, the pressure ($P(\vec{r},t)$) wave obeys the homogeneous WAVE EQUATION in three-dimensional space $\nabla^2 P(\vec{r},t) - (1/v_s^2)(\partial^2/\partial t^2) P(\vec{r},t)$, with v_s the speed of SOUND (see Figure B.5).

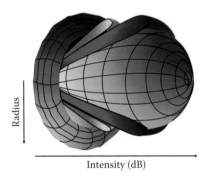

Radius

Intensity (dB)

Figure B.5 Power profile of baffled piston, used, for instance, in naval sonography.

Bainbridge, Kenneth Tompkins (1904–1996)

[atomic, nuclear] A physicist from the United States who was instrumental in the development process and theoretical analysis of the CYCLOTRON. Kenneth Bainbridge's research on the mass differences between isotopes provides the evidence to support the theoretical concept of the mass–energy equivalence as proposed by ALBERT EINSTEIN (1879–1955) (see Figure B.6).

Figure B.6 Kenneth Tompkins Bainbridge (1904–1996). (Courtesy of Los Alamos National Laboratory, Los Alamos, New Mexico.)

Bakelite®

[biomedical, solid-state] Polyoxybenzylmethylenglycolanhydride, one of the first synthetic PLASTICS. Bakelite is strong and temperature resistant to high temperatures. The chemical basis is thermosetting phenol formaldehyde resin. Bakelite is formed resulting from an elimination reaction of formaldehyde with phenol, combined with strengtheners such as wood pulp or CELLULOSE. It is a composite material designed in 1907 and commercially introduced in 1939. Due to its mechanical, thermal, and chemical properties, bakelite is used in various consumer items (e.g., rotor for distributor in car), poker chips, billiard balls, and chess pieces, as well as in medical devices in its early days, and still is to a lesser degree. The name was derived from the Belgian inventor LEO HENRICUS ARTHUR BAEKELAND (1863–1944) (see Figure B.7).

Figure B.7 Picture of a cook pot with bakelite handle.

Ballistic motion

[general] The trajectory associated with projectiles that are ejected under explosive force (e.g., bullet and metallic glowing cinders produced under welding) or under mechanical impulse (e.g., baseball). The trajectory of the ballistic PROJECTILE is a product of the resultant of the forces acting on the object. GRAVITY being the constant force acting in vertical direction, whereas only a force acts on release in a generalized direction, based on the planned target range. The ballistic projectile will provide a parabolic track that obeys the path outlined by the initial escape VELOCITY ($\vec{v_i}$, with horizontal projection $v_{ix} = v_i \cos\theta$ [neglecting fluid resistance] and vertical projection $v_{ix} = v_i \sin\theta$) providing the horizontal displacement $s_x = v_{ix}t$ as a function of time t, and the vertical DISTANCE $s_y = v_{iy}t + (1/2)gt^2$ with a maximum height when released in upward direction, for example, cork form champagne bottle (g is the GRAVITATIONAL ACCELERATION); the time is derived from the duration in which the vertical MOTION is suspended (hits the GROUND/target) (see Figure B.8).

Figure B.8 Ballistic motion of a cannonball.

Balmer, Johann Jakob (1825–1898)

[atomic, computational, general, nuclear, solid-state] A computational physicist and mathematician from Switzerland. In 1885, Johann Balmer defined hydrogen excitation spectra by the BALMER SERIES, where the transition FREQUENCY (v) spectrum is defined as $v = cR\left[(1/2^2) + (1/n^2)\right]$, where $R = 10,967,758\,m^{-1}$ is the RYDBERG CONSTANT, $c = 299,792,458\,m/s$ is the speed of light, and $n = 3, 4, 5, \ldots$ is an integer depicting the excitation levels. The balmer series emission lines (i.e., ATOMIC LINE SPECTRA) are all in the visible spectrum (see Figure B.9).

Figure B.9 Johann Jakob Balmer (1825–1898).

Balmer series

[atomic, general, nuclear] Hydrogen excitation spectra are described by JOHANN JAKOB BALMER (1825–1898) in 1885. The electron transitions within the HYDROGEN ATOM generate a set of discrete emission wavelengths with FREQUENCY (ν) spectrum defined as $\nu = cR\left[\left(1/2^2\right)+\left(1/n^2\right)\right]$, where $R = 10,967,758\,\text{m}^{-1}$ is the RYDBERG CONSTANT, $c = 299,792,458\,\text{m/s}$ is the speed of light, and $n = 3, 4, 5, \ldots$ is an integer depicting the excitation levels. The Balmer series emission lines (i.e., ATOMIC LINE SPECTRA) are all in the visible spectrum, also known as Balmer formula (see Figure B.10).

Figure B.10 Balmer lines in the visible spectrum representative of hydrogen gas emission. (Courtesy of Jan Homann.)

Balmer thermometer

[astronomy/astrophysics] An indicator of stellar ACTIVITY by means of the ENERGY spectrum expressed by the number (N) of neutral H atoms in the excited ($n = 2$) state available to produce Balmer lines. This THERMOMETER expresses the stellar temperature in KELVIN (see Figure B.11).

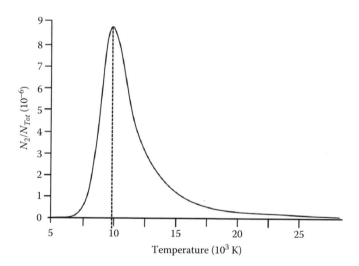

Figure B.11 Balmer thermometer emission energy spectrum.

Bar magnet

[general] It is the simplest construction for a MAGNET and has a well-defined NORTH and south pole on the opposite sides of the long section of the bar. The MAGNETIC FIELD lines curve around the long side of the bar in all directions forming a set of concentric elongated doughnuts in outline. When placed underneath a paper sheath, the cross-sectional outline of the field lines will become evident. Note that the magnetic field lines always form closed loops (see Figure B.12).

Figure B.12 Bar magnet with paperclips attached on the north- and south pole, connecting the magnetic field loop.

Bardeen, John (1908–1991)

[electronics, solid-state] A physicist from the United States, known for his pioneering efforts in semiconductor PHYSICS next to his work in SUPERCONDUCTIVITY. John Bardeen received a Nobel Prize in Physics in 1956 for his work on transistors, shared with WALTER HOUSER BRATTAIN (1902–1987) and WILLIAM BRADFORD SHOCKLEY (1910–1989), and he received another Nobel Prize in Physics in 1972 regarding superconductivity, shared with LEON NEIL COOPER (1930–) and JOHN ROBERT SCHRIEFFER (1931–). In 1947, Bardeen discovered the transistor function while working at the Bell laboratories (see Figure B.13).

Figure B.13 John Bardeen (1908–1991). (Courtesy of the Nobel Prize files.)

B

Barkla, Charles Glover (1877–1944)

[atomic, nuclear] A physicist from Great Britain. In 1911, Barkla revealed his discovery of the emission lines in the X-ray spectrum resulting from ATOMIC EMISSION transitions, which are sharp lines. These lines are superimposed on the broad BREMßTRAHLUNG spectrum. Charles Barkla received the Nobel Prize in Physics for his work in 1917. Charles Barkla also discovered the fact that X-ray had POLARIZATION, similar to ordinary visible light, allowing the extrapolation of the ELECTROMAGNETIC SPECTRUM. The work of Barkla also illustrated the correlation between the characteristic LINE SPECTRUM and the RADIATION consisting of PARTICLE emission during Röntgen radiation (see Figure B.14).

Figure B.14 Charles Glover Barkla (1877–1944). (Courtesy Library of Congress, George Grantham Bain Collection, Washington, DC.)

Barn (the unit)

[atomic, nuclear] A cross-sectional unit in nuclear dimensions, $1\,\text{barn} = 10^{-28}\,\text{m}^2$. The unit expresses the PROBABILITY of a specific nuclear reaction taking place in terms of cross-sectional area.

Baroclinic waves

[fluid dynamics, geophysics] Atmospheric waves that migrate and incorporate horizontal thermal gradients and variability of FLOW as a function of altitude, essentially caused by the changing CORIOLIS FORCE with altitude. In order to account for variability in properties with altitude, a new reference frame is introduced to allow for the use of standard WAVE EQUATION. The "virtual" altitude (z^*) is corrected by means of PRESSURE (P, against a reference $_{00}$) and temperature (T) as $z^* = H_{00}\ln(P_{00}/P)$, where $H_{00} = RT_{00}/g$ is the equivalent height of the column, $R = 8.3144621(75)$ J/Kmol is the universal GAS constant, and g is the GRAVITATIONAL ACCELERATION. The vertical velocity for atmospheric flow is defined as $w^* = Dz^*/Dt = -(H_{00}/P)\varpi_T$, where ϖ_T is the vorticity and $D/Dt = [(\partial/\partial t) + (\vec{v}\cdot\nabla)]$, $\vec{v}$ is the velocity vector. For the equivalent VELOCITY POTENTIAL ($\Psi_v^* = \varphi_v/f_{C0}$, φ_v the velocity potential), the equation of MOTION is now defined as $[(\partial/\partial t) + u_g(\partial/\partial x) + v_g(\partial/\partial y)]\nabla^2\Psi_v^* + \beta_C(\partial/\partial x)\Psi_v^* = f_{C0}[\exp(z^*/H_{00})](\partial/\partial z)\{w^*\exp[-(z^*/H_{00})]\}$, where f_{C0} is the geostrophic vorticity, and β_C is the Coriolis parameter, with the longitude and LATITUDE velocity components, respectively, $v_g = (\partial/\partial x)\Psi_v^*$ and $u_g = -(\partial/\partial y)\Psi_v^*$. In thermodynamic perspective, the temperature (T) effects are represented as $(\partial/\partial z^*)\Psi_v^* = RT/f_{C0}H_{00}$, where H_{00} is the equivalent "pressure" height. The generalized solutions are provided in the format $\Psi_v^{*'} = [\text{Const}\times e^{-(z^*/2H_{00})}e^{i(\omega_i t + k_x x + k_y y + \kappa_z z^*)}]$, where intrinsic angular frequency $\omega_i = 2\pi\nu_i$ (ν_i is the intrinsic frequency), k_x is the wave number for the longitudinal direction, k_y is the wave number as a function of latitude, with boundary condition $\kappa_z^2 = v_{BV}^2 / f_{C0}^2 \left[\beta_C / \left(u_0 + (\omega_i/k) \right) - \left(k_x^2 + k_y^2 \right) \right] - \left(1/4 H_{00}^2 \right)$, where $v_{BV} = \sqrt{\left[(g/\theta_\rho)(\partial \theta_\rho / \partial z^*) \right]}$ is the BRUNT–VÄISÄLÄ BUOYANCY FREQUENCY, θ_ρ is the local temperature, as a function of DENSITY (ρ). The baroclinic wave describes the situations pertaining to conditions that do not apply to BAROTROPIC WAVES, also known as baroclinic Rossby waves (*also see* BAROTROPIC WAVE *and* LAMB WAVE) (see Figure B.15).

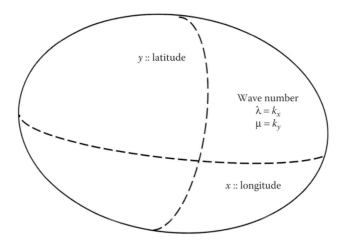

Figure B.15 Reference for the nomenclature used in the baroclinic and barotropic wave pattern definition.

Barometer

[energy, fluid dynamics, general, geophysics] A device used to measure the ATMOSPHERIC PRESSURE. The first barometer was constructed by EVANGELISTA TORRICELLI (1608–1647) in 1643. The barometer uses the principle of COMMUNICATING VESSELS to measure the height of a column of mercury with density ρ_{Hg} to match the height of a column of AIR with density ρ_{air}, the air density declines with altitude and hence the MEASUREMENT performs a virtual integration (see Figure B.16).

Figure B.16 Picture of barometer indicating the pressure and the anticipated weather conditions.

Barometric law

[fluid dynamics, geophysics] *See* LAW OF THE ATMOSPHERES.

Barotropic waves

[fluid dynamics, geophysics] Large-scale atmospheric WAVE patterns that are generated as a result of the changing CORIOLIS FORCE with altitude. In a barotropic Rossby wave, the total VORTICITY (ϖ_T) is conserved: the sum of the geostrophic vorticity (ϖ_g) and the MAGNITUDE contribution by the earth's vorticity ($f_C = f_{C0} + \beta_C y$; β_C is known as the Coriolis parameter). Barotropic waves have an essential influence on the midlatitude weather development by means of forming the large-scale CIRCULATION based on its crests and troughs. The changing Coriolis parameter as a function of altitude (y) ($\beta_C = \partial f_C / \partial y$) creates a restoring force in the horizontal direction, with ensuing horizontal oscillations in pressure fronts generating the barotropic waves. The WAVE EQUATION provides the DISPERSION for the wave with wave number $k = 2\pi/\lambda$, where λ is the wavelength, and the pattern has an intrinsic angular frequency $\omega_i = 2\pi v_i$ (v_i is the intrinsic frequency), defining the PHASE VELOCITY as a function of LATITUDE (ℓ) as $\omega_i/k = \beta_C / [k^2 + \ell^2 + (f_{C0}/gH)]$, where g is the GRAVITATIONAL ACCELERATION and H is the height of the atmospheric column. For stationary waves, the wavelength is derived to be represented as $\lambda_{Bs} = 2\pi\sqrt{(\bar{u}/\beta_C)}$, where $\bar{u}$ is the migration/DRIFT VELOCITY perturbation; for instance, at latitude 50°N with $\bar{u} = 10\,\text{m/s}$, this provides $\lambda_{Bs} = 5179$ km. Waves that have a shorter WAVELENGTH ($\lambda < \lambda_{Bs}$) will travel west to east, and at greater wavelengths ($\lambda > \lambda_{Bs}$), they will migrate east to west. As a reference, nonstationary barotropic waves at midlatitude will travel eastward at a rate of approximately 6°longitude/day. The barotropic wave pattern is indicated on global weather maps represented by the 50 kPa pressure surface (which equates to the geopotential), with wavelength stretching 1000 km or more. Expanding the wave equation to three dimensions yields the influence of a stratified ATMOSPHERE and wave propagation in the vertical direction $\omega_i/k = -\beta_C / [k^2 + \ell^2 + (f_{C0}^2/v_{BV}^2)(\partial_m^2 + C_{Eckart}^2)]$, where ∂_m represents the dampening coefficient, v_{BV} is the BRUNT–VÄISÄLÄ BUOYANCY FREQUENCY, $C_{Eckart} = [1/2\rho(z)][\partial\rho(z)/\partial z] + g/v_s^2$ defines the Eckart coefficient, a scale of the ratio of the moment of INERTIA for the different wave aspects; angular—translational (a parameter carried over from QUANTUM mechanical analysis with respect to molecular interaction) ($\rho(x, y, z)$ the perturbation to the density as a function of location and height, v_s is the speed of SOUND). Barotropic waves are often forced by the local topography, Rocky Mountains, Andes Mountains, Himalayas, and so on. Generally, barotropic wave theory assumes that there is no vertical velocity for the wave (the AIR is moving), a "standing wave" approach. Including vertical propagation is discussed under BAROCLINIC WAVES, also known as barotropic Rossby wave or Rossby wave. The name CARL-GUSTAF ARVID ROSSBY (1898–1957) is associated with the atmospheric waves due to the initial discovery and theoretical description by this Swedish meteorologist and scientist in 1919. A similar process in found in the oceanic fluctuations where the FLOW velocity at great depth is in the order of 2 m/day, and a time constant is in the order of years (*also see* BAROCLINIC WAVES) (see Figure B.17).

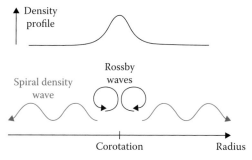

Figure B.17 Barotropic Rossby wave in atmospheric jet-stream evolution.

Barrier

[nuclear] *See* COULOMB BARRIER, FISSION BARRIER, *and* POTENTIAL BARRIER.

Bartlett, James Holley (1904–2000)

[chemical, dynamics, general, nuclear] A scientist from the United States who contributed to the description and definition of the DONNAN EQUILIBRIUM.

Baryon

[atomic, general, quantum] Heavy subclass of the PARTICLE group HADRON, examples are NEUTRON and protons. Baryons can be "strange" and "charmed." Baryons obey FERMI–DIRAC STATISTICS and have half-integral SPIN, also known as FERMIONS. They are the complementary particles of MESONS, which obey BOSE–EINSTEIN STATISTICS and have zero or integral spin, also known as BOSONS. Masses of the known baryons as well as mesons range from one-seventh of that of the PROTON for the pi MESON, extending 10 times the proton mass. Baryons are subject to the STRONG NUCLEAR FORCE. The known baryons are $p, n, \Delta, \Lambda, -$. In 1964, both MURRAY GELL-MANN (1929–) and GEORGE ZWEIG (1937–) provided evidence that baryons (as well as the other hadron group: mesons) are not primary ELEMENTARY PARTICLES. Baryons are composed of quarks, the smallest elementary particle currently known and verified.

Baryon number

[atomic, general, quantum] In order to satisfy the conservation principle for BARYONS, the BARYON NUMBER $B = 1$ was introduced with respect to the antibaryon ($B = -1$), whereas nonbaryons have $B = 0$.

Basal metabolism

[biomedical, thermodynamics] The average ENERGY consumption rate for the HUMAN BODY at rest (maintaining equilibrium, constant temperature, etc.), keep in mind that the metabolic rate is a function of height, weight, AGE, and current and prior ACTIVITY: 72 calories/h = 301.45 J/h. At this rate, the daily recommended consumption comes to approximately 2000 calories, as listed as a reference on several food packaging with regard to nutritional guidelines.

Basic forces of nature

[energy, general, nuclear, quantum, thermodynamics] Force can be classified into four types—STRONG NUCLEAR FORCE, weak nuclear force, electromagnetic force (its subdivision is ELECTROWEAK FORCE), and gravitational force.

Basilar membrane

[acoustics, biomedical, computational, general] SENSORY structure in the COCHLEA of the EAR that converts a mechanical VIBRATION to ELECTRONIC impulses. The conversion is performed by a long stretch of hairs lining the length of the cochlea construction, resembling a snail shell. The hairs are part of the ORGAN OF CORTI.

When the hair bends and applies shear strain to the MEMBRANE, it generates an ACTION POTENTIAL as described in the entry for the ORGAN OF CORTI (see Figure B.18).

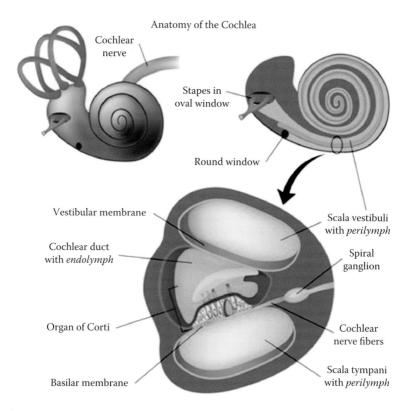

Figure B.18 Basilar membrane in the cochlea of the ear.

Bat

[acoustics, biomedical] The only mammal capable of sustained flight, with the use of webbed wings. The WING of the bat is remarkably distinguished from those of birds and pterosaurs, both are not mammals. Bats generally rely on acoustic TELEMETRY for hunting and flight guidance. Bats are of the biological order Chiroptera, with the following Greek reference, contraction of χειρ (phonetic: cheir), which translates roughly as hand, and πτερον (phonetic: pteron), meaning wing. Certain bats use ACOUSTIC probing at frequencies (ν) reaching in excess of 200 kHz. The guidance by ACOUSTICS is also used by submarines. Based on general sensing constraints, the RESOLUTION associated with wavelength-based detection cannot generally exceed half the shortest WAVELENGTH (λ_{min}), which at 200 kHz equates to $\text{Res} = (1/2)\lambda_{min} = (1/2)(v_{sound}/\nu) = (1/2)\left(340.29\text{m/s}/2\times10^5\,\text{s}^{-1}\right) = (1/2)1.7015\times10^{-3}\,\text{m} \cong 850\,\mu\text{m}$, where $v_{sound} = 340.29\,\text{m/s}$ in AIR at standard temperature and pressure. In comparison, human VISION is limited by both the RAYLEIGH CRITERION and the density of RODS and CONES in the FOVEA. On average, the eyes can distinguish two point sources with angular resolution of $0.3\,\text{arc min} = 0.0000872\,\text{rad}$ apart, resulting in a spatial resolution at the center wavelength 550 nm applying a focal length of 22 mm to an object placed in the NEAR POINT of the HUMAN EYE (0.25 m) yielding approximately 88.65 μm resolution on average. Alternatively, the bat is used to strike a ball in the sports baseball and cricket (see Figure B.19).

Figure B.19 Picture of a representative bat.

Battery

[energy, general] A reversible electrical ENERGY storage device. Electrical energy can be stored primarily by chemical means, which can be released under reversed conditions. When applying an electrical current to the battery, the internal structure and composition converts the electrical energy into a form of energy that will hold the potential energy for a finite lifetime, and the process can be reversed to release an electrical current for later use. Certain batteries hold only potential electrical energy based on the production process and will only deliver current under a fixed electrical voltage. For instance, a lead battery for a car operates on the following principle: $PbO_2 + 2H_2SO_4 + Pb \leftrightarrow PbSO_4 + 2H_2O + PbSO_4$, where the ANODE is made from PbO_2, while the CATHODE is made from porous lead (Pb), with WATER (H_2O) and sulfuric ACID (H_2SO_4) as catalyst and SOLUTE. In the lead battery, the electric potential (ε_{emf} or V_{EMF}) primarily depends on the sulfuric acid concentration. The electrical potential of the battery can be described by a function derived from the GIBBS FREE ENERGY as $\Delta\varepsilon_{emf} = \Delta\varepsilon_{emf}^0 - (RT/nF_{Helmholtz})\ln Q_r$, also known as the NERNST EQUATION or NERNST POTENTIAL, where ε_{emf} is the electromotive TRANSMEMBRANE POTENTIAL, $R = 8.314462175\,J/Kmol$ is the universal GAS constant, and T is the temperature. This equation also provides the potential difference generated in a galvanic CELL as a function of the ION concentrations in both compartments during equilibrium. In this equation, Q_r states the ratio of the respective compartmental ionic concentrations (see Figure B.20).

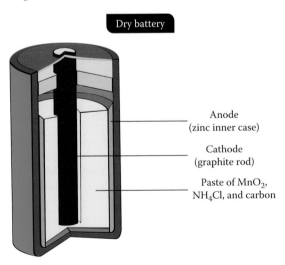

Figure B.20 Picture of a C-cell battery cut open, revealing a graphite core electrode.

Bayes, Thomas (1701–1761)

[biomedical, computational] A statistician and mathematician from Great Britain. Bayes is most known for his posthumously published work on "inverse probability," used to predict the verification of a phenomenon, known as the Bayes' theorem.

Bayes' theorem

[biomedical, computational] A PROBABILITY theory pertaining to the occurrence of an event (or data point) (E_{data}) with respect to a scientific hypothesis ($P(H_i)$) related to another event (H_i). The principle of combining the recently acquired data with experimental observations recorded previously was introduced in the mid-eighteenth century by THOMAS BAYES (1701–1761) to provide a mechanism of verification. The theorem is sometimes used to validate the PERCEPTION of an observation and hence verify the assumptions made that define a condition. This process can also be applied prior to performing the experiment. The BAYES' THEOREM is also used in MEDICINE to ascertain the probability of the presence of a medical condition based on laboratory findings and patient examinations. In medicine, the derivation is not necessarily based on computational processes, rather on likelihood from documented and personal experiences. Many events/observations with associated hypothesis and data points involved in a single phenomenon will converge to yield the probability of the occurrence of a specific phenomenon. Mathematically, this is defined as the probability of the correctness of the hypothesis under the observation of specific experimental data $P(H_i|E_{\text{data}}) = P(H_i)P(E_{\text{data}}|H_i)/\sum_i P(H_i)P(E_{\text{data}}|H_i)$, where $P(E_{\text{data}}|H_i)$ represents the probability of obtaining realistic observations under the assumption H_i.

Bayesian statistics

[biomedical, computational] A prediction and verification mechanism of action based on the BAYES' THEOREM.

Bazin equation, open channel

[fluid dynamics] Considering the free FLOW in a water channel, defined as having one exposed surface to open AIR, there are several factors to consider. Water channels can be traced back to at least 300 BC with respect to the Roman aqueducts, stretching 16.5 km. Water channels are frequently deep with rough WALL and high flow rate and hence a high REYNOLDS NUMBER that ensures turbulent flow. Under the high flow rate and rough surface on the walls of the channel, the ROUGHNESS ($\alpha_{\text{roughness}}$, found in tables; shaved wood surface: $\alpha_{\text{roughness}} = 0.06$, concrete: $\alpha_{\text{roughness}} = 1.30$, canal bed with grass and rocks: $\alpha_{\text{roughness}} = 2$) becomes the determining factor in the boundary flow. The "wetted perimeter" (s_w) of the channel is the contiguous length stretching from one edge across the bottom surface up to the opposing edge at the free surface. Consider that the channel is uniform over its length ℓ, with shear stress at the wall σ_s, the FRICTION must balance the gravitationally induced flow (under incline ANGLE θ), for LIQUID with density ρ flowing through a CROSS SECTION A, since the liquid is not accelerating, as $\rho g \ell \sin\theta = \sigma_s s \ell$. The shear stress can be approximated as a function of the flow velocity v as $\sigma_s = f_f \rho(v^2/2)$, where f_f is the frictional coefficient. This allows the flow velocity to be written as $v = \sqrt{2g\sin\theta/f_f} = c_f\sqrt{(d_h\sin\theta)}$, since the angle is small $\tan\theta \cong \sin\theta$. The mean hydraulic depth can be defined as $d_h = A/s_w$. In channel flow, the flow velocity coefficient (c_f) is expressed by the Bazin equation $c_f = 87/[1+(\alpha_{\text{roughness}}/\sqrt{d_h})]$ (also see CHEZY EQUATION).

Beam forming

[acoustics, optics] A linear parameter estimation process used to define the incident ANGLE of a ray/beam with the specific associated distribution of MAGNITUDE as a function of angle and location. The beam-forming process is designed to represent a profile that will provide the optimal configuration for specific RESOLUTION imaging applications. In certain cases, the beam profile may need to be altered in the process of SIGNAL acquisition. In OPTICS, a laser beam may be formed to produce a specific spot shaped by means of lenses. One specific optical beam-forming mechanism is LENS-AXICON FOCUSING. In ACOUSTICS, the use of certain arrangements of sources (PHASED ARRAY) or single source with ORIFICE (damped piston) can provide a specific emission profile in angular distribution and in spatial outline.

Bearden, Joyce Alvin (1903–1986)

[atomic, nuclear] A physicist from the United States who worked on the identification of atomic structures by means of X-RAY probing. Bearden listed the characteristic lines for a range of ELEMENTS. He was a student of ARTHUR HOLLY COMPTON (1892–1962).

Beat frequency

[acoustics, general, imaging, mechanics, optics] The frequency difference between two frequencies that are closely separated (*see* BEATS).

Beats

[acoustics, general, imaging, mechanics, optics] The superposition of two waves that are slightly off on the WAVELENGTH (λ) produces an INTERFERENCE that has an optimum at the difference between the two wavelengths. The BEAT FREQUENCY ($v = \omega/2\pi$) is observed at the low frequency, fading in and out with the period of the beat frequency next to the base frequencies. The superposition principle provides the mechanism to describe the effect as $\sin(\omega_1 t) + \sin(\omega_2 t) = 2\sin\left[(\omega_1 t + \omega_2 t)/2\right]t\cos\left[(\omega_1 t - \omega_2 t)/2\right]$, creating the mean angular frequency $\omega_1 t + \omega_2 t/2$, at small difference, to be AMPLITUDE modulated at angular frequency $\Delta\omega = (\omega_1 - \omega_2)$. In case the two wavelengths are separated beyond a certain threshold (greater than 0.01% for ACOUSTICS and greater than 0.001% for ELECTROMAGNETIC RADIATION), the two waves are perceived as separate waves. In acoustics, this can, for instance, be heard for two diesel trains running at almost equal revolutions per minute, but at less than a few Hz off (~at a fraction of a Hz off) to SOUND as the mean frequency for the two trains becoming louder and fading out at the beat frequency (see Figure B.21).

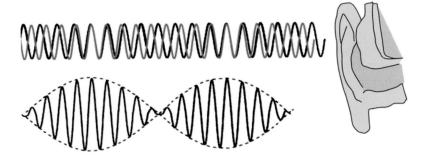

Figure B.21 Graphical illustration of the beat frequency concept.

Beattie–Bridgeman equation of state

[thermodynamics] A thermodynamic relation between PRESSURE (P), VOLUME (V), and TEMPERATURE (T) as a refinement to the VAN DER WAALS EQUATION. The Beattie–Bridgeman equation of state was introduced by James Alexander Beattie (1895–1981) and OSCAR Cleon Bridgeman (1897–1967) in 1928. The Beattie–Bridgeman equation of state has the following format: $P = (RT/V)\left[1 - \left(c'/VT^3\right)\right]\{1 + (B_0/V)[1 - (b'/V)]\} - \{(A_0/V^2)[1 - (b'/V)]\}$, where A_0, B_0, and a', b', and c' are constants that represent the specific properties of the substance and are listed in look-up tables for the respective boundary conditions and gasses in question.

Becquerel, Antoine Henri (1852–1908)

[atomic, general, nuclear] A scientist from France dedicated to FLUORESCENCE. His contributions to the definition of nuclear RADIATION and RADIOACTIVITY were helpful after the discovery of X-RAY radiation by WILHELM CONRAD RÖNTGEN (1845–1923) (see Figure B.22).

Figure B.22 Antoine Henri Becquerel (1852–1908) in the laboratory performing experiments with magnets.

Becquerel (Bq)

[general] The System International unit of RADIATION. The Becquerel defines the number of nuclear decays per SECOND for a quantity of ISOTOPE ($1\,\text{Bq} = 1\,\text{s}^{-1}$), named after ANTONIE HENRI BECQUEREL (1847–1922).

Bel

[acoustics, general, theoretical] A logarithmic SIGNAL scale unit used to express the relative MAGNITUDE of a signal (ACOUSTICS, ELECTRONICS, optical) with respect to a reference value $bel = \log_{10} I/I_0$. In acoustics, the auditory limit is generally the reference point at $I_0 = 1.0 \times 10^{-12}\,\text{W/m}^2$ (associated pressure: a pressure of $P \cong 1.0 \times 10^{-5}\,\text{Pa}$). For HEARING, the scale of LOUDNESS is an indication of the number of factors of ten above the threshold of hearing, ranging to $\sim 120\,\text{dB}$ as the threshold of hearing with power density of $I = 1.0\,\text{W/m}^2$ and a pressure of $P \cong 30\,\text{Pa}$. More commonly known under the unit DECIBEL (dB, incorporating the 10-factor): β, in $dB = 10 \log_{10}(I/I_0) = 20 \log_{10}(A/A_0)$, for AMPLITUDE A. It is named after the scientist ALEXANDER GRAHAM BELL (1847–1922) who initiated the signal qualification process in acoustics.

Bell, Alexander Graham (1847–1922)

[general] A scientist and inventor from Scotland, Great Britain. Alexander Graham Bell provided standardization for electronic design and ACOUSTICS based on his research and industrial involvement. He received the first patent for the telephone in 1876. He was one of the founding fathers of the National Geographic Society. He provided a range of resources and training for the deaf and deaf-mutes, as well as HEARING aid development and processes to assist in communication (see Figure B.23).

Figure B.23 Alexander Graham Bell (1847–1922). (Courtesy Bibliothèque et Archives Canada.)

Bell, George Howard (1897–1985)

[biophysics, chemical, mechanics] A physiologist and biochemist from the United States who contributed to the theoretical definition of whole MUSCLE strength.

Bell, George Irving (1926–2000)

[biomedical, chemical, computational, mechanics] A biomedical engineer, physicist, and biologist from the United States. Apart from Bell's efforts on the development of the thermonuclear weapon concept and associated NEUTRON transport, he developed theoretical models related to MUSCLE contraction by means of the description of ADHESION STRENGTH based on ligands in the ACTIN–MYOSIN interaction, next to the models of the MAGNITUDE of immunological events, and played an significant role in the Human Genome project, heading up the Center for Human Genome Studies.

Benedetti, Giovanni Battista (Gianbattista) (1530–1590)

[general] A scientist and mathematician from Italy (Venice, Venetian States) who in 1585 recognized that INERTIA is a phenomenon that can maintain an object in MOTION unless impeded by an external force. This statement was before the work by GALILEO GALILEI (1564–1642) and the laws of motions by SIR ISAAC NEWTON (1642–1727). Benedetti's mathematical efforts focused on Euclid problems. Giovanni Benedetti also worked out a theory of free fall, explaining gravitational acceleration.

Benedict, Francis Gano (1870–1957)

[biomedical] A physiologist, chemist, and nutritionist from the United States. Francis Benedict developed a CALORIMETER and successfully measured metabolic rate; he designed and built a spirometer to determine oxygen consumption. This was all instrumental in the development of our current understanding of the human METABOLISM and scheduling an effective diet. Francis Benedict, with assistance from JAMES ARTHUR HARRIS (1880–1930), produced a heuristic formula for ENERGY expenditure in 1919. The formulation for men and women was different, respectively, male expenditure in calories/day $= \left(66.473 + 13.752 \times \text{weight}\right) + 5.003 \times \text{height} - 6.775 \times \text{age}$ and female expenditure in calories/day $= \left(655.095 + 9.563 \times \text{weight}\right) + 1.85 \times \text{height} - 4.676 \times \text{age}$, known as the Benedict–Harris equations.

Benedict–Webb–Rubin equation of state

[thermodynamics] An empirical EQUATION OF STATE based on the initial work by JOHANNES DIEDERIK VAN DER WAALS (1837–1923) and the modification introduced in the BEATTIE–BRIDGEMAN EQUATION OF STATE, as performed by MANSON BENEDICT (1907–2006), GEORGE B WEBB (?–?), and LOUIS C. RUBIN (?–?) published in 1940. The Benedict–Webb–Rubin equation of state links the PRESSURE (P), VOLUME (V), and TEMPERATURE (T) of specific substances with the use of coefficients and constants that are listed in look-up tables for a broad range of media, expressed as $P = (RT/V) + \left(1/V^2\right)\left[B_0'RT - A_0' - \left(C_0'/T^2\right)\right] + \left(1/V^3\right)\left(b''RT - A_0'\right) + \left(a''\alpha^*/V^6\right) + \left(1/V^3\right) \{c''\left[1 + \left(\gamma^*/V^2\right)\right]/T^2\} e^{-\gamma^*/V^2}$, where A_0', B_0', C_0', and a'', b'', c'', α^*, and γ^* constants that represent the specific properties of the substance.

Berkelium ($^{247}_{97}$Bk)

[atomic, nuclear] A radioactive ELEMENT. BERKELIUM is an actinide, discovered in 1949, artificially created at the University of California, Berkeley, California.

Bernoulli, Daniel (1700–1782)

[computational, fluid dynamics, general] A mathematician from the Netherlands who moved to Switzerland for his professional career. Daniel Bernoulli conceived the equation that described the flow ENERGY balance known as the BERNOULLI'S EQUATION. He also introduced the mathematical format of the series function, later referred to as the BESSEL FUNCTIONS, due to the involvement from FRIEDRICH WILHELM BESSEL (1784–1846). Daniel was a nephew of JACOB BERNOULLI (1654–1705) (see Figure B.24).

Figure B.24 Daniel Bernoulli (1700–1782), engraved by Johann Jakob Haid.

Bernoulli, Jacob (1654–1705)

[computational] A mathematician from Switzerland. His work made significant contributions in solving separable differential equations. His efforts in PROBABILITY theory and enumeration were published posthumously in 1713 providing proof of mathematical solutions as well as his introduction of the LAW OF LARGE NUMBERS.

Bernoulli number (B_n)

[computational, thermodynamics] A sequence of discrete rational numbers defined through a process of an exponential generating function expressed as $x/(e^x - 1) = \sum_{n=0}^{\infty} (B_n x^n / n!)$. These numbers are invaluable in the series expansions of trigonometric functions, specifically pertaining to number theory and analysis. The number sequence was introduced by JACOB BERNOULLI (1654–1705).

Bernoulli number of the second kind

[computational] A recursive equation, $(b_n = \int_0^1 (x)_n \, dx)$, used in mathematical modeling and NUMERICAL equation solving, where $(x)_n$ is a coefficient known as a falling factorial that has the parameters expressed by an exponential generating function $E(x)$ as $E(x) = \sum_{k_i=0}^{\infty} (k_i! / a_{k_i})$ $x^{k_i} = x/\ln(1+x) = 1 + (1!/2)x - (2!/6)x^2 + (3!/4)x^3 + \cdots$, where a_{k_i} is a coefficient and k_i is an integer.

The BERNOULLI NUMBER OF THE SECOND KIND has been defined in many different ways over the centuries after its original conception, published posthumously with respect to the work by JACOB BERNOULLI (1654–1705), in 1713.

Bernoulli's equation

[fluid dynamics, thermodynamics] An equation designed by DANIEL BERNOULLI (1700–1782) with respect to the ENERGY content of inviscid FLUID flow, in particular the use of CONSERVATION OF ENERGY. The Bernoulli's equation links the gravitational attraction on a volume of medium at elevation h, with density ρ while in FLOW, to the kinetic energy density ($KE = (1/2)\rho v^2$, v the velocity) and the applied PRESSURE (P) over a DISTANCE with different conditions as $P_1 + (1/2)\rho v_1^2 + \rho_1 g h_1 = P_2 + (1/2)\rho v_2^2 + \rho_2 g h_2$, ($g$ is the GRAVITATIONAL ACCELERATION) generally considering an INCOMPRESSIBLE FLUID, $\rho_2 = \rho_1$, also referred to as the Bernoulli's theorem.

Bessel, Friedrich Wilhelm (1784–1846)

[astronomy, atomic, computational, fluid dynamics, general, geophysics, quantum] A mathematician and astronomer from Germany. Friedrich Bessel redefined the series of equations outlined by DANIEL BERNOULLI (1700–1782), known as BESSEL FUNCTIONS. Bessel's observations and calculations with respect to the Hailey's COMET provided significant detail in the orbital trajectory in 1804. Friedrich Bessel also provided the location in the order of 3222 stars, as well as the DISTANCE from the SUN to the nearest STAR, using an experimental method referred to as parallax (see Figure B.25).

Figure B.25 Friedrich Wilhelm Bessel (1784–1846), painting by Christian Albrecht Jensen.

Bessel beam

[optics] A nondiffracting beam profile described by a first-order Bessel function ($J_0\left(k\theta_r r\right)$, where $\theta_r = \arcsin\left(n_{\mathrm{lens}} \sin\theta_A\right) - \theta_A$ is the DIVERGENCE ANGLE defined by the axicon angle (θ_A) and the index-of-refraction of the axicon material n_{lens}, r is the radial location, and $k = 2\pi/\lambda$, where λ is the wavelength. The radiance of this beam is constant as a function of position, measured in transverse direction, in the path of propagation. One specific application is found in LENS-AXICON FOCUSING, specifically with respect to high-energy pulsed laser RADIATION.

Bessel functions

[atomic, computational] A mathematical series formulation introduced by DANIEL BERNOULLI (1700–1782) and generalized by FRIEDRICH WILHELM BESSEL (1784–1846). The Bessel functions are canonical ($y(x)$) EIGENFUNCTION solutions to the Bessel differential equation $x^2\left(d^2 y/dx^2\right)+x\left(dy/dx\right)+\left(x^2-v^2\right)y=0$, where v is a complex number, indicating the order of the Bessel function. The complex number can take on any value; however, the most relevant solutions are for the integer and half-integer values of v. Bessel functions of the first kind are defined by $J_v(x)=\sum_{m=0}^{\infty}(-1)^m/m!\,\Gamma(m+v+1)(x/2)^{2m+v}$, where $\Gamma(m+v+1)$ is the gamma function ($J_1(x)$ represents a first-order Bessel function of the first kind). The Bessel functions of the first kind are finite at the origin under boundary conditions v integer or v real and positive, with the additional condition that when approaching zero for negative nonzero v. One useful tool is the fact that the Bessel function can be developed in a Taylor series around the origin ($x=0$). Bessel functions of the first kind for integer $\alpha=n; n\in\mathbb{N}$ can also be formulated as trigonomic integrals $J_n(x)=(1/\pi)\int_0^{\pi}\cos(n\xi-x\sin\xi)d\xi=(1/2\pi)\int_0^{\pi}e^{i(n\xi-x\sin\xi)}d\xi$, where ξ is the ANGLE in the complex plane, with the following examples of specific special solutions: $J_0(x)=(1/\pi)\int_0^{\pi}\cos(x\sin\xi)d\xi=(1/\pi)\int_0^{\pi}\cos(x\cos\xi)d\xi$, $J_v(x)=\left[(1/2)x\right]^v\big/\pi^{1/2}\Gamma\left[v+(1/2)\right]\int_0^{\pi}\cos(x\cos\xi)\sin^{2v}\xi d\xi$, and $J_n(x)=(1/\pi)\int_0^{\pi}\cos(x\sin\xi-n\xi)d\xi$. An example of the Taylor development is shown in the TERMINAL VELOCITY entry, the torsional oscillations entry, and DIVERGING WAVES entry to name but a few examples. For the case $\alpha=$ integer, the following equality arises: $J_{-n}(x)=(-1)^n J_n(x)$, indicating that the solutions are interdependent; this introduced the Bessel functions of the second kind. Bessel functions of the second kind are combining solutions for the first kind to form: $Y_v(x)=J_v(x)\cos(vx)-J_{-v}(x)/\sin(vx)$, which converges from $v=n; n\in\mathbb{N}$ to $Y_n(x)=\lim_{v\to n}Y_v(x)$. Special solutions for the second kind of zero order are $Y_0(x)=(4/\pi^2)\int_0^{\pi/2}\cos(x\cos\xi)\{\gamma+\ln(2x\sin^2\xi)\}d\xi$. Bessel functions of the third kind are also referenced as HANKEL FUNCTIONS $H_v^{(1)}=J(x)+iY_v(x)=i\csc(v\pi)\{e^{-iv\pi}J_v(x)-J_{-v}(x)\}$ and $H_v^{(2)}=J_v(x)-iY_v(x)=i\csc(v\pi)\{J_{-v}(x)-e^{-iv\pi}J_v(x)\}$. Many solutions can be presented under specific conditions and can be found in SOLUTION books such as *Handbook of Mathematical Functions* by Milton Abramowitz and Irene A. Stegun. Bessel functions form attractive solutions to many complex mathematical and practical problems.

Beta decay (β⁻)

[atomic, general, nuclear, solid-state] Two types of DECAY can be distinguished: intrinsic SPIN of electron and NEUTRINO are parallel (GAMOW–TELLER DECAY) or antiparallel (FERMI DECAY). The BETA PARTICLE is similar to an electron and is primarily the result of the breakdown of an ATOM with a PROTON deficiency or excess in neutrons. Due to the ENERGY configuration, the NEUTRON is separated into a proton and an electron or beta particle [β⁻] as well as an ANTINEUTRINO ($\bar{v}$): $n_{neutron}=p_{proton}+\beta^-+\bar{v}$. Beta particles are emitted in a continuous energy spectrum, however, with an upper limit. During beta decay, electrons are emitted at relativistic speeds, and classical theory will be insufficient to explain the transmission TUNNELING through the nuclear potential energy BARRIER. This so-called penetrability is described by the relativistic FERMI FUNCTION. The beta-decay spectrum is directly correlated to the transition PROBABILITY: $(1/h)|\mathcal{M}|^2(dN_{tot}/dQ_{\beta^-})$, dominated by the density of states of the radio ISOTOPE formulated as $dN_{tot}/dQ_{\beta^-}=(p_{e^-})^2\delta(p_{e^-})(p_{\bar{v}})^2(dp_{\bar{v}}/dKE_{\bar{v}})(L^6/4\pi^4\hbar^6)$, with $\hbar=h/2\pi$, where $h=6.62606957\times10^{-34}$ m²kg/s is the PLANCK'S CONSTANT, N_{tot} is the total number of isotopes, $Q_{\beta^-}=KE_{e^-}+KE_{\bar{v}}$ represents the beta energy, $|\mathcal{M}|$ is determinant of the matrix defined by the WAVE function (ψ) ($\mathcal{M}=\int\psi_D^*\psi_{e^-}^*\psi_{\bar{v}}^*\Delta V\psi_p dxdydz$), $L=Mvr$ defines the angular momentum for the mass M at DISTANCE r with velocity v, p_i is the respective momentum for the various particles (p positron; $\bar{v}$ antineutrino; e^- electron; β^- beta particle), and $KE_{\bar{v}}$ is the kinetic energy for the antineutrino.

Beta particle

[atomic, nuclear, quantum] A charged PARTICLE emitted from the NUCLEUS that has a mass and charge equal in MAGNITUDE to those of the electron.

Beta plus decay

[atomic, energy, nuclear] In contrast to beta DECAY, the beta-plus PARTICLE is mostly similar to an electron apart from being opposite in charge and is primarily the result of the breakdown of an ATOM with a NEUTRON deficiency or excess in protons. The beta-plus particle release is also called as POSITRON DECAY. Due to the ENERGY configuration, the neutron is separated into a neutron and a positron or beta-plus particle [β^+] as well as an ANTINEUTRINO ($\bar{v}$): $p_{proton} = n_{neutron} + \beta^+ + \bar{v}$. Beta-plus particles are emitted in a continuous energy spectrum, however, with an upper limit, which resembles regular beat decay.

Betatron

[atomic, nuclear] A PARTICLE ACCELERATOR that applied a gradient in the magnetic flux, continuously increasing in the path of the particle. The gradient is established as a result of a periodic OSCILLATION in the field. The time-dependent oscillation provides a voltage gain due to the principles outlined in the FARADAY'S LAW. The accelerator works especially well for electrons (e), hence the name. The rate of gain in particle momentum ($p = mv = eBr$; m is the particle mass and v is the velocity) as a result of the applied FLUX ($\Phi_B = AB$; B is the MAGNETIC FIELD strength and A is the cross-sectional area of the accelerator tube) is $dp/dt = (e/2\pi r)(d\Phi_B/dt)$, where r is the radius of the orbit. The VACUUM requirements for the accelerator are better than 10^{-5} Torr. The electron in orbit gains several hundred electrovolt with each revolution. The accelerated electrons produce X-RAY in the order of $100\,\text{R/min}$ at $1\,\text{m}$ while operating in the range 25–35 MeV. Large betatrons operating in excess of 320 MeV can produce mesons.

Betelgeuse

[astronomy, general] An exceptionally bright STAR that is located in the Orion constellation. Betelgeuse has been the topic of many galactic observations, specifically the fact that due to its brightness (radiance), it can be investigated by the MICHELSON INTERFEROMETER, as well as the INTERFEROMETER improved by ROBERT HANBURY BROWN (1916–2002) and RICHARD Q. TWISS (1920–2005). The use of the Michelson interferometer, for instance, allowed for the determination of the diameter of the star (see Figure B.26).

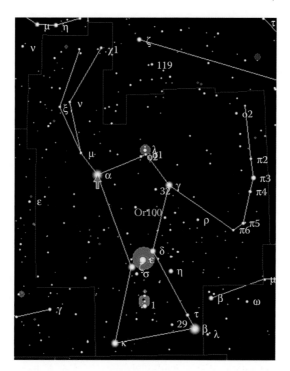

Figure B.26 Picture of the star Betelgeuse in the constellation Orion at position Alpha Ori.

B

Bethe, Hans Albrecht (1906–2005)

[astronomy/astrophysics, atomic, nuclear] A physicist from France. Hans Bethe received the Nobel Prize in PHYSICS for his work on explaining and theoretically defining nuclear reactions in 1967. Bethe provided insight into the nuclear ENERGY production in stars (see Figure B.27).

Figure B.27 Hans Albrecht Bethe (1906–2005). (Courtesy of Los Alamos National Laboratory, Los Alamos, New Mexico.)

Bevatron

[atomic, nuclear] SYNCHROTRON, a PARTICLE ACCELERATOR at Lawrence Berkeley Laboratories, University of California, Berkeley, California, in operation since 1949. Originally, it was designed to accelerate protons, but has been used to accelerate ionized particles as heavy as uranium. The acceleration was produced by a discrete upscaling of the potential difference gradually to 10 MeV (see Figure B.28).

Figure B.28 First charged particle tracks observed in liquid hydrogen bubble chamber at the Lawrence Berkeley Laboratories Bevatron in the 1930s.

B-field

[general] *See* MAGNETIC FIELD.

Big Bang theory

[astronomy, atomic, general] A theoretical concept of the origin of the UNIVERSE as expressed by ALEXANDER FRIEDMANN (1888–1925) and refined by GEORGE GAMOV (1904–1968). The model of George Gamov assumes that the universe started out as an immensely dense assembly of neutrons, which would be stable in this configuration. Following an implosion, the neutrons were expelled in an explosion that formed other nuclides, grouping to generate the ELEMENTS we now know (listed in the PERIODIC TABLE OF ELEMENTS) or those we still anticipate to find. The big bang event supposedly took place 15×10^9 years ago. Within the first 100 s supposedly, based on ENERGY laws, the first light nuclei were formed, starting out from elementary indistinguishable particles to leptons, quarks, bosons, and photons within the first 1×10^{-34} s. The elementary 1×10^{-43} s on initiation would have been controlled by the "grand unification theories"-based interaction forces, predating the electroweak interaction, forming the leptons and the like (see Figure B.29).

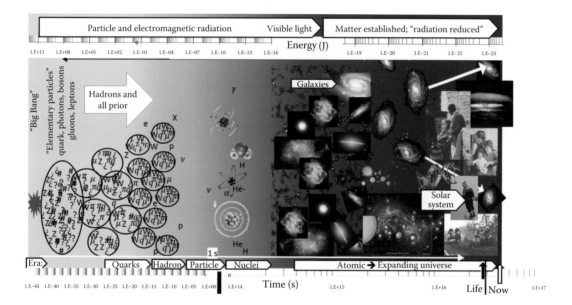

Figure B.29 The hypothetical sequence of events associated with the "Big Bang" process.

Binding energy

[atomic, nuclear, solid-state] The total ENERGY required to separate an ATOM of any ELEMENT with atomic number Z and mass number A into Z individual hydrogen (^{1}H) atoms as well as N neutrons: $A = Z + N$. This primarily pertains to the splitting of the NUCLEUS into its respective individual nucleons: protons and neutrons, not just the IONIZATION process. The total binding energy for hydrogen ^{2}H is 2.224574 MeV; for carbon ^{12}C, it is 92.16239 MeV, compared to ^{14}C with 105.28522 MeV; and for uranium ^{235}U, it is 1783.881 MeV. The greater the binding energy, the more stable the element, particularly the binding energy per NUCLEON. The binding energy per nucleon has an apparent peak with $8.7\,\text{MeV} = 1.394 \times 10^{-12}\,\text{J}$ per binding at $A \cong 60$, closest element cobalt ^{59}Co (one of the "magic nuclei"). On a related item, the binding energy per PARTICLE is generally different for protons versus neutrons (approximate same mass); however, they are the same for deuterium, although the expelled PROTON will have kinetic energy. The end-product of any FISSION will release the energy difference expressed by the sum of the masses of the fissure components subtracted from the mass of the original atom based on the mass–energy equivalence: $E = mc^2$, with the speed of light $c = 2.99792458 \times 10^8$ m/s, in addition to accounting for any kinetic energy of the constituents as well as PHOTON emission and elementary particle release.

Bingham, Eugene Cook (1878–1945)

[fluid dynamics, mechanics] A chemist and physicist born in the United States. Bingham made significant mathematical and experimental contributions to the understanding of RHEOLOGY and deformable FLOW (see Figure B.30).

Figure B.30 Eugene Cook Bingham (1878–1945). (Courtesy of Lafayette University, Easton, Pennsylvania.)

Bingham number (Bm = ($\tau_\gamma L / \eta v$) = ($\sigma_\gamma / \eta \text{Bm} \dot{\gamma}_0$))

[fluid dynamics, mechanics] A representation of the yield-stress ratio to viscous stress within a deformation process, where τ_γ is the shear stress, L is the "YIELD LENGTH" or characteristic length, η is the VISCOSITY, v is the velocity, σ_γ is the residual stress in the FLOW that stopped, and η_{Bm} is the BINGHAM PLASTIC VISCOSITY. This parameter relates to VISCOPLASTIC liquids and amorphous PLASTICS (e.g., Bingham plastics).

Bingham plastic

[fluid dynamics, mechanics] A FLOW medium with viscometric flow response to applied stress, nonlinear with a threshold stress (σ_0) above which flow is initiated $\sigma_{\text{stress}} - \sigma_0 = \eta \dot{\gamma}$, where $\dot{\gamma} = (dv_x / dx) + (dv_y / dy)$ is the rate of shear combining the directional flow velocity (v_i) gradients in all direction (*Note*: $\dot{\gamma}$ is the temporal derivative of the shear and the stress is $\sigma_{\text{stress}} = F_{xy} / A_{xy}$, the force F_{xy} is parallel to a surface A_{xy}) and η is the VISCOSITY. More formally, the Bingham plastic stress is defined as $\sigma_{\text{stress}} = \sigma_\gamma \, \text{sgn}(\dot{\gamma}) + \eta_{\text{Bm}} \dot{\gamma}_0 (\dot{\gamma} / \dot{\gamma}_0)$, where η_{Bm} is the BINGHAM PLASTIC VISCOSITY and $\text{sgn}(\dot{\gamma})$ return the NUMERICAL sign of the SHEAR RATE value $\dot{\gamma}$ (i.e., –; 0; or +). Some media are SHEAR THICKENING and

SHEAR THINNING under flow. At the point where the threshold is reached in flow reduction, the flow will stop ($\dot{\gamma} = 0$); however, a residual stress (σ_y) will remain (see Figure B.31).

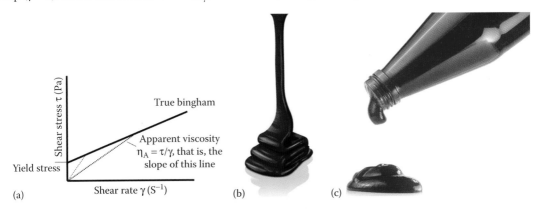

(a)

(b)

(c)

Figure B.31 (a) Graphical representation of the behavior of a Bingham plastic. (b) Example of chocolate as a Bingham fluid and (c) tomato ketchup as a Bingham fluid.

Bingham plastic liquid

[biomedical, fluid dynamics] A material under strain with an associated SHEAR STRESS yield value. When the VISCOELASTIC material exceeds a certain value, it will FLOW as a FLUID (e.g., tomato ketchup and toothpaste). The BINGHAM PLASTIC fluid has a RHEOLOGY that is non-Newtonian but is independent of time (not changing during the process apart from under the influence of the MAGNITUDE of the applied shear rate) (see Figure B.31).

Bingham plastic shear stress

[fluid dynamics, mechanics] A nonlinear description of the shear stress for a medium that deforms under FLOW, defined by $\sigma_{Bm} = \sigma_{yield}$.

Bingham plastic viscosity (η_{Bm})

[biomedical, fluid dynamics] The response of the shear stress under the influence of an applied shear rate described as $\eta_{Bm} = \mu_p + \left[\tau_y / \sqrt{\left[(\partial v_r / \partial z)^2 + (\partial v_\phi / \partial z)^2 \right]} \right]$, where μ_p is the plastic VISCOSITY, τ_y is the parallel shear stress, v_r is the radial velocity, v_ϕ is the angular velocity of FLOW, and z is the direction of flow. A Bingham FLUID has an inherent offset from the origin to overcome the yield stress in order to accurately characterize the material RHEOLOGY. The linear behavior of the shear stress as a function of the shear rate provides the slope in the form of the Bingham plastic viscosity. The shear stress (τ) is a function of the viscosity and the shear rate $\dot{\gamma}$ is given as $\tau = \tau_0 + \eta_{Bm} \dot{\gamma} Y'$, where Y' is the elastic modulus and $\dot{\gamma}$ is the shear rate. Apparent viscosity $\eta_{Bm} = \tau / \dot{\gamma}$ is given by slope of the line.

FLUID	η(mPa s)–ALTERNATIVELY (cP)
Water	1
Coffee cream	10
Vegetable oil	100
Honey	10,000
Asphalt	100,000

Note: cP, centiPoise.

Bioelectrode

[biomedical] An ELECTRODE with specific design to collect signals from within a CELL either by means of a needle filled with ELECTROLYTE FLUID or by an array of electrodes to collect the electrical DEPOLARIZATION wave front as a function of location through three-dimensional interpolation, such as a "sock" with a multitude of electrodes that fits around the HEART for accurate transmural "imaging" of the depolarization pattern during the systolic event and using volume conduction to locate the exact origin of any irregularities in the depolarization pathway. The irregularities in the ventricular depolarization can identify MUSCLE tissue that has been damaged either by lack of oxygen (CIRCULATION problems in coronary vessels) or by viral or parasite (e.g., Chagas disease) infections with inflammatory response (see Figure B.32).

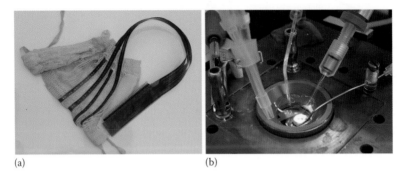

(a)   (b)

Figure B.32 (a) Example of a bioelectrode. (b) A special kind of electrode called a patch-clamp electrode for electrolyte-mediated recording of ionic potentials.

Bioluminescence

[biomedical, chemical, solid-state] Light emission resulting from OXIDATION process in biological systems. One preeminent example is the firefly, and several deep-ocean creatures also emit light. The firefly uses the light SIGNAL to attract attention during its mating ritual. There is also evidence that in mammals, in certain cases, cells may communicate with each other through light pulsation. Additional examples of bioluminescence are found in bacteria and fungi. Another description of this process is CHEMILUMINESCENCE, based on the chemical foundation, although the chemistry is controlled from biological origin. During the bioluminescence ACTIVITY, the biological entity releases specific chemical(s) that are oxidized mediated through a catalyst that employs an enzymatic process. The respective molecule released for oxidation by oxygen under ENERGY supplied by adenosine hate is referred to as LUCIFERIN, while the catalyst enzyme is branded luciferase. The firefly produces a yellow–green light in pulsating format (see Figure B.33).

(a)   (b)

Figure B.33 (a) Image of bioluminescence for a firefly and (b) bioluminescence of algae.

Biomedical optics

[astronomy/astrophysics, biomedical, computational, optics] Theoretical treatise of interaction of ELECTROMAGNETIC RADIATION with biological tissue. The biological media are mixed within the biological entity, creating an inhomogeneous three-dimensional structure where all constituents have specific optical characteristics that are representative of the particular tissues. The optical characteristics are represented by several parameters: absorption coefficient, SCATTERING coefficient, scattering anisotropy factor, and index of REFRACTION, each of which is a function of temperature, wavelength, and density, in turn linked to water content. The light propagation through the turbid biological media can be time dependent, specifically based on periodic biological events affecting the optical characteristics as well as depending on the format of light delivery, pulsed or continuous. The theoretical description of light propagation is derived from earlier work on galactic light propagation, interacting with dust in outer space as well as dust (i.e., various chemical configurations of particulates within a range of sizes) and gasses in the ATMOSPHERE at the location of observation. The early work by ARTHUR SCHUSTER (1851–1934) and KARL SCHWARZSCHILD (1873–1916) published in 1905 and 1906, respectively, provided the time-dependent EQUATION OF RADIATIVE TRANSFER that can be used to describe the three-dimensional propagation of light in turbid media such as biological tissues. The work by Schuster and Schwarzschild was refined in 1918 by FRIEDRICH KOTTLER (1886–1965). The final refinements were made by SUBRAHMANYAN CHANDRASEKHAR (1910–1995) published in 1947. The time-dependent angular and spatial photon ENERGY rate distribution is described by the light power incident on a cross-sectional area flowing within a SOLID ANGLE and is designated by the parameter radiance: $\vec{L}(\vec{r},\vec{s},t)$ [W/sr.cm^2]. In this notation, the vector $\vec{r}$ denotes the position of the location where the radiance is quantified, $\vec{s}$ is the direction vector for the PHOTON migration, and all as a function of time is t (*see* EQUATION OF RADIATIVE TRANSFER) (see Figure B.34).

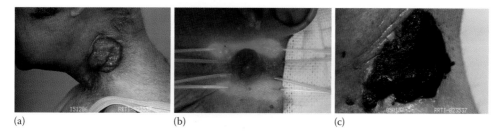

(a)   (b)   (c)

Figure B.34 (a–c) The use of laser to interact with an injectable photochemical that causes cell death in biomedical optics. The cancer treatment of this large cancer growth requires that the illumination is provided from the inside of the tissue volume to overcome the shallow light penetration.

Biophysics

[general] A branch of PHYSICS that integrates physics, biology, chemistry, and ENGINEERING to understand the workings of physiological phenomena as well as the MECHANICS of the ANATOMY and the operational mechanism of action for the various senses. The senses for animals may be different and supplemental to human senses: smell, taste, hearing, sight, touch, and pain (pain is considered a separate sense). Several animals can sense magnetic and electric fields, for instance, sharks for prey identification, and platypus for navigation, next to birds for their migratory path. The BLOOD flow and corrections to FLOW by means of atherectomy or artificial HEART replacement are a FLUID DYNAMICS aspect of biology that are critical for survival. The biological field stretches out from mechanics to fluid dynamics, THERMODYNAMICS, OPTICS, and chemistry. In biology, most physical phenomena are not consistent between individuals and they change over time (COMPLIANCE/time dependence). One specific note on pain: when a pain SIGNAL is processed in the somatosensory area of the parietal lobe of the brain, the location of the source for the pain is registered. Although pain originates from an internal organ, the source of pain may be elusive and misdirected. The action-potential activation of the insular cortex in the frontal lobe of the brain provides an immediate motivation to reverse the cause. In case the frontal lobe has areas that are damaged, one can

feel the pain, but is not motivated to counteract it or remove from the source. The periaqueductal gray area of the brain (midbrain) has the ability to close down the pain gate, thus alleviating the sensation of pain by chemical means. Chemicals such as endorphins affect the periaqueductal gray area by increasing its gate-blocking activity. Endorphins are produced naturally in the brain. Similar externally produced analgesics, such as morphine, also affect the periaqueductal gray area. Additional endorphins can be produced by the pituitary and adrenal glands. These glands release hormones under the influence of external stimuli. One specific example is the phenomenon of stress-induced suppression of pain (analgesia).

Biopotential

[biomedical] An electrical potential generated as the result of a transmembrane gradient in chemical concentration across the cellular MEMBRANE for nerves and MUSCLE as well as other cells for communication and data transfer. The MAGNITUDE of the biopotential is derived with the NERNST EQUATION.

Biot, Jean-Baptiste (1774–1862)

[general] A French physicist. The work of Biot provided the first documented and theoretical proof and definition of OPTICAL ACTIVITY by solutes and solids. The most well-known contributions of Jean Biot are in ELECTRODYNAMICS, specifically his work with FÉLIX SAVART (1791–1841). In 1820, Biot and Savart provided the description of the force on a MAGNET at a DISTANCE from a current in a wire yielding a MAGNETIC FIELD of its own, known as the BIOT–SAVART LAW, also referred to as AMPÈRE'S LAW, more specifically using concepts introduced by ANDRÉ MARIE AMPÈRE (1775–1836) (see Figure B.35).

Figure B.35 Jean-Baptiste Biot (1774–1862).

Biot number (Bi $= h_{conv}L/k_{cond}$)

[biomedical, thermodynamics] A dimensionless number applied to HEAT TRANSFER. The concept was introduced by JEAN-BAPTISTE BIOT (1774–1862) around 1804 describing the heat transfer conduction within a solid in relation to the convection. The BIOT NUMBER provides the equivalence of conduction and convection with respect to a characteristic length (L), convection coefficient (h_{conv}, convection heat transfer coefficient), and THERMAL CONDUCTIVITY (k_{cond}) of the medium. The BIOT NUMBER is derived using the FOURIER CONDUCTION LAW and NEWTON'S LAW OF COOLING. One ordinary example of the association of this number is that smaller objects cool faster than larger objects. Another implication is that for large objects, the temperature gradient must be very large, specifically for large BIOT NUMBER.

Biot number, mass transfer ($Bi_m = k_{cond} L / D_{AB} = h_{conv}{}^{fluid} L / D_{AB} \delta_m$)

[general, mechanics, thermodynamics] A dimensionless number relating the mass transfer to HEAT TRANSFER, particularly involving PHASE transitions. The BIOT MASS-TRANSFER NUMBER describes the ratio of "internal diffusion resistance" over the "external diffusion resistance," where k_{cond} is the THERMAL CONDUCTIVITY of the fluid, D_{AB} indicates the mass DIFFUSIVITY.

Biot–Savart law

[biomedical, computational, electronics, general] Definition of the MAGNITUDE and direction of the MAGNETIC FIELD ($\vec{B}$) resulting from a CURRENT (I) in a length of wire $\Delta \ell$ at a DISTANCE r (i.e., radius of loop of the magnetic field line) from the core of the wire $\vec{B}(r)_x = \mu_m \left[I \left(\overrightarrow{\Delta \ell} \times \vec{r} \right)_x \middle/ 4\pi r^3 \right]$, where $\mu_m = \mu_{mr}\mu_{m0} = \mu_{mr} \times 4\pi \times 10^{-7}$ N/A^2 is the magnetic permeability, represented by the magnetic permeability of VACUUM ($\mu_{m0} = 4\pi \times 10^{-7}$ N/A^2) multiplied by the relative magnetic permeability (μ_{mr}), at any location x in the length of the wire. This definition was devised by JEAN-BAPTISTE BIOT (1774–1862) and FÉLIX SAVART (1791–1841) in 1820. This principle was initially investigated in relation to the forces that act on a MAGNET as a result of any combination of nearby currents (*Note*: superposition principle).

Birefringence

[general, optics] Material property. The fact that a ray of light originating from one source (which consists of a finite or infinite number of point sources) traverses the medium in two separate paths. In solids, a medium can have two optical axis that are at an ANGLE to each other. In liquids, the same phenomenon can be achieved through an externally applied electric field, which polarizes the molecules to provide two axis with different index of REFRACTION. The artificially induced birefringence plays a critical role in the KERR EFFECT. The Kerr CELL can be made to change transmissivity on a very short timescale, in sync with the changing applied external electric field. The Kerr effects for liquids have a timescale of less than picoseconds, depending on the MAGNITUDE of the applied field and the constituents of the LIQUID (see Figure B.36).

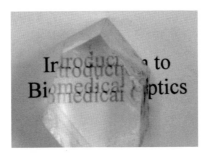

Figure B.36 The birefringence in a crystal, splitting the original text in separate images.

B

Birge, Raymond Thayer (1887–1980)

[atomic, chemical, nuclear, quantum] A physicist and chemist from the United States. Raymond Birge introduced certain concepts that may be considered QUANTUM chemistry. He was a firm supporter of the BOHR ATOMIC MODEL. His work on the spectral shifts of diatomic molecules, specifically the spectral emission of nitrogen when placed under conditions that create a high degree of DISPERSION, created insight into the ENERGY configuration of the electron orbit in the diatomic nature. His work provided details on the methods of establishing the HEAT OF DISSOCIATION for such diatomic molecules as oxygen, nitrogen, and carbon monoxide using the spectral emission profile. He also provided the most accurate values for constants such as the RYDBERG CONSTANT and Planck's constant (see Figure B.37).

Figure B.37 Picture of Raymond Thayer Birge around 1920. (Courtesy of Bancroft Library, University of California Berkeley, Berkeley, California.)

Bjork–Shiley mechanical heart valves

[biomedical] An artificial HEART VALVE introduced in 1968 by VIKING OLOV BJÖRK (1918–2009), a Swedish physician and cardiac surgeon, and DONALD PEARCE SHILEY (1920–2010), an engineer from the United States, to replace malfunctioning (not closing properly) bicuspid valves (aortic valve) leading from the left ventricle into the AORTA. The valves open and close due to a fluid-dynamic pressure gradient as a result of the contraction and relaxation of the ventricular MUSCLE of the heart (see Figure B.38).

Figure B.38 Picture of equivalent Bjork–Shiley mechanical heart valve.

Black, Joseph (1728–1799)

[atomic, general, thermodynamics] A physician from Scotland, Great Britain. Joseph Black was the first person to formulate the concepts of LATENT HEAT and SPECIFIC HEAT. He also discovered the gaseous compound carbon dioxide. He was born in France and resided in his country of nationality Scotland, UK. He introduced the concept of CALORIMETRY and caloric value in 1760. He is a British pioneer in THERMODYNAMICS and introducer of the concept of "the MATTER of heat." The quantity of HEAT was placed in contrast to the known designation of TEMPERATURE, later translated as CALORIC VALUE, as introduced by Black in 1760. JAMES WATT (1736–1819) was the instrument maker at Black's university, the University of Glasgow during his time. Black also recognized that carbon dioxide was produced during the metabolic process and was expelled during exhalation. His work is directly linked to the introduction of the ZEROTH LAW OF THERMODYNAMICS.

Black body

[general, nuclear, optics, thermodynamics] A radiative emissive object. A black body readily absorbs and emits thermal RADIATION, with a certain degree of efficacy. The surface absorption efficiency is defined by the spectral absorbance, which is a fractional proportion of the incident ENERGY with respect to the absorbed energy. A perfect absorber has a spectral absorbance of 1: $\alpha_\lambda (T) = 1$ (as a function of temperature (T)), defining 100% efficiency. METAL surfaces are generally not perfect absorbers, neither are other polished surfaces. The emission is correspondingly represented by the EMISSIVITY, ranging from $0 \leq \varepsilon_\lambda (T) \leq 1$. A body that has the maximum absorbing efficiency generally also has equally effective emission of ELECTROMAGNETIC RADIATION due to the thermal agitation of charges at or near the surface (temperature is proportional to the internal kinetic energy). For instance, aluminum foil has $\varepsilon_\lambda (T) = 0.15$, compared to $\varepsilon_\lambda (T) = 0.9$ for carbon and $\varepsilon_\lambda (T) = 0.98$ for exposed human SKIN (considered a perfect black body—no makeup or sunscreen). The thermal agitation of charges in particular presents the process of accelerated charges which, by definition, will produce electromagnetic radiation (as well as absorption under the same concept: CONSERVATION OF ENERGY). The emission process is described by WIEN'S DISPLACEMENT LAW. A highly effective absorber/emitter is referred to as a perfect black body. The spectral profile of a perfect black body follows PLANCK'S LAW of radiative emission. The ideal perfect black body is a box with a rough internal surface (preferably tinted black, but not a requirement) and a small hole; the hole presents the black body, confining all the heat (see Figure B.39).

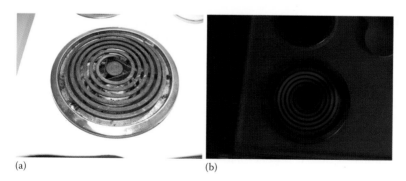

(a)   (b)

Figure B.39 (a and b) Picture of black body, cold and emitting when hot.

Black-body emission

[astronomy, energy, general, nuclear, optics, solid-state] The RADIANT POWER (P_{rad}) per unit area emitted from a body with EMISSIVITY of a "black body" (emissive object) as a function of emission FREQUENCY (ν) expressed as $P_{rad} = \left[4\pi h\nu^3 / c^2 \left(e^{h\nu/kT} - 1 \right) \right] \Delta\nu$, where h is the PLANCK'S CONSTANT, k is the BOLTZMANN CONSTANT, and T is the TEMPERATURE (in KELVIN).

Blackett, Patrick Maynard Stuart, Baron (1897–1974)

[astronomy/astrophysics, nuclear] An English physicist who's groundbreaking work on cloud chambers helped understand the disintegration of isotopes and the trajectory of charged particles. The cloud chamber was instrumental in recognizing the charge state of COSMIC RAYS. Blackett's interests also included the investigation of PALEOMAGNETISM. He received the Nobel Prize in Physics in 1948 for his work on defining cosmic rays and charge–ray interaction and his experimental verification of annihilation RADIATION. He frequently worked closely with GIUSEPPE ("BEPPO") PAOLO STANISLAO OCCHIALINI (1907–1993). He was a pupil of ERNEST RUTHERFORD (1871–1937) in the Cavendish (Cambridge, UK) laboratory (see Figure B.40).

Figure B.40 Picture of Baron Patrick Maynard Stuart Blackett (1897–1974) in approximately 1950.

Black hole

[astronomy/astrophysics, general, quantum, thermodynamics] A galactic phenomenon that apparently captures all mass and ELECTROMAGNETIC RADIATION in its vicinity and will not reemit. The theoretical basis lies in the fact that the mass of the phenomenon is so large that the gravitational force is enormous and interacts even with the electric and MAGNETIC FIELD energies. Keep in mind that GRAVITY is one of the weakest primary forces in the PHYSICS models. The electronic attraction between a PROTON and an electron is on the order of 10^{40} times stronger than their interrelated gravitational interaction. Mathematically, the interaction of gravity with MATTER is described as the coupling of the gravity vector with the ENERGY-momentum TENSOR of matter. An indication of the phenomenological minimum requirements would be an object with diameter of 15 km and 5 times the mass of the SUN ($m_{black\ hole} = 1.0 \times 10^{31}$ kg), yielding a density $\rho_{black\ hole} \geq 10^{19}$ kg/m^3. The presence of black holes has been identified based on the gravitational influences on nearby bodies as well as diversions in stellar RADIATION patterns. The MILKY WAY also has a verified black hole it its center. Due to the mechanism, the black hole will continue its growth; the black hole in the center of the Milky Way is presumably composed of millions of stars, planets, and "DARK MATTER." The total mass of a black hole is defined to be confined within the Schwarzschild radius ($R_s = 2GM/c^2$, where $G = 6.673 \times 10^{-11}$ Nm^{2}kg^{-2} is the gravitational constant, M is the mass of the object, and $c = 299,792,458$ m/s is the speed of light) (see Figure B.41).

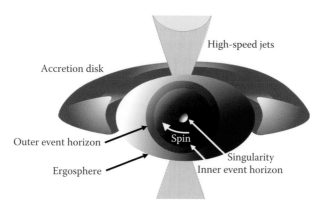

Figure B.41 The configuration of a black hole, based on the efforts from Dr. Luka Č. Popović, Dr. Predrag Jovanović, and Dr. Dragana Ilic, all from the Astronomical Observatory, Belgrade, Serbia.

Black-holes Colliding

[astrophysics, relativistic] Ripples in SPACETIME, time perturbations (*see* TIME PERTURBATION, COLLIDING BLACK HOLES, GRAVITATIONAL WAVES, *and* EINSTEIN EQUATIONS).

Blake number ($B\ell = v\rho/\eta(1-\varepsilon)A_s$)

[energy, fluid dynamics] In FLOW through an AGGREGATION of solids (e.g., rock bed) a nondimensional expression to relate the ratio of the inertial to the viscous force. In the representation of the Blake number, the following parameters apply: v is the flow velocity, ρ is the FLUID density, η is the dynamic VISCOSITY, ε is the "bed porosity" or void fraction, and A_s is the characteristic or "bed-specific" surface area. The number is used in momentum transfer as a generalization of the REYNOLDS NUMBER.

Blanc's law

[atomic, condensed matter] The total ION MOBILITY velocity achieved by a mixture of ions moving through a gaseous medium when an external unit electric field is applied μ_{ion}. For a mixture, this translates into $\mu_{ion} = \left(\sum_{i=1}^{n} f_i/\mu_{ion,i} \right)^{-1}$, where f_i is the fractional value for constituent i and $\mu_{ion,i}$ is the respective ion mobility.

Blasius, Paul Richard Heinrich (1883–1970)

[computational, fluid dynamics] A German physicist, student of Ludwig Prandtl (1875–1953), who in 1908 described the separation of a boundary layer under certain flow conditions. Additional work by Blasius provided the expression for pipe flow resistance in terms of the Reynolds number in 1911. Blasius is a contemporary of Theodore von Kármán (1881–1963) (see Figure B.42).

Figure B.42 Picture of Paul Richard Heinrich Blasius (1883–1970).

Blasius' theorems

[fluid dynamics] A flow of a medium with density ρ in the z-direction around a body will exert a vector force in two dimensions ($\vec{F} = F_x - iF_y$) on this body that is dependent on the flow velocity potential (Φ_{flow}) and the contour of the body (C_{contour}), expressed as $\vec{F} = i\rho/2 \int_{\text{contour}} (d\Phi_{\text{flow}}/dz)^2 \, dz$.

Blatt, John Markus (1921–1990)

[computational, nuclear] A theoretical nuclear physicist and mathematician from Austria. Blatt contributed to the theoretical description of elastic electron collisions as well as superconductivity.

Bloch, Felix (1905–1983)

[atomic, biomedical, nuclear, quantum] A physicist from Switzerland. The contributions of Felix Bloch range from establishing the quantum theory of solids to the de Broglie equivalent Bloch waves for electrons. He also provided significant information leading to the description of the magnetic moment for neutrons as well as the Bloch equations that describe the evolution of the magnetic moment with respect to nuclear magnetization, specifically beneficial for the development of MAGNETIC RESONANCE imaging. He shared the Nobel Prize in PHYSICS with EDWARD MILLS PURCELL (1912–1997) from the United States for their nuclear magnetic work, specifically nuclear magnetic resonance (see Figure B.43).

Figure B.43 Picture of Felix Bloch (1905–1983).

Bloch equations

[atomic, biomedical, nuclear] The nuclear magnetization definitions with respect to nuclear MAGNETIC RESONANCE phenomena observed and described by FELIX BLOCH (1905–1983). Using the GYROMAGNETIC RATIO (with respect to the NUCLEUS, the gyromagnetic ratio is $\gamma_N = eg_N/2m_p = g_N\left(\mu_N/\hbar\right)$, where g_N is the nuclear g-factor, m_p is the PROTON mass, $\hbar = h/2\pi$ is the Planck's constant with $h = 6.626070040(81) \times 10^{-34}$ m^{2}kg/s (Js), $\mu_N = e\hbar/2m_p c$ is the nuclear MAGNETON, and c is the speed of light) and nuclear magnetization $\vec{M}(t) = [M_x(t), M_y(t), M_z(t)]$, the magnetic moment of an atomic nucleus that results from the MAGNETIC DIPOLE associated with the SPIN of the protons and neutrons. The Bloch equations for a nucleus exposed to an external time perturbed MAGNETIC FIELD (in the z-direction) expressed by $(\vec{B}(t) = [B_x(t), B_y(t), B_0 + \Delta B_z(t)])$ are written as follows:

$$dM_x(t)/dt = \gamma_N \left[\vec{M}(t) \times \vec{B}(t)\right]_x - \left(M_x(t)/T_t\right),$$

$$dM_y(t)/dt = \gamma_N \left[\vec{M}(t) \times \vec{B}(t)\right]_y - \left[My_x(t)/T_t\right], \text{ and}$$

$$dM_z(t)/dt = \gamma_N \left[\vec{M}(t) \times \vec{B}(t)\right]_z - \left[M_z(t) - M_0/T_\ell\right],$$

where T_ℓ and T_t represent the longitudinal and transverse RELAXATION TIME (which will tend to infinity when no relaxation is obtained).

Bloch's theorem

[atomic, biomedical, nuclear] The ENERGY bands in a LATTICE structure have solutions to the SCHRÖDINGER EQUATION $(\Delta \phi_{\vec{k}}(\vec{r}) + (2m/\hbar)\left[KE_{\vec{k}} - V(\vec{r})\right]\phi_{\vec{k}}(\vec{r}) = 0$, where $KE_{\vec{k}}$ represents the kinetic energy for a PARTICLE with mass $m \cong \hbar^2\left(\partial KE_{\vec{k}}/\partial \vec{k}\right)$, $\hbar = h/2\pi$, with $h = 6.62606957 \times 10^{-34}$ m^{2}kg/s the Planck's constant) for a periodically varying potential $(V(\vec{r}))$ and have solutions $(\phi_{\vec{k}}(\vec{r})$, where $\vec{k}$ represents k-SPACE) of the form $\phi_{\vec{k}}(\vec{r}) = e^{-i\vec{k}\cdot\vec{r}} U(\vec{r})$, where $U(\vec{r})$ represents the internal energy for the crystal lattice with periodicity in $\vec{r}$. This format applies specifically to SEMICONDUCTORS, requiring determination of $KE_{\vec{k}}$ primarily in the first BRILLOUIN ZONE only. The eigenstates of the SOLUTION for an electron in a crystalline lattice are forming Bloch waves, confirming the electronic band structures (formation of electrons in "shells," with the specific VALENCE and CONDUCTION BAND in reference to the energy structure). The concept was introduced by FELIX BLOCH (1905–1983). It is also known as Bloch WAVE.

Blondlot, Prosper-René (1849–1930)

[atomic, nuclear] A physicist from France known for his controversial and unsubstantiated definition and discovery of so-called N-ray in 1903. The N-ray refers to the hypothetical *neutron radiation*, shortly after the introduction of X-RAY RADIATION by WILHELM CONRAD RÖNTGEN (1845–1923). The failed confirmation of the hypothetical V-ray by his peers created a cautionary attitude in the scientific world toward research based on biased opinion in order to prove a concept, skewing the methods and results.

Blood

[biomedical, fluid dynamics] A biological FLUID that transports oxygen, carbon dioxide, nutrients, and waste products. The oxygen transport relies on the use of hemoglobin enclosed in the red BLOOD CELL. Blood is, due to its particulate structure, a non-Newtonian LIQUID, specifically a shear-thinning liquid (pseudoplast). One may sometimes consider blood as a thixotropic fluid due to the delay experienced for ROULEAUX FORMATION and the response to changes in shear rate. The process of rouleaux formation is the clumping of red-blood cells to form large amalgamations (French for *roll* of blood, resembling a roll of coins) potentially under the influence of abnormal proteins. Rouleaux formation may be considered a rheological abnormality, but it does occur also during pregnancy and as a result of certain FLOW conditions. However, the viscous behavior for blood is a function of the vessel diameter, specifically the flow in capillaries forms a *train* of RED blood cells that are transported as disks that touch the WALL in circumference, balancing the forces, migrating as *parachutes* with significant FRICTION due to indentation of the wall of the vessel. Blood as a transport vehicle for oxygen is perfectly adjusted for the low oxygen content of the Earth's ATMOSPHERE. The oxygen SATURATION CURVE, describing the binding efficacy of oxygen to hemoglobin, is optimized for low oxygen partial pressure. Considering a partial pressure for oxygen of approximately 20.9% O_2 to be 20.9×760 mmHg $\cong 160$ mmHg $= 2.133 \times 10^4$ Pa, the oxygen saturation for blood is relatively level down to 20 mmHg. The blood *consumption* (VO_2) as well as the minute VENTILATION in the alveoli (V_A) determines the drop in partial pressure from inspired to alveolar step in the lung exchange (generally from 150 mmHg to 100 mmHg; *Note*: 20% of oxygen content at an ATMOSPHERIC PRESSURE of 760 mmHg yields a partial pressure for oxygen of free AIR of ~152 mmHg). The partial pressure difference between the oxygen level in the alveoli and the blood in the artery just inside the lung, opposite to the MEMBRANE of the alveoli, is relatively small (~5 mmHg). In the CIRCULATION, starting from arterial stream and tissue exchange, the partial pressure of the oxygen in the blood drops from 95 mmHg to 40 mmHg on the venous side. The arterial to venous oxygen content is a function of the oxygen consumption per minute, the blood flow, and the hemoglobin dissociation curve. The brain controls the RESPIRATION based on the carbon dioxide level only, due to the fact that it affects the pH of the blood (i.e., dissolved provides carbonic ACID, BOHR EFFECT) and can hence be detected by chemical sensors in the blood stream on the vessel wall. In the alveoli of the lungs, there are chemical "transporters" that facilitate the exchange of oxygen and carbon dioxide. Alternatively, carbon monoxide binds permanently to hemoglobin, hence reducing the oxygen exchange, specifically with respect to cigarette smoke. Nitrogen, as the majority constituent in atmospheric air, is also dissolved in the blood and consequently in tissue. The nitrogen content in the tissue will increase dramatically for deep-sea divers due to the fact that the applied external pressure directly affects the partial pressure of the dissolved gasses in tissues. When the deep-sea divers approach the surface of the water too quickly, the tissue will not have time to gradually release the dissolved nitrogen in the blood stream and will release it quickly in GAS form, creating BUBBLES that form emboli that can result in restrictive blood flow in critical anatomical positions, including the HEART and the brain, causing serious illness and potentially death. The decompression disease is often referred to as CAISSON DISEASE or *the bends* (see Figure B.44).

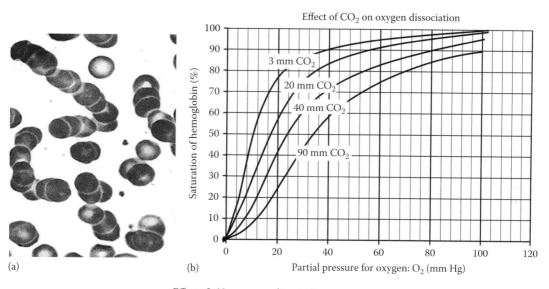

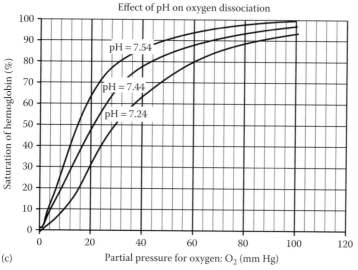

Figure B.44 Blood flow of RBCs in the blood vessel: (a) clumping together of the primary components of blood, the RBCs—rouleaux formation and (b) graphical illustration of the influence of the carbon dioxide concentration in the blood on the oxygen release by the hemoglobin in the RBCs. (c) graphical illustration of the influence of the blood acidity on the oxygen release by the hemoglobin in the RBCs. *(Continued)*

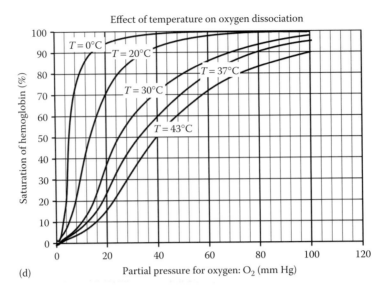

Effect of temperature on oxygen dissociation

(d)

Partial pressure for oxygen: O_2 (mm Hg)

Figure B.44 (Continued) Blood flow of RBCs in the blood vessel: (d) graphical illustration of the influence of the solvent temperature on the oxygen release by the hemoglobin in the RBCs.

Blood pressure

[general] The heart's ventricular contraction provides a pressure (SYSTOLIC PRESSURE) that results in BLOOD flow through the arteries. The COMPLIANCE of the arteries accommodates the sudden increase in blood volume in a restrictive CIRCULATION system, allowing for the pressure to be sustained by the vascular compliance and the surrounding tissue volume that is "pushed aside" as a result of the blood flow. The AORTA is the first vessel attached to the left ventricle leading into the whole body, apart from the pulmonary artery carrying oxygen poor blood from the right ventricle to the lungs. The aortic connection to the brachial artery in the arm provides a relatively reduced pressure profile with respect to the aorta. The brachial artery is accessible in the elbow of the arm and is used for blood pressure measurements by means of the SPHYGMOMANOMETER. The mechanical contraction of the ventricle has a periodic pattern, known as the *heart rate*, providing systolic pressure during contraction and DIASTOLIC PRESSURE during relaxation. The blood pressure in the arm is measured by means of a pressure cuff that will totally close off the brachial artery when inflated to a randomly high pressure in excess of 250 mmHg, the statistically determined highest pressure in the general population. When the pressure cuff pressure is gradually reduced (letting the AIR escape from the bellows at a controlled rate), at one point the pressure in the cuff will match the interstitial pressure applied to the tissue surrounding the artery during systolic blood flow. At this point, a small amount of blood will seep through the narrow opening in the brachial artery, creating a fluttering SOUND similar to allowing air to escape from a balloon by squeezing the opening. The sound made by the blood being forced through the narrow opening generating a tissue VIBRATION was recognized by the Russian physician NIKOLAI SERGEYEVICH KOROTKOV (1874–1920) in 1905. The "Korotkov sounds" are identified by placing a stethoscope in the pit of the elbow. Once the pressure is reduced further, at one point the blood will FLOW freely, referenced as the diastolic pressure. The rate at which the pressure in the cuff is released provides the accuracy and RESOLUTION in the blood pressure diagnosis (slope of the curve, where the curve is composed of

B

data points acquired at each heartbeat). The two values, systolic and diastolic pressure, are used for clinical identification of problems and health issues associated with either the HEART or vasculature. The normal range is lower than 120/80 mmHg (for a 20-year-old healthy subject); however, these values depend on AGE, where the accepted systolic pressure increases with age. High blood pressure still remains a topic of clinical discussion, and generally, the accepted upper limit for a problematic cardiovascular system is greater than 140/90 mmHg, with specific warning signs for systolic pressure greater than 160 mmHg and diastolic pressure greater than 100 mmHg. The pressure differences as well as ABSOLUTE PRESSURE gradually decline with resistive (vascular diameter) and capacitive (vascular compliance, stretch) DISTANCE with respect to the heart (*also see* WINDKESSEL) (see Figure B.45).

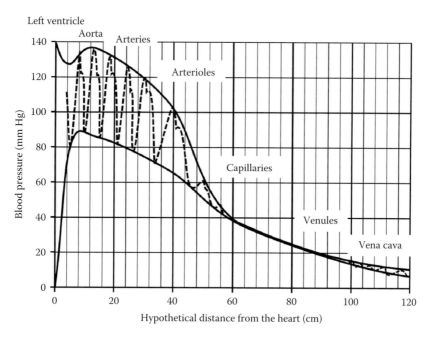

Figure B.45 Graphical representation of the idealized blood pressure in the healthy (average) human circulation with respect to the distance to the heart and the local vessel diameter, depicted by the envelope of the pulsatile flow. The pulsatile (temporal) profile is indicated by a dotted line. The temporal profile also illustrates the dispersion resulting from acoustic impedance based on the vascular compliance and viscous flow constraints.

Blue shift

[astronomy/astrophysics, computational, mechanics, optics] A shift in the spectral line value for verified atomic transitions for specific ELEMENTS to apparently shorter wavelengths due to a relative decrease in proximity between PHOTON emission source and observer. This process is described under *Doppler shift*. The phenomenon relates to the galactic movements of stars with respect to EARTH. The absorption (or emission) lines generally used are the transitions for hydrogen (*see* **BALMER LINES**).

B-meson

[atomic, general, solid-state] As the B-meson was originally thought to DECAY only to produce either a MUON or a tau, new discoveries have shown a broader range in energies. The charged leptons may indeed not all be equal. The decay processes measured are spread over an ENERGY band that is greater than two times the standard deviation based on the original model (see Figure B.46).

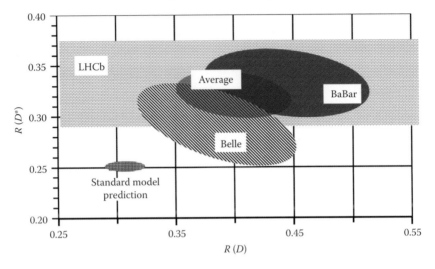

Figure B.46 The experimental results from various groups with regard to the "branching ratios" (R) in relation to the two observed distinct decay processes (D and D*) in reference to the standard single B-meson model and the average. The unitless branching ratios, representing the decay mechanism of action for the respective B-mesons, observed by the Bell group at the High Energy Accelerator Research Organization, KEK, Tsukuba, Ibaraki 305-0801 Japan, respectively, the Stanford BaBar collaboration (BaBar = B/B-bar system of mesons) at the SLAC National Accelerator Laboratory, Menlo Park, CA (Stanford Linear Accelerator Center), and the LHCb (Large Hadron Collider beauty) collaboration at CERN, European Laboratory for Particle Physics, CH-1211, Genève 23, Switzerland.

Bode, Hendrik Wade (1905–1982)

[commmunication, electronics, mathematics] A physicist and electrical engineer from the United States.

Bode diagram

[electronics, general] A graphical representation of the frequency response for the MAGNITUDE of a complex quantity of a system (process) as a function of either frequency or PHASE. In electronic applications, the correlation between the complex impedance $Z(f) = \sqrt{(X_c^2 + R^2)}$ and the phase ANGLE $\varphi(f) = -\mathrm{arctg}(X_c/R)$ (R is the RESISTANCE and X_c is the capacitive reactance) is called a Bode diagram, named after the efforts from HENDRIK WADE BODE (1905–1982) (see Figure B.47).

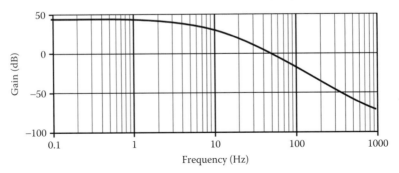

Figure B.47 Bode diagram for a complex electronic alternating current circuit response.

Bodenstein number (Bs = $vL/D_{v,a}$)

[fluid dynamics] An equation applied to chemical reaction calculations with respect to DIFFUSION, where v is the velocity, L is the characteristic length, and $D_{v,a}$ is the axial DIFFUSIVITY. It can also be considered as the inverse of the PÉCLET NUMBER.

Bohr, Christian Harald Lauritz Peter Emil (1855–1911)

[biomedical] A physiologist from Denmark, father of the physicist NIELS HENRIK DAVID BOHR (1885–1962). Christian Bohr is known for the BOHR EFFECT with respect to the oxygen exchange for BLOOD with respect to the partial pressure of the other constituents of the blood; published in 1902.

Bohr, Niels Henrik David (1885–1962)

[atomic, general, nuclear] A physicist from Denmark. In 1913, Niels Bohr introduced a new concept that the angular momentum of the electrons in the HYDROGEN ATOM has distinct discrete values, equaling $n\hbar$, where $\hbar = h/2\pi$ with $h = 6.62606957 \times 10^{-34}$ m^{2}kg/s the Planck's constant. The associated radius for the respective Bohr ORBITS is $r_{Bohr} = n^2 (0.0529\,\text{nm})$, where $n = 1,2,3,\ldots$. Niels Bohr received the Nobel Prize in PHYSICS for his work on the atomic model in 1922 (see Figure B.48).

Figure B.48 Niels Henrik David Bohr (1885–1962), anno 1922. (Courtesy of Nobel Prize Organization, Stockholm, Sweden.)

Bohr atom

[atomic, general, nuclear, solid-state, thermodynamics] An ATOM consists of two main components: NUCLEUS and orbiting "PLANET" electrons. This model was developed in parallel with the RUTHERFORD ATOMIC MODEL. Using the information provided by the experimental observations by ERNEST RUTHERFORD (1871–1937) along with JOHANNES "HANS" WILHELM "GENGAR" GEIGER (1882–1945) and ERNEST MARSDEN (1889–1970), and later again supported by HENRY (HARRY) GWYN JEFFREYS MOSELEY (1887–1915), Bohr was able to postulate that the electron-binding ENERGY levels (E_n) are ordered as a function of the "position" relative to the nucleus (identified by the PRINCIPAL QUANTUM NUMBER n, which delineates the DISTANCE to the nucleus: the radius increases by n^2) as follows: $E_n = -\left(2\pi^2 m_e e^4 Q^2 / h^2\right)\left(1/n^2\right)$, where m_e is the electron mass, e is the ELECTRON CHARGE, h is the PLANCK'S CONSTANT, and $Q = k_0 Z$ is the ATOMIC NUMBER (Z)

multiplied by the Coulomb force constant (k_0). Based on Planck's observations with regard to the quantization of ELECTROMAGNETIC RADIATION ($E = h\nu$, where h is the Planck's constant and ν is the frequency of light) in addition to the PHOTOELECTRIC EFFECT described by Einstein, in 1908 Niels Bohr formulated his interpretation of the atomic structure. Bohr's atomic model assumptions are as follows: (1) electrons orbit outside the nucleus in discreet stable revolutions $F_{centripital} = \left(mv^2/r \right) = \left(e^2/r^2 \right) = F_{electrostatic}$, where m is the mass, v is the ORBITAL VELOCITY, e is the electron charge, and r is the radius to the core of the atom (the radius of the orbit of the electron); (2) the electronic orbit angular momentum (m_θ) has discreet values $m_\theta = mvr = nh/2\pi$, where n, an integer, is the principal QUANTUM NUMBER; (3) light is emitted or absorbed in electronic orbital transitions as quantum jumps, each with photonic energy content equal to the difference in orbital energy of the electron $h\nu = E_{initial} - E_{final}$. Combining these postulates yields the following expression for the orbit configuration: $r_n = n^2 \left(h^2/4\pi^2 me^2 \right)$, also known as BOHR MODEL OF THE ATOM.

Bohr effect

[biomedical] The principle that the binding of oxygen is affected by both the carbon dioxide concentration and the acidity. The statement made by Christian Bohr in 1902 described that an increase in pH will influence the oxygen affinity of hemoglobin resulting in an increased binding, whereas an increase in carbon dioxide (CO_2) will result in a release of oxygen. Keeping in mind that the additional information became available that dissolved carbon dioxide forms carbonic ACID, hence increasing the pH.

Bohr magneton

[atomic, energy, nuclear, optics] An expression for electron MAGNETIC DIPOLE moment $\mu_B = e\hbar/2m_e c$, where $\hbar = h/2\pi$ with $h = 6.62606957 \times 10^{-34}$ m^{2}kg/s the Planck's constant, m_e is the electron mass, $e = 1.60217657 \times 10^{-19}$ C is the electron charge, and $c = 2.99792458 \times 10^8$ m/s is the speed of light. Additionally, the Bohr magneton provides an indication of the intrinsic MAGNETIC FIELD associated with the ELECTRON SPIN of an electron in orbit in a BOHR ATOMIC MODEL with MAGNITUDE $\vec{m} = 9.274 \times 10^{-24}$ Am^2. The orientation of the magnetic moment in this case is in the opposing direction of the angular momentum due to the NEGATIVE CHARGE of the electron.

Bohr method

[biomedical, fluid dynamics] In RESPIRATION, the volume capacity of the lung is measured using this method. The method specifically determines the physiological dead space of the lung total expired volume, named after CHRISTIAN HARALD LAURITZ PETER EMIL BOHR (1855–1911). The physiological dead space (V_D) is the lung volume that is not receiving waste CO_2 from the BLOOD as it flows through the lung vasculature. The AIR expired from the lung is collected over several breaths. The collected expired air is analyzed for P_{e,CO_2}, the partial pressure of carbon dioxide. Subsequently, a collection is made of air forcefully expelled at end expiration, the alveolar air sample, providing the alveolar P_{CO_2}: P_{A,CO_2}. The physiological dead space (V_D) is derived from $V_T P_{e,CO_2}$ ($V_T - V_D$) P_{A,CO_2}, where V_T is the total lung volume.

Bohr model of the atom

[atomic] *See* **BOHR ATOM**.

Bohr radius (r_{Bohr})

[atomic, general] A radius of "allowed" ORBITS in the hydrogen atomic model, as defined by the discrete fits $r_{Bohr} = n^2 \left(0.0529 \text{ nm} \right)$, where $n = 1,2,3,\dots$. This concept was introduced by NIELS HENRIK DAVID BOHR (1885–1962) in 1913, with preliminary expressions in 1908.

Boltzmann, Ludwig Eduard (1844–1906)

[atomic, mechanics, nuclear, quantum] A mathematician and physicist from Austria. Ludwig Boltzmann is best known for his contributions to the statistical description of the behavior and characteristics of gasses and the molecular movement, that is, statistical MECHANICS. Boltzmann's kinematic model predicted the ATOM, but this was not verified until the 1920s after his untimely death (he committed suicide under the assumption that his life's efforts were a waste because no practical evidence could be provided at the time), partially based on the COMPTON SCATTERING experimental results, hence making Ludwig Boltzmann's work a seminal effort. In 1884, Ludwig Boltzmann provided the theoretical argument to verify the black-body RADIATION in collaboration with the work provided by the physicist JOSEPH STEFAN (1835–1893) from Slovenia, in 1879, expressed as the STEFAN–BOLTZMANN LAW (see Figure B.49).

Figure B.49 Ludwig Eduard Boltzmann (1844–1906) taken approximately in 1902.

Boltzmann constant (k_B)

[atomic, fluid dynamics, mechanics, nuclear, quantum] $k_B = R/N_A = 1.38065 \times 10^{-23}$ J/K $= 1.3806488 \times 10^{-23}$ m^{2}kg/s^{2}K, where $R = 8.3144621(75)$ J/Kmol is the universal GAS constant and $N_A = 6.022137 \times 10^{-23}$ J/K is the Avogadro's number. This constant is named after LUDWIG EDUARD BOLTZMANN (1844–1906).

Boltzmann distribution

[mechanics, thermodynamics] A PROBABILITY distribution ($p(E)$) for the states of a system at ENERGY E that is operating at a temperature T in thermal equilibrium $p(E) = (1/P^B)e^{-E/k_BT}$, where $P^B = \sum_i e^{-E_i/k_BT} = \sum_j g_d(E_j)e^{-E_i/k_BT}$ is a NORMALIZATION function over all the states with respective energy E_i, referred to as the PARTITION FUNCTION, $k_B = 1.3806488 \times 10^{-23}$ m^{2}kg/s^{2}K is the Boltzmann constant, $g_d(E_j)$ refers to the DEGENERACY of the respective energy states, and e^{-E_j/k_BT} is the Boltzmann factor. The Boltzmann distribution can also be written with respect to the kinetic molecular theory written in the format of PARTICLE distribution (quantity N and mass m) with respective velocities ($\vec{v}$) as $dN/N = (m/2\pi k_BT)^{1/2} e^{-m\vec{v}^2/2k_BT} d\vec{v}$. The probability distribution was generalized by LUDWIG EDUARD BOLTZMANN (1844–1906) in 1871, based on the prior work by James Clerk Maxwell (1831–1879) in 1859. It is also referred to as the Gibbs distribution, the Maxwell–Boltzmann distribution, and the Boltzmann's distribution law (see Figure B.50).

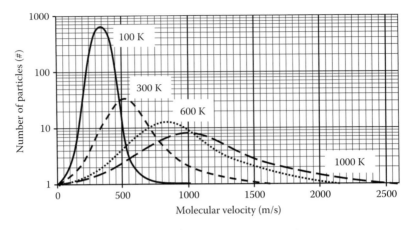

Figure B.50 Boltzmann distribution curve, for respectively four different (arbitrary) temperatures.

Boltzmann equilibrium

[biomedical, computational, general] A statistical thermal equilibrium for a kinetic gas. The energy exchange for an isolated system is in equilibrium.

Boltzmann H-theorem

[computational, energy, mechanics, nuclear, quantum] An expression for the distribution of velocity vectors ($\vec{v}$) in a dilute gas with respect to the thermodynamic entropy (S). In a dilute gas with finite volume elements d^3V, such that one only needs to consider binary collisions, the velocity distribution of the molecules is described by $H^B = \int f(\vec{v}.t) \log f(\vec{v}.t) d^3V$, where the velocity distribution under equilibrium is defined by the Maxwellian function $f(\vec{v}.t) = (m/2\pi kT)^{3/2} \exp(-mv^2/2kT)$. The equilibrium Maxwellian velocity distribution depends on the thermal energy ($k_B T$, where $k_B = 1.38066 \times 10^{-16}$ J/K is the Boltzmann constant and T is the local temperature in Kelvin) and kinetic energy ($KE = mv^2/2$, where m represents the mass of the gas molecules). When not in equilibrium, the velocity distribution (H^B) decreases monotonically with time.

Boltzmann transport equation

[atomic, general] Semi-classical mechanical description of the migration of particles in a six-dimensional volume for particles at position $\vec{r}$ with momentum $\vec{p} = \hbar\vec{k}$ ($\vec{k}$ is the wave number; $|\vec{k}| = 2\pi/\lambda$, λ is the wavelength; $\hbar = h/2\pi$ with $h = 6.62606957 \times 10^{-34}$ m^{2}kg/s the Planck's constant). The Boltzmann transport equation describes the rate of change for the parameters of a system defined by the function $f(\vec{k}, \vec{r}, t)$ as a function of time (t), subjected to flow and scattering as $(d/dt) f = [\nabla f \cdot (d\vec{k}/dt)]_{\vec{k}} + [\nabla f \cdot (d\vec{r}/dt)]_{\vec{r}} + (\partial/\partial t) f_{\text{scatt}}$, where the last term identifies the contributions from scattering. The function for a Fermi gas (e.g., electrons in conductor) as a function of energy ($\epsilon_{\vec{k}}$) and temperature (T) is defined as $f_0^F(\epsilon_{\vec{k}}, T) = 1/\{\exp[(\epsilon_{\vec{k}} - \mu_c)/k_B T] + 1\}$, where μ_c is the chemical potential and $k_B = 1.3806488 \times 10^{-23}$ m^{2}kg/s^{2}K is the Boltzmann coefficient. For bosons (e.g., phonons), the function adheres to the Bose–Einstein relation as $f_0^b(\epsilon_{\vec{k}}, T) = 1/[\exp(\epsilon_{\vec{k}}/k_B T) - 1]$. When there is no external magnetic or electric field, the system will reach an equilibrium situation, with representative

PARTICLE distribution f_0. The QUANTUM mechanical equivalent TRANSPORT EQUATION was defined in 1928 by FELIX BLOCH (1905–1983). The transport equation was introduced by LUDWIG EDUARD BOLTZMANN (1844–1906) in 1872 (*also see* TRANSPORT EQUATION).

Boltzmann's distribution law

[atomic] *See* BOLTZMANN DISTRIBUTION (also known as Boltzmann's law).

Bond graph

[fluid dynamics, general, mechanics] Graphical representation of the parameters and conditions for a dynamic system, which may be subject to change. Specifically, the ENERGY and work as well as momentum with respect to the DYNAMICS of the system are outlined in blocks in order to solve the individual components of the system in a recursive and interactive manner, based on a block-diagram format. One specific example of a dynamic system that benefits from the use of the Bond graph is the hemodynamic CIRCULATION in the HUMAN BODY. The BLOOD circulation is subject to the periodic MOTION (pressure function) of the HEART (left ventricle) with subsequent vascular COMPLIANCE and vascular branching followed by convergence (arteries to arterioles to capillaries [organs] and converging to venules and veins back to the right side of the heart) (see Figure B.51).

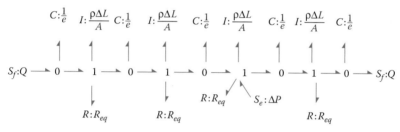

Figure B.51 Bond diagram for a complex and convoluted flow system. Each segment can be solved individually, adhering to the boundary conditions being equal between each sequential tube-flow segment.

Bond number ($Bo = g(\rho_\ell - \rho_v)L^2 / \gamma_s$)

[biomedical, fluid dynamics] Dimensionless number describing the range in MAGNITUDE for the ratio between the gravitational force on an object (or buoyance for a BUBBLE) and SURFACE TENSION. In the definition, the following parameters are used: γ_s is the surface tension, L is the characteristic length, g is the GRAVITATIONAL ACCELERATION, and ρ_i is the density of the LIQUID (ℓ) or "submerged" object (v). This number can be used to scale a phenomenon such as the CAVITATION from an object falling into a liquid medium creating cavitation and the potential formation of a water JET as the process develops over time. In another application, bubbles rising through a liquid column are categorized by Bond number. When the Bond number is small, the gravitational force can be neglected in the remaining theoretical analysis (e.g., $Bo \ll 1$) when compared to the surface tension effects as could be the case for a bubble, depending on the boundary conditions. Additionally, undershooting a certain Bond number threshold, the bubbles will not rise in a vertical column (e.g., $Bo < 0.842$). The Bond number can also be used to describe the boundary conditions of the process related to the way a red BLOOD CELL interacts with the CAPILLARY wall in comparison to passing through an artery.

Born, Max (1882–1970)

[atomic, general, nuclear, quantum, thermodynamics] A physicist from Germany. Max Born is most known for his contributions to QUANTUM MECHANICS, and he received the Nobel Prize in PHYSICS for his work on the interpretation of the Schrödinger WAVE equations in 1953, shared with the German nuclear physicist WALTHER WILHELM GEORG BOTHE (1891–1957) (see Figure B.52).

Figure B.52 Max Born (1882–1970).

Born wave function

[nuclear] NORMALIZATION of the WAVEFUNCTION ($\Psi(x,y,z,t)$ *see* **SCHRÖDINGER EQUATIONS**) yielding the PROBABILITY of finding a PARTICLE in a particular VOXEL $dx\,dy\,dz$ defined as $\Psi^*\Psi = |\Psi(x,y,z,t)|^2\,dxdydz$.

Born–Oppenheimer approximation

[thermodynamics] A computational approximation for molecular interactions based on the assumption that atomic nuclei have infinite mass and the atomic masses are fixed in space for location (*see* **ADIABATIC APPROXIMATION**).

Bose condensate

[computational, energy, fluid dynamics, mechanics, quantum, thermodynamics] Collection of IDENTICAL PARTICLES (particularly bosons) that, when under low temperature conditions (in general below a critical temperature; specifically, approaching ABSOLUTE ZERO Kelvin), act as a single entity, surrendering their individual identities as part of an ISOLATED SYSTEM. Bose–Einstein condensation is considered a PHASE TRANSITION. Under Bose–Einstein condensation, the PARTICLE density (ρ_N, yielding an atomic separation $\sim \rho_N^{-1/3}$) and the temperature (T) are related through the DE BROGLIE WAVELENGTH ($\lambda_{dB} = \sqrt{(2\pi^2\hbar/mk_bT)}$), Boltzmann coefficient $k_B = 1.3806488 \times 10^{-23}\,\mathrm{m^2kg/s^2K}$, $\hbar = h/2\pi$ with $h = 6.62606957 \times 10^{-34}\,\mathrm{m^2kg/s}$ the Planck's constant) as $\rho_N\lambda_{dB}^3 = 2.612$, representative of the lowest energy QUANTUM STATE. The internal energy (U_{BE}) for a GAS with particulate mass m, as a Bose–Einstein condensate $U_{BE} = 4\pi\hbar^2(\rho_N\mu_{s,a}/m)$, with $\mu_{s,a}$ the characteristic SCATTERING length for s-wave interaction, in the order of $2-5$ nm when considering alkali atoms.

Bose liquid

[computational, energy, fluid dynamics, mechanics, quantum, thermodynamics] BOSON GAS very close to ABSOLUTE ZERO temperature reaching a state of MATTER where the majority of bosons are occupying the lowest QUANTUM STATE. The quantum effects at this point reach MACROSCOPIC scale (*see* BOSE CONDENSATE).

Bose–Einstein statistics

[atomic, nuclear] QUANTUM-mechanical description of a system with symmetric WAVEFUNCTION, for which the constituents can freely be distributed over the ENERGY levels (ϵ_i) for the system. The symmetric wavefunction places the requirement of integer SPIN (i.e., bosons). The system can be described by the PROBABILITY distribution of particles n_i, which also serves as a representation of the occupation number for the respective (i) energy states, at the temperature T. The energy levels are considered with finite width (hence the "at will" distribution, with $g_{\epsilon i}$ nondegenerate individual energy quantum levels. The probability distribution under Bose–Einstein statistics for the respective energies is $P_{BE} = \prod_i [(n_i + g_{\epsilon i} - 1)!/n_i!(g_{\epsilon i} - 1)!] = \prod_i (g_{\epsilon i}{}^{n_i}/n_i!)$, with the total number of constituents $N = \sum n_i$ and respective total energy $E = \sum n_i \epsilon_i$. The probability distribution for the occupation of states is now $n_i = g_{\epsilon i}/C_{BE}{}^{-1}e^{\epsilon_i/k_B T}$, where $k_B = 1.3806488 \times 10^{-23}$ m^{2}kg/s^{2}K is the Boltzmann coefficient.

Boson

[atomic, general] Elementary PARTICLE (GLUON) named after the physicist SATYENDRA NATH BOSE (1894–1974) who provided the critical information with respect to the description and discovery. Bosons provide the vehicle for the strong interaction between quarks. The boson has integer SPIN values (in contrast to fermions with half-integer spin, named after ENRICO FERMI [1901–1954]). Specific vector bosons are W^+, W^-, and Z particles as well as the PHOTON.

Boson gas

[thermodynamics] A system of particles that satisfies the Bose–Einstein condensate criteria (see Figure B.53).

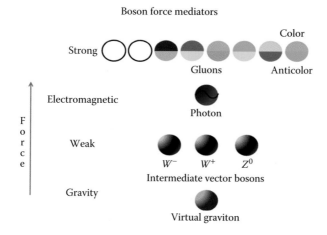

Figure B.53 Boson.

Bothe, Walter Wilhelm Georg (1891–1957)

[atomic, nuclear, thermodynamics] A German scientist who worked on the nuclear program and shared the Nobel Prize in PHYSICS in 1954 with MAX BORN (1882–1970).

Boulton, Matthew (1728–1809)

[thermodynamics, mechanics] A Scottish engineer, partner of JAMES WATT (1736–1819). Together they produced steam engines used to drain mines. Later applications of steam engines produced by Boulton and Watt were in the original field of Boulton, coin stamping (see Figure B.54).

Figure B.54 Matthew Boulton (1728–1809), painting by Carl Frederik von Breda, c. 1792.

Boundary layer

[fluid dynamics] A concept introduced by LUDWIG PRANDTL (1875–1953) in 1905. Fluid-element layer of molecules from the surrounding FLUID covering the surface of a solid body subjected to a fluid in rest or in MOTION that is not subjected to the same parameters as the whole LIQUID. Basically, the boundary layer is a transition from the fluid to the mechanical conditions of the object. Outside the boundary layer, the INVISCID FLOW is behaving similar to all the fluid, with perturbation for localized vector direction based on the dimensions of the object obstructing the flow (see Figure B.55).

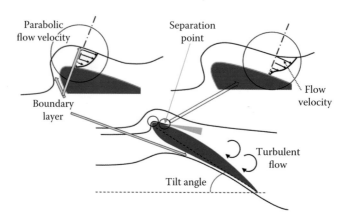

Figure B.55 Boundary layer aspect of flow, for instance, with respect to an airfoil (wing). The detachment of the boundary layer forms the onset of turbulent flow.

Boundary layer, main flow

[fluid dynamics] A transition layer between fixed surface and main flow region of uniform flow velocity. The velocity at the surface is zero due to FRICTION and ADHESION, the layer with a high gradient in velocity is the boundary layer, until the region is reached where the FLOW is uniform through the remaining volume of flow. On the face of the body, the flow velocity is zero with absolutely no slip. For this reason, owing to the effect of friction the flow velocity near the WALL varies continuously from zero to uniform velocity. In other words, it is found that the surface of the body is covered by a coat comprising a thin layer where the velocity gradient is large. This layer forms a zone of reduced velocity, causing vortices, called a WAKE, to be cast off downstream of the body.

Boyle, Robert (1627–1691)

[atomic, chemical, energy, general, nuclear] A scientist from Ireland. The work of Robert Boyle on gases followed by the efforts of the French scientist JOSEPH LOUIS GAY LUSSAC (1778–1850) formed the gateway to modern GAS theory and the IDEAL GAS LAW. Additional contributions were in the fields of pneumatics and chemistry (see Figure B.56).

Figure B.56 Robert Boyle (1627–1691).

Boyle's gas law

[general, fluid dynamics, thermodynamic] *See* **BOYLE'S LAW**.

Boyle's law

[atomic, fluid dynamics, mechanics, nuclear] For an isothermal process, the pressure–volume relationship can be defined as PV = constant. During an ADIABATIC PROCESS (T ≠ constant), the pressure will drop faster with increasing volume than for an isothermal process and a correction is required, which has been established empirically as PV^{γ} = constant, where the exponential factor γ depends on the process and the medium. For a monoatomic GAS, γ = 1.67, while for a diatomic and polyatomic gas, the values are, respectively, γ ≅ 1.4 and γ ≅ 1.3. The "IDEAL GAS LAW" still hold true. The correlation between volume and pressure at constant temperature was described earlier by EDME MARIOTTE (1620–1684) from France in 1650, hence also known as Mariotte's law. Since it was introduced by ROBERT BOYLE (1627–1691), it is also known as BOYLE'S GAS LAW (see Figure B.57).

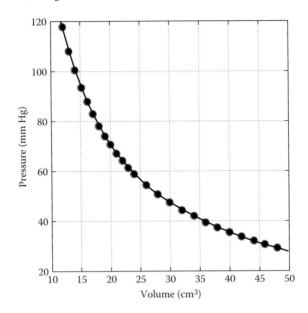

Figure B.57 Boyle's law in graphical form using the data as published by Robert Boyle.

Boyle–Charles' law

[fluid dynamics] An IDEAL GAS LAW based on the observations of ROBERT BOYLE (1627–1691) and refined by the French scientist JACQUES ALEXANDRE CÉSAR CHARLES (1746–1823). Jacques Charles recognized that for a fixed quantity of dilute gas (closed volume) under constant pressure, the volume is directly proportional to the applied ABSOLUTE TEMPERATURE (in KELVIN) (*see* IDEAL GAS LAW).

Boyle–Gay Lussac's law

[fluid dynamics, general, thermodynamic] Equilibrium conditions between VOLUME (V), PRESSURE (P), and TEMPERATURE (T) for a dilute GAS: $PV = nRT$ or $P_1V_1/T_1 = P_2V_2/T_2$, where $n = N/N_A$ represents the number of MOLE of medium (N is the number of molecules and N_A is the Avogadro's number) and R is the universal gas constant.

Bragg, William Henry, Sir (1862–1942)

[general] A scientist from Great Britain. He is an experimentalist on crystal diffraction and a contributor to the BRAGG'S EQUATION with his son WILLIAM LAWRENCE BRAGG (1890–1971).

Bragg, William Lawrence, Sir (1890–1971)

[atomic, general, nuclear] A scientist from Great Britain. He is an experimentalist on crystal diffraction and a contributor to the BRAGG'S EQUATION with his father SIR WILLIAM HENRY BRAGG (1862–1942) (see Figure B.59).

Figure B.59 Picture of the team at the laboratory of William Lawrence Bragg (1890–1971), c. 1931.

Bragg plane

[atomic, nuclear, solid-state] SIR WILLIAM LAWRENCE BRAGG (1890–1971) modeled a LATTICE constructed with parallel planes of symmetry, separated by a periodically reproduced DISTANCE, in various sectional directions, not specifically parallel to the surface plane of the lattice. The planes can be identified from diffraction experimentation using X-RAY diffraction based on the BRAGG'S EQUATION. Plane of a lattice structure bisects one reciprocal lattice vector (wavevector: $k = 2\pi/\lambda$, where λ is the wavelength), defined in the RECIPROCAL SPACE (*also see* **BRAVAIS LATTICE** *and* **MAX THEODOR FELIX VON LAUE**) (see Figure B.58).

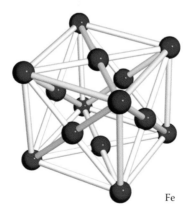

Fe

Figure B.58 The Bragg plane in a crystal structure.

Bragg's diffraction

[atomic, nuclear] *See* **BRAGG'S EQUATION**.

Bragg's equation

[nuclear] A mechanism of action for diffraction of ELECTROMAGNETIC RADIATION proposed by the father-and-son team Sir WILLIAM HENRY BRAGG (1862–1942) and Sir WILLIAM LAWRENCE BRAGG (1890–1971)

based on the preliminary experimental work by the German physicist Max Theodor Felix von Laue (1879–1960). The location with respect to the DIFFRACTION GRATING for constructive INTERFERENCE between diffracted rays is given by the equation $2d \sin(\theta_m) = m\lambda$, where d is the GRATING spacing (the regular spacing between the atoms in the crystal structure), m is the diffraction order number ($m = 1, 2, 3, \ldots$), θ_m is the diffraction ANGLE or angle of observation, and λ is the wavelength of the electromagnetic radiation (e.g., X-RAY and light), also referred to as BRAGG's LAW.

Brahe, Tycho {Tyge Ottesen Brahe} (1546–1601)

[astronomy/astrophysics, general] A scientist and nobleman born in Denmark. Tycho Brahe made detailed observations on the planetary motions, which helped JOHANNES KEPLER (1571–1630) formulate his THREE LAWS OF PLANETARY MOTION. He made considerable improvements on the optical quality of TELESCOPE construction and obtained significant details on events in the UNIVERSE. He discovered a SUPERNOVA (i.e., birth of a new STAR) in 1572. He also observed the trajectory of a passing COMET in 1577, providing critical information about gravitational forces. Nevertheless, his solar model was still GEOCENTRIC, based on the teachings of NICOLAUS COPERNICUS (1473–1543) and CLAUDIUS PTOLEMY (approximately AD 90–168) (see Figure B.60).

Figure B.60 Painting of Tyge Ottesen Brahe [Tycho Brahe] (1546–1601) by Eduard Ender.

Bravais lattice

[condensed matter, solid-state] A crustal structure with various configurations with respect to the centering of the LATTICE. In atomic structure, a total of fourteen (14) Bravais lattices can be defined based on their geometric composition, per respective lattice system: triclinic; monoclinic: base-centered and simple cubic monoclinic; orthorhombic: base-centered, simple orthorhombic, face-centered, and body-centered orthorhombic; rhombohedral; tetragonal: body-centered and simple cubic rhombohedral; and cubic: simple cubic, face-centered cubic, and body-centered cubic. The rocksalt (sodium chloride: halite) structure is an octahedral configuration, forming a three-dimensional checkerboard pattern. The zinc blende structure has tetrahedral coordinates but is not fully defined as such. The structure of the zinc blende is similar to the diamond structure, although without the regular atomic structure, instead alternating types of atoms at the lattice corners. The planes of organization within the lattice are defined through their respective Miller index (see Figure B.61).

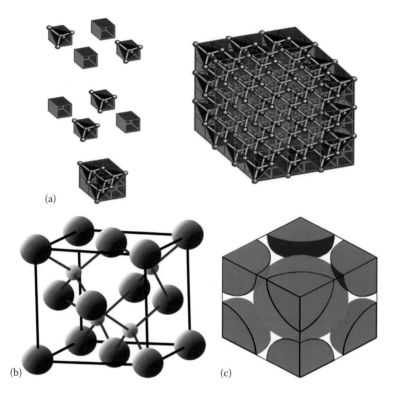

(a)

(b)   (c)

Figure B.61 (a–c) Illustrations of the structure involved in Bravais lattice configurations.

LATTICE SYSTEMS (7)	BRAVAIS LATTICES (14)			
Triclinic	P			
	$\alpha, \beta, \gamma \neq 90°$			
Monoclinic	P	C		
	$\beta \neq 90°$ $\alpha, \gamma = 90°$	$\beta \neq 90°$ $\alpha, \gamma = 90°$		

(*Continued*)

B

LATTICE SYSTEMS (7)	BRAVAIS LATTICES (14)			
Orthorhombic	P	C	I	F
	$a \neq b \neq c$	$a \neq b \neq c$	$a \neq b \neq c$	$a \neq b \neq c$
Tetragonal	P	I		
	$a \neq c$	$a \neq c$		
Rhombohedral	P			
	$\alpha = \beta = \gamma \neq 90°$			
Hexagonal	P			
Cubic	P (pcc)	I (bcc)	F (fcc)	

Bremβtrahlung

[atomic, biomedical, general, nuclear] *See* **Bremsstrahlung**.

Bremsstrahlung

[atomic, biomedical, general, nuclear] Radiation created by the impact of high energy electrons on a medium with high molecular weight. The electrons are accelerated in a negative manner, creating radiation with energy equivalent to the electron impact, resulting in X-ray radiation. The specific X-ray spectrum created is a function of the atomic structure of the target. The maximum energy E released as radiation $E = h\nu$, where $h = 6.6260755 \times 10^{-34}$ Js is the Planck's constant and ν is the frequency of the radiation, will equate to the kinetic energy of the electron $E_{kin} = (1/2)mv^2 = e*V$, where V is the voltage applied to induce acceleration of the electron with charge e and mass m moving with final velocity v. Note that the frequency in this case will be the upper limit ν_{max}, yielding a minimum X-ray wavelength (as pertaining to the maximum resolution defined by the Rayleigh criterion) as $\lambda_{min} = c*\left(h/e*V\right)$.

Brewster, David (1781–1868)

[general] A physicist from Great Britain. David Brewster provided particular information with regard to light polarization, specifically known as Brewster's law or the Brewster's angle (see Figure B.62).

Figure B.62 David Brewster (1781–1868) c. 1850.

Brewster's angle (θ_p)

[general] An angle of incidence for light reflecting from the interface between two media with index of refraction: incident from medium with index n_1 and reflecting at medium with index n_2 resulting in propagated light with linear polarization in the direction perpendicular to the plane of incidence (for a perfectly flat, semi-infinite medium, the polarization is thus parallel to the plane of interface).

The minimum angle (θ_p) of incidence required to provide perfectly polarized light is defined as $\tan \theta_p = n_2/n_1$. One specific application is found in the Nicol PRISM (see Figure B.63).

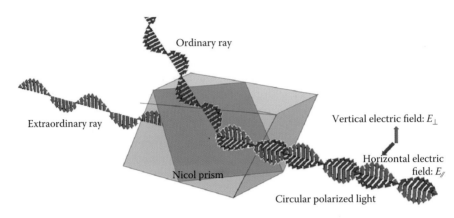

Figure B.63 Brewster angle in Nicol prism providing orthogonal polarization of the split beams.

Brewster's law

[general] *See* **BREWSTER'S ANGLE.**

Brillouin zone

[computational, electronics, quantum, solid-state] A concept used with respect to "density of states" relating to electron charges in solids and liquids. On a crystal structural level, a primary configuration can be identified that defines the outline of a primitive unit CELL that confines all the symmetry properties of the entire crystal in repetition, named the Wigner–Seitz cell. For the crystal, the first Brillouin zone defines the Wigner–Seitz cell with respect to the reciprocal atomic LATTICE in k-SPACE in a BRAVAIS LATTICE (k represents the wavevector: $k = 2\pi/\lambda$, where λ is the wavelength). The regularity of the atomic lattice provides the framework for the eigenvalues of the WAVE EQUATION solutions, that is, electronic eigenstates. Traveling electron waves (i.e., Bloch wave) through the lattice with WAVELENGTH (modulation) matching the periodicity of the polyhedron crystal lattice structure are in the first Brillouin zone. On an acoustic basis, the wave is referred to as a vibrational QUANTA, a phonon. Additional reference to the FERMI SURFACE will be required for full definition of the Bloch wave, where the Fermi ENERGY is the upper primary limiting factor. The second Brillouin zone may be reached from within the first Brillouin zone by means of crossing a single BRAGG PLANE only.

Brinkman, Henri Coenraad (1908–1961)

[fluid dynamics] A scientist from the Netherlands. Henri Brinkman is best known for his work on VISCOSITY, specifically viscous dissipation. The work of Henri Brinkman was based on the Darcy equation from the French engineer Henry Philibert Gaspard Darcy (1803–1858).

Brinkman number ($Br = \eta v^2 / \kappa_{cond}(T - T_0)$)

[fluid dynamics] A number that indicates the viscous dissipation, where v is the velocity, v is the VISCOS-ITY, $\kappa_{cond} = 75k/64r^2 \sqrt{kT/\pi m}$ is the THERMAL CONDUCTIVITY of the FLUID k the Boltzmann constant, r is the radius of the hard sphere that is in FLOW, m is the molecular mass (*Note*: the universal GAS constant: $R = k/m$), T is the temperature of the boundary (WALL), and T_0 is the bulk fluid temperature. This phenomenon generally applies to the CONTINUUM (*see* **KNUDSEN NUMBER**).

Brinkman number, modified ($Br^* = (T - T_0)_{iso}/(T - T_0)$)

[fluid dynamics] A number that indicates the viscous dissipation, where T is the temperature of the boundary (WALL) and T_0 is the bulk FLUID temperature, and iso denotes isothermal conditions while all objects are moving at identical bulk VELOCITY (*also see* **BRINKMAN NUMBER**).

British thermal unit (Btu)

[general] A unit of heat or work. In heat, the Btu represents the ENERGY requirement to raise one (1) British pound of WATER ($1\,lb = 453.59\,g$) by one (1) degree Fahrenheit: $1\,Btu = 1055.05585\,J$.

Brown, Robert (1773–1858)

[biomedical, nuclear, solid-state] A scientist, biologist, and botanist from the United Kingdom (specifically Scotland). His primary observations leading to his description of BROWNIAN MOTION were performed on moving cells under a MICROSCOPE, specifically related to his work on fertilization (as well as pollination) and the movement of sperm. He observed particle MOTION in liquids, such as pollen in water, apparently randomly moving about. This was later described as molecular and particulate collisions by ALBERT EINSTEIN (1879–1955). The phenomenon can sometimes be observed during fine snow-dust precipitation (not real snowflakes, equivalent to suspended frozen mist) under the absence of wind while there is a LIGHT SOURCE providing contrast (e.g., the SUN; causing reflections that can be traced). The thermal gradients in the AIR near buildings can provide a MACROSCOPIC TURBULENCE, moving the floating frozen ice specks in collision-based random orientation (see Figure B.64).

Figure B.64 Robert Brown (1773–1858), approximately 1855.

Brownian motion

[atomic, biomedical, computational, nuclear, solid-state] A STOCHASTIC PROCESS of random movement of particles in colloidal suspension (in liquids or gasses, and even on atomic and molecular level) as a result the thermal MOTION (~kinetic ENERGY) of the particles in the suspending medium. It has been postulated that ELEMENTARY PARTICLES may also perform Brownian motion, subject to nuclide collisions. It was first described by ROBERT BROWN (1773–1858) in 1827. The principle is also applied in NUMERICAL theory, referred to as "particle theory." The PARTICLE interaction was described on an interactive gravitational scale using the gravitational length: $\ell_g = k_B T / m^* g$, where $k_B = 1.3806488 \times 10^{-23}$ m^{2}kg/s^{2}K is the Boltzmann constant, T is the local temperature, m^* is the mass of a GAS molecule, and g is the GRAVITATIONAL ACCELERATION.

Brunt–Väisälä buoyancy frequency of the atmosphere

[astronomy/astrophysics, fluid dynamics, geophysics] OSCILLATION frequency (v_{BV}) of a column of AIR around an equilibrium position in altitude z_0 during adiabatic MOTION, provided as the result of BUOYANCY of a LIQUID, at ABSOLUTE TEMPERATURE with respect to a reference temperature under constant pressure: $T/T_0 \sim \theta/\theta_0$ ($\theta = T\left(P/P_0\right)^{\kappa_g}$, $\kappa_g \equiv R/c_p = \gamma_c/\left(1-\gamma_c\right)$, c_p is the SPECIFIC HEAT under constant pressure, $\gamma_c = c_p/c_v$, where θ defines the potential temperature), yielding $v_{BV}^2 = g\left[d\left(\ln \theta_0\right)/dz\right]$, with g the GRAVITATIONAL ACCELERATION. Under these conditions, the rate of change in height (time: t) of a reference point in the column of FLUID (i.e., infinitesimal parcel of air of cross-sectional area A and height dz) and DENSITY (ρ, with steady-state density ρ_0) are correlated as $d^2z/dt^2 = -g\left[\left(\rho-\rho_0\right)/\rho\right]$, where the fluid obeys the hydrostatic equation with respect to pressure P as $dP/dz = -\rho g$, with the pressure as a function of height ($H = RT/g$, GAS constant: $R = 8.3144621(75)$ J/Kmol). The frequency is named after the two scientists who, independently, introduced the concept: the Swedish scientist Vilho VÄISÄLÄ (1889–1969) and British meteorologist DAVID BRUNT (1886–1965). The full perturbation at a location in the column of fluid is described as $\left[\left(D^2/Dt^2\right) + v_{BV}^2\right]w + (1/\bar{\rho})(D/Dt)\left\{\left[\left(\partial/\partial z\right) + \left(g/v_s^2\right)\right]P\right\} = 0$, where v_s is the speed of SOUND, $(D/Dt) = (\partial/\partial t) + \bar{U} \cdot \nabla \sim (\partial/\partial t) + \bar{u}(\partial/\partial x)$, $\bar{u}$ is the average FLOW velocity in horizontal direction, and w is the vertical displacement VELOCITY (*also see* **HOUGH FUNCTIONS, THERMOSPHERE,** *and* **LAMB WAVE**).

B-scan

[acoustics] A brightness scan used to describe a specific ULTRASONIC IMAGING technique. Ultrasounds are defined as mechanical SOUND waves having frequencies higher than the audible range. All sounds that can be detected by the human EAR are called audible, thus their frequencies are in the audible range. Audible sound frequencies range from approximately 20 Hz to 20 kHz. During a person's development, the highest detectable frequency progressively decreases. A baby ear can detect sound waves with frequencies as high as 20 kHz; however, an adult can't detect higher than 17 kHz. Moreover, research showed that the fetal ear can detect sounds up to 40 kHz. Many instances showed that animals can use sound waves in more advanced ways than human do. Bats and dolphins are known to be able to detect frequencies up to, respectively, 30 kHz and 158 kHz and showed a fairly high detection RESOLUTION that enables them to navigate and maneuver around obstacles. ULTRASOUND was first characterized in animals and shown to be mostly generated in the larynx. The first ultrasound generating device, Galton whistle, was developed in the 1880s. This device permitted a continuous change in frequency from audible to ultrasonic. This concept is still used for silent dog whistle. Wedge resonators, widely used in the alimentary industry for emulsification, employ a jet of AIR or LIQUID to induce a flexible wedge to flexural OSCILLATION. Another ultrasound GENERATOR used in the alimentation industry is Sirens. It generated ultrasounds by forcing a GAS through punctured rotating disks. Even though the waves generated are not pure sine WAVE, they are widely used for emulsification. The other methods to generate sound waves with very high frequencies are based on material that can change shape upon change in the electromagnetic characteristics of the external environment. Two characterized materials have been shown to have such property: magnets and piezoelectric crystals. MAGNETOSTRICTION is defined as the shape change of an oscillating MAGNET under the action of an external MAGNETIC FIELD. This change of shape is due to the rearrangement of the molecules composing the magnet. Magnetostriction has been first described by JAMES PRESCOTT JOULE (1818–1889) in 1847. Magnetostricitve devices are extensively used in industry. However, it is difficult to manufacture devices that generate frequencies higher than 500 kHz. The brothers PIERRE CURIE (1859–1906) and PAUL-JACQUES CURIE (1856–1941) discovered piezoelectric crystals in 1880. These crystals have the property to generate electrical charge in their surface when they are under mechanical compression forces. Moreover, these crystals changed their shape when a voltage is applied. The application of an AC voltage with a very high frequency will generate a sound wave with the same frequency. Piezoelectric material can be made from barium titanate and lead zirconate titanate. The most commonly used medical ultrasound devices are based on piezoelectric ceramic. The back layer absorbs back RADIATION thus giving the TRANSDUCER directionality. The piezoelectric ELEMENT is wired to an electric circuit. Actual transducer is controlled by microprocessors and composed of many active ELEMENTS, each consisting of a controlled piezoelectric element. The beam formed will have two zones: near field (Fresnel region) and the far field (Frauhofer region). While the near field is mostly composed of waves traveling in parallel, waves traveling in the far field are divergent. It is possible to use the diffraction properties of sound waves to focus the wave front in a smaller area. An acoustic LENS is generally used for focusing, making the near field a little convergent. This will lead to an increase in the DISTANCE that the front wave will travel before becoming divergent and losing its RESOLUTION.

Physical concepts

Propagation

ACOUSTIC WAVES arise when the particles, of which a medium is composed, are supplied with ENERGY and are driven to vibrate. The VIBRATION of these particles will be transmitted to neighboring molecules due to an energy transfer. This repeated process of particles oscillating and transferring their energy to neighboring particles leading to their subsequent OSCILLATION with the same frequency is characterized as propagation. Even if a SOUND wave can propagate over relatively long distances, the molecules forming the medium stay in the same position. This propagation depends on the physical properties of the medium: v is the velocity of propagation of the sound wave; Y_K is the Young's elastic modulus; and ρ is the density of the medium. A WAVE propagation classical equation relates velocity of propagation, FREQUENCY (v), and WAVELENGTH (λ): $v = v\lambda$. Given that the frequency is source dependent, one can deduce that the wavelength will be dependent on the acoustic properties of the medium.

Reflection and refraction

During its propagation, a sound wave will move through different media. At the interface between two acoustically different media, the sound wave can be significantly affected. Many secondary waves will be generated, one of which is the reflected wave. The ANGLE of REFLECTION is equal to the angle of incidence. The portion of the sound wave that will go through the interface is called the refracted wave. When the interface is smaller than λ, the molecules composing the interface act as point sources and will radiate spherical waves. This effect is called SCATTERING and will be discussed in the next paragraph. The ratio of reflected to refracted sound waves is dependent on the acoustic properties of both media. These properties are characterized by the acoustic impedance which is $Z = \rho v$.

Attenuation

The total energy of a sound wave will progressively decrease due to wave attenuation. Two physical concepts are related to the energy dissipation: The acoustic conductance of a medium will determine its absorption levels. Most biological media are VISCOELASTIC in which a portion of the total MECHANICAL ENERGY will be transformed to heat resulting in a net loss of energy; during the passage of the sound wave through an interface, the sound could be scattered. The scattering implies loss of energy $I = I_0 e^{-\mu z}$, where z is the depth, $\mu = \mu_a + \mu_s$ with the attenuation factor due to absorption (μ_a) and scattering (μ_s).

Mechanical and cavitational effects

First order

The propagation of a WAVE through a media results in MOTION and acceleration of particles (high peak PARTICLE acceleration = 22.452 g; g is the GRAVITATIONAL ACCELERATION). Damages may result when different portions of the same structure are subjected to different forces so that it is twisted and torn.

Cellular membrane "fatigue" (red blood cell, RBC)

Autolysis of erythrocytes may be the result of repeated exposure to ultrasonically induced hydrodynamics shear stress. Ultrasounds may liquefy thixotropic structure including the mitotic/meiotic spindles.

Cavitation

Oscillatory ACTIVITY of highly compressible bodies such as GAS BUBBLES or "cavities" can be enhanced if the same dose of continuous wave ultrasonic power was administered in the form of relatively long pulses. This will result in violent oscillatory behavior and can disrupt cells such as white BLOOD cells (WBC), platelets, erythrocytes, and epithelial cells and may cause blood coagulation dysfunction.

Macromolecular effects

Several studies have been conducted on different cells' DNA; the results showed no evidence for single- or double-strand breakage in the DNA. There are no obvious effects of ultrasound waves on enzymes; a CELL is more likely to be destroyed before its enzymatic content is degraded. Ultrasounds have a wide range of application in the medical field, both for diagnostic and for therapeutic purposes. ULTRASONIC IMAGING represents a wide array of use as a primary modality or associated with other medical imagining techniques. Like most of the medical imaging techniques, ultrasonic imaging is mainly a TOMOGRAPHY modality since it visualizes sections of the HUMAN BODY. However, in many of its modalities, a third dimension is integrated which is time. One of the most appealing benefits of ultrasonic imaging is its ability to visualizes soft tissues with a fairly good RESOLUTION. One of the first and most important imaging techniques was X-ray-based imaging. This technique produced very poor resolutions on soft tissues. Hence, ultrasonic techniques represented a major advance in medical imaging by allowing visualization of "non-visualizable" tissues. Ultrasonic imaging uses very fast spreading, and its vulgarization is due to three main reasons: (1) It allows real-time imaging. Ultrasonic imaging added time as a third or fourth dimension. (2) It doesn't use ionizing RADIATION and has no major side effects. And (3) it provides qualitative as well as quantitative data. Quantitative data are very important for monitoring the progress of a disease under specific therapeutic conditions and doesn't rely on the subjective analysis of the radiologist.

Resolution

Spatial resolution

It is defined as the ability of a particular system to discriminate between two closely separated scatters. This concept arises from the nature of the detected RADIATION. Spatial resolution is generally referred to LATERAL RESOLUTION, which is the x coordinates of an ULTRASOUND-generated IMAGE and would be directly correlated with the width of the SIGNAL. It is dependent on an f# factor, which is equal to the focal DISTANCE divided by the APERTURE size. While the aperture size is directly related to the number of active ELEMENTS of the TRANSDUCER (by active elements we mean the number of piezoelectric elements that are controlled by their specific electrical wiring), the focal distance is directly dependent on the depth of the focal area. Many medical imaging research companies have focused on increasing the depth of the focal area without losing resolution and not only resolution at the focal depth but also homogeneity of resolution all along the image. The evolution of transducers has been always correlated with the addition of active elements and the increase in complexity of the MICROPROCESSOR that controls the electrical signals that activate these elements. But increasing active elements is physically constrained and researcher needed another way to enhance resolution that doesn't involve aperture. This was achieved by creating multiple focal systems.

Homogeneity of resolution

Since each focal zone will have appreciatively equivalent resolutions, a high resolution at all depths is mainly achieved by having a wide APERTURE system. The main limitation of such design is related to the image processing needed to integrate all the images detected of each focal area. Computing limitations and also the need for a fast real-time imaging system limited this approach. Nevertheless, this design has been implemented by many imaging companies and showed great results.

Contrast resolution

It is defined as the ability of a system to discriminate between two acoustically similar scatters. In the early 1940s, ULTRASONIC IMAGING design was thought to be done like X-rays by detecting transmitted ultrasounds. This showed a very low contrast resolution that didn't extract any relevant data. Pulse-echo technique based on analyzing the reflected ultrasounds rather than the transmitted has been proposed and immediately showed much better results. Contrast resolution was much better. The intensity of the reflected SOUND wave depended on the impedance of both media. This demonstrates that even a fairly small difference in acoustic impedances between both media will still generate enough change in intensity to be detected.

Ultrasonic imaging modality

B-mode

It represents a 2D SLICE through a portion of ANATOMY. This modality is widely used in obstetrics to visualize the fetus development in uterus. To acquire the IMAGE, the technician needs to move the TRANSDUCER along the line with a constant velocity. The computer will integrate all the data detected at all the focal distances and will form a 2D image. The classical B-mode will involve the visualization of a sectional plane $z = z_0$ by the liner movement of the transducer in the x direction along that plane. At each time t, a scan will occur, and the image will reflect the sequence of scans starting at t until time $t = t_0 + n\Delta t$, $T = T_0 + n\Delta t$.

Doppler mode

The DOPPLER effect states that changes in the DISTANCE beam–receptor will affect the frequency of the WAVE perceived by the receptor $v = v_0 \left[v_s / (v_s - v_o) \right]$. $f = f_0 * c/(c - v)$, given v is the frequency perceived by the receptor, v_0 is the frequency of the source, v_s is the velocity of SOUND, and v_o is the velocity of the moving beam or receptor. This effect can be used to image movements of acoustic scatters within the HUMAN BODY. Doppler imaging is widely used to visualize the movement of BLOOD. It is used both for peripheral arteries and veins but also for cardiopathologies. Doppler imaging measures blood velocity and detects blood FLOW direction. Two main ways to visualize Doppler imaging are as follows: A COLOR code imaging coupled with a B-mode shows the blood velocity and is very useful to detect regurgitation or abnormal blood flows. A plot of blood velocity along a peripheral artery can detect stenosis and other pathologies related to vascular degeneration.

Harmonic imaging

Concept

Superficial tissue structure, for example, SKIN, fat, and MUSCLE, produces pulse distortion and aberration. The sound wave emitted by the TRANSDUCER is a perfect sine wave; however, the reflected wave that will be detected by the transducer is a much distorted wave. This WAVE includes not only the FUNDAMENTAL WAVE but also many other waves, one of which is the second harmonic wave whose frequency is twice the frequency of the emitted wave. During the year 1980, Muir and Carstensen first demonstrated the conventional imaging array that produced significant nonlinear effects. Many research projects have been done to explore the nonlinear effects of ultrasounds. Today, major medical ULTRASOUND companies sell harmonic imaging systems based on conventional array/system technology. A better understanding of harmonic waves will push toward a new generation of transducers and systems that are more specific to harmonics and will lead to better RESOLUTION. Acoustic nonlinearity is due to the pressure density (constitutive) behavior, meaning that high pressures are correlated with higher densities and lead to higher SOUND propagation velocities. In contrast, low pressures are correlated with lower densities and lead to lower propagation velocities. Most of the nonlinear characteristics of ultrasounds have been studied through computer simulations and based on mathematical modeling. One of the most recent mathematical models is the quasilinear model. A nonlinear equation, which describes the propagation sounds in soft tissues, is defined by the pressure fluctuations ($P(\vec{r},t)$) as a function of location ($\vec{r}$) and time: $\nabla^2 P(\vec{r},t) - (1/v^2) \left[\partial^2 P(\vec{r},t) / \partial t^2 \right] + LP(\vec{r},t) = \epsilon \left[\partial^2 P^2(\vec{r},t) / \partial t^2 \right]$, here v is the speed of sound, L is a LINEAR OPERATOR that accounts for loss, and ϵ is a small parameter and is very small in soft tissue (10^{-3}). The QUASILINEAR approximation is based on a perturbation of the first equation: $P(\vec{r},t)=P_1(\vec{r},t)+\epsilon P_2(\vec{r},t)++O(\epsilon^2)$. Thus, $\nabla^2 P_1(\vec{r},t)-(1/v^2)[\partial^2 P_1(\vec{r},t)/\partial t^2]+LP_1(\vec{r},t)=0$ and $\nabla^2 P_2(\vec{r},t)-(1/v^2)[\partial^2 P_2(\vec{r},t)/\partial t^2]+LP_2(\vec{r},t)=\epsilon[\partial^2 P_1^2(\vec{r},t)/\partial t^2]$, where P_1 represents the fundamental field and P_2 is the second harmonic wave, assuming that $\epsilon^2 P_1=O(\epsilon^2)$.

Generation

In addition of being naturally generated, second harmonic waves can be generated by using contrast agents. Contrast agent will lead to a much homogenous second harmonic wave with a higher intensity. Contrast agents are still under extensive research. Properties of an ideal contrast agent are as follows: it should be nontoxic, easily eliminated, administered intravenously, pass easily through microcirculation, physically stable, and acoustically responsive: stable harmonics.

Advantages

Harmonic signal is narrower and have less side lobes. This will significantly increase the spatial resolution. Since the harmonic waves are generated inside the tissue, they will travel only once through the tissue to reach the transducer and thus will be less deformed. Harmonic imaging has been proven to increase resolution and to enable more enhanced diagnostic capabilities (see Figure B.65).

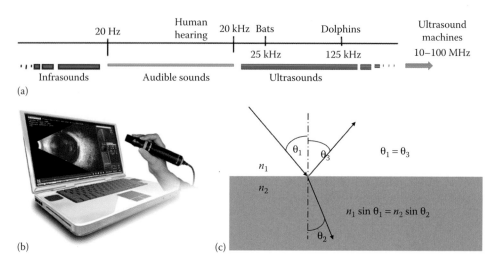

Figure B.65 (a) B-scan sound spectrum background, (b) portable ultrasound B-scan probe and signal processing and display unit from Optos, and (c) representation of interface between two media (e.g., air [above] and water [below]).

Bubble

[mechanics, thermodynamics] Volume of VAPOR or GAS surrounded by a LIQUID surface, respectively, flexible solid surface of a medium with SURFACE TENSION, either as a shell in a gaseous medium or alternatively confined within a liquid medium. The bubble is formed under equilibrium forces of surface TENSION ($F_s = 2\pi r\sigma_s$) and pressure force applied to the opposing hemispheres ($F_p = PA = P\pi r^2$, where A is the cross-sectional area and r is the radius of the spherical bubble): $\sigma_s = Pr/2 = (\partial U/\partial\alpha_A)_{S,V,n_i} = (\partial E_H/\partial\alpha_A)_{S,V,n_i}$ (U is the internal ENERGY of the medium at the surface, α_A is an infinitesimal surface area, E_H is the Helmholtz free energy, S is the entropy, V is the volume, and n is the quantity of constituent i). Bubble VIBRATION is the deformation of a bubble under the influence of inherent or external conditions. The forces on the bubble are governed by the Bond number: $Bo =$ oscillation force/surface tension $= (kr)^2 \varphi_A^2 (\rho_o/\rho_i)$ ($k = 2\pi/\lambda$ is the wave number for wavelength λ and φ_A is the AMPLITUDE of the VELOCITY POTENTIAL for the surface), the REYNOLDS NUMBER: $Re =$ inertial force/viscous force, and the Weber number: $We =$ inertial force/surface tension. Many different deformations are possible and for convenience only a radial EXPANSION/collapse is described. A free OSCILLATION of a bubble (or droplet for that MATTER) has several harmonic frequencies ($v = \omega/2\pi$; ω the ANGULAR FREQUENCY) (order n) defined by $\omega_n^2 = n(n+1)(n-1)(n+2)/r^3[n\rho_o + (n+1)\rho_i]$, where ρ_o is the density of the outer FLUID (or the density of the "SKIN" of the bubble) and ρ_i is the density of the inner fluid. Forced bubble oscillations can be induced by acoustic modulations and are referred to as initiated by the primary Bjerknes force. Conservation of momentum provides the outline for the volumetric extension (with velocity $\vec{v}$) as $(\partial\vec{v}/\partial t) + \vec{v} \cdot \nabla\vec{v} = -(1/\rho)\nabla P + (1/\rho)\nabla \cdot \vec{\sigma}_s + (1/\rho)\vec{F}_B + (1/\rho)\vec{F}_s + A(2\pi v)^2 \cos(2\pi v t)$ with $\sigma_s = \eta_{visc}(\nabla\vec{v} + [\nabla\vec{v}]^T)$ for NEWTONIAN FLUID with VISCOSITY η_{visc}, the shear stress TENSOR; external forces $\vec{F}_B$; and oscillating resonant force $A(2\pi v)^2 \cos(2\pi v t)$. Generally, the oscillations will be nonuniform (chaotic), abandoning the perfect spherical design. The bubble shape design can be classified under Legendre shapes, based on the LEGENDRE POLYNOMIAL series extension for the radius $r = (r_0/2) + \sum_{i=0}^{\infty} A_i(t) p_i(\cos\theta)$, where $p_i(\cos\theta)$ is the oscillatory harmonic of order i with AMPLITUDE $A_i(t)$ (*also see* CAVITATION *and* LAPLACE LAW) (see Figure B.66).

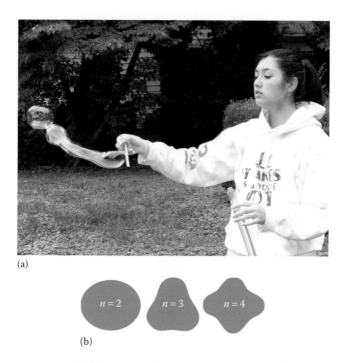

(a)

(b)

Figure B.66 Illustrations of bubble. (a,b) Examples of several Legendre bubble shape configurations.

Bubble chamber

[general, nuclear] A sealed vessel designed to provide a visual impression of the trajectory of an electrically charged PARTICLE. The chamber is either filled with a saturated GAS (i.e., cloud chamber) or superheated transparent LIQUID. On the passage of the ION, the interaction with the medium generates a trail of condensation droplets or BUBBLES in a liquid. The application of an external electric or MAGNETIC FIELD provides the means of "steering" the particle in MOTION (acting as a current) under the influence of Lorentz force. The first cloud chamber was constructed in 1912 by CHARLES THOMSON REES WILSON (1869–1959) from Scotland, UK, for which he received the Nobel Prize in Physics in 1927 (shared with ARTHUR HOLLY COMPTON [1892–1962]). The bubble chamber evolved later as introduced by DONALD ARTHUR GLASER (1926–2013) from the United States in 1952 (for which he received the Nobel Prize in Physics in 1960) (see Figure B.67).

Figure B.67 The trajectory of a charged particles exposed to a magnetic field passing through a liquid, forming a bubble track.

Buckminster Fuller, Richard (1895–1983)

[general, mechanics] An engineer, inventor, and architect from the United States. Richard Buckminster Fuller designed a multi-point interconnect that can support a "hollow structure" with minimal use of material while providing maximal strength (see Figure B.68).

Figure B.68 Richard Buckminster Fuller (1895–1983).

Buckyball

[chemical, general, mechanics] Geometric configuration described by RICHARD BUCKMINSTER FULLER (1895–1983), also known as a geodesic dome. The buckyball has surprising mechanical stability and the concept is also applied to towers for the inherent strength and reduced requirements for construction materials. In chemistry there are several molecular configurations that use the atomic interconnects) covalent bonds) forming a "spherical" structure, for instance Buckminsterfullerene, a fullerene molecule with the formula C60, which displays a fused-ring structure, resembling a cage or the outline of a soccer ball the C60 structure has 60 carbon atoms that are arranged in twenty hexagons and twelve pentagons. The carbon ATOM at each vertex provides the structural integrity of each polygon. Buckminsterfullerene was discovered in 1985 and is named after the architect that provided the mechanical concept in the early 1940s out of aluminum tubes with 5-point JOINTS and covered by a plastic-vinyl skin, in icosahedron (polyhedron with 20 triangular faces, 30 edges and 12 vertices) configuration (see Figure B.69).

Figure B.69 The "buckyball" in chemistry and engineering: Biosphere dome.

Buffer, Chemical

[chemical] Solution mixture selection with the primary purpose to reduce the fluctuations in HYDRO-GEN concentration as a result of interactions with other chemicals. In general, the buffered solution has a relatively stable pH. The stable pH guarantees a controlled environment to study the interaction with the hydrogen ION (i.e., H^+). The buffer results from an interactive molecular SOLUTE coexistence, where the constituents are capable to transfer protons rapidly and reversibly. Generally, the buffer is a mixture of ingredients that maintains a chemical equilibrium by means of a constant conversion of MOLECULES to IONS and in reverse. In this manner, the CHEMICAL POTENTIAL of the respective constituents is directly correlated with the chemical potential of all the chemical species in the respective chemical reactions. The premise of the solution is to make it impervious to REDOX POTENTIAL of the ingredients, TEMPERATURE, ACIDITY changes due to variations in external conditions as well as PRESSURE. The most common type of chemical reactions is dominated by the hydronium ion (H_3O^+) as a source for PROTONS (i.e., H^+), that is, the ACID–base buffer. The equilibrium condition in the chemical reaction between the acid concentration ($[AH^+]$) and base concentration ($[B]$) with respect to the conjugated base in solution (e.g., $AH^+ + H_2O \rightleftarrows B + H_3O^+$) is defined by an equilibrium constant or dissociation constant $K^D = [H_3O^+][B]/[AH^+][H_2O]$. The acidity of the buffer, defined as the concentration of the buffer component ($[H_3O^+]$), is a direct function of the dissociation constant and the concentration of the added acid only ($[AH^+]$), not the acidity of the acid (*Note*: hydrochloric acid has a much lower pH than vinegar), as expressed by $\left[H_3O^+\right] \cong \sqrt{K^D\left[AH^+\right]}$.

Buffer, Electronical

[electronics] An electronic circuit designed to minimize the AMPLITUDE (A) and FREQUENCY (ν) content of the SIGNAL transfer between a high IMPEDANCE (Z) source and a low impedance LOAD. Generally, the buffer consists of an amplifier with unit AMPLIFICATION ($M_A = 1$), wired in a circuit to provide infinite input impedance ($Z_{in} = \infty$) and zero output impedance ($Z_{out} = 0$). In this manner, the source is not subjected to an ENERGY drain, since in theory there is no current emerging from the source.

Bulk modulus (*B*)

[dynamics, mechanics] A deformation parameter based on the elastic potential defined as the unidirectional stress (σ_n) per unit strain (ϵ_n) where the transverse modes are not confined (i.e., ϵ_2 and ϵ_3 are allowed to change): $B = -V\left(dV/dP\right)$, where V is the volume of the medium and P is the pressure.

Bulk modulus (*B*~M~)

[general, mechanics] The measure of compressibility of a VOLUME (V) of substance, defined as the ratio of the change in stress $\Delta F/A$ over the fractional change in volume when subjected to a uniform compression force on all surfaces with constant area simultaneously $B_M = \left(\Delta F/A\right)/\left(\Delta V/V\right)$, for convenience the substance will be a cube. For liquids and solids, the bulk modulus will be constant in approximation; however, for a GAS the bulk modulus will be a function of the applied pressure.

Bullialdus, Ismaël (1605–1694)

[general] An astronomer and scientist from France. Ismaël Bullialdus provided supporting evidence for the ideas posted by NICOLAUS COPERNICUS (1473–1543), GALILEO GALILEI (1564–1642), and JOHANNES KEPLER (1571–1630). Supposedly, he postulated the inverse square law for gravitational attraction. Additionally, he presumably hypothesized that the PLANETARY ORBITS should be ellipses, not circular (Bullialdus' conical hypothesis) (see Figure B.70).

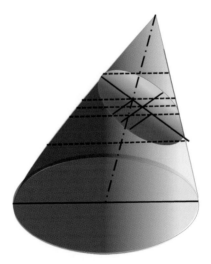

Figure B.70 The Bullialdus' conical hypothesis.

Bunsen, Robert Wilhelm Eberhard (1811–1899)

[general, optics] A physicist and chemist from Germany. In collaboration with GUSTAV ROBERT KIRCHHOFF (1824–1887), they discovered that the spectral absorption of atmospheric constituents provided specific patterns, representative of the ENERGY transitions. These observations were used to explain the FRAUNHOFER LINES in the solar spectrum, discovered by JOSEPH VON FRAUNHOFER (1787–1826) in 1814. Bunsen also investigated the spectral emission lines of several ELEMENTS under elevated temperatures and discovered caesium (1860) and rubidium (1861). He also devised the "Bunsen burner" for use in his chemical experimentation, still widely used up to this day for heating beakers and test tubes with chemical reagents (see Figure B.71).

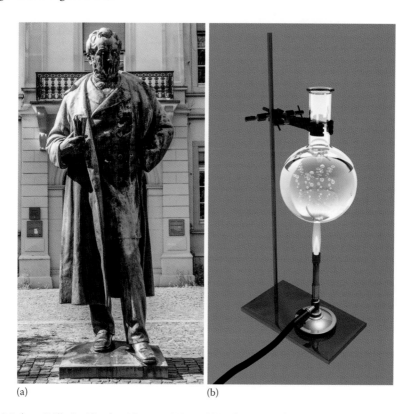

(a)   (b)

Figure B.71 (a) Robert Wilhelm Eberhard Bunsen (1811–1899). (b) Bunsen burner, used in most general laboratory experiments.

Bunsen–Roscoe rule

[chemical, thermodynamics] Based on CONSERVATION OF ENERGY constraints, it can be shown that the MAGNITUDE of the product resulting from a PHOTOCHEMICAL REACTION is directly proportional to the administered total dose of ENERGY, irrespective of the exposure time over which the energy dose is delivered. It is derived by ROBERT WILHELM EBERHARD BUNSEN (1811–1899) and HENRY ENFIELD ROSCOE (1833–1915).

Buoyancy

[fluid dynamics, general] [ref. mechanics] (syn.: Float, Archimedes' principle) {use: biomedical, statics} An object submerged in a FLUID medium (LIQUID or GAS) will experience an apparent weight that is its own weight reduced by the weight of the volume of the displaced fluid. This phenomenon was first described by the Greek physicist ARCHIMEDES (287–212 BC) in approximately 214 BC, while taking a bath. SURFACE TENSION will affect the total outcome of the apparent weight. The Archimedes principle can be used to derive the DENSITY (ρ) of an unknown object. Additionally, the relative composition of components can be established, such as the amount of body fat or the gold content. The upward (buoyancy-) force is defined as $F_b = \rho_l V_l g$, where V_l is the volume and ρ_l is the density of the displaced fluid. Technically, the forces are vectors, and hence, a negative sign means against the direction of GRAVITATIONAL ACCELERATION, also known as Archimedes' principle (see Figure B.72).

Figure B.72 Buoyancy principle. Row-boat floating due to displaced water providing buoyant upward force.

Buys Ballot, Christophorus Henricus Diedericus (1817–1890)

[fluid dynamics, general, geophysics, meteorology] A Dutch physicist, chemist, mathematician, and meteorologist known as Buys Ballot, after whom the correlation between the direction of the wind and the horizontal pressure-gradient *lines* was named, specified by hemisphere as the BUYS-BALLOT LAW. For his work on geophysics and natural phenomena, he was appointed as the first director of the newly established Royal Dutch Meteorological Institute (KNMI: Koninklijk Nederlands Meteorologisch Institute), a research institute for weather phenomena, still in use as the official Dutch weather observatory and weather forecasting facility. He observed that the airflow around low-pressure systems in the northern hemisphere is counterclockwise and clockwise around high-pressure systems. Additional observations made by Buys Ballot involve the banking (rising higher on one side) of the airflow to the right in the northern hemisphere and to the left in the southern hemisphere under a force described under CORIOLIS FORCE. The Coriolis force described the MOTION influenced by an object in rotation. Buys Ballot made the first steps to the formation of the World Meteorological Organization (WMO) by the formation of the International Meteorological Committee in 1873 (see Figure B.73).

Figure B.73 Christophorus Henricus Diedericus Buys Ballot (1817–1890), drawing made in 1857.

Buys Ballot law

[fluid dynamics, general, geophysics, meteorology] In atmospheric sciences, the AIR FLOW will travel around a high (or low) pressure system in counterclockwise on the northern hemisphere and in clockwise direction on the southern hemisphere. This can be interpreted by a person standing on the northern hemisphere with his/her back toward the wind that the low pressure area will be on the left hand side. This concept was introduced by CHRISTOPHORUS HENRICUS DIEDERICUS BUYS BALLOT (1817–1890) in 1857.

Cadherins

[biomedical, chemical] In biology there is cell-to-cell interaction, which can be divided into four types of receptors: cadherins, selectins, INTEGRINS, and IMMUNOGLOBULIN SUPERFAMILY. Cadherins rely on the abundance of Ca^{2+} ions to provide a platform for temporary ADHESION. The adhesion is homotype, using the cadherin molecules to bind to each other spanning the extracellular space.

Cadmium [Cd]

[nuclear] ELEMENT: $^{112}_{48}$Cd. Semiconductor material.

Caisson disease

[biomedical, fluid dynamics, general] Decompression disease, formation of embolisms resulting from the release of GAS BUBBLES from tissue into the bloodstream, primarily nitrogen, when the external pressure applied to the body is reduced too quickly. This phenomenon is connected to OSMOTIC PRESSURE, the partial pressure associated with dissolved gases, which can occur during scuba-diving expeditions on return to the sea level, while surfacing too rapidly. Originally the disease was associated with the exodus from a closed compartment, a caisson (when returning from underwater efforts while enclosed in the caisson under compressed AIR). One example is the use of caissons while working on the New York Brooklyn Bridge in the 1870s. The compressed air was necessary to compensate for the external hydrostatic pressure applied by the resting water column, where each 10 m of water corresponds to approximately 1 atmosphere, or 101.325 kPa, also called "the bends" or "divers disease" (see Figure C.1).

Figure C.1 Decompression chamber used to mitigate and treat decompression sickness after deep-sea diving, used by the United States Navy. (Courtesy of Jayme Pastoric, Mass Communication Specialist 2nd Class, US Navy.)

Calcification

[atomic, biomedical] The ADHESION of calcium to surfaces of implanted devices in biological organisms. Another form of calcification is in the formation of plaques in the lumen of BLOOD vessels, making it tougher and harder than fatty plaques. Other mineralization effects of calcium are in diseased tissues, causing the biological material to become more rigid and hence fragile for tearing and fracture. During osteoporosis, bone calcium level decreases due to metabolic inequilibrium and a deficiency in the supply chain, primarily due to eating habits. Calcium deposition on HEART valves gradually evolves into hardening of the valves, resulting in a diminished efficacy of the PUMP function due to regurgitation and seepage (see Figure C.2).

Figure C.2 Calcification of a water heater element due to "hard water" with calcium content.

Calcium [Ca]

[atomic, biomedical] ELEMENT used in the formation of bone and teeth as well as in the communication between cells ($^{40}_{20}$Ca). Biological CELL have special calcium channels that carry the calcium ION (Ca^{2+}) charges for the creation of an ACTION POTENTIAL in combination with sodium and potassium ions. The calcium ion electropotential is described by the NERNST EQUATION. The calcium ions and atoms also play a role in the Gibbs free ENERGY calculations for biological cells. Calcium is a major catalyst in SKELETAL MUSCLE contraction. Another implication of calcium is in the formation of plaque in the BLOOD vessels. About 1% of calcium is dissolved in blood, and the remaining 99% is in cellular structures and muscular processes.

Californium [$^{251}_{98}$Cf]

[nuclear] Radioactive unstable ELEMENT, classified as a METAL in the element category of Actinide with a half-life of 898 years. Californium was synthesized at the University of California, Berkeley, in the 1950. The metal is artificially generated as a result of colliding CURIUM [$^{247}_{96}$Cu] with ALPHA PARTICLES [$^{4}_{2}$He^{2+}]. Other isotopes have a much shorter half-life with aggressive RADIATION. The disintegration process of Californium generates NEUTRON emission and also produces ALPHA DECAY, which results in the production of isotopes of Curium. Californium possesses a simple hexagonal crystal structure.

Calipers

[computational, general] Measuring device that is designed for analog measurements of size with relative accuracy, generally in the order of $0.01 - 0.05\,mm$. In geometry this refers to a mechanism of calculating the configuration of a polygon, expressed by its ANTIPODAL POINTS forming pairs and vertices. The computational rotating caliper respectively aligns a "caliper line" with the edges of a polygon, thus defining the corners or antipodal pairs while successively moving the caliper line around the polygon. Also used to describe the disk-break pressure clamp for the wheels of an automobile or motorcycle, also called "calliper" (see Figure C.3).

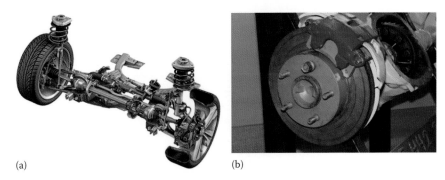

(a)   (b)

Figure C.3 (a) Brake calipers on a Mercedes automobile front axle and (b) close up of disk brake calipers. (Courtesy of Daimler AG.)

Callen, Herbert Babar (1919–1993)

[computational, thermodynamics] Physicist from the United States who made significant contributions to the theoretical formulation of thermodynamic concepts.

Calley, John (1663–1717)

[energy, mechanics] Physicist and engineer from the United Kingdom, coinventor of the modern concept of the steam ENGINE in 1712 with THOMAS NEWCOMEN (1663–1729). The other, more well-known, inventor related to the stem engine is JAMES WATT (1736–1819).

Caloric

[general, thermodynamics] Hypothetical concept of heat as a FLUID, named CALORIC, introduced in 1809 by Horatio Gates Spafford (1778–1832) in his *General Geography, and Rudiments of Useful Knowledge: In Nine Sections [Illustrated with an Elegant Improved Plate of the Solar System. A Map of the World. A Map of the United States]*, propounds the idea of heat being a subsystem of MATTER that is indestructible and self-repellent is incorrect. This concept was intended to replace KINETIC THEORY, the idea that atomic/molecular MOTION is the foundation for the transfer of heat and was not introduced until COUNT RUMFORD (BENJAMIN THOMPSON

[1753–1814]) described the equivalence of work and rising temperature in 1798. However, the principle of heat and "MOLECULAR MOTION" was introduced in a rudimentary form by PLATO (427–347 BC) and later reaffirmed by GALILEO GALILEI (1564–1642) (see Figure C.4).

Figure C.4 Horatio Gates Spafford (1778–1832), thermodynamics engineer who introduced the caloric concept in 1809.

Caloric theory

[general, thermodynamics] The concept of a physical medium that transfers ENERGY, also referred to as theory of CALORIC. The fluidic medium involved in the energy transfer in question was named "caloric," supposedly indestructible (*see* CONSERVATION OF ENERGY) and self-repellent (this statement comes back in a different format with respect to a concept loosely connected to energy: ENTROPY).

Caloric value

[general, thermodynamics] Concept of HEAT TRANSFER introduced by Professor JOSEPH BLACK (1728–1799) in 1760. This concept is based on Black's theoretical interpretation of the ENERGY transfer between two systems; he also introduced the concept and unit of Calorie into the process of raising the temperature of a body.

Calorie (cal)

[biomedical, general, thermodynamics] The amount of HEAT necessary to raise the temperature of 1 g of water by 1°C (from 14.5°C to 15.5°C) under ATMOSPHERIC PRESSURE (1 atm = $1.013 \times 10^5 \, \text{N/m}^2 = 1.013 \times 10^5 \, \text{Pa}$). Abbreviated unit: cal = 4.186 J; kilocalorie is expressed as Cal (*also see* HEAT CAPACITY).

Calorimeter

[general, thermodynamics] Device used to determine the CALORIC VALUE of a substance, insulated to avoid exchange of HEAT with the exterior media (see Figure C.5).

Joule's apparatus

Figure C.5 A calorimeter.

Calorimetry

[biomedical, energy, thermodynamics] Indication of the interchange of ENERGY: energy loss by hot object = energy gain by colder object. Additionally, in biological applications calorimetry describes the process of converting one form of energy, specifically the CALORIC VALUE of fuel (i.e., food) involved in METABOLIC activity, for example, body temperature, CELL division, and MUSCLE action. In general, thermodynamic applications calorimetry is an indication of the energy consumption of pyrotechnic activity, combustion as well as PHASE transitions (e.g., melting, vaporization, and general heating and cooling, such as the Carnot process). The process of melting and vaporization introduces the LATENT HEAT of FUSION and the LATENT HEAT OF VAPORIZATION.

Calpain

[biomedical, chemical] Protein in the proteolytic enzyme family, calcium-dependent with symmetric peptidase core. This type of protein is found in the connective tissues of the brain, EYE LENS, and SKELETAL MUSCLE. The proteolytic nature identifies the use of peptide bonds for binding to amino acids, forming a POLYPEPTIDE chain.

Calvin cycle

[biomedical, chemical, thermodynamics] Photosynthetic carbon reaction in biological cellular METABOLISM ENERGY transfer. The "dark reactions" pathway in PHOTOSYNTHESIS. In this chemical reaction pathway, the free energy obtained from cleaving phosphor bonds within ATP is applied to fix CO_2 and chemically reduce, in order to form CARBOHYDRATE. The enzymes and their intermediates involved in the Calvin cycle

C

are located in the cellular chloroplast stroma. The stroma forms a compartment that is equivalent to the mitochondrial matrix. This metabolic cycle was characterized by Melvin Ellis Calvin (1911–1997), with the support from James Bassham (1922–2012) and Andrew Benson (1917–), for which Melvin Calvin received the Nobel Prize in Chemistry in 1961 (see Figure C.6).

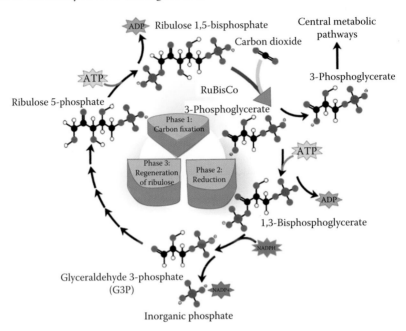

Figure C.6 The Calvin cycle. (Courtesy of Mike Jones.)

Camber

[fluid dynamics] The asymmetric surface contours of the upper and lower surfaces of an AEROFOIL. Because of a difference in camber between the upper and lower surface, the air-flow velocity over the top will be greater than under the bottom, providing a pressure difference between the top (lower) and the bottom (higher) based on the Bernoulli equation. The definition of camber ensures the maximum LIFT COEFFICIENT for the object in-flight. This principle is defined numerically by the line connecting the front and the tail end of the WING described by the CAMBER LINE $(Z(x))$ with a path length for the AIR to travel: for the upper portion, $Z_{up}(x) = Z(x) + (1/2)T(x)$, where $T(x)$ is the "thickness function," which varies with location (x) from front to rear, and for the bottom of the aerofoil, $Z_{low}(x) = Z(x) - (1/2)T(x)$. The thickness curve can be of any formulations, for instance $T(x) = -(t_r/0.2)\left[-a_1 x^{1/2} + b_1 x + c_1 x^2 - (c_1 - 1)x^3 + d_1 x^4\right]$, where t_r is the thickness ratio. In supersonic design the camber can also be negative (see Figure C.7).

Figure C.7 Camber of model plane wing on profile.

Camber line

[fluid dynamics] On an AEROFOIL the camber line indicates the partition line separating the upper and lower curved surfaces of the WING. The CAMBER of the respected curved surfaces provide the in-flight LIFT. An aerofoil that curves upward at the tail end of the wing has a chamber line that curves up as well. An example of a curved chamber line is $Z(x) = D_2 \left[b_2 x - c_2 x^2 + (c_2 - 1)x^3 \right]$, which varies with location (x) from front to rear and a and b are constants.

Camera

[general, optics] Optical instrument used for collecting still images or moving footage on a SOLID-STATE or digital format recording mechanism. The early cameras used photographic film to record single still images for still PHOTOGRAPHY and a sequence of still images in high frame-rate acquisition for movies. Later photography relied on a CHARGE-COUPLED DEVICE (CCD) array to record matrix ELEMENTS of grayscale or red–green–blue (RGB) values in electronic format with a variable dynamic range and in a range of PIXEL magnitudes. The camera has a simple or complex set of lenses (or only a single LENS, for low-end quality imaging) that correct for distortions and COLOR aberrations resulting from diffraction. The early mode solid-state camera used silver-chloride-coated plastic strips for the photographic effect. The silver chloride oxidizes and is later processed by chemical interactions to form a fixed chemical structure that covers the surface of the PHOTOGRAPHIC PLATE. This mechanism will result in a negative IMAGE, since more light will result in an increase of the OXIDATION process, making the chemical darken respectively. The final process in the solid-state design requires the production of a print on a paper that is chemically treated to respond to the exposure to light in a similar manner, this presents the hues in the same relative distribution as when observed in real life. The image formation system uses a thin lens combination with a range of lenses with various optical configurations to provide wide-angle or zoom image capture possibilities. The image formation relies on magnification for both capture and hard-copy reproduction (see Figure C.8).

(a)   (b)

Figure C.8 (a) Picture of an old-fashioned Kodak camera. (b) Picture of a new age digital camera.

C

Camera obscura

[general, optics] Imaging device using a pinhole as the Fourier LENS that provides an inverted spatial DECONVOLUTION of a light pattern generated by a source distribution on a screen or PHOTOGRAPHIC PLATE. The IMAGE has an optimal RESOLUTION depending on the diameter of the pinhole depending on the Rayleigh diffraction mechanism and the WAVEFRONT. The image formation is based on the PARAXIAL IMAGE FORMATION approach. The pinhole image formation was first described by ARISTOTLE (384–322 BC) (see Figure C.9).

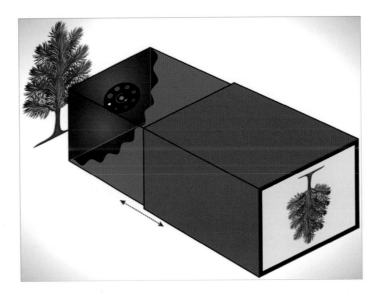

Figure C.9 Camera obscura.

Canal rays

[nuclear] In gas-filled tube with ANODE and CATHODE this represents the ions released from the anode, and are called anode ray or canal ray. The POSITIVE IONS migrate toward the anode under the applied electric field. The canal rays migrate in the opposite direction from the cathode (electron) ray. The exact nature of the positive canal ray depends on the GAS and hence the ions formed by surrendering any number of electrons at the positive anode. The discovery of the positive PARTICLE stream was made in 1886 by EUGEN GOLDSTEIN (1850–1930).

Canal theory of the tides

[fluid dynamics, geophysics] Canals have a specific tidal FLOW pattern due to FORCED OSCILLATION induced by the ebb and flood conditions. In some places the riverbed topography can create spectacular flow conditions, primarily resulting from the changes in propagation velocity with respect to long versus short waves.

One example is the surf in the Kampar River in Sumatra and another is in Girdwood, Alaska, with a succession of TIDAL WAVES running for 8 km along the riverbed (see Figure C.10).

Figure C.10 Canal wave examples. (a) Sumatra, Indonesia. The "Seven Ghosts" wave can have a crest of up to 3 m and travels the distance of the Kampar River for up to 50 km, (b) Sumatra; tidal bore, (c) Kampar River; Bono wave, and (d) Alaska. (Courtesy of Bono Surf.)

Canal waves

[fluid dynamics] *See* TIDAL WAVES.

Cancellous bone

[biomedical] Spongy bone with internal structure similar to a honeycomb ("trabecular") found for instance in ball JOINTS, pelvis (ilium), skull (cranium), and the scapula (i.e., shoulder blade). Cancellous bone constitutes approximately 20% of the bone structure in the human ANATOMY. The Young's modulus for cancellous bone is $E_Y = 1\,\mathrm{GPa}$, whereas the CORTICAL BONE found in the skeletal structure such as the tibia has a Young's modulus of $E_Y = 18\,\mathrm{GPa}$. Cancellous bone is rich in bone marrow and is an essential supplier of red BLOOD cells.

Canonical ensemble

[computational, thermodynamics] A thermodynamic arrangement with two components: one component is a system at constant temperature and the second system a heating bath to maintain the temperature in system 1.

The ensemble will still have fluctuations in ENERGY (E). The ensemble consists of a fixed number of particles (N), a constant volume (V), and at constant temperature (T), which defines the CANONICAL ENSEMBLE at $N;V;T$. Note that the temperature is related to ENTROPY (S) as $(1/T) = (\partial S/\partial E)$. The distribution of the Hamiltonian (H_s) and the Entropy (S_s) for the ensemble is defined by a normalized BOLTZMANN DISTRIBUTION, or canonical energy distribution $(\rho_c(\{q_i\},\{p_i\})$, where $\{q_i\},\{p_i\}$ represent the DEGREES OF FREEDOM for the system) as

$$\rho_c(\{q_i\},\{p_i\}) = \left[1 \Big/ \int \prod_{i=1}^{3N} \exp\left(H_s(\{q_i\},\{p_i\})/k_b T\right)\right] \exp - \left[H_s(\{q_i\},\{p_i\})/k_b T\right]$$

where $k_b = 1.3806488 \times 10^{-23}$ m^{2}kg/s^{2}K is the Boltzmann coefficient.

Canonical transform

[computational] When compared to Hamiltonian transform (with operator H_{tr}) and Lagrangian formulation, the canonical transform (with operator C_{tr}) implicitly does not have time as a parameter. The canonical transform requires an additional factor to conform to a Hamiltonian expressed as $H_{tr} = C_{tr} + (\partial F_{tr}/\partial t)$, where F_{tr} is a generating or transform function.

Canonical turbulent boundary layer

[computational, fluid dynamics, geophysics] Elaborating on the work by LUDWIG PRANDTL (1875–1953) on boundary layers published in 1904, the features in a zero pressure gradient in the streamlined high Reynolds number FLOW of an incompressible medium are described in a PIPE where the flow phenomena are not affected by WALL curvature or SURFACE ROUGHNESS. In the formation of the canonical turbulent boundary layer the conditions also mandate that there are no obstruction upstream. The high REYNOLDS NUMBER $(\text{Re} > 350-730 \text{ range})$ in the shear layer will be the instigator for canonical condition. One way of achieving the conditions for canonical turbulent boundary layer flow are through "shock" flow, jets of high velocity.

Capacitance (C)

[biomedical, electronics, general] Electrical capacitance [C] specific to alternating current [AC] conditions; in general the CAPACITANCE of a device changes from direct current (steady-state) to alternating current operations. Specifically, the RESISTANCE, inductance, reactance, and capacitance under alternating current become complex; FREQUENCY $(\nu = \omega/2\pi$, where ω is the ANGULAR FREQUENCY$)$ dependent; and parameters that have crossover properties, meaning the parameters are no longer linear and singular defined. Consider the current (i), a sinusoidal function with time (t), $= I_{max} \cos(\omega t)$, where I_{max} represents the maximum current (i.e., AMPLITUDE). For the charging process of a CAPACITOR this current now transforms into a function of the capacitance (C) and the applied voltage (V_{max}), $i = \omega C V_{max} \cos(\omega t)$, which translates to a current–voltage relationship defined by $V = (1/\omega C)I$, which yields a new parameter identified as the reactance, $X_c = 1/\omega C$, a derived capacitance. The frequency dependence will induce retardation in the current with respect to the driving voltage and can lag in PHASE. Indication of the amount of CHARGE (Q) that can be "steady-state" stored on a capacitor when a VOLTAGE (V) is applied, $C = Q/V$. The value of the capacitance is expressed in Farad [F], in actuality the value of capacitors in general use has a denomination of pF or μF. For a plate capacitor the capacitance is a function of the surface area (A), and the separation between the plates (d), $C = [\varepsilon(N-1)A]/d$, where ε is the DIELECTRIC permeability and N the number of plates while $E = V/d = \sigma_{elec}/\varepsilon = Q/A\varepsilon$ indicates the electric field between the plates, with σ_{elec} the plate charge density. Because of the repulsive force between like charges, the charge distribution on a CONDUCTOR or semiconductor will be uniform, which does not hold true for an INSULATOR. Any object has a complex impedance, for instance the capacitance of a horse with respect to soil is approximately $150-200\,pF$, the hooves act as the insulation between "GROUND" or better, the dielectric separating the two sides of the capacitor. The storage of ELECTRIC CHARGE has many experimental observations culminating in the final description. Two concurrent but separate events are notable in this historical development, around 1745 the world was pursuing the collection of what was thought to be "electric fluid," now known as electric current. Both PIETER VAN MUSSCHENBROEK (1692–1761) and EWALD JURGENS VON KLEIST (1700–1748) performed experiments that illustrated the storage and respective discharge of charge in a container, mediated by the conductive interface of a wet hand holding the device, providing the means to discharge.

Capacitance, membrane

[biomedical, chemical, electronics] *See* MEMBRANE CAPACITANCE.

Capacitive reactance (X_c)

[biomedical, electronics, general] Because of time lag under alternating current charge distribution in an electronic circuit the capacitive charging requires introduction of a new complex parameter representing RESISTANCE. As described under CAPACITANCE (AC), $V = (1/\omega C)I$, which introduces the parameter defined as the capacitive reactance, $X_c = (1/\omega C)$.

Capacitive spectroscopy

[biomedical, electronics, imaging] PARTICLE counting under microfluidic conditions, specifically when the channel diameter is relatively large with respect to the particle diameter, put severe constraints on the equipment design. Polarizable micropillars can be used to provide a LIGAND-based capacitive system that specifically responds to certain molecular configurations. For instance, the use of polydimethylsiloxane can allow for such capacitive coupling counting system design. When an external ELECTRIC FIELD (E) is applied the capacitance (C) changes under the influence of the passage of the quantity of specific molecules. The current in the capacitive system will adhere to the EQUATION OF CONTINUITY, $\nabla \cdot J = 0$, where J is the current density, and OHM's LAW (constitutive Ohm's law), $J = (\sigma + i\omega\varepsilon_r\varepsilon_0) E$, with ε_r the RELATIVE PERMITTIVITY and $\varepsilon_0 = 8.854187817 \times 10^{-12}$ F/m the absolute permittivity of VACUUM, σ the conductivity per unit length of medium, and $\omega = 2\pi v$ the ANGULAR FREQUENCY of alternating applied voltage with frequency v. The impedance (i.e., capacitance) can be calculated and plotted as a function of frequency in a BODE DIAGRAM (also referred to as a Bode plot) to derive the capacitive spectrum and hence the chemical composition of the FLUID passing through the "lab-on-a-chip." This sensing technology is also referenced under IMPEDANCE SPECTROSCOPY.

Capacitor

[biomedical, electronics, general] Device capable of storing and releasing electrical charge, expressed in Farad (F). The capacitor design uses two conductors that are spaced apart by an insulating medium, such as GLASS, paper, polycarbonate, polyester (PET film), polystyrene, polypropylene, PTFE or teflon, silver mica, electrolytes, POLYMER, printed circuit board, and VACUUM. Next to conductors semiconductor materials are also used. A charge applied on a capacitor will be equal but opposite in value for the two respective plates. The capacitance (C) of the capacitor represents the ability to hold CHARGE (Q) and is defined by the electric field between the two charge layers or the electrical POTENTIAL (V) associated with this electric field, $C = Q/V$. Capacitors are made with various media, such as METAL plates separated by AIR (tuning capacitor, used in early RADIO design), ELECTROLYSIS (oxidized metal foil emerged in conducting gel), and ceramics. The ENERGY stored on a capacitor is the result of transferring a charge over an average potential $(1/2)V$, applying the capacitance yields $E = (1/2)CV^2$. Placing capacitors in parallel equate to increasing the surface area for the capacitive charge storage thus adding the individual capacities, whereas in series the total charge is equal for each capacitor and the combined capacitance equates: $C_{equivalent} = C_1 C_2/(C_1 + C_2)$ or $(1/C_{equivalent}) = (1/C_1) + (1/C_2) + (1/C_i)$. A capacitor discharges its electrical charge while the voltage applied (V_{suppl}) over the external resistor (R) (i.e., impedance: $X_{electric}$) will drop accordingly, resulting in a time DECAY of the current (I) (i.e., charge released per unit time: t) expressed as $dQ/dt = I = C(dV/dt) = C(dIR/dt) = RC(dI/dt)$; solving for the charging phenomenon or respectively the time release of charge yields $Q(t) = CV_{suppl}\left[1 - e^{-RC/t}\right]$; or discharge, $Q(t) = Q_0 e^{-RC/t}$, or for the charging current $I(t) = (V_{suppl}/R)e^{-RC/t}$. When the capacitor is charged and discharged resulting from an alternating current $(I_{cap}(t) = I_{max}\sin(\omega t))$, where ω is the ANGULAR FREQUENCY indicating the change in direction of the current through the capacitor (I_{cap}) per unit time; there will be a lag between the voltage and the current of precisely 90° or $\pi/2$ in PHASE, the current leading over the voltage with respect to time (see Figure C.11).

Figure C.11 Capacitor.

Capacitor, parallel plate

[electronics, general] *See* **PARALLEL PLATE CAPACITOR**.

Capacitor, radio

[general] Variable capacitance plate CAPACITOR used in early design FM and AM RADIO receiver and analog television receivers. The value of the capacitor determines the RESONANCE FREQUENCY of the receiving circuit, hence locking on to one specific radio transmitter SIGNAL only.

Capillarity

[fluid dynamics, thermodynamics] The surface interaction between a fine gauge hollow tube inserted into a LIQUID surface and the resulting surface shape and level height with respect to the surrounding free liquid inside a narrow column. The height of the liquid surface inside the tube with respect to the level liquid surface is a function of the DENSITY (ρ) of the liquid and the "adhesive forces." For instance, the height (h) as a function of tube diameter (d) for the following liquids is as follows: mercury falls below the surface $h = -10/d$, water $h = 30/d$, or ALCOHOL $h = 11.6/d$. The surface curvature will make an ANGLE θ with the WALL of the tube that is directly dependent on the TENSION (F_T) between the liquid and the surface material of the tube (or the lining of the tube, e.g., OIL), $h = 4F_T\cos\theta/\rho gd$, where g is the GRAVITATIONAL ACCELERATION (see Figure C.12).

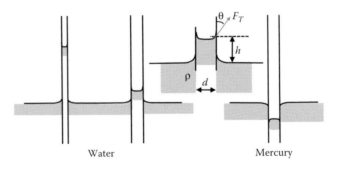

Water   Mercury

Figure C.12 Capillarity.

Capillary

[biomedical, biophysics, general] Small diameter tubing that results in hemodynamic conditions where residual forces become apparent in the FLOW mechanism of action as well as under steady-state conditions revealing the influence of forces between the WALL and the medium. One specifically notable situation is BLOOD flow in capillary systems of the circulatory vasculature under mechanical contraction of the HEART. The CAPILLARY FLOW is confined by the fact that the RED blood cells will flow through with minimal gradient force, which entails that the red blood cells line up as stacked parachutes, resulting in what is known as ROULEAUX FORMATION. In case the red blood cells are tilted there will be a gradient in force making the CELL straighten out with the edge of the circumference of the disk-shaped cell in cross-sectional contact with the wall of the tube, that is, capillary vessel. The full-circumference contact of the red blood cells results in a full mechanical contact with high FRICTION and frequently resulting in both vascular and red blood cell deformation, emphasizing the VISCOELASTIC properties of blood flow.

Capillary action

[fluid dynamics] (Greek: *capilla*: hair, i.e., hollow tube) Surface forces in the operational environment between a FLUID and a solid that are dominated by either cohesive or adhesive forces. The SURFACE TENSION for a certain LIQUID adheres to the WALL and wets the surface because of a polar attraction greater than the internal Van der Waals forces of the liquid. This phenomenon may cause the fluid to rise inside a narrow tube. The radius of curvature of the MENISCUS of the liquid in a tube with radius R is defined as $r = R/\cos\theta$, where θ is the ANGLE with the wall outside the liquid. The height (h) with which the liquid may rise depends on the density of the liquid ρ, expressed as the pressure of the column of liquid under gravitational attraction (g the GRAVITATIONAL ACCELERATION): $P = \rho g h$ compared to the force per unit area resulting from the surface tension over the curved surface, $2(\sigma_{surface}/r)$, yielding $h = 2\sigma_{surface}\cos\theta/\rho g R$.

Capillary flow

[fluid dynamics] FLOW is expressed as the integral of the velocity of medium over the cross-sectional area of a conduit with diameter $2R$. The velocity is the result of the pressure gradient ($P_1 - P_2$) over a length of a passage way (L), while the velocity can be a function of the radius (r) to the center of the conduit, $Q = (2\pi/4\eta)[(P_1 - P_2)/L]\int_0^R [R^2 - (r')^2]rdr'$, known as POISEUILLE'S LAW.

Capillary number (Ca)

[biomedical, fluid dynamics] Ratio of viscous force to elastic force, $Ca = \mu\dot{\gamma}a/hE$, when applied to vascular WALL where μ is the VISCOSITY, $\dot{\gamma}$ is the shear rate, a is the radius, h is the MEMBRANE thickness, and E is the membrane Young's modulus. For large diameter vessels the shear rate can still be large near the wall. The shear rate on the surface of the wall concomitantly leads to deformation of red BLOOD cells, and attributes to the formation of a flow-free layer at the vessel wall.

Capillary number (Ca = η_v/γ_s)

[fluid dynamics] γ_s SURFACE TENSION, η viscosity, $\boldsymbol{v}$ velocity. In the description of CAPILLARY FLOW, the capillary number ties together with the Bond number to describe conditions where the viscous stress between a BUBBLE and the WALL will be ruled by VISCOUS LUBRICATION. Under capillary flow transporting a bubble the FRICTION forces across the surface of the bubble can vary locally. Under certain conditions the bubble may be elongated and hence in close contact with the wall of the tube, which may be identified by a low, nonzero capillary number. Under these circumstances the surface of the bubble can be divided in regions that are subject to specific force diagrams, respectively locally unique from tip to tail. The capillary number may additionally influence approximations and reductions in the NAVIER–STOKES EQUATION (see Figure C.13).

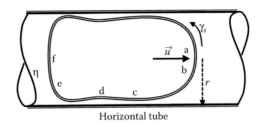

Horizontal tube

Figure C.13 The relation for the capillary number in bubble growth.

Capillary waves

[energy, fluid dynamics, mechanics] Ripples occurring at an interface between two liquids. One circumstance is the RIPPLE wave resulting from wind on the surface of a LIQUID. Because of the size or MAGNITUDE, CAPILLARY wave grow and extinguish rapidly when the source (i.e., wind) engages or disengages. The WAVE process is confined primarily by the SURFACE TENSION (T_s). The breaking wave causes the FLUID pressure to become discontinuous at the surface in the crest. Assuming two curves are forming the crest, each with respective radius R_i, the discontinuity in the fluid surface PRESSURE (P) can be defined as $P - P' = T_s[(1/R_1) + (1/R_2)]$, where P and P' represent the pressure at the surface on either side of the crest. In the case of negligible influence of GRAVITY the VELOCITY POTENTIAL (ϕ) of the wave propagation in the x-direction with AMPLITUDE in the y-direction can be represented as $\phi_y = Ce^{ky} \cos(kx)\cos(\psi t + \varphi)$ and $\phi_y' = C'e^{-ky} \cos(kx)\cos(\psi t + \varphi)$, respectively, where C and C' are constants and $k = 2\pi/\lambda$ the wave number, ψ the imposed velocity, and φ a PHASE shift. This translates in a varying pressure under external force defined for one side respectively, $(P/\rho) = (\partial\phi_y/\partial t) = (\psi^2 a/k)e^{ky} \cos(kx)\sin(\psi t + \varphi)$, in case of a fluid density ρ. The surface deformation velocity is defined by $\psi = \sqrt{[T_s k^3/(\rho + \rho')]}$, where ρ' the density on the "opposite side of the crest." The wavelength is not a singular value, but rather forms a spectrum, ranging from low to high frequencies. The higher frequencies will be subject to DISPERSION and will cause the sharp crest to "break" eventually (on large scale this corresponds to braking wave at the beach in Hawaii, for instance). The driving force and the associated ENERGY of motion (KE) will determine the wavelength spectrum based on the velocity potential, $KE = (1/2)\rho\int_0^\lambda \phi_y(\partial\phi_y/\partial y)_{y=0}dx - (1/2)\rho'\int_0^\lambda \phi_y'(\partial\phi_y'/\partial y)_{y=0}dx$ for waves at the interface of two fluids. Under capillary WAVE MOTION the surface tension has an overpowering force influence on the wave description, acting as a restoring force. Generally, water waves with a wavelength less than 0.0173 m can be considered CAPILLARY WAVES. Both the PHASE VELOCITY and group velocity increase with decreasing wavelengths. The group velocity for capillary waves is greater than the phase velocity. At the crossover wavelength the phase velocity has a minimum value, approximately 0.231 m/s in a water–air fluid surface interface. Capillary waves can generally be recognized as small ripples, and result from a fresh wind blowing over smooth water. These waves have curved crowns and v-shaped troughs.

Capillary waves are extinguished by molecular "VISCOSITY." For GRAVITATIONAL WAVES, in contrast, generally the crests are sharp and the troughs are rounded (see Figure C.14).

Figure C.14 Capillary wave.

Carathéodory, Constantin (1873–1950)

[computational, thermodynamics] Mathematician from Germany. The work of Dr. Carathéodory made significant changes in the computational approach to thermodynamic axiomas. In axiomatic thermodynamics, the SOLUTION mechanism for the SECOND LAW OF THERMODYNAMICS relies on how to interpret the geometric configuration of a specific differential equation and the appropriate solutions, known as the PFAFFIAN. Carathéodory made significant strides in formulating thermodynamic laws in proper mathematical expressions. Carathéodory made statements with respect to the second law of thermodynamics concerning the mathematical definition of thermal concepts for a system (S_y), such as TEMPERATURE (T) and heat ($Q = \Delta U + W$, with the change of internal ENERGY U and the work W performed), described in mechanical terms of displacement (i.e., momentum: $p = mv$, with mass m and velocity v), volume (V), and PRESSURE (P). Each system will consist of different phases (e.g., LIQUID, solid), where the total energy of the system is the sum of the energy constituents of the various phases $\varepsilon = \sum \varepsilon_i (V_i, p_i, m_{ik})$, where k denotes the respective constituents, for the respective PHASE i. This has analogies with LORD KELVIN's (1824–1907) definition of temperature as the average kinetic energy (E) of a system, $E = (1/2)mv^2 = (3/2)k_bT$, where $k_b = 1.3806488 \times 10^{-23}$ m^{2}kg/s^{2}K the Boltzmann constant.

Carbohydrate

[biomedical, chemical] Molecular structure principally consisting of carbon, hydrogen, and depending on the full macromolecular definition, including oxygen as an organic compound. Also referred to as saccharide. Examples of carbohydrates are sugars, wheat, and lactose. The carbohydrate designation is further subdivided in monosaccharides (e.g., GLUCOSE), polysaccharides (e.g., starch), disaccharides (e.g., lactose, sucrose), and oligosaccharides (e.g., inulin, also found specifically in AB BLOOD and on the MEMBRANE of cells where they form a marker in CELL to cell communications).

Carbon ($^{12}_{6}$C)

[atomic, biomedical, chemical, imaging] Organic ELEMENT. Part of molecular structures such as graphite and diamond as well as carbohydrates and organic fuels. Major atomic building block in organic structures, biological organism are composed of molecular chains containing carbon in a large percentage. This stable element has by definition the ATOMIC WEIGHT of 12.00000 atomic mass units (u). Stable carbon consists of 6 protons, 6 neutrons, and 6 electrons. Additionally, carbon isotopes are used in imaging applications since they will seamlessly integrate with the biological structure. Some of the available isotopes are ^{11}C and ^{14}C (see Figure C.15).

(a) (b) (c) (d)

Figure C.15 Carbon is the building block of organic structures, as well as a fundamental ingredient for construction. (a) Log cabin made completely out of wood; (b) a tree; (c) a chameleon, biological life form; and (d) graphite tip of a pencil and a diamond earring.

Carbon radionuclide (^{11}C)

[biomedical, chemical, imaging] This unstable carbon ISOTOPE has a half-life of 20.4 min and is used in positron emission TOMOGRAPHY as a tracer in order to label physiological processes. The DECAY process of ^{11}C is by one positron.

Carbon radionuclide (^{14}C)

[biomedical, chemical] Major atomic building block in organic structures, biological organism are composed of molecular chains containing carbon in a large percentage. The DECAY process of ^{14}C is by two electrons with a half-life of $\tau_{1/2} = 5,730$ years. The RADIOACTIVE ISOTOPE ^{14}C occurs in normal living carbon structures as a common "cousin" with a ratio of 1.3×10^{-12} to stable carbon-12 (same as the atmospheric ratio), and is used as a standard tool for radioactive dating of organic structures. During normal METABOLISM the standard ratio of ^{14}C to ^{12}C is maintained while after CELL death the ^{14}C will start to decay. Determining the remaining ratio provides the quantitative tools for radioactive dating.

Carbon-14 dating

[general, nuclear] Carbon ISOTOPE; 14C is produced in the upper ATMOSPHERE as 14N is bombarded by COSMIC RAYS. (*Note*: standard carbon has ATOM mass 12.) The 14C drops to the EARTH where it is absorbed by plants and animals. The 14C levels in an organism are constant throughout the organism's life (since it continuously adds and removes 14C through nutrition and RESPIRATION). When an organism dies it can no longer replenish its 14C levels and the 14C begins to DECAY. Using the fact that the half-life for 14C is 5730 years, RADIOACTIVITY levels of 14C are measured and the level of decay from the original value is used to estimate the organism's AGE. This type of radioactive dating only applies to objects that have carbon as a base (primarily a metabolic base), not stones and such. There are however limitations to the accuracy and time range (see Figure C.16).

Figure C.16 Carbon dating of human skull.

Cardiac output

[biomedical, fluid dynamics] BLOOD flow principle recognized by ADOLF EUGEN FICK (1829–1901) as the total volume ejected by the contracting HEART per minute of time, expressed in liters per minute. The cardiac output (CO) is thus a function of both the volume of the ventricles and the heart rate (HR): heart rate multiplied by stroke volume (SV): CO = HR × SV. The heart ejects a volume of blood based on the muscular function of the cardiac MUSCLE as well as the FLUID dynamic constraints resulting from the vascular RESISTANCE and COMPLIANCE in the whole body CIRCULATION. The circulation system comprises the AORTA on the left ventricular side and the pulmonary artery on the right ventricular side, branching out into a web of arteries, arterioles, and capillaries, flowing under a combination of venules and veins, leading back to the left and right atrium by means of the pulmonary veins from the lungs (carrying oxygen-rich blood) and the hollow VEIN (superior vena cava), carrying oxygen-depleted blood, respectively. During exercise the cardiac output can increase by more than fivefold, depending on the training and general health of the individual.

Cardiac pacemakers

[biomedical] *See* PACEMAKER.

Carnot, Nicolas Léonard Sadi (1796–1832)

[computational, thermodynamics] French physicist who provided essential insight on the practical applications of the SECOND LAW OF THERMODYNAMICS in 1824 and may be credited as one of the cultivating founders of the essence of thermodynamics. Carnot's most remarkable asset is the theoretical description of a process that relies on the exchange of heat and the performance of work. The process has four (4) stages in an ideal mechanism; isothermal EXPANSION during delivery of heat (Q_H) at high temperature, followed by adiabatic expansion, with the third-stage isothermal compression, with heat exchange (i.e., loss) at low temperature heat (Q_L); and the fourth phase ADIABATIC COMPRESSION, ending up in the exact same energetic point in pressure and volume (PV plot) as where it left off. WORK (W) is performed during the isothermal expansion of stage 1 as well as the adiabatic expansion of stage 2. INERTIA or other stored ENERGY mechanism provides for the stroke in stage 3. Since the described Carnot process is ideal there is a reality associated with the efficiency of working devices and hence an associated efficiency: e_c = output of useful energy/delivered energy for the process = $W/Q = (Q_H - Q_L)/Q_H$. Examples of the efficiency of specific devices are as follows: electric MOTOR 50%–95%; GAS furnace for domestic heating 70%–85%; hydrogen–oxygen FUEL CELL 60%; nuclear power-plant 30%–35%; incandescent lamp 5% (see Figure C.17).

Figure C.17 Nicolas Léonard Sadi Carnot (1796–1832).

Carnot coefficient

[thermodynamics] Identification parameter for the distribution of the ENERGY "consumption" of a system, divided in two components: work performed on the external system and the disposal of the entropy received from both the external heat source and the entropy generated by the ENGINE. The CARNOT ENGINE primarily operates based on internal temperature (T_{in}) and the external temperature that drives the process (T_{out}), yielding the Carnot coefficient, $\eta_{Carnot} = (T_{out} - T_{in})/T_{out}$. This provides the ENTROPY (S) generated by an irreversible process as $S_{irr} = \eta_{Carnot}(Q_{out}/T_{in})$, where Q_{out} represents the energy transferred out of the system.

C

Carnot cycle

[energy, general, thermodynamics] (syn.: work) {use: cooling, heating, thermodynamics} A four-stage process where a GAS is expanded and contracted in two isothermal stages and two ISENTROPIC stages in interchangeable order. The Carnot cycle and CARNOT ENGINE principle were conceived by NICOLAS LÉONARD SADI CARNOT, a French scientist (1796–1832). The principles of the thermodynamics and the associated efficiency of ENERGY exchange processes and resulting work were first described in 1824 (see Figure C.18).

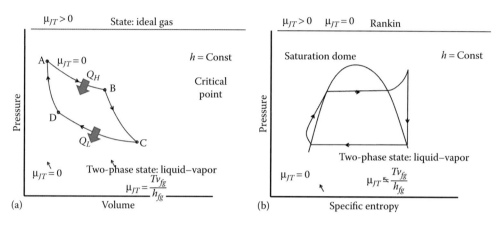

Figure C.18 (a,b) Pressure–volume diagram for a Carnot cycle at a specific temperature.

Carnot engine

[energy, thermodynamics] (syn.: work) {use: cooling, heating, thermodynamics} The Carnot engine consists of a cylinder and a piston, the compartment sealed off by the piston is filled with GAS and the reservoir is exposed to either low or high temperature resulting in a heat exchange process. Alternating the heat exchange (cooling/heating) results in the periodic EXPANSION and contraction of the sealed-off gas, moving the piston up and down.

Carrel, Alexis (1873–1944)

[biomedical] Vascular surgeon from France, who invented the mechanism of BLOOD oxygen exchange outside of the HUMAN BODY for which he received the Nobel Prize in PHYSIOLOGY and MEDICINE in 1912. This mechanism was adapted and incorporated in the first HEART–lung machine developed by Dr. JOHN HEYSHAM GIBBON (1903–1973). Together with CHARLES LINDBERGH (1902–1974) Carrel developed the first heart–lung machine in 1935.

Carrier

[biomedical] Protein chains that facilitate the transport of molecules through the biological CELL MEMBRANE at specific receptor sites. The respective individualized protein chains are highly specialized in the selection of molecules to transport. This transport system is classified as facilitated DIFFUSION, an active process.

Carrier wave

[general] Electromagnetic or acoustic WAVE in analog format that can be frequency or AMPLITUDE modulated to encrypt information for transmission to a remote receiver location. In digital format the mechanism of transfer is in bit rate, not in modulation. For television the transfer mechanism in the United States (in 2009) and in various other countries has changed from frequency modulation to digital encoding (using the same carrier wave for the respective channels/television stations). The carrier wave can be chosen to promote long DISTANCE propagation, whereas the modulation can be in the audible range or in a range that can be analyzed with some kind of spectral mechanism of action after the carrier wave has been removed by a demodulation scheme. In ELECTROMAGNETIC RADIATION amplitude modulation is

used in AM RADIO (150 kHz–26 MHz), whereas frequency modulation is used in FM radio (87.5 MHz–108 MHz). General FM radio is transmitted in stereo. RADIO stations perform the audio encoding for the transmission across the AIR ways to the audience. The STEREO modulation is obtained at two separate audio frequencies: v_{left} and v_{right} (i.e., multiplexing) with respect to the pilot tone (v_p), expressed as $\{0.9([(v_{left}+v_{right})/2]+[((v_{left}-v_{right})/2)\sin(4\pi v_p t)])+0.1\sin(2\pi v_p t)\}\times 75$ kHz. The audio SIGNAL in FM radio transmission is between 30 Hz and 15 kHz, defined as HiFi, or high-fidelity. The HiFi standard was introduced in 1966 to set the standards for accuracy in frequency transfer (reproduction of the input signal to the consumer) and low signal-to-noise ratio. The converted multiplexed signal is recombined in a decoder, which is part of the RADIO RECEIVER unit, and the output is transferred to an amplifier and relayed to speakers for the listening pleasure. Amplitude modulated radio waves generally have a short range, up to 200 km, whereas frequency modulated waves reaching into very short waves can travel extreme distances due to REFLECTION from the IONOSPHERE ranging from the opposite side of the world.

Cartesian coordinates

[general, mechanics] Coordinate system named after RENÉ DESCARTES (1596–1650), using two planes (two-dimensional) or three orthogonal planes (three-dimensional [coordinates: x,y,z]).

Cassini, Giovanni Domenico (1625–1712)

[astronomy, general, geophysics] Mathematician, astronomer, and engineer from Italy, and geophysicist. He completed the work of JEAN PICARD (1620–1682) on the longitude and LATITUDE angular division scale of the globe for cartography. In addition he discovered the split in Saturn's ring which was aptly named after him. He presented a thorough theoretical description of the precision of the axis of rotation of the MOON as well as the rotation period of several planets. Giovanni Cassini is known for his contributions on the description of our SOLAR SYSTEM, specifically the discovery of four satellites orbiting SATURN. Cassini also discovered the division of the rings of Saturn (not a continuous belt), which are named in his honor: Cassini Division. As an engineer Cassini worked on water management (FLOW control for rivers; specifically the river Po in Italy, known for its devastating flooding) as well as set out a topographical map of France in 1670, to be completed by his grandson, after the continuation by his son, in 1789. Cassini's name was also used for the (to date) 10 year mission to Saturn, as well as the in situ investigation by the National Aeronautics and Space Administration, powered by the Cassini spacecraft (see Figure C.19).

Figure C.19 Giovanni Domenico Cassini (1625–1712).

Cassini, Jacques (1677–1756)

[astronomy, general] French physicist, mathematician, and astronomer, son of GIOVANNI DOMENICO CASSINI (1625–1712). Cassini provided a detailed mathematical description of curvature of various closed loop shapes, primarily oval, and irregular shapes in the general outline of figure eight. These contracted oval shapes are sometimes referred to as cassinoids.

Cassini state

[astronomy, geophysics] A system is defined as residing in a Cassini state when it is in COMPLIANCE with the following three laws. The MOON revolves the EARTH in a 1:1 spin-orbit ratio, meaning that the same side of the Moon always faces the Earth. On second observation by GIOVANNI DOMENICO CASSINI (1625–1712) he found that the PRECESSION of the moon, describing a cone, is executed in a plane that intersects the elliptic orbit as a circle. The third conclusion from Cassini's observations was that the plane carved by the intersect from the normal to the rotational elliptic plane around the Earth with the plane described by the normal to the orbital plane around the SUN will contain the orientation of the rotational axis of the Moon as a function of location. These laws are known as the CASSINI'S LAWS. This applies in the scope of the processes to the mean orientation of a synchronously locked SATELLITE as well.

Cassini's laws

[astronomy, astrophysics, geophysics] Three sets of PLANETARY MOTION rules described by GIOVANNI DOMENICO CASSINI (1625–1712) in 1693, with mathematical geometric derivations of the position and orientation of the MOON with respect to EARTH. The same rules have been applied in setting the configuration of geostationary ORBITS for earth's satellites. With modifications these laws also appear to apply to the obliqueness of VENUS as well. The moon's axial rotation according to these laws is constrained as follows: (1) The Moon spins at a uniform rate that matches its mean orbital rotation rate; meaning the same side of the Moon always faces the Earth. (2) The normal to the moon's equatorial plane subtends a fixed ANGLE with the normal to the ecliptic plane as $\theta = 1.59°$. (3) The normal to the moon's equatorial orbital plane, the normal to general orbital plane of the Moon, and in addition to the normal to the ecliptic lunar plane form three coplanar vectors that are orientated in such a manner that the vector describing the ecliptic lunar plane is confined by the other two planes.

Cassiopeia

[astronomy, astrophysics, general] The location of the Cassiopeia constellation is found in reference to the NORTH STAR (POLARIS) (Polaris is fixed in our view of the northern sky), which is part of the URSA MINOR (LITTLE DIPPER), with URSA MAJOR (BIG DIPPER) to one side and Cassiopeia of the lower other side (the relative position will change with the seasons as the rotational axis of the earth pivots [PRECESSION]); to the right in SUMMER and to the left of Polaris in WINTER. The shape of Cassiopeia is more or less a flattened "W" (see Figure C.20).

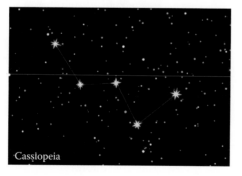

Figure C.20 Outline of the constellation Cassiopeia.

Casson, N. (twentieth century)

[fluid dynamics] Relatively unknown fluidic behavior engineer who published his model of viscous behavior in 1959, known as the CASSON MODEL, or the fluids as Casson fluids.

Casson model

[biomedical, fluid dynamics, general] Non-Newtonian FLOW model of colloid suspensions such as BLOOD, dough, and chocolate under low shear rate. The VISCOSITY is defined as $\eta_{\text{visc}} = (\sigma_{s,0}/\dot{\gamma}) + \sqrt{(k_C \sigma_{s,0}/\dot{\gamma})} + k_C$, where $\sigma_{s,0}$ is the material yield stress, $\dot{\gamma}$ the shear rate, and k_C the Casson constant or Casson rheological constant. Note that based on the FÅHRAEUS–LINDQVIST EFFECT the determination of the apparent viscosity is a function of the device and methods used to derive the viscosity. Specifically there are four primary device mechanisms in use to determine the viscosity: two concentric rotating cylinders; rotating disk with respect to stationary plane separated by a LIQUID thickness; rotating cone with respect to stationary plate; and measure of flow rate through a CAPILLARY tube (*also see* VISCOSITY MODEL). It is also known as Casson equation.

CAT scan

[general] Computerized axial TOMOGRAPHY, also known as CT scan or X-ray tomography. X-ray IMAGE reconstruction resulting from multiple sequential or parallel transmission measurements at incremental ANGLES completing a 360° circumferential scan. The computational component is in the calculation of the intersects of lines from various (multiple and incongruent) angles for identification of points in the two-dimensional dissection plane that have equal attenuation, providing the triangulation for the location of the anatomical features for the image reconstruction. The axial component is added by moving the scanning array in the axial direction by discrete stages, or through the use of multiple X-ray sources and detector arrays that are stacked in a scanning belt. Once all the measurements have been acquired from 360° circumferential scans as well as axial increments, a three-dimensional reconstruction is performed by means of a computer program that renders a visualization. The visualization can be selected by choosing slices at azimuthal and ZENITH angular orientations in order to identify regions with X-ray density that is different from the NORMAL DISTRIBUTION peak or that are sharply distinguished from the neighboring volumes (see Figure C.21).

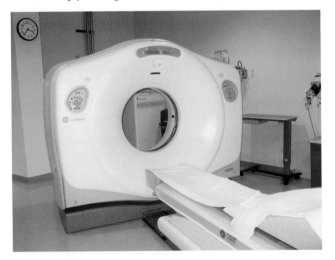

Figure C.21 CT scanner X-ray machine.

Cathode

[biomedical, chemical, electronics, general] Negatively charged ELECTRODE, usually accompanied by the positively charged ANODE, as introduced by MICHAEL FARADAY (1791–1867). The word "cathode" comes from the Greek word for "the direction in which the SUN sets." Both the cathode and anode submerged in a SOLUTION of ACID, SALT, or base will divide the constituents of the respective electrolyte(s) based on the sign of the charge. The positive (CATION) and negative (ANION) charged ions will diffuse respectively to the cathode and anode under the influence of the applied electric field resulting from an external electrical potential source (e.g., BATTERY, chemical reaction), which is required to be stationary in order to induce the ionic migration.

When direct current (consisting of electrons) between the electrodes is applied to pure water the ELECTRIC CHARGE gradient between the cathode and the anode will result in IONIZATION of the water $\left(4H_2O \rightleftarrows 2HO^- + 2H^+ + 2H_2O \rightleftarrows 2(H_3O)^+ + 2HO^- \rightleftarrows 4O^{2-} + 8H^+ \leftrightarrow 4O_2 + 8H_2\right)$, which is a poor CONDUCTOR. The formation of the $(H_3O)^+$ ION is temporary, and is due to the inherent attraction between the polarized water molecule and the hydrogen ion with POSITIVE CHARGE (cation). Upon reaching the respective anode and cathode, the respective cation and anion will lose their charge in exchange with the electrodes and form oxygen (O_2) and hydrogen molecules (H_2). The ELECTROLYSIS of water can be enhanced by the addition of an ACID, such as carbonic acid, sulfuric acid, nitric acid, or perchloric acid. The speed of the electrolysis will greatly depend on the strength of the acid (acidity) expressed in pH. Electrode will be at negative electrical potential, with respect to the anode which is at a positive electrical potential. A heated cathode will emit electrons, forming CATHODE RAYS. The electron emission will provide a free space current, with current density J, which can be calculated with the use of Richardson's law, $J = (1 - \tilde{r})(4\pi e m_e k_b^2 / h^3) T^2 e^{-(\phi_{elec}/kT)}$, also referred to as DRIFT current density or displacement current density; defined as a function of temperature T, using Boltzmann coefficient $k_b = 1.3806488 \times 10^{-23}$ $m^2 kg/s^2 K$; electron charge equivalence, $e = 1.60217657 \times 10^{-19}$ C; ϕ_{elec} the work function; electron mass, $m_e = 9.10939 \times 10^{-31}$ kg; $h = 6.62606957 \times 10^{-34}$ $m^2 kg/s$ the Planck's constant, and $\tilde{r}$ the mean REFLECTION coefficient for the material surface ($\tilde{r} \cong 0.5$ for tungsten). The concept of displacement current also applies to the time-dependent MAXWELL EQUATIONS (see Figure C.22).

Figure C.22 Cathode ray vacuum tube.

Cathode glow

[atomic, nuclear] Low-pressure GLOW DISCHARGE first described in 1675 by the French astronomer JEAN PICARD (1620–1682). It wasn't until 1870 that SIR WILLIAM CROOKES (1832–1919) provided a formal description of the cathode glow, and introduced the investigational ELECTRODE tube, the CROOKES TUBE. The Crookes tube was the predecessor to the modern neon light signs.

Cathode ray tube (CRT)

[biomedical] VACUUM tube with a CATHODE that emits electrons and a hollow ANODE to provide a PROJECTILE electron (electron gun) that can be deflected by electrodes before it impacts a fluorescent screen. CRT was the elementary component in old-fashioned televisions and OSCILLOSCOPE screens. The monitors predating the flat-panel display, PLASMA screen, and LED screen are referred to as CRT screens. CRT screens are bulky and heavy due to the large amount of GLASS and ELECTRONICS involved. Additionally due to the construction of the CRT mechanism of action these devices are consuming a significant amount of electrical ENERGY, with inherent generation of heat (see Figure C.23).

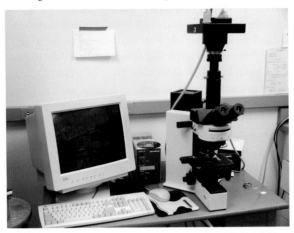

Figure C.23 CRT computer monitor attached to a camera mounted on an optical microscope.

Cathode rays

[atomic, nuclear] Electron beam, emitted from the CATHODE in a VACUUM TUBE or CROOKES TUBE (after SIR WILLIAM CROOKES [1832–1919]).

Cation

[chemical, electronics] ION that is attracted to the CATHODE, hence carrying a net POSITIVE CHARGE. The process of ion DIFFUSION applies in particular to ELECTROLYSIS and atmospheric AIR IONIZATION coronas.

Cauchy, Augustin-Louis (1789–1857)

[general] Mathematician, physicist, and engineer from France. Cauchy provided significant details on the description of stress and strain on MICROSCOPIC scale based on the initial MACROSCOPIC definitions by ROBERT HOOKE (1635–1703) more than a century prior in 1676. Another mathematical contribution by Cauchy is the SOLUTION method based on the residues of the developed series with respect to a function (see Figure C.24).

Figure C.24 Drawing of Augustin-Louis Cauchy (1789–1857).

Cauchy method of residues

[computational, fluid dynamics, general] Mechanism of complex function analysis named after AUGUSTIN-LOUIS CAUCHY (1789–1857). The "residue" theorem applies to solving integrals over closed curves in the complex domain. Consider a function (f) that is analytic and holomorphic inside the curve of the domain carved out by the integral loop as well as on the points of the curve of the closed integral path, but is not analytic in the points inside the curve within the complex domain $z_j = x_j + iy_j$: excluded points $z_1, z_2, \ldots, z_n$ (i.e., singular points). The function can be developed in a Laurent series as: $f(z) = \sum_{n=-\infty}^{\infty} a_n (z - z_j)^n$, where the coefficient of a_{-1} with respect to $1/(z - z_0)$ is the *residual* of the function $f(z)$ in z_0; $\text{Res}[f, z_0] = a_{-1}$. For instance the Laurent function for $f(z) = e^{i/z}$ around the point z_0 can be written as $f(z) = 1 + (2^1/1!)(1/z) + 2^2(1/2!z^2) + 2^3(1/3!z^3) + \cdots$ with $\text{Res}[f, z_0] = a_{-1} = 2$. The Cauchy method of residues is described by the algorithm: $\int_C f(z)dz = 2\pi i \sum_{k=1}^{n} \text{Res}[f, z_j]$, where the residue of the function f at any point z_j is $\text{Res}[f, z_j]$. This is also tied to the Cauchy integral $f(a) = 1/2\pi i \oint f(z)/(z-a)dz$ (see Figure C.25).

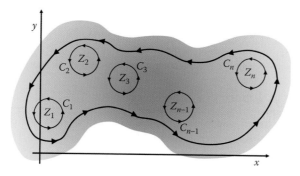

Figure C.25 The Cauchy residue method.

Cauchy number (C = ρv²/E)

[engineering, fluid dynamics] This number identifies regimes in FLUID FLOW WHERE compressibility is a factor of influence, named after a contributor in the field AUGUSTIN-LOUIS CAUCHY (1789–1857). This number describes the relative importance of inertial and elastic forces under compressible flow conditions; in fact the ratio of the inertial forces to the compressible forces. The Cauchy number is a function of the fluid DENSITY (ρ), as well as the flow VELOCITY (v) and the elastic modulus of the fluid medium (E). The Cauchy number is closely related to the MACH NUMBER and is often expressed as the Mach number to the second power.

Cauchy strain tensor

[fluid-dynamics, mechanics] $\varepsilon = \nabla \vec{v} + (\nabla \vec{v})^T \sim E$, where E is the LAGRANGIAN FINITE STRAIN TENSOR and $\vec{v}$ the deformation or FLOW velocity (solid respectively FLUID).

Cauchy–Poisson wave problem

[computational, fluid dynamics] Mathematical problem introduced in 1815 by AUGUSTIN-LOUIS CAUCHY (1789–1857) and SIMÉON DENIS POISSON (1781–1840) regarding the WAVE propagation on a (slightly compressible) LIQUID at the surface, with finite depth (h). The displacement follows from the local PRESSURE (P) and DENSITY (ρ) as $P/\rho = (\partial\phi/\partial t) - gz + F(t)$, under the condition that the Laplace derivative of the VELOCITY POTENTIAL (ϕ) equals zero $\nabla^2\phi = 0$, z the vertical displacement, x and y the propagation directions, g the GRAVITATIONAL ACCELERATION, and $F(t)$ the time-perturbed influence function (e.g., storm). As boundary conditions the displacement pressure will satisfy $\xi = 1/g[\partial\phi/\partial t]_{z=0}$, as well as $\partial\xi/\partial t = -[\partial\phi/\partial t]_{z=0}$. The SOLUTION can be written with the use of a Bessel function of zeroth order ($J_0(k\varpi)$, $\varpi = r\sin\theta$; the surface rotational displacement, wave with curve of radius r, respectively $z = -r\cos\theta$), using the wave number $k = 2\pi/\lambda$, where λ represents the wavelength, provided $\phi = gt/2\pi \int_{-\infty}^{\infty} \{k - (gt^2/3!)k^2 + [(gt^2)^2/5!]k^3 - \cdots\} e^{kz} J_0(k\varpi) dk \approx (gt/2\pi)\{[P_1(\cos\theta)/r^2] - (gt^2/3!)[2!P_2(\cos\theta)/r^3] + [(gt^2)^2/5!][3!P_3(\cos\theta)/r^4]\}$, after developing the pressure in a series, while the displacement satisfies $\xi = t/2\pi\rho\varpi^3 \{1 - [(1)^2(3)^2/5!][(gt^2)2/\varpi] + (1^2 3^2 5^2/9!)[(gt^2)^4/\varpi] - \cdots\}$, which for $(gt^2/2\varpi) \gg 1$ yields: $\xi = (gt^3/2^{7/2}\pi\rho\varpi^4)\sin(gt^2/4\varpi)$. Note that the WAVEFRONT can be described by $r = t\sqrt{(gb)}$.

Cauchy–Riemann equations

[computational, fluid dynamics] Based on the Cauchy integral $f(a) = (1/2\pi i)\oint f(z)/(z-a)dz$, and the fact that holomorphic functions are analytic it follows $f''(a) = n!/2\pi i \oint f(z)/(z-a)^{n+1}dz$. For an arbitrary disk ($\mathbb{D}$) within the closure ($\mathbb{C}$) of the integrand it can be shown for $\chi \in \mathbb{D} \subset \mathbb{C}$, for which $f(\chi) = 1/2\pi i \int_{\mathbb{D}} [f(z)/(z-\chi)]dz + 1/2\pi i \int \int_{\mathbb{D}} (\partial f(z)/\partial z')[dz \wedge dz'/(z-\chi)]$, where $\wedge$ is the "logical" "and" for the integration constraints to the complex function. The latter integral provides the tools to solve the Cauchy–Riemann equations with respect to two variables, $u(x, y)$ and $v(x, y)$, as expressed in the thesis by the German mathematician Georg Friedrich Bernhard Riemann (1826–1866) in 1851, based on the work by JEAN LE ROND D'ALEMBERT (1717–1783) published in 1752, LEONARD EULER (1707–1783) published in 1779, and with finishing touches by AUGUSTIN-LOUIS CAUCHY (1789–1857) in 1814. The Cauchy–Riemann equations are expressed as $= \partial v(x, y)/\partial y$ and $\partial u(x, y)/\partial y = -\partial v(x, y)/\partial x$, with respect to the complex number $z = x + iy$ as $f(x + iy) = u(x, y) + iv(x, y)$, a holomorphic function.

Cavendish, Henry, Lord (1731–1810)

[general] Scientist, chemist, physicist, and British nobleman of French origin, one of the wealthiest men of his days. Cavendish derived the universal gravitational constant in 1728 as $6.75 \times 10^{-11} \text{ Nm}^2/\text{kg}^2$, compared to current day accepted standard, $6.67259 \times 10^{-11} \text{ Nm}^2/\text{kg}^2$ (see Figure C.26).

Figure C.26   Lord Henry Cavendish (1731–1810). (Courtesy of George Wilson.)

Caveolae

[biomedical] Location of actively formed invagination in the CELL MEMBRANE to support the ENDOCYTOSIS processes. The process results in the formation of vacuole or vesicles used to transport solid or LIQUID material intended for consumption or processing.

Cavitation

[biomedical, fluid dynamics, general] Sudden decrease in volume (collapse) of EXPANSION (e.g., mechanical deformation or vapor BUBBLE; NUCLEATION) that exerts forces on the surrounding media and materials. Generally the formation of bubble progresses at a slower rate than the collapse, providing a unique mechanism of mechanical interaction. Cavitation may result from interaction of ULTRASOUND pressure wave with soft materials (i.e., initiated by mechanical forces), specifically biological media. Additional occurrences of cavitation can be described resulting from thermal ablation processes, specifically short duration events, such as pulsed laser ablation (i.e., initiated by THERMAL EXPANSION). The cavitation process is accompanied by shear waves that will obey the WAVE EQUATION with associated propagation of effects. Depending on the material properties the wave can be critically damped ranging to undamped. Cavitation can provide mechanical effect as well as chemical changes. The collapse of a spherical bubble can be associated with a KINETIC ENERGY (KE) that is a function of the rate of change over TIME (t) of the bubble radius ($r = r(t)$) expressed as $\dot{r} = dr(t)/dt$, with respect to the initial radius R_0 with associated internal pressure P_1, defined using the density $\rho = \rho(t)$ as $KE = 2\pi\rho r^3\dot{r}^2 = (4/3)\pi P_1\left(R_0^3 - r^3\right)$. The true nature of the bubble collapse is more intricate, specifically when involving ionized PLASMA created during an ablation process (e.g., laser vaporization) or when condensation occurs during the size reduction of the bubble. During condensation the density of the VAPOR bubble changes and hence other CONSERVATION LAWS need to be included in the theoretical evaluation process. The rate of change in bubble diameter can be described as a wave phenomenon. Assuming the special case of adiabatic expansion/collapse the rate of change is captured by a surface velocity equivalent (i.e., $v_s' = \sqrt{(P_1/\rho)}$; "constant" parameter with the dimensions of velocity) in second-order derivative ($\ddot{r} = d^2r/dt^2$): $r\ddot{r} + (3/2)\dot{r}^2 = v_s'^2\left(R_0/r\right)^{3\gamma_a}$, where γ_a represents the adiabatic factor (defined as the ratio of the SPECIFIC HEAT [$c_{sp} = Q/m\Delta T$, describing the HEAT (Q) required to change the TEMPERATURE (T) of a MASS (M) by 1 K] of the GAS in two phases of the RAREFACTION vs. expansion processes), which ties pressure and density together as $P/P_0 = \left(\rho/\rho_0\right)^{\gamma_a}$. The rate of heat exchange of the bubble to the surrounding LIQUID medium is primarily subject to the NUSSELT NUMBER (Nu) of the system. The FORCE (F) exerted by the cavitation process at any specific point in the collapse is directly proportional to the momentary change in kinetic energy ($W = \Delta KE$, i.e., the work) over the DISTANCE traveled by the surface of the bubble captured by incremental delta in radius (Δr), $W = F\Delta r$ (see Figure C.27).

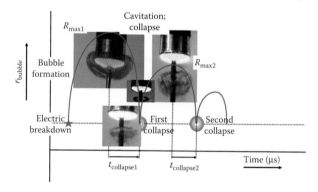

Figure C.27 Cavitation process for bubble generated by laser heating resulting from nanosecond laser pulse.

Cavitation number ($\sigma_c = 2\,(P - P_v)/\rho v_{flow}^2$)

[energy, fluid dynamics] Ratio of "static pressure" gradient in relation to KE density, where P is the local pressure, P_v the fluid VAPOR PRESSURE, v_{flow} the FLOW velocity (average), ρ the fluid density. In addition to PIPE flow, especially in a curved vessel, this will also be very useful in determining the impact of shape and angular velocity in PROPELLER action. Even though this may resemble the Euler number the functionality is entirely different.

CCD

[computational, electronics, quantum, solid-state] *See* CHARGE-COUPLED DEVICE.

CCD camera

[electronics, fluid dynamics, optics, quantum, solid-state] Charge-coupled device semiconductor structure from which the top layer has been removed which makes the charge layer light sensitive and can be used in array form to capture images that are limited in RESOLUTION by the number of CCD unites in the array or matrix format (see Figure C.28).

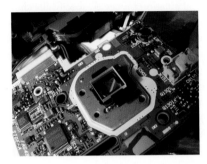

Figure C.28 Picture of digital camera using CCD element for registration of light spectrum and radiance.

CCT

[general, optics] Correlated COLOR temperature. The COLOR appearance of LIGHT emitted by a source, primarily a lamp. The color of source light is related to the EMISSION from a reference source heated to a particular TEMPERATURE, measured in KELVIN (K). The correlated color temperature classification for a LIGHT SOURCE, however, does not give information on its specific SPECTRAL DISTRIBUTION, more so on the appearance of the color of objects illuminated by the source. Two lamps may visually appear to be the same color, while the respective visual experience of object colors can be significantly different (*also see* VISION) (*see* Figure C.29).

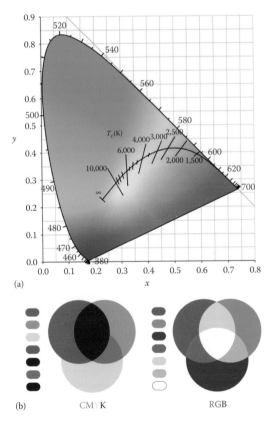

Figure C.29 Correlated color temperature map: (a) CIE 1931 x, y chromaticity space. (b) Color chart outlining the difference between the CMYK standard (cyan–magenta–yellow–black), for instance, used in dyes with a subtractive character, and the RGB (red–green–blue) standard.

Celestial bodies

[astronomy/astrophysics, general, mechanics] Any object in galactic space, ranging from planets (e.g., EARTH) to galactic dust and stars. The escape velocity for an object from any celestial body with mass M is the minimum speed (v_e), in perpendicular direction, required to overcome the gravitational attraction of the body with radius R and gravitational constant G and can be defined as $v_e = \sqrt{(2GM/R)}$. Note that the release from higher altitude (e.g., from a plane at cruising altitude) changes the radius (r) factor to the location specifics, which becomes $\sqrt{2}$ times the orbital equilibrium VELOCITY ($v_{\tan}$): $v_{\tan} = \sqrt{(GM/r)}$ for stationary orbit, where centripetal (tangential) force balances the gravitationally confined orbital force (see Figure C.30).

Figure C.30 A full moon (celestial body) in the eastern sky.

Cell

[biomedical, chemical, energy, solid-state] There are several types of cells to be defined: primarily we are familiar with the biological composition of single- and multicell organisms and secondarily the charge pump cell in a BATTERY needs to be considered. The biological cell has a lipid–protein dynamic bilayer forming the CELL MEMBRANE that encloses the intercellular FLUID, which may contain the NUCLEUS with genetic information and physically adjusts to external SHEAR FORCE and strain, reinforcing weak spots as well as providing a mechanism to actively move the cell by means of changing the configuration of the enclosure. Cells can grow and form semipermanent adhesions to other structures and other cells. The ameba is an example of a single-cell organism, whereas the species *Homo sapiens* has a multitude of specialized cells that form organs, supporting structures, muscles as well as a highly configurable communications network. For a battery the electrical potential that can be achieved is a function of the materials used, the HALF-CELL potential. In biology the half-cell potential is derived from the NERNST POTENTIAL equation for the specific ions in place at either side of the cell membrane: $\varepsilon_{ion} = (RT/FZ_{ion})\ln([Ion]_{out}/[Ion]_{in})$, where $R = 8.3144621$ J/Kmol is the GAS constant, $F = 9.64853399 \times 10^4$ C/mol is the FARADAY CONSTANT, Z_{ion} the ionic charge, $[Ion]_i$ represent the concentration of a specific ION, respectively, on the outside or inside of the cell membrane and T the ABSOLUTE TEMPERATURE. At the body temperature of 37°C, the equation yields $\varepsilon_{ion} = 61.5 \times \ln([Ion]_{out}/[Ion]_{in})$, which provides for chlorine with respective intracellular concentration $[Cl^-]_{in} = 9.0$ mM, and extracellular concentration $[Cl^-]_{out} = 125.0$ mM; $\varepsilon_{Cl^-} = -70$ mV, which can be experimentally verified. The galvanic half-cell potential is $\varepsilon_{ion} = RT/F \ln(a_{ion}/\sqrt{(P_{ion}/P^0)})$, where $a_{ion} = \exp[(\mu_i - \mu_i^{\ominus})/RT]$ is the "ACTIVITY" of the ion SOLUTION with the chemical potential $\mu_i = (\partial G/\partial N_i)_{T,P,N_{j \neq i}}$ (also known as the partial molar ENERGY) where G represents the Gibbs free energy and N_i the particulates for the respective constituents under the operational conditions (concentration $[ion]$, pressure $[P]$, and temperature $[T]$) and with respect to a chosen standard state $\mu_i^{\ominus}$, representing the partial solution as if it responds as an ideal chemical solution, P_{ion} is the partial pressure of the ionic solution respectively gas, expressed in Pascal, and P^0 is the standard pressure 10^5 Pa. The half-cell potential can also be defined in terms of acidity (pH) of the solution as $\varepsilon_{ion} = -(\Delta G^{\ominus}/nF) - (0.05916/n)\log\left(\{a_A\}^a \{a_B\}^b / \{a_C\}^c \{a_D\}^d\right) - (0.05916h/n)\text{pH}$, where the number of released electrons is described by n as in the chemical process: $aA + bB + n[e^-] + h[\text{H}^+] \rightleftarrows cC + dD$, describing the range of chemical constituents and their respective quantities in the reaction (see Figure C.31).

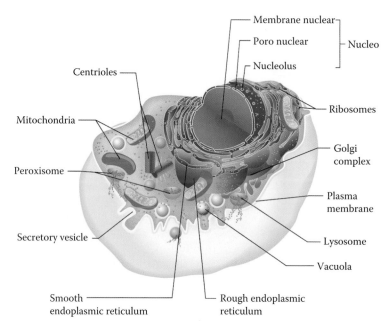

Figure C.31 Biological cell.

Cell, electrically responsive

[biomedical, electronics] In biological media there are certain cellular configurations that allow the individual cells to respond to electrical influences that are extremely small in MAGNITUDE. One specific example is the sensory organ in the head of certain sharks that can detect electric field strengths as small as $10^{-6} N/C$. In comparison the background RADIATION in outer space is $\sim 3 \times 10^{-6} N/C$, the electric wiring in the average household produces field strengths in the order of $10^{-2} N/C$, and LIGHTNING produces approximately $10^4 N/C$. The sensory PERCEPTION of the SHARK provides capabilities to detect MUSCLE contractions from several kilometers DISTANCE.

Cell composition and response to radiation

[biomedical, nuclear] Cells respond to RADIATION depending on the type of radiation (α [Helium ion]; β [electron]; γ [EM radiation]; etc.) and dose. Additionally, radiation exposure is cumulative, leading to a slow build-up before catastrophic response is achieved. Damage can be affected on the CELL MEMBRANE and on the DNA, as well as total cell death. Damage may be reversible or irreversible. Although the initial deposition of the ENERGY may be very short (order of 10^{-17} s), the large-scale implications are directly proportional to the ENERGY of the applied radiation and the mechanism of action: PARTICLE or ELECTRO-MAGNETIC radiation. The biological change in specific cells as a result of radiation may occur after a latent period. The duration of the delay in time depends on initial and cumulative dose and may vary, depending on the type (i.e., energy) of the radiation ranging from minutes to years. One specific radiation response may be an altered communication behavior, changes in the chemical responses.

Cell junction

[biomedical, chemical, fluid dynamics, mechanics] CELL to cell interaction (e.g., communication) is mediated by cell junctions that provide a mechanism for protein binding (e.g., ligands) and define the response to deformation. Cell junctions are regions in the CELL MEMBRANE of a biological medium that are enriched with specific proteins that provide a means of interaction (signal exchange, electrical or chemical, and potentially optical) between two adjacent cells. They also provide a mechanism to sustain SHEAR STRESS (σ_s), which is of particular importance in the GASTROINTESTINAL TRACT (GI tract) and BLOOD vessels. The shear stress is a function of the VISCOSITY (η) of the FLUID in MOTION, the FLOW (Q), and the radius (r), $\sigma_s = 4\eta Q/\pi r^3$. The RESISTANCE to shear stress characterizes the specific cell junctions, in particular for the lumen in question. Three types of cell junctions can be identified: (1) desmosomes, (2) tight junctions, and (3) gap junctions. Desmosomes (zonulae adherens) are located in tissues that are under the influence of shear stress, for instance the SKIN and GI tract. The desmosomes are proteins that form a strong intercellular bond as well as a strong bond to the basal LAMINA (extracellular matrix [ECM] layer that is formed by secretion from the epithelial cells). Desmosomes provide a regenerative/reconstructive mechanism to the cell membrane under deformation, in particular with respect to the organization of the cytoskeleton. Tight junctions on the other hand are found in tissue providing a specific exchange function, such as the KIDNEY (nephron cells), the endothelial cells in the blood vessels (nutrients, oxygen, waste exchange) as well as the nutrient resorption by the enterocytes in the GI tract (intestines). Tight junctions have specific functions to prevent the movement of proteins in the cellular membrane, in particular those that serve an apical function (chemical sorting function, for instance found in the Golgi complex [also known as Golgi apparatus]; named after the Italian scientist and physician Camillo Golgi [1843–1926]). Gap junctions are found for instance in the ORGAN OF CORTI in the cochlea of the EAR, supporting the SIGNAL transduction in response to deformation by movement of the hair cell. Gap junctions are composed of six connexin proteins (Cx) that form hemichannels, also known as connexons. The connexons will align for proper sharing of cytoplasm between neighboring cells. The sharing mechanism is based on small molecules (<1000 Da) and ions providing both electrical and chemical signal transduction through "messengers." Generally, the GAP JUNCTION transfer of molecules and ions does not require ENERGY; however, the HEARING effort is certainly facilitated by the ATPase activity and recycling of K^+ ions. The gap junction provides a selection mechanism for which ions may pass by means of POLARIZATION of the connexons and the molecular orientation. In addition to the interaction with adjacent cells by means of the cell junction, the cell junctions also support communication with ECM. ECM is one of the key components for multicellular organisms and is composed of a collection of extracellular molecules that is purposely secreted by cells and provides both biochemical and structural support. ECM is composed on any or multiple of the following constituents: collagens, fibronectin, LAMININ, as well as vimentin and vitronectin. Cells are attached to the ECM by means of INTEGRINS (transmembrane receptor molecule acting as a bridge between cells and between cells and the ECM). The integrins provide a chemical response to shear stress (deformation), which is specific for the transmembrane receptor kind. Finally, the cell junction provides interaction with the cellular scaffolding, also known as cytoskeleton, which is an integral part of the cellular structure and support the intracellular transport as well as support cell division. The cellular scaffolding comprise three kinds of filaments: microfilaments, intermediate filaments (IF), and microtubules. The microfilaments form a support mechanism by integrating with the double-helix ACTIN chains in the vicinity of the cellular membrane. The microtubules provide the transportation support. All components of the cellular scaffolding support the shape and in particular the resistance to deformation and tension. The MEMBRANE resilience to deformation can be defined by the membrane stiffness (Ei) and the deformation length L as $F = 3Ei/L^2$. The function of the cytoskeleton in this regard is the restructuring of the cell membrane to compensate for tension by means of actin (F-actin, a component of the cytoskeleton), IFs, and microtubule filaments. Specifically, the IFs provide a membrane "motor function," providing molecular realignment and thickening of the cell membrane in compensation for the applied deformation. The adjustments are made in steps in the order of 8 nm with local force applied by the cytoskeleton in response to the cell junction efforts in the order of 6 pN.

Cell membrane

[biomedical] Lipid–protein bilayer that encloses the cellular PLASMA and in most cases a NUCLEUS containing the biological DNA. The MEMBRANE is a living structure that can adjust to chemical challenges as well as mechanical influences resulting from both internal and external forces. The CELL MEMBRANE accommodates chemical and particulate transport through ENDOCYTOSIS next to facilitated gated transport through pores. Mechanically, the cell can modify its cell membrane to become resilient to local stress and strain resulting from external influences on both MICROSCOPIC and MACROSCOPIC scale. The macromolecular structure provides an elastic configuration that supports an active lifestyle. External forces can be found resulting from high-frequency vibrations, as well as pressure gradients resulting from local osmotic pressures as well as gravitational influences. Additional external forces may result from FLUID FLOW resulting in shear stress. Internal forces include the chemical configuration making up ACTIN and MYOSIN to form cytoskeletal cables that orchestrate cellular movements as well as morphological changes. Other external influences include hydrostatic pressure and ADHESION from neighboring cell, next to growth. A free (nonenclosed) body of LIQUID is subject to LAPLACE LAW, whereas the cell membrane forms an equilibrium by means of both stretch and modification. The TENSILE FORCE (F_t) on the membrane for this situation is expressed as $F_t = K_a \left(\Delta A / A_0 \right)$, where ΔA is the increase in surface area for the bilayer with respect to the original area A_0 and K_a is the area EXPANSION constant (range: 10^{-1}–10^1 N/m), where the tension 3×10^{-3} N/m $< F_t < 3 \times 10^{-2}$ N/m. With surface area expansion the membrane thickness (d) will change proportionally, $\Delta A / A_0 = \Delta d / d_0$, where the response to shear stress of the membrane describes it as an elastic solid (see Figure C.32).

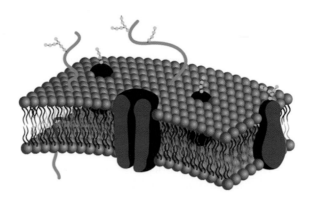

Figure C.32 Artist impression of the design of a biological cell membrane.

Cellophane

[biomedical] Regenerated CELLULOSE ACETATE, used as a filtration mechanism in the early days of dialysis MACHINES (~1930s), following the recurring infection rate when using peritoneal membranes (obtained from the abdominal lining).

Cellular automata

[computational] System definitions with an abstract foundation that are used in discrete mathematics problem solving. These concepts are equivalent to the field concept in field equations in general PHYSICS. Another analogy may be drawn with respect to the wavelet concept in SIGNAL processing, which is both place and time confined. Under this computational approach, the space is divided into a uniform matrix format that contains array parameters which are called the "cells." Each CELL is associated with a state variable defined in digital format. For each cell, time progresses in incremental steps, not continuous, and frequently relies on a "look-up" table as the provider for the definition of the state. The fact that the system can be solved for a finite number of steps in time with respect to a finite segment of the system lends this mechanism ideal for parallel computing, and for that MATTER for cluster computing, since all system segments are defined to be noninteractive, each with the same local and uniform sets of rules.

Cellular signaling

[biophysics, chemical, energy] Biological cells communicate with neighboring cells as well as with the autonomic and PARASYMPATHETIC NERVE SYSTEM to regulate their own metabolic and reproductive ACTIVITY as well as balance with the organ or tissue they are an integral part of. In addition, the cells regulate their natural chemical and energetic balance by secretion of chemical signals as well as electronic charges which interact with its cellular MEMBRANE and the extracellular conditions. The extracellular conditions can involve the secretion of hormones or the opening or closing of ports on the cell's own or neighboring cell's membranes (PHAGOCYTOSIS, PINOCYTOSIS, active processes as well as the ACTION POTENTIAL specifically). Some of the signaling processes involve soluble factors, primarily proteins, other are ionic in nature. Some of the signaling processes are identified by their specific mechanism as follows: AUTOCRINE, HORMONAL, PARACRINE, and SYNAPTIC SIGNALING.

Cellulose

[biomedical, chemical] Artificially constructed as well as naturally occurring polymer. POLYMER materials that are degradable and biocompatible, made from β-D-glucose (6-carbon sugar) monomers, that is, polysaccharides, are called cellulose. The long molecular configuration provides cellulose with its inherent strength. Cellulose is the main component forming the cellular matrix of plants CELL walls. Cotton is a more specific example of naturally occurring material composed of cellulose. This material was used in artificial lung devices developed in the 1950s and 1960s to generate a SEMIPERMEABLE MEMBRANE from the transport of oxygen to BLOOD in an extracorporal system. Additional applications are found in drug delivery, sutures, tissue ENGINEERING, and bandages and wound dressings. Cellulose can be reconstituted with chemical treatment to break lignin bonds and reformed with the aid of alkali media or bisulfites, followed by cross-linking the cellulose fibers to form paper (see Figure C.33).

Figure C.33 Raw cotton with greater than 90% cellulose content.

Cellulose acetate

[biomedical, chemical] Acetate ester formed from CELLULOSE, which can be reconstituted to form CELLOPHANE. Cellulose acetate was first produced in 1865. Cellulose was made to react with acetic anhydride to form cellulose acetate (one of the foundation materials used in the production of photographic emulsions). Several applications are available under commercial names. Other examples of applications of cellulose acetate are in textiles (often in blends with other materials, such as cotton and NYLON), cigarette filters, and frames for EYE glasses. Cellulose acetate can be spun into strands for weaving, for instance, acetate rayon.

Cellulose nitrate

[biomedical, chemical] Chemical result of the treatment of CELLULOSE with sulfuric ACID and nitric acid. Chemical component used in the construction of extremely thin microporous membranes, for instance used to encapsulate artificial biologically active components such as sensing chemical ELEMENT transducer lining with the chemical ligands separated from the TRANSDUCER. Other applications are in paint (lacquer) and explosives. Cellulose nitrate can be used to produce TRANSPARENT film sheets and light-sensitive "transparent" sheets and roles used in cameras. With AGE, chemicals degrade, culminating in the release of acidic by-products such as nitric oxide, makes the cellulose nitrate unstable and highly flammable, it will even burn under water. It is also known as NITROCELLULOSE, guncotton, flash cotton, or flash paper.

Celsius, Anders (1701–1744)

[general, thermodynamics] Astronomer from Sweden, generally known for his introduction of the CELSIUS SCALE used in temperature MEASUREMENT. Anders Celsius followed the inspiration of CARLOS RENALDINI [RINALDINI] (1615–1698) who performed his temperature measurements at least 50 years prior. Another contemporary metrologist was DANIEL GABRIEL FAHRENHEIT (1686–1736), competing in the English culture with the Fahrenheit scale (see Figure C.34).

Figure C.34 Painting of Anders Celsius (1701–1744).

Celsius scale

[general, thermodynamics] Generally accepted measure for temperature (T) in household use, established by ANDERS CELSIUS (1701–1744) as the total range of observable expressions of average molecular KE defined as temperature ($T = (2/3k)\bar{E}$, where $k_b = 1.3806488 \times 10^{-23}\,\text{m}^2\text{kg/s}^2\text{K}$ is the Boltzmann constant) in relative format as introduced in 1741. Sometimes one may still find Celsius defined as CENTIGRADE. The principle of measuring temperature by means of liquid EXPANSION can be attributed to the Italian scientist Ferdinand II de' Medici, the grand duke of Tuscany (1610–1670), dating back to approximately 1654.

The temperature scale generally uses three points as reference to define the temperature. The Celsius scale uses the boiling point of water at 1 atmosphere pressure ($T_{boil} \equiv 100°C$); the PHASE TRANSITION point from solid to LIQUID for H_2O recognized as the MELTING POINT, also referenced as the FREEZING point ($T_{melt} \equiv 0°C$); and additionally the triplet point for water is expressed with respect to the Kelvin temperature scale as $T_{Triplet} \equiv 273.16\,K$. The Celsius scale is the range of temperatures from melting point to boiling point subdivided in 100 incremental steps. Other temperature scales are the Fahrenheit scale introduced by GABRIEL DANIEL FAHRENHEIT (1686–1736) in 1717 (still used in the United States) as well as the absolute Kelvin scale, next to scales from past exercises; Réaumur (defined by the French scientist RENÉ-ANTOINE FERCHAULT DE RÉAUMUR [1683–1757]) and RANKINE (defined by the Scottish scientist WILLIAM JOHN MACQUORN RANKINE [1820–1872]). The THERMOMETER used to express the Celsius scale uses either ALCOHOL or mercury in a closed tube, relying on linear THERMAL EXPANSION ($\Delta\ell = \alpha(T_2 - T_1)$, where α is the linear expansion coefficient and ℓ the length of the medium) of the medium to indicate a MAGNITUDE when lining up with a calibrated scale according to the Celsius definitions. The first thermometer invented is attributed to GALILEO GALILEI (1564–1642) dated circa 1595, using GAS expansion to raise or lower the hydrostatic pressure in a tube; hence raising or lowering the liquid level of the open container medium connected to the gas-filled tube. Other thermometer mechanisms from early stages of thermodynamics work are based on gas or liquid expansion principles. Later mechanisms of action used for the determination of temperature use PLANCK'S LAW and rely on the black-body RADIATION captured by optical spectroscopic techniques. ABSOLUTE TEMPERATURE SCALES are expressed in Kelvin or Rankin, both referencing the point of lowest temperature with no MOLECULAR MOTION as the zero point. A useful conversion between degree Fahrenheit (T_F) and degree Celsius (T_C) is $T_C = (5/9)(T_F - 32)$.

Celsus (second century AD)

[biomedical] Egyptian scientist and medic. Author of an encyclopedia of MEDICINE called the "DE MEDICINA."

Center of gravity

[biomedical, mechanics] The virtual point in the volume of a body where for convenience the forces are apparently acting upon. A body in equilibrium requires that the sum of the forces through the center of gravity align with the normal force. In case the normal force and sum of acting forces do not line up, the system forms a torque that will result in rotation. The center of gravity may be defined for a specific subsection of a total system, for instance the foot of a human, in reference to the entire HUMAN BODY.

Center of mass (r_{CM})

[general, mechanics, nuclear] The point inside the outline of a body of a system where all mass can be considered to be concentrated. For instance, for an air-filled rubber ball all the mass is in the rubber shell whereas the center of mass is in the center of the ball. The location of the perceived single mass in a reference frame of choice as summarized over all the masses ($\sum_i m_i = M$) (each respective mass [m_i] constituent identified by i) as a function of each respective location (r_i) contributed from a system of particles (regardless of size and structure; including atoms and molecules) or concurrently the location of the combined effect of all the masses of the constituents of a device with various components and materials. The definition of the center of mass is formulated as $r_{CM} = \sum_i r_i m_i / \sum_i m_i$. The center-of-mass will directly influence the determination of the moment of INERTIA for an object. The center of mass will also be the point where all forces (F_i) are considered to be acting on the object providing the acceleration to the center of mass ($\overrightarrow{a_{cm}}$) in NEWTON'S SECOND LAW, $\sum \vec{F} = m\overrightarrow{a_{cm}}$. The location of the center of mass and dimensions of a body directly determine the moment of inertia for that body, specifically if the body is nonuniform and has various geometric extensions of certain composition and materials that hold mass. The location (R) of the center of mass in a coordinate system is defined by the components of an object, each with respective mass (m_i) and respective DISTANCE to an origin (r_i) in a coordinate system, defined by the total mass ($M = \sum_i m_i$) as $R = \sum_i m_i r_i / M$. The relative position (r^*) of each of the components (i) can now be defined for a specific subsection (k) as

$r^*_k = r_k - R$; similarly the velocity is defined by $\vec{v}^*_k = \vec{v}_k - \vec{V}$, where V is the velocity vector for the system. Under relativistic conditions (approaching the speed of light [c], which applies to nuclear conditions), the velocity needs to be corrected by means of the LORENTZ TRANSFORMATION, which translates in the momentum ($p = mv$) formulation as follows: $\vec{p}^*_k = \vec{p}_k - [(\gamma_{cm} e_k/c) - \gamma_{cm}^2/(\gamma_{cm} + 1)(\vec{\beta} \cdot \vec{p}_k)]\vec{\beta}$, where the quantum ENERGY is $e_k = m_k c^2/\sqrt{(1 - \{\vec{v}_k/c\}^2)}$ and "inertial energy" $\vec{\beta} = \sum_i \vec{p}_k c/e_i$ and $\gamma_{cm} = \sum_i e_i/E^*$. In this the QUANTUM energy is defined as $E^* = \sqrt{[(\sum_i e_i)^2 - (\sum_i \vec{p}_i)^2 c^2]}$. In the Lorentz transformation the momentum components that are orthogonal to the parameter $\vec{\beta}$ are conserved, while the parallel components are modified. This last description can be applied under various conditions, including COMPTON SCATTERING (see Figure C.35).

Figure C.35 A high jumper is manipulating her center of mass to enhance the high jump level when performing a pole vault jump during the different vaulting phases.

Centered difference

[computational, fluid dynamics] Computational technique used to isolate extremes by using the half-intervals ($\Delta h/2$) in the first derivatives of a function $f(f'(x_i) = [f(x_i + (\Delta h/2)) - f(x_i - (\Delta h/2))/\Delta h] + O(\Delta h))$ when derived over incremental step size Δh, the error is expressed by $O(\{\Delta h\}^2)$. The slope connecting these two points and respective accuracy can be expressed in (and is directly dependent on) the increment step size. Increasing this into a Taylor series the derivative has a second component: $f'(x_i) = -f(x_i + (2\Delta h/2)) + 8f(x_i + (\Delta h/2)) [-8f(x_i - (\Delta h/2)) + f(x_i - (2\Delta h/2))/12(\Delta h/2)] + O(\{\Delta h\}^2)$ the second-order difference now becomes $f''(x_i) = [f(x_i + (\Delta h/2)) + f(x_i) - f(x_i - (\Delta h/2))/\{(\Delta h/2)\}^2] + O(\Delta h)$ and $f''(x_i) = [-f(x_i + (2\Delta h/2)) + 168f(x_i + (\Delta h/2)) - 30f(x_i) - 16f(x_i - (\Delta h/2)) + f(x_i - (2\Delta h/2))/12(\Delta h/2)] + O(\{\Delta h\}^2)$, which uses the centered differences around x_i. This SOLUTION method is frequently used in computational FLUID DYNAMICS, primarily since the NUMERICAL solution needs to be solved in a finite ELEMENT configuration due to the complexity of the local boundary conditions.

Center-seeking forces

[general, mechanics] Any object with mass (m) moving in a curved trajectory (e.g., circular, elliptic motion) with certain tangential VELOCITY (v) will be forced to follow the curved path with radius of curvature (r) under the influence of a CENTRIPETAL FORCE. The centripetal force (F_c) is directed toward the center of the curvature with MAGNITUDE $F_c = ma_c = mv^2/r$; this force may result from FRICTION (tires on the pavement), normal force (barrel racing), or TENSION (e.g., rope) (see Figure C.36).

Figure C.36 Merry-go-round with center-seeking force.

Centigrade

[thermodynamics] *See* CELSIUS.

Centimeter

[general] Metric system. Fractional unit of DISTANCE/length, 1/100 of the SI unit meter.

Centripetal force

[general, mechanics] Circular MOTION obtained by balancing forces, determining the MAGNITUDE of the CENTRIPETAL FORCE maintaining the object in revolution, $F_C = mv^2/R$, where m is the mass of the PROJECTILE, v the velocity of motion, and R designates the radius of curvature of the circular motion. Force is required to keep a moving mass travel in a circular path and should be directed toward the axis of the circular path.

Cepheid variables

[astrophysics, energy, mechanics] Named after the prototype Delta Cepheid. Indication of the mechanical properties of stars, specifically the oscillatory behavior.

CERN synchrotron

[atomic, mechanics, nuclear, quantum] High-energy PARTICLE ACCELERATOR located in Switzerland, at the French border, near the city of Geneva. The particle accelerator is formed in a large circular loop with radius R. An applied voltage alternates with a single period determined by the condition $T = 2\pi m/qB$. The particle accelerator uses a MAGNETIC FIELD (B) to generate a force $\left(F = qvB = m\left(v^2/R\right)\right)$ on a charged particle with charge q and mass m to induce acceleration resulting in a maximum velocity $v_{max} = qBR/m$, with KE upon exit from the CYCLOTRON $(1/2)\,mv^2 = q^2B^2R^2/2m$ (see Figure C.37).

(e)

Figure C.37 CERN ("Conseil Européen pour la Recherche Nucléaire," or European Council for Nuclear Research) facilities and operations in Meyrin, Switzerland: (a) outline of the 27 km diameter facilities of the European Organization for Nuclear Studies with beam guide (Large Hadron Collider: LHC) as insert. (Courtesy of CERN, Geneva, Switzerland.) (b) Compact Muon Solenoid (CMS) inner tracker barrel consisting of three layers of silicon modules placed at the center of the CMS (LHC) experiment, guiding 14 TeV proton–proton collisions, the silicon that is used will be able to withstand the powerful magnetic field and high doses of radiation without damage. (Courtesy of Maximilien Brice, CERN, Geneva, Switzerland.) (c) Calorimeter used measure the energy of particles that are produced during collision of protons, close to the axis of the beam, chilled inside a cryostat to provide optimal operational conditions for the detector. This is part of the LHC used to verify the Higgs boson and provide an insight into the origin of dark matter. (Courtesy of Claudia Marcelloni, CERN, Geneva, Switzerland.) (d) Simulation of particle trajectories. (Courtesy of CERN, Geneva, Switzerland.) (e) The World Wide Web was conceived at CERN by the scientist Sir Timothy John "Tim" Berners-Lee (1955–) from Great Britain in 1989, the use of the Internet made the transport of large amounts of data easy and convenient. (Courtesy of Rory Cellan-Jones, BBC, London, UK.)

Chadwick, James (1891–1974)

[atomic, general, nuclear] Physicist from Great Britain. In 1914 Chadwick described the release of electrons with different energies, introducing the concept of RECOIL ENERGY to the atomic and nuclear interaction from PROJECTILE PARTICLE interaction. Chadwick worked with Rutherford on the definition of the nuclear composition and in 1932 described the release of the neutron (n) resulting from the bombardment of beryllium (Be) with ALPHA PARTICLES (α), $_2^4\alpha + {_4^9}\text{Be} \rightarrow {_6^{12}}\text{C} + {_0^1}n$, also producing CARBON (see Figure C.38).

Figure C.38 James Chadwick (1891–1974).

Chain reaction

[chemical, general] Any chemical or nuclear process in which some of the products of the process are instrumental in the continuation or magnification of the process.

Chandrasekhar, Subrahmanyan (1910–1995)

[astronomy/astrophysics, biomedical, energy, general, optics] Indian mathematician who theoretically described the RADIATIVE TRANSFER of ENERGY within stellar ATMOSPHERES. The radiative transfer theory was toward the end of the twentieth century successfully adapted for biomedical purposes in light tissue interaction—EQUATION OF RADIATIVE TRANSFER (see Figure C.39).

Figure C.39 Subrahmanyan Chandrasekhar (1910–1995).

Chandrasekhar mass

[astronomy/astrophysics, energy, quantum] The gravitational mass of a STAR where GRAVITY will overcome the Fermi ENERGY and collapses, $M \sim 1.44 M_{\odot}$, where $M_{\odot}$ represents the solar mass. Stellar mass less than the Chandrasekhar mass will not possess enough energy to produce neutrons, thus preventing the formation of a NEUTRON STAR by means of the constraints to the electron DEGENERACY. At a mass less than the Chandrasekhar mass the star will only be able to produce a WHITE DWARF. This also defines the theoretical upper limit for the mass of a stable white dwarf. The electron density (N_e) for the star defines the Fermi energy of that star, $E_F = (5/3)(1/N_e)C\left(N_e^{5/3}/R\right)$, where R defines the equilibrium radius for the star and C is a constant for the star (based on its history and size).

Characteristic X-ray

[atomic, nuclear] *See* **X-RAY, CHARACTERISTICS**.

Charge (q)

[electronics, general] Excess or shortage of electrons in an atomic or molecular system that produces either negative or POSITIVE CHARGE. On a nuclear level the excess protons can provide a positive charge. The concept of two different charges was officially introduced by BENJAMIN FRANKLIN (1706/1705?–1790) in 1747; however, the concept of charge was known and has been described as far back as ancient Greek philosophers and old Chinese documentation dating back well before Christ on the Christian/western calendar. It is also referred to as ELECTRIC CHARGE.

Charge conjugation

[computational, relativistic, thermodynamics] Under a LORENTZ TRANSFORMATION all particles are transformed into their respective antiparticles. The operator is discontinuous and applies only to RELATIVISTIC–QUANTUM–MECHANICAL situations. The charge conjugation follows the pattern outlined for space INVERSION and time reversal.

Charge density wave

[atomic, solid-state, thermodynamics] Periodic charge density fluctuations in a low-dimensional METAL LATTICE structure. Because of an interaction between electron density in free FLOW and phonon propagation an electron density WAVE pattern is generated with a periodic charge density modulation that has a rhythmic intermittent lattice distortion superimposed, causing a charge density wave pattern. Each perturbation is associated with the Fermi wave vector. The phonon spectral pattern that results is referred to as the KOHN ANOMALY. The periodic PHASE transitions induced by the electron charge density fluctuations generate a wave pattern with a wavelength $\lambda_{\text{charge}} = \pi/k_F$, where k_F is the FERMI WAVE VECTOR. The wavelength may be commensurate with the LATTICE CONSTANT, however not so when the lattice has only partially filled electron bands. When the lattice density perturbations are disorganized, the PARTICLE gap occurring at the FERMI LEVEL will create an INSULATOR.

Charge per unit area (σ_e)

[general] Total charge that is distributed evenly on a "two-dimensional" object per unit area.

Charge per unit length (λ_e)

[general] Total charge that is distributed evenly on a "one-dimensional" object per unit length.

Charge-coupled device (CCD)

[computational, electronics, quantum, solid-state] Capacitive semiconductor material structure that can be used to temporarily store ELECTRIC CHARGE. The storage time can be controlled by external circuitry and can be used to form a delay line (time delay) and concurrently acts as a "bucket memory," temporarily storing the electronic information and transferring it over to the next CCD in the circuit. The stored information is primarily binary, but there are also analog applications.

Charles, Jacques Alexandre César (1746–1823)

[general] Scientist and physicist from France. Charles' work was in the behavior of ideal gasses with pre–IDEAL GAS LAW definitions. Cotemporary of JOSEPH LOUIS GAY-LUSSAC (1778–1850) (see Figure C.40).

Figure C.40 Jacques Alexandre César Charles (1746–1823). (Courtesy of the United States Library of Congress, Washington, DC.)

Charles's law

[atomic, nuclear] GAS LAW introduced by JACQUES ALEXANDRE CÉSAR CHARLES (1746–1823) with respect to the VOLUME (V) and temperature (T) of an IDEAL GAS expressed as V/T = constant for constant pressure and number of moles. It is also known as the GAY-LUSSAC LAW.

C

Charm

[general] Classification of ELEMENTARY PARTICLES, specifically quarks. In reference, an electron can be identified by spin-up or spin-down. In the construction of rudimentary particles, mesons consist of a QUARK and an antiquark, whereas a BARYON will be composed of three quarks. With the broad variety of baryons and mesons the type of quark defines the final product. Next to the fact that various quarks have fractional charge the charm provides another identification. The charm can be "up (u)" with charge $u :+(2/3)e$ (rest ENERGY 360 MeV), electron charge MAGNITUDE: $|e| = 1.60217657 \times 10^{-19} C$, "down ($d$)" with charge $d :-(1/3)e$ (rest energy 360 MeV), "charmed (c)" with charge $c :+(2/3)e$ (rest energy 1500 MeV), "strange (s)" with charge $s :-(1/3)e$ (rest energy 540 MeV), "top (t)" with charge $t :+(2/3)e$ (rest energy 173,000 MeV), and "bottom (b) with charge $b :+(2/3)e$ (rest energy 5000 MeV). The respective antiquarks: $\bar{u}, \bar{d}, \bar{c}, \bar{s}, \bar{t}, \bar{b}$, have opposing charge to the "regular" quark and identical rest energy. The quark charm principle was introduced by MURRAY GELL-MANN (1929–), for which he received the Nobel Prize in Physics, and GEORGE ZWEIG (1937–) in 1963 (*see* QUARK) (see Figure C.41).

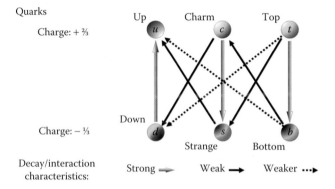

Figure C.41 Concept of "charm" pertaining to elementary particles, quarks in particular.

Châtelier–Braun principle

[chemical, thermodynamics] The principles of the changes to the chemical equilibrium introduced by the French chemist HENRY LOUIS LE CHÂTELIER (1850–1936) and independently by the German physicist and inventor Karl Ferdinand Braun (1850–1918). A system defined by the number of moles of the constituents of the medium, the pressure, and volume and temperature when subjected to change will affect the other parameters as well as the chemical balance of the constituents. In this situation an EXOTHERMIC REACTION will be forced to reverse its reaction when the system temperature is artificially increased, hence reducing the reaction constant. Alternatively, a temperature increase for an endothermic reaction will be encouraged to proceed in the direction that requires ENERGY, thus increasing the equilibrium reaction constant. Similarly changes in pressure or volume will force the chemical reaction in the direction that complies with the changing boundary conditions. This principle is also known as Le Châtelier principle, or as LE CHÂTELIER–BRAUN PRINCIPLE. Also referred to as CHÂTELIER–BRAUN INEQUALITIES, or Le Châtelier theorem.

Chebyshev, Pafnuty Lvovich [Пафну́тий Льво́вич Чебышёв] (1821–1894)

[computational, fluid dynamics] Mathematician from Russia. Chebyshev described several stochastic principles and introduced the basic concepts of number theory providing basic mathematical foundations, as well as PROBABILITY theory fundamentals.

Chebyshev inequality

[computational] In a PROBABILITY distribution mostly all values of a series of data are near the average value. Introduced by PAFNUTY CHEBYSHEV (1821–1894).

Chemical compound

[chemical] Pure substance composed of two or more ELEMENTS combined in a fixed and definite proportion by weight.

Chemical energy

[chemical, general, solid-state] Chemical reaction that delivers electrical ENERGY with electrical current in the form of free electrons. The electrochemical property is a direct result of the HALF-CELL phenomenon. This principle applies in particular to batteries. Alkaline batteries are producing nonreversible chemical reactions (disposable), whereas lithium-ion (i.e., LITHIUM) batteries (used in mobilephones and laptop computers) are rechargeable and the chemical reaction is reversed to the point where the electromotive force (V_{emf}) will be maintained for a long duration. The alkaline BATTERY is for instance based on the chemical reaction between zinc (Zn) and manganese dioxide (MnO_2), whereas the electrodes are potassium hydroxide (i.e., the alkaline). In the lithium-ion battery, lithium ions actually migrate between the CATHODE and the ANODE, hence producing a current externally; during the recharging process the ions are forced back by the applied external electric field. Other battery materials include lead-ACID, used in automobiles and hospital equipments. The charge carrying capability of a battery and the chemical reaction is expressed in ampere-hours, or charge multiplied by time. Another chemical energy application is in fuel cells, relying on OXIDATION of an ELECTROLYTE into free electrons, for example, hydrogen–oxygen and ALCOHOL and fossil fuel-based systems. The first battery (VOLTAIC PILE) was conceived by the Italian physicist ALESSANDRO VOLTA (1745–1827) in 1800 (see Figure C.42).

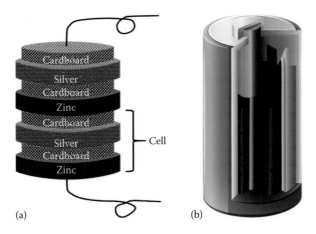

Figure C.42 (a) Chemical energy expressed by voltaic pile and (b) "dry cell."

Chemical gradient

[biomedical] The relative difference in concentration over a DISTANCE within specific constituents of a mixture (GAS, LIQUID, or SOLID-STATE: n- and p-structures). One specific application is in the use of REVERSE osmosis in water purification (*also see* CHEMICAL POTENTIAL) (see Figure C.43).

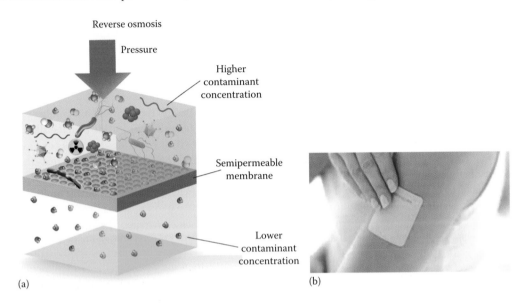

Figure C.43 (a) Chemical gradient in reverse osmosis water filtration. (b) Chemical gradient for nicotine patch used by individuals who are in the process of smoking cessation.

Chemical potential (μ_i)

[biomedical, thermodynamics] The partial derivative change in free ENERGY (F) per change in atomic and molecular (N) content, respectively, of the constituent $\mu_i = (\partial F/\partial N)_{T,V} = G(T,P,N)/N$ under constant temperature (T) and constant volume (V), incorporating the definition of the Gibbs free energy (G). Two systems in the same volume under identical conditions with the same ith constituent and in equilibrium will have the same chemical potential. There are four different types of chemical potential that can be discriminated: (1) thermodynamic chemical potential, (2) chemical potential, (3) electronic chemical potential, and (4) in relativistic concepts. The thermodynamic concept is used to describe the state in a system under certain conditions. For instance water will evaporate when the water temperature is above the boiling point, thus migrating from LIQUID to GAS state with an associated lower chemical potential in the gas/VAPOR state in comparison with the liquid state at that specific temperature. The rise in temperature will initiate the reordering from the higher to the lower potential. Similar change of state or equilibrium situations can be depicted for chemical reactions such as the OXIDATION of butane as described by the chemical reaction $2C_4H_{10} + 13O_2 \xrightarrow{\text{flame}} 8CO_2 + 10H_2O$ with respective chemical potentials and the release of energy described by $E = 2\mu_{gas,C_4H_{10}}C_4H_{10} + 13\mu_{gas,O_2}O_2 - 8\mu_{gas,CO_2}CO_2 + 10\mu_{gas,H_2O}H_2O$ at the combustion temperature. The change in internal energy per change in unit state for each of the constituents i is written as $\mu_i(S,V,n) = (\partial U/\partial n_i)_{S,V,n}$, for $i = 1, 2, \ldots, r$, where U is the internal energy, V the volume of the mixture, S the entropy of respective state, n the number of states, and r the total number of constituents. Chemical potential of SOLVENT with respect to constituent i is expressed as $\mu_i = \mu_{ii}(T,p) + RT \ln y_i$, where $\mu_{ii}(T,p) = \mu_i(T,p,y_1,y_2,\ldots y_r)$ or $\mu_{ii}(T,p_{ii}) = h_{ii}(T,p_{ii}) - Ts_{ii}(T,p_{ii})$ is the chemical potential of the pure solvent in the mixture at temperature T and pressure p, and y_i is the fractional component of each respective ingredient, with $\lim_i y_i \to 1$ yielding the solvent factor for constituent i and $R = N_A k = 8.3145$ J/molK is the universal gas constant, $N_A = 6.022 \times 10^{23}$ molecules/mol the Avogadro number and $k = 1.38066 \times 10^{-23}$ J/K molecule the Boltzmann constant. In this configuration the following variable are used: $h_{ii}(T, p_{ii})$ the specific enthalpy of system ii and $s_{ii}(T, p_{ii})$ the respective specific entropy with p_{ii} the partial pressure of constituent i. Under chemical equilibrium at a GROUND STATE p_0 the chemical potential of a genuine constituent will equal the Gibbs free energy of reaction for that particular constituent, $\mu_{ii}(T, p_0) = g_{ii}(T, p_0)$. The latter can describe the conversion of CHEMICAL ENERGY into electrical energy for a BATTERY, for instance. The use in true chemical potential applies to biomedical PHYSICS and relates to the CHEMICAL GRADIENT. The chemical potential describes the difference in chemical concentration across a cellular MEMBRANE creating an electrical potential gradient according to the respective NERNST POTENTIAL for the constituents at either side of the membrane. The electronic chemical potential falls under theoretical chemistry as applied to density functional or energy functional theory. The electronic chemical potential is the functional partial derivative of the electrochemical density functional with respect to the system state defined as the electron density, which is expressed as $\mu_i(r) = [\partial E(\rho)/\partial \rho(r)]_{\rho=\rho_{ref}} = V_{ext}(r) + [\partial F(\rho)/\partial \rho(r)]_{\rho=\rho_{ref}}$, where ρ is the density functional $(E(\rho) = \int \rho(r)V(r)d^3r + F(\rho)$, where $F(\rho)$ is the universal functional which identifies the KE of the electrons in the CHEMICAL COMPOUND and $V_{ext}(\rho)$ is the combined influence of nuclear electrostatic potential and the electric influences of electric and magnetic fields external to the ATOM or molecule resulting from surrounding chemicals and outside influences, also referred to as the external potential). Since the electronic chemical potential is based on the electron density, this potential is equivalent to the electron negativity of the ATOM, as a MATTER of fact the sum of the electron affinity (EA) and the IONIZATION POTENTIAL (IP) for the atom. This commodity is also referred to as the Mulliken potential, $\mu_{Mulliken}(r) = -\chi_{Mulliken}(r) = -(IP + EA)/2 = [\partial E(N)/\partial N]_{N=N_0}$. Last but not least, in relativistic physics the term chemical potential relates the states of FUNDAMENTAL PARTICLES analogous to the thermodynamic potential but with the following inherent and unique characteristics relativistically there will be no enthalpy of state for fermions and bosons, but each elementary PARTICLE contributes to the total internal energy of the system into which it is introduced or from which it escapes. The potential describes the tendency of ELEMENTARY PARTICLES to migrate out of energetic regions with higher chemical potential. The "relativistic" chemical potential is associated with the WAVE mechanical description of the elementary particles by the Boltzmann equation in QUANTUM physics and possesses a higher potential, which relates to a higher particle density. The elementary particle can be described as a gas of fermions and bosons. When just considering the fermions, the electrical potential is $\mu_F = kT \ln(z) = kT \ln(e^{\epsilon_F/kT} - 1)$, where is the fugacity, ϵ_F is the Fermi energy, or $\mu_F = (\partial E(N)/\partial N)_s = (\partial E_{orbital}(N)/\partial N)_s + (\partial E_{electrostatic}(N)/\partial N)_s$, where both terms represent the change in energy of the system if one particle is added under constant entropy (S). For an IDEAL GAS the chemical potential is linked to the PRESSURE (P) as $[\partial \mu_i(N)/\partial P]_T = V = RT/P$, or equivalently $[\partial(\mu_i/RT)/\partial \ln P]_T = 1$ (also see FERMI ENERGY).

Chemiluminescence

[biomedical, chemical, solid-state] Light produced as the result of a chemical reaction. The most well-known CHEMILUMINESCENCE reaction is oxidation–reduction, specifically produced during a fire. A more intricate oxidation–reduction reaction is the CHEMICAL ENERGY transfer that results in an excited SINGLET STATE, which has a limited lifetime and degrades with the emission of light. A strictly chemical reaction describing just such formation of a singlet state is in the reduction reaction of an amino-based derivative with a base and oxygen-based compound that can release the oxygen for OXIDATION, producing nitrogen and (high-energy) blue light. Other available mechanisms that produce light under a variety of ENERGY transfer principles are BIOLUMINES-CENCE (e.g., firefly), ELECTROLUMINESCENCE (e.g., LIGHTNING), radioluminescence (e.g., scintillation), and thermoluminescence (every object with a temperature above ABSOLUTE ZERO kelvin emits light with a wavelength range directly inversely proportional to the ABSOLUTE TEMPERATURE; gloving heating coil on electric stovetop or, indirectly, the incandescent lightbulb) (see Figure C.44).

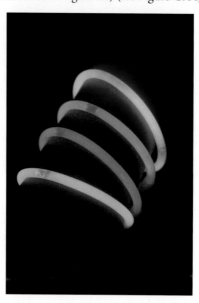

Figure C.44 Chemiluminescence example producing light in a "glow stick," for instance, used during deep-sea diving. Also used as decorative jewelry.

Chézy, Antoine de (1718–1798)

[fluid dynamics] Engineer from France. In 1769 Chézy performed experiments on the FLOW RESISTANCE in open channels such as the Seine River near Paris and the Courpalet Canal. This information provided the heuristic Chézy formulas. The Chézy quantity (i.e., coefficient), C_{chezy}, varies from about $30\,\text{m}^{1/2}/\text{s}$ for small rough channels to $90\,\text{m}^{1/2}/\text{s}$ for large smooth channels (see Figure C.45).

Figure C.45 Antoine de Chézy (1718–1798).

Chézy's formula, open channel

[fluid dynamics] Heuristic expression for FLOW velocity in an open canal. The flow velocity in an OPEN CHANNEL is defined as $v = C_{\text{chezy}}\sqrt{(m\theta_{\text{incl}})}$, where m is the mean hydraulic depth, θ_{incl} the inclination ANGLE of the riverbanks, and C_{chezy} the Chézy's flow velocity coefficient. The Chézy's flow velocity coefficient can be approximated by the formula derived by the Swiss engineers Emile-Oscar Ganguillet (1818–1894) and Wilhelm Rudolf Kutter (1818–1888) (GANGUILLET–KUTTER EQUATION, 1869): $C_{chezy} = [23 + (1/n) + (0.00155/\theta_{incl})]/[1 + [23 + (0.00155/\theta_{incl})](n/\sqrt{m})]$, where n defines the conditions of the surface of the WALL of the flow and is found listed in tables for specific materials (e.g., polished wood: $n = 0.010$ to 0.013; river with rocks imbedded in the shores and bottom as well as grass and growth: $n = 0.035$ to 0.05).

Chilton–Colburn j-factor for heat transfer

[fluid dynamics, thermodynamics] *See* COLBURN J-FACTOR, HEAT TRANSFER.

Chilton–Colburn j-factor for mass transfer

[fluid dynamics, mechanics, thermodynamics] *See* COLBURN J-FACTOR, MASS TRANSFER.

Chip

[electronics, solid-state] Silicon SEMICONDUCTOR crystal substrate designed to perform select electronic functions using N-TYPE and P-TYPE materials, also known as integrated circuit (see Figure C.46).

Figure C.46 Chip.

Chromaticity

[biomedical, chemical, optics] Definition of perceived COLOR space, also defined under CIE standards, based on the organization that has standardized the concept of color "Commission International de Illumination" (see Figure C.47).

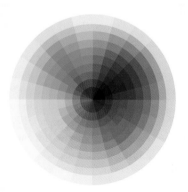

Figure C.47 Chromaticity color chart.

Chromatograph

[electronics, general, solid-state] Analytical device to assist in the analysis of the composition of media in a process called CHROMATOGRAPHY. Also used as gas–LIQUID chromatograph, or liquid–gas chromatograph, gas chromatograph, and gel permeation chromatography (see Figure C.48).

Figure C.48 Gas chromatograph equipment for analytical processing based on spectral attenuation profile of chemical constituents.

Chromatography

[electronics, general, solid-state] Analytical process designed to resolve the composing constituents of a mixture. Chromatography is applied to chemical analysis. The applications of chromatography range from analysis of petrochemical, environmental, pharmaceuticals, cosmetics ("fragrance"), food- (e.g., safety, flavor) and water quality, and in therapeutic diagnostics as well as for analysis of biological specimen submitted for instance during urine analysis or drawn BLOOD, next to forensics. Chemical chromatography uses the process of DIFFUSION when subjected to a bed of resin particles, with different diffusion coefficients for different SOLUTE molecule sizes. The resin is generally arranged in a column, analyzing the mobile PHASE of the respective components with high degree of RESOLUTION. The "steady-state" medium, affecting the separation process, is referred to as the stationary phase of the mixture. The separation process is arranged in a column. In GAS chromatography the mobile unit is in gas form. Gas chromatography is more popular for volatile components. The analysis relies on the optical spectral attenuation from transmitted light, providing specific representative absorption peaks at representative wavelengths (actually in chromatography one relies on the chromatography "wave numbers," $k' = 1/\lambda$; note that the regular WAVE number is defined as $k = 2\pi/\lambda$). Different chromatography techniques are available for specific applications. Chromatography can define molecular chains and atomic ingredients. Different mechanisms of action involved in the chromatographic analysis are, for instance, ADSORPTION chromatography, affinity chromatography, ION-exchange chromatography, gel chromatography, and partition chromatography, each with its own mechanism of molecular separation. Specifically, the mechanisms described influence the retardation process associated with the respective DIFFUSION processes. For instance gel chromatography uses polyacrylamide and other types of gels to act as sieves. In affinity chromatography the use of a chemical matrix (for instance, a biomolecular matrix used to identify antigens or antibodies) provides a binding process used for separation. Specific analyses are available for the identification, for instance, of alkaloids, amino acids, nucleic acids, proteins, steroids, and vitamins.

Chromophore

[biomedical, general, optics] Dye constituent that provides the essence of COLOR of a molecular composite medium observed or measured with optical devices. The chromophore is either a separate molecule or a component of a larger molecule and gains color when irradiated by a broad band LIGHT SOURCE, ideally WHITE LIGHT. The chromophore will absorb selected ranges of the source bandwidth, while the remaining spectral band(s) are re-emitted by means of shallow SCATTERING processes, which yields the visible color spectrum distribution (see Figure C.49).

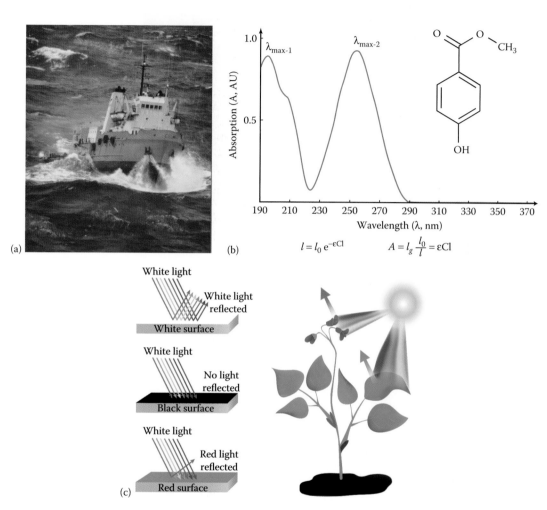

(a)

(b)

$$I = I_0\, e^{-\varepsilon Cl} \qquad A = I_g\, \frac{I_0}{I} = \varepsilon Cl$$

(c)

Figure C.49 (a) The orange color on the hull of the supply boat for oil rigs is the result of the choice in chromophores in the paint used, absorbing all but the "orange" color band of the incident white light. Whereas the "blue" color of the water is the result of Raleigh scatter of the incident white light. (b) Absorption lines for the specific chromophores of the chemical compound paraben. (c) color of object based on backscattered/back-reflected portion of the incident white light; remaining part of spectrum is absorbed by the chromophores inclosed in the (surface) of the object. A black object inherently absorbs all incident light, a white object reflects all incident light. The full spectrum is perceived as white.

Chromosphere

[astronomy/astrophysics, geophysics] Solar layer above the PHOTOSPHERE, with a temperature greater than 4200 K, and a thickness for our SUN is approximately 2000 km, containing primarily helium and hydrogen. The chromosphere is best investigated during a full total eclipse, exposing just the chromosphere (see Figure C.50).

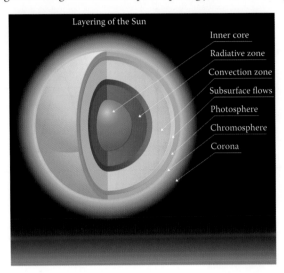

Figure C.50 Layers of the Sun, including the chromosphere.

CIE

[general, optics] Commission Internationale de l'Eclairage; The International Commission on Illumination; technical, scientific organization implementing standards and information exchange related to light, LUMI-NESCENCE, color and spectrophotometry, VISION and imaging as well as photobiology. CIE was founded in 1913 and is a nonprofit organization that acts independently providing advice on standardization. Although operating on a voluntary basis, CIE is recognized by ISO (International Organization for Standardization) as an international standardization institute. The committee is seated in Vienna, Austria.

CIE 1931 color space

[biomedical, general, optics] The definition of COLOR perception by the International Commission on Illumination (CIE) initially released in 1931. The CIE 1931 color space provides a mathematically defini-tion of COLOR spaces associated with the virtual XYZ color space as introduced. In the color space, Y means brightness, Z is proportional to blue stimulation, and X is an interpreted representation of the RED sensitivity curve of the CONES of the HUMAN EYE. The XYZ color space is also referred to as the TRISTIMULUS VALUES. XYZ can be confused with red–green–blue (RGB) cone responses even if X and Z are roughly equivalent to RED and blue, respectively. However, in the CIE XYZ color space, these values are not equal or similar to the S, M, AND L RESPONSES OF THE HUMAN EYE, but can be considered derived values (*also see* VISION).

CIE color space

[general, optics] Graphical representation of the hues and radiance of COLOR in a three-dimensional space, defined in "brightness" (L), hue and chrome (a: red-green; b: yellow-blue), expressed in the CIE $L * a * b$ color space, yielding the three axes. This scientific analysis overcomes subjectivity. For instance, when analyzing the color of teeth by a dentist for reconstructive dental work the ambient light, or examination light color temperature, will influence the observations, as will the personal history for the observer (good night rest, etc.; color blindness). The CIE color analysis uses spectrophotometric instrumentation for NUMERICAL determination. The color representation in EUCLIDEAN SPACE is defined based on the respective measurements and plotted according to the "E-value" as $E = \sqrt{(L^2 + b^2 + c^2)}$. In case only hue and SATURA-TION is displayed (two-dimensional graph), the space is defined as $Cab = \sqrt{(a^2 + b^2)}$ (see Figure C.51).

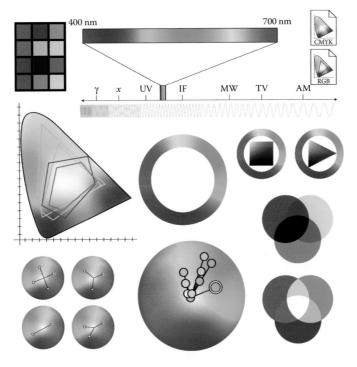

Figure C.51 CIE color charts and related tools for spectral identification and standardization.

Circulation

[fluid dynamics] The average FLOW velocity (i.e., wind velocity) on a path looping around the object in a flow pattern (e.g., WING), where flow against the path direction has a negative value and flow directions perpendicular to the path ($\vec{s}$) are not considered. Circulation is defined as the steady-state closed loop integral of the vorticity within the enclosed path. Where, the vorticity is the degree of "rotation" experienced in flow VELOCITY ($\vec{v}$) as a function of location ($\vec{r}$) and time (t), which is the "curl" of the velocity at a given place. This will not consider initiation and termination processes. This yields for the circulation $\oint_C \vec{\varpi}(\vec{r},t)\,d\vec{s} = \oint_c \nabla \times \vec{v}(\vec{r},t)\,d\vec{s}$. A flow at rest will have circulation zero. The WAKE of the flow in, for instance, the airstream around a wing has zero circulation, which also describes the VORTEX at the tip of the wake, assuming the loop integral is encompassing a large enough volume of space. The wake and wing-tip vortex have no viscous forces (no boundary layers) and hence the circulation is zero.

Circulation (Γ_{circ})

[fluid dynamics] The average FLUID FLOW velocity (i.e., wind velocity) on a path looping around the perturbation (i.e., WING), and can be represented as the closed loop integral of the VORTICITY ($\vec{\varpi}(\vec{r},t)$) within the enclosed path. $\Gamma_{circ} = \oint_C \vec{\varpi}(\vec{r},t)\,d\vec{s} = \oint_c \nabla \times \vec{v}(\vec{r},t)\,d\vec{s}$. Where the vorticity is the degree of "rotation" experienced in flow VELOCITY ($\vec{v}$) as a function of location ($\vec{r}$) and time (t), which is the "curl" of the velocity at a given place. The circulation at the boundary of the phenomena will be finite, the outer (i.e., horizontal) edges of the wings create a VORTEX that is at the edge of the WAKE.

Circulation function

[fluid dynamics] Arbitrarily closed loop over the FLOW around an object, $\oint_S [v_x(dx/ds) + v_y(dy/ds) + v_z(dz/ds)]\,ds$, in a Cartesian system (x, y, z) with flow velocity $\vec{v} = (v_x, v_y, v_z)$ over a loop S with path steps s tangential to the loop.

Circulatory system

[biomedical, fluid dynamics, general] Blood flow loop in a biological system; HEART, arteries, arterioles, capillaries, venules, veins, and back to hearts. The circulatory system has an oscillatory component (induced by the pumping rhythm of the heart) on the arterial side that dampens the AMPLITUDE (both velocity and pressure) in the growing impedance with reduction in FLOW lumen associated with increasing number of branching vessels. The circulatory system is defined by CONSERVATION OF MASS. In addition to the pumping mechanism of the heart, the veins have a flow mechanism that is supported by encapsulation from skeletal muscles. The SKELETAL MUSCLE function is important, especially when considering that a large number of people will faint when standing perfectly still for extended periods of time. An easy exercise to maintain CIRCULATION in the veins of the LEG is to periodically flex and release the leg muscles (see Figure C.52).

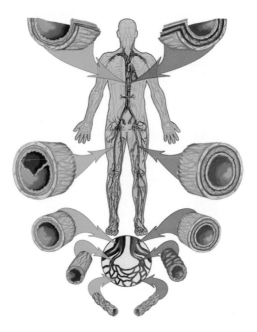

Figure C.52 Blood circulation diagram. Venous blood in blue and arterial (oxygen rich blood) blood in red.

Clapeyron, Benoît Paul Émile (1799–1864)

[thermodynamics] Physicist and engineer from France. Clapeyron made several recommendations for thermodynamics and defined many conditions as a pioneer in the new field of thermodynamics. Clapeyron defined the two-phase mixture, for instance, steam and boiling WATER (*also see* CLAUSIUS–CLAPEYRON RELATION) (see Figure C.53).

Figure C.53 Benoît Paul Émile Clapeyron (1799–1864).

Clausius, Rudolf Julius Emanuel (1822–1888)

[general, mechanics, thermodynamics] German physicist and scientist. Clausius' father was a contemporary of JAMES PRESCOTT JOULE (1818–1889) and LORD KELVIN (a.k.a. WILLIAM THOMSON: 1824–1907), while he himself was a contemporary of NICOLAS LÉONARD SADI CARNOT (1796–1832). The work of Carnot inspired Clausius to formulate his own interpretation of HEAT TRANSFER, different from that expressed by Lord Kelvin in the SECOND LAW OF THERMODYNAMICS as: There is no single isolated process that results in the transfer of heat from a body at low temperature to a body at higher temperature. Meaning that there are multiple processes at a stage to accomplish one goal, in the refrigeration this equates to work performed with losses due to the work itself (generating heat of its own), while transferring heat against a gradient. In 1865 Clausius introduced the concept of ENTROPY (see Figure C.54).

Figure C.54 Rudolf Julius Emanuel Clausius (baptized name: Rudolf Gottlieb; 1822–1888).

Clausius–Clapeyron relation

[thermodynamics] $dP/d\tau_T = L/\tau_T \left(\rho_V^{-1} - \rho_L^{-1}\right) = L/\tau_T \Delta v_a$, where P is the VAPOR PRESSURE, τ_T the temperature, ρ_V the density of the vapor phase, ρ_L the density of the LIQUID phase, L the LATENT HEAT OF VAPORIZATION ($L \equiv \tau_T \left(S_g - S_\ell\right) = dQ$, where S_i the entropy of the GAS and liquid phases, respectively, and dQ the heat added to the system), and v_a the volume occupied by one ATOM or molecule of the constituent. Note the derivative is not simply as in the EQUATION OF STATE for the gas, but assumes the gas and liquid to coexist. Generally we can assume that the GAS PHASE of an atomic gas (v_{ag}) occupies a greater volume than in the liquid phase ($v_{a\ell}$) and $\Delta v_a = v_{ag} - v_{a\ell} \cong v_{ag} = V_g/N_g$, with V_g the gas volume of the constituent and N_g the number of molecules or atoms in the gas state, respectively.

Climatology

[energy, fluid dynamics, general, geophysics] Field of PHYSICS dealing with physical phenomena in the lower part of EARTH's ATMOSPHERE directly related to the weather and providing warnings for people in the path of inclement weather (see Figure C.55).

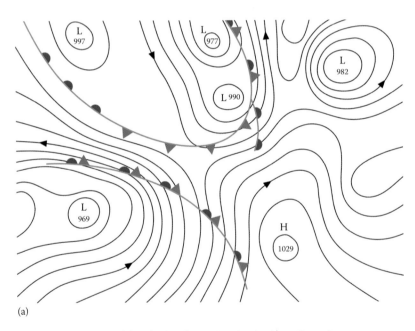

(a)

Figure C.55 Climatological tools: (a) cold and warm front migration in three dimensions   (Continued)

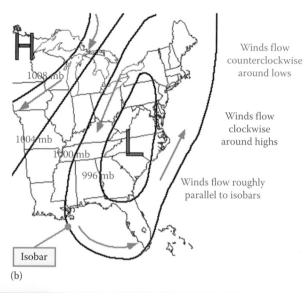

Winds flow
counterclockwise
around lows

Winds flow
clockwise
around highs

Winds flow roughly
parallel to isobars

Isobar

(b)

(c)

Figure C.55 (Continued) Climatological tools: (b) drawing of isobars on a section of the global map. (Courtesy of University of Illinois, Department of Atmospheric Sciences, IL.) and (c) mobile meteorological station. (Courtesy of CBS KCNC-TV Denver, CO.)

C/M Ratio

[electromagnetism] Expression used in frequency modulation synthesis, where M is the modulator frequency and C the CARRIER frequency. Frequency modulation synthesis is a tool used in spectral analysis.

Coalsack Nebula

[astrophysics] Dark "cloud" observed in the band of the Milky Way galaxy, dating back to millennia, documented by the Inca's and early European observers (documented in 1499 by Spanish explorer Vicente Yáñez Pinzón [1462–1514]), next to Australian Aborigines approximately 40,000 years ago (WALL drawings). Recent discoveries show that the interstellar "dust" cloud of particles is so dense that it blocks virtually all background light (blocks better than 90%) at an azimuth in the "southern" portion of the MILKY WAY.

The Coalsack Nebula is located between 610 and 790 light-years distance, in the constellation Crux ("The Southern Cross") and spans approximately 50 light-years at its widest expansion. The Coalsack Nebula is also featured in the original science fiction television series *Star Trek*, the "Immunity Syndrome" episode (1968). The configuration of the Coalsack Nebula can be derived from the equations on galactic light scattering introduced by Karl Schwarzschild (1873–1916), based on cumulative information acquired over the years, and most recently by observations made by the MPG/ESO 2.2-m telescope at European Southern Observatory's La Silla Observatory in Chile, which is one of the driest areas in the world as well as extremely remote, located 150 km northeast of La Serena on the periphery of the Atacama Desert in Chile. The dust spread-out in the nebula provides significant attenuation for transmitted light from remote stars. The dark spot has been known and described for a considerable time. The aboriginal population of Australia described it as the head of an emu. The Incas in South America circulated the story that it was the god Ataguchu who kicked a hole in the band of stars (now known as the Milky Way) in a fit of anger (see Figure C.56).

Figure C.56 "Coalsack" in the constellation Crux, depicted by the dark "hole." The Crux constellation is up and to the right, in the shape of a cross. (Courtesy of Dr. Poshak Gandhi.)

Coefficient of drag

[fluid dynamics, mechanics] Coefficient defining the influence of the retarding effects resulting from friction and viscosity with respect to a body moving through a fluid. The coefficient of drag (D_{flow}) can be found in tables for fluids and surface conditions and is correlated to the drag force (F_{drag}) as $F_{drag} = D_{flow} \rho A \left(v^2/2 \right)$, with v the velocity of the object in the fluid (e.g., water, air), ρ the fluid density, and A the effective cross-sectional area of the object in motion.

Coefficient of friction ($C_f = \sigma_s/(1/2)\rho v^2$)

[fluid dynamics, mechanics] Surface shear stress expressed as dimensionless number defined by the stress over the kinetic energy density, where ρ the density of the fluid, v velocity, and σ_s the stress. Also defined as the ratio of shear stress (σ_s) with respect to normal stress (σ_n), $C_f = \sigma_s/\sigma_n$.

Coefficient of friction, kinetic (μ_{kin})

[mechanics] The ratio of the FRICTION force (F_f) to the NORMAL FORCE ($F_n = mg\cos(\theta)$, where θ is the ANGLE with the normal to the surface) for an object in MOTION. An object being dragged over a surface will experience KINETIC FRICTION, such as a sled sliding down a snowy hill (see Figure C.57).

Figure C.57 Sliding down a ski slope with limited kinetic friction.

Coefficient of friction, rolling (μ_{rol})

[mechanics] The ratio of the FRICTION force (F_f) to the NORMAL FORCE ($F_n = mg\cos(\theta)$, where θ is the ANGLE with the normal to the surface) for an object in MOTION. The friction force needs to be overcome to maintain uniform rolling motion at uniform VELOCITY (see Figure C.58).

Figure C.58 Deformation of an automobile tire resulting in rolling friction. Rear axle with tire cut open illustrating room for deformation. (Courtesy Daimler AG, Stuttgart, Germany.)

Coefficient of friction, static (μ_{stat})

[mechanics] The ratio of the FRICTION force (F_f) to the NORMAL FORCE ($F_n = mg\cos(\theta)$, where θ is the ANGLE with the normal to the surface) for an object at rest. This would apply to an automobile tire on the road under normal steady-state velocity and acceleration without slip. Also a box resting on a slope will experience static friction (see Figure C.59).

Figure C.59 Person standing in fixed location on Lombard Street incline in San Francisco due to static friction.

Coefficient of kinetic friction

[mechanics] *See* COEFFICIENT OF FRICTION, KINETIC.

Coefficient of kinetic friction (μ_K)

[general, mechanics] Relation between the resistive force (F_K) and NORMAL FORCE (F_N) for a body in uniform MOTION while experiencing a force with a component parallel to the free surface the body rests on, $F_K = \mu_K F_N$.

Coefficient of rolling friction

[mechanics] *See* COEFFICIENT OF FRICTION, ROLLING.

Coefficient of rolling friction (μ_R)

[general] FRICTION coefficient for the dissipative force resulting from deformations observed, for instance, for a tire on the road surface or a waltz rolling out asphalt (*see* COEFFICIENT OF FRICTION, ROLLING).

Coefficient of static friction

[mechanics] *See* COEFFICIENT OF FRICTION, STATIC.

Coefficient of static friction (μ_S)

[general, mechanics] Relation between the resistive force (F_S) and NORMAL FORCE (F_N) for a body at rest while experiencing a force with a component parallel to the free surface the body rests on. As long as the parallel component of the force does not exceed the frictional force ($F_f = \mu_s F_N$) the body will remain at rest, $F_S \leq \mu_s F_N$.

Cohesion

[general, solid-state, thermodynamics] Binding force between molecules of a substance. Contrast to ADHESION, which applies to MACROSCOPIC entities.

Colburn j-factor, heat transfer ($j_H = (h_0/c_p G_m)_s (c_p \eta / \kappa)_s^{2/3} (\eta_w/\eta)_s^{0.14} = St Pr^{2/3}$)

[fluid dynamics, thermodynamics] Dimensionless number illustrating the ratio of heat transfer to crossflow REYNOLDS NUMBER, where s describes the conditions of FLOW on the shell or WALL, h_0 the film heat-transfer coefficient, $G_m = W_s/S_m$ the mass velocity, specifically pertaining to the free flow area bridging adjacent tubes, with W_s the mass flow rate at the wall and S_m the minimum free flow through either crossflow areas, c_p the SPECIFIC HEAT under constant pressure, $c_p \eta$ the specific heat under constant flow at wall FRICTION, κ the THERMAL CONDUCTIVITY, η the VISCOSITY of the FLUID, η_w the viscosity at the wall, St the Stanton number, and Pr the Prandtl number. Sometimes also referred to as CHILTON–COLBURN J-FACTOR FOR HEAT TRANSFER.

Colburn j-factor, mass transfer ($j_m = St_m Sc^{2/3}$)

[fluid dynamics, mechanics, thermodynamics] $j_m = (\kappa_m/v)(\eta/\rho D_{dif}) = St_m Sc^{2/3}$, dimensionless number representing the ratio of FRICTION forces to DIFFUSION forces, where $\kappa_m = mole/A \times t \times [a]$ is the mass transfer coefficient, with A the area, t time, $[a]$ the concentration of constituent "a," v the velocity of mass transport (either FLOW or diffusion), D_{dif} diffusion coefficient, ρ the density, St_m the Stanton mass transfer number, and Sc the Schmidt number. Sometimes also referred to as CHILTON–COLBURN J-FACTOR FOR MASS TRANSFER. In case there is transfer of momentum, the Colburn mass transfer j-factor will be equal to the Colburn HEAT TRANSFER j-factor, which is only the case when there is no FORM DRAG. The conditions of no form drag apply under certain specific circumstances to flat plates and inside straight conduits.

Colliding black holes

[astrophysics, relativistic] Generate ripples in time referenced as gravitational space-time perturbations (*see* TIME PERTURBATION, COLLIDING BLACK HOLES; GRAVITATIONAL WAVE; *and* EINSTEIN EQUATIONS).

Colloid osmotic pressure

[biomedical, chemical, thermodynamics] The osmotic pressure of a SOLUTION as if the PARTICLE solution would be a GAS with the same volume and temperature as the solution. The osmotic pressure can be expressed by the IDEAL GAS LAW, $\Pi_{osm} = nRT/V = [C]RT$, where $R = 8.135\,J/molK$ is the gas constant, T the local temperature in Kelvin, n is the number of moles of solution, and $[C]$ the concentration (*also see* STARLING'S LAW *and* VAN'T HOFF LAW).

Color

[computational, energy, general, nuclear, quantum, relativistic, solid-state, thermodynamics] Also referenced as color force. This label is associated with the FOUR FORCES. Designation in the STRONG NUCLEAR force that acts on a property defined as "color," which has three states: r, g, b (also referenced as: RED, green, and blue). This in comparison with the ELECTROMAGNETIC FORCES that act on charges (positive and negative) and the GRAVITATIONAL FORCE which acts on mass. Basic constituents red, green, and blue can form all other colors visible to the HUMAN EYE. Depending on the mechanism the colors can be additive (e.g., light, where all three form WHITE LIGHT) or canceling or subtractive (e.g., paint; a mixture of all three colors yields black).

Comet

[astronomy/astrophysics, mechanics] Celestial object consisting of frozen medium and solids, leaving a trail of GAS and charged particles, the comet's tail. The tail emanates from the head, or coma of the comet.

The principal content of the comet is supposedly found in the NUCLEUS of the coma. Note that the tail of charged particles has a different ANGLE to the comet's trajectory than the VAPOR cloud tail; forming two tails. This is different from an ASTEROID, which has no tail and consists of solids only. One of the more noted comets is HALLEY'S COMET, last seen in 1986. Halley's comet (initially Comet 1982i; renamed Comet 1986III) moves on its orbit around the SUN, traveling to the edges or our SOLAR SYSTEM, every 75–76 years, depending on our own position with respect to the observation in our orbit around the Sun. Another notable comet is Hale–Bopp (astronomy nomenclature: Comet C/1995 O1, "discovered" in 1995), which has been described dating back to millennia and has a period of 2537 years, and Hyakutake. So far a total of 10 comets have been identified in our solar system. The direction of the gas/vapor tail is indicative of the SOLAR WIND. Comets are considered to be essential constituents of the formation of our solar system; hence are bound to carry information about the history of the creation of the solar system (see Figure C.60).

Figure C.60 A comet.

Communicating vessels

[general, fluid dynamics] The FLUID surface of open vessel that are connected by free-flowing passages will be at equal level under equilibrium, independent of the shape and size of the respective vessels (see Figure C.61).

Figure C.61 Communicating vessels.

Compliance

[biomedical, general, mechanics] (C_{compl}) indication of elastic nature of a material or composition (e.g., amalgamation of multiple components such as multilayer or mash/webbing and "solid" filler). The phenomenon is in particular applicable for an distensible tube, in which case the tube expands with increasing applied PRESSURE (P), making the compliance a direct function of the change in VOLUME (V) as a result of the change in pressure: $C_{compl} = \Delta V/\Delta P$.

Compressibility factor (Z$_{com}$)

[thermodynamics] Parameter indicating the deviation from the IDEAL GAS LAW, and ideal gas conditions. The compressibility factor, $Z_{com} = PV/nRT$, will equal 1 for an IDEAL GAS (*also see* VAN DER WAALS EQUATION OF STATE).

Compressibility of gas in fluid

[fluid dynamics] The rate of change in volume of a dissolved GAS under compression of the solvent FLUID: $B_g = \gamma P/\propto$, with P the externally applied pressure (in Pascal), $\gamma = c_p/c_v$ the ratio of the specific heats of the gas, and $\propto$ the volume fraction of the undissolved gasses.

Compression waves

[acoustics, atomic, biomedical, fluid dynamics, geophysics] Transfer WAVE in liquids, solids, and gasses. For compression waves in fluids see both ACOUSTICS and SOUND.

Compression wave, propagation

[acoustics, atomic, biomedical, fluid dynamics, geophysics] Any solid, LIQUID, or GAS can be compressed by an external force, resulting in relaxation based on the bulk modulus, which culminates in a (damped-) OSCILLATION in the form of a longitudinal elastic compression wave. This compression wave can be at the atomic level. In fluids the medium responds differently than in solids and the WAVE is called an acoustic wave. In general the compression oscillation mode is also the direction of propagation. The oscillatory medium is identified by alternating rarefactions and condensations (*also see* LAMB WAVE). The speed of propagation (v) in sea water is a direct function of depth, salinity, and temperature and can be defined in first-order approximation as $v = \sqrt{B/\rho}$, where B is the bulk modulus and ρ the local density. The propagation velocity at the water surface at certain wavelength and temperature is approximately 1.4×10^3 m/s, while at a depth of 5 km this becomes 1.5×10^3 m/s while an increase of 1°C results in an increase of 3.7 m/s and an increase in salinity by 1% increases the speed of wave propagation by 1.2 m/s. In comparison the speed of propagation in solids is defined by $v = \sqrt{Y/\rho}$, where Y is the Young's modulus and ρ the local density. For EARTHQUAKE COMPRESSION, wave propagation velocity is 1.3×10^4 m/s, approximately 10× greater. Compression waves are also used in ULTRASONIC IMAGING, relying on the differences in speed of propagation or modulus of tissue constituents in organs (see Figure C.62).

Figure C.62 Example of atmospheric compression wave. The expansion (lowering the local pressure) in a linear direction forms condensation (cooling during expansion), expressed by parallel rows of clouds.

Compton, Arthur Holly (1892–1962)

[atomic, biomedical, nuclear] Scientist from the United States who provided empirical proof of the momentum as well as ENERGY phenomena associated with ELECTROMAGNETIC RADIATION next to that of particles, published in 1922. Based on the energy observations and the prevailing CONSERVATION LAWS the energy loss of a colliding electrons generates a PHOTON with the energy balance at hand as described by the COMPTON EFFECT. Arthur Compton received the Nobel Prize in Physics in 1927 for his description of the ATOMIC ENERGY explained by the Compton effect (see Figure C.63).

Figure C.63 Arthur Holly Compton (1892–1962). (Courtesy of Noble Foundation, Stockholm, Sweden.)

Compton effect

[atomic, biomedical, condensed matter, energy, nuclear] Description of the ENERGY loss of a PHOTON involved in a collision with free electrons. The conservation of momentum forms the basis of the determination of the wavelength of the re-emitted photons as follows: $\vec{p}_{i,photon} = \vec{p}_{elec} + \vec{p}_{s,photon}$, where $p_{i,photon} = h/\lambda$ is the momentum of the incident photon ($h = 6.6260755 \times 10^{-34}$ Js the Planck's constant, λ the wavelength of the incident ELECTROMAGNETIC RADIATION with energy $E_{i,photon} = h\nu$, ν the frequency of the RADIATION), the momentum of the scattered photon $p_{s,photon}$, with momentum of the electron as $p_{elec} = m_e c$ with rest energy: $E_{elec} = m_e c^2$, however the electron needs to be considered in relativistic terms which yields $E_{elec}^2 = p_e^2 c^2 + m_e^2 c^4$, based on the CONSERVATION OF ENERGY: $E_{i,photon} + m_e c^2 = E_{s,photon} + E_{elec}$. The increase in wavelength from the incident photon with wavelength λ_i to the emitted photon with wavelength λ_s is due to the collision loss, which can be written as $\Delta\lambda = \lambda_i - \lambda_s = h/m_e c(1 - \cos\theta)$, where θ represents the ANGLE OF SCATTERING with respect to the incident direction and $\Delta\lambda$ represents the COMPTON WAVELENGTH (see Figure C.64).

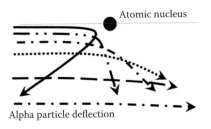

Figure C.64 Compton effect. The angular deflection of a beam of alpha particles nearing an atomic nucleus.

Compton scattering

[atomic, biomedical, condensed matter, energy, nuclear] *See* **Compton effect**.

Compton wavelength

[atomic, biomedical, condensed matter, energy, nuclear] The increase in wavelength from the incident PHOTON with wavelength λ_i to the emitted photon with wavelength λ_s is due to the collision loss, which can be written as $\Delta\lambda = \lambda_i - \lambda_s = h/m_e c(1 - \cos\theta)$, where θ represents the ANGLE of SCATTERING with respect to the incident direction. In molecular PHYSICS the interaction between two protons with the release of a PION ("π") (i.e., π-MESON), with mass m_π, is represented by the potential equation: $V = \overline{\overline{\tau_a}} \cdot \overline{\overline{\tau_b}} g_\pi^2 (m_\pi/2M) \big[(1/3)\big(e^{-\mu_c r}/r\big)\big(\overline{\overline{\sigma_a}} \cdot \overline{\overline{\sigma_b}}\big) + \big[(1/3) + (1/\mu_c r) + (1/\mu_c^2 r^2)\big]\big(e^{-\mu_c r}/r\big)S_{ab}\big]$, where $1/\mu_c = \hbar/m_\pi c = 1.4$ fm is the pion Compton wavelength ($\sqrt{G\pi^2/\hbar c} =$ is the pion–nucleon coupling constant, $\hbar = h/2\pi$, where $h = 6.6260 \times 10^{-16}$ Js is Planck's constant, $\overline{\overline{\sigma_a}}$ and $\overline{\overline{\sigma_b}}$ are the PAULI SPIN MATRICES for the respective nucleons a and b, $\overline{\overline{\tau_a}}$ and $\overline{\overline{\tau_b}}$ are the respective isospin matrix in 2×2 format for NUCLEON a and b, r the nucleonic interspacing DISTANCE, c the speed of light, g_π the pion ENERGY, and $S_{ab} = 3\big(\overline{\overline{\sigma_a}} \cdot \vec{r}\big)\big(\overline{\overline{\sigma_b}} \cdot \vec{r}\big) - \big(\overline{\overline{\sigma_a}} \cdot \overline{\overline{\sigma_b}}\big)$ the TENSOR operator; the intrinsic SPIN can also be characterized as $\overline{S^{(a)}} = \overline{\overline{\sigma_a}}\hbar/2$.

Computational fluid dynamics (CFD)

[biomedical, computational, fluid dynamics, thermodynamics] FLUID DYNAMICS problems solved by advanced and extensive mathematical analysis. Early on (prior to computers) NUMERICAL solutions were attempted by hand dating back to the eighteenth century, referred to as theoretical fluid dynamics, in contrast to experimental fluid dynamics. The intricacy of the theoretical approach extended to complex and extensive problem statements with lengthy periods (e.g., months) spent on working out the solutions. More recent examples are the work by LEWIS FRY RICHARDSON (1881–1953) in 1910 on Laplace equations for HEAT TRANSFER and RICHARD COURANT (1888–1972), KURT OTTO FRIEDRICHS (1901–1982), and HANS LEWY (1904–1988) in 1928 for their work on hyperbolic partial differential equations. The culmination of these efforts came to flourish with the more widespread availability of computers in the 1960s and 1970s shown by the 1973 work of W. Roger Briley (twentieth century) and H. McDonald (twentieth century), and the work of Richard M. Beam (twentieth century) and R.F. Warming (twentieth century) in 1976 and 1978 solving Navier–Stokes and Euler algorithms (see Figure C.65).

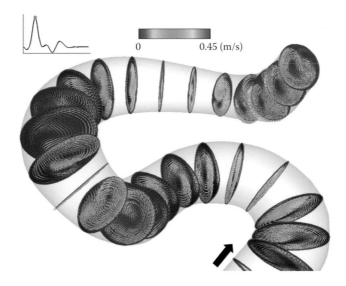

Figure C.65 A computational analysis of fluid flow in a tube.

Condensation number (Co = $g\rho^2 \Delta H_{vap} L^3 / \kappa\eta\Delta T$)

[thermodynamics] The ratio of molecular condensation on a surface to the total number of molecules in contact with the surface, where ΔH_{vap} is the LATENT HEAT OF VAPORIZATION, g GRAVITATIONAL ACCELERATION, L the characteristic length, ΔT temperature gradient, κ the THERMAL CONDUCTIVITY (Kappa), η VISCOSITY, and ρ the density.

Condenser

[electromagnetism, general] Old terminology for CAPACITOR, a device that can store ELECTRIC CHARGE.

Conduction band

[atomic, electronics, general] Under the influence of an external FORCE (e.g., ELECTRIC or KINETIC), ELECTRONS can be moved to a higher ENERGY ORBIT, removed from the VALENCE BAND (with low BINDING ENERGY) to an ENERGY configuration that allows the electron to move away from the host ATOM's NUCLEAR BINDING FORCE as a FREE ELECTRON. The energy configuration is referenced as the conduction band. Under room TEMPERATURE (~300 K) the kinetic energy of electrons is approximately KE = 0.026 eV $\approx 4.1 \times 10^{-21}$ J. In comparison the gap energy (energy gap between valence band and conduction band) for a SEMICONDUCTOR such as germanium or silicon is approximately 1.1 eV, whereas in common CONDUCTORS, such as IRON and copper for instance, the conduction band overlaps the valence band, allowing electrons to move freely between neighboring copper atoms. When the THERMAL ENERGY can bring the VALENCE electron in the conduction band the phenomenon of ELECTRICAL CONDUCTION is enabled and an electric CURRENT can be established.

Conductor

[atomic, general] Materials made of ELEMENTS with specific ELECTRON configuration. An ATOM has the electrons arranged according to the BOHR ATOMIC MODEL with electrons in specific allowed ENERGY LEVELS or ORBIT SHELLS. In comparison with either INSULATOR or SEMICONDUCTOR the electron configuration is more energetic than in these two. In an INSULATOR the CONDUCTION BAND containing free electrons is unpopulated and separated by a FORBIDDEN GAP from the VALENCE BAND, whereas in conductor configuration the valence band is overlapping the conduction band, hence populating the conduction mechanism of action with free electrons.

Cones

[biomedical, chemical, optics] Anatomical feature in the EYE responsible for the electrochemical conversion of ELECTROMAGNETIC RADIATION in the PERCEPTION of three elementary colors that in combination with relative intensity will generate virtually infinite number of COLOR combinations. The various color pattern definitions are, for instance, defined by the CIE system and described in greater detail under CHROMATICITY (see Figure C.66).

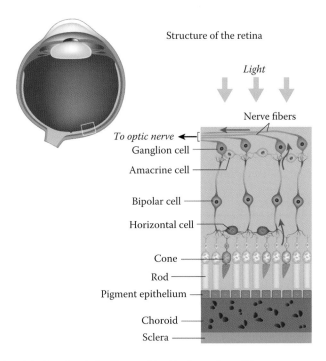

Figure C.66 Cones (color vision) and rods (radiance vision) in the retina of the eye.

Conservation laws

[atomic, general, thermodynamics] The following conservation principles/laws can be distinguished: CONSERVATION OF ANGULAR MOMENTUM, CONSERVATION of atomic nuclei, conservation of BARYONS, conservation of leptons, CONSERVATION OF CHARGE, conservation of current, CONSERVATION OF ELECTROMAGNETIC ENERGY, CONSERVATION OF ENERGY, CONSERVATION OF LINEAR MOMENTUM, CONSERVATION OF MASS, CONSERVATION OF MASS–ENERGY, conservation of MECHANICAL ENERGY, and conservation of momentum. Each conservation principle will be outlined at the respective location.

Conservation of angular momentum

[fluid dynamics, general, nuclear] The angular momentum is the product of the ANGULAR VELOCITY (ω) and the moment of INERTIA (I_{inertia}) of the object ($L = I_{\text{inertia}}\omega$), when no external forces are involved the TOTAL ANGULAR MOMENTUM of a system will be conserved. When a rotating system disintegrates (without exchange of ENERGY), the sum of the angular momentum as vectors will combine to provide the same angular momentum as for the initial structure. Conservation of angular momentum also applies to nuclear and atomic transitions, setting boundary conditions and selection rules for transitions, in direct combination with CONSERVATION OF ENERGY (see Figure C.67).

Figure C.67 Conservation of energy (conversion of potential energy to kinetic energy) and conservation of angular momentum illustrated by ball rolling down incline, which will continue its rolling path when a level plane is reached.

Conservation of charge

[atomic, fluid dynamics, general, nuclear] Charge cannot be destroyed or altered. Combining a PARTICLE with POSITIVE CHARGE and a particle with NEGATIVE CHARGE as part of the same system will have as a system the net charge for the two particles, as will the combined particle.

Conservation of electromagnetic energy

[energy, general, optics] Based on the MAXWELL EQUATIONS the continuity in ELECTROMAGNETIC RADIATION is defined by $(\partial u_e/\partial t) + \nabla \cdot \vec{S} + \vec{J} \cdot \vec{E} = 0$, where $\vec{E}$ is the ELECTRIC FIELD, $u_e = 1/8\pi\left(\left|\vec{E}\right|^2 + \left|\vec{B}\right|^2\right)$ the ENERGY density per unit volume V, t is time, $\vec{J}$ is the CURRENT DENSITY, $\vec{S} = \sqrt{\varepsilon_0\mu_0}/4\pi\left(\vec{E} \times \vec{B}\right)$ the POYNTING VECTOR, $c = \sqrt{\varepsilon_0\mu_0}$ the speed of light, $\vec{B}$ is the MAGNETIC FIELD, $\vec{n}$: the normal to the surface: A, of the enclosed space, and $\vec{J} \cdot \vec{E}$ represents the work per unit time per unit volume performed as a function of the ELECTROMAGNETIC FIELD interaction.

Conservation of energy

[general] The principle that ENERGY can neither be created nor destroyed, and therefore the total amount of energy in the UNIVERSE is constant. This law of classical PHYSICS is modified for certain nuclear reactions (*see* CONSERVATION OF MASS–ENERGY).

Conservation of linear momentum

[general, mechanics] Specifically applying to lossless collisions (no external net forces, specifically no FRICTION, no *permanent* deformation), the momentum $\vec{p}$ as a vector will be conserved when transferred to a different set of objects, or adding object to the MOTION: $\vec{p} = \sum_n^{\Omega} m_i \vec{v}_i$, for a system with constituents with mass m_i, each traveling with respective velocity v_i; where $\Delta\vec{p} = 0$ (see Figure C.68).

Figure C.68 During the game of pool (billiards) the principle of conservation of momentum is the driving principle however; there are friction losses involved during the rolling action of the billiard balls.

Conservation of mass

[thermodynamics] In principle, MASS (m) cannot be destroyed. The exception will be nuclear reactions where mass can be converted into ENERGY based on ALBERT EINSTEIN's (1879–1955) energy principle: $E = mc^2$, where $c = 2.99792458 \times 10^8$ m/s is the speed of light. In FLOW the conservation principle applies to gasses and liquids, where the mass is captured by the DENSITY (ρ) multiplied by the VOLUME (V), $m = \rho V$. In chemical reactions the conservation of mass is represented by the reaction equation; no atoms are lost in the forming or dissociation of molecules and residue atoms. The concept of flow is a primary example of the conservation of mass principle (see Figure C.69).

Figure C.69 Conservation of mass illustrated.

Conservation of mass–energy

[general, mechanics] The principle that both mass and ENERGY combined are conserved based on the principle that energy and mass are interchangeable in accordance with the equation: $E = mc^2$, where E is the energy, m is the mass, and c is the velocity of light.

Continuity equation

[fluid dynamics, general, nuclear] The general conservations laws: CONSERVATION OF MASS, conservation of momentum, and CONSERVATION OF ENERGY, as well as CONSERVATION OF ANGULAR MOMENTUM, next to CONSERVATION of charge.

Continuum

[fluid dynamics] For instance a KNUDSEN NUMBER of $Kn \leq 0.1$ sets the boundary conditions where the FLUID FLOW can be treated as a continuous medium, with the MACROSCOPIC parameters such as PRESSURE, TEMPERATURE, VOLUME, VELOCITY, and DENSITY. More confined is the regime $0 < Kn \leq 0.1$, describing diffusive slip-flow in the continuum.

Convectively coupled Kelvin wave (CCKW)

[fluid dynamics, mechanics, meteorology] Atmospheric NONDISPERSIVE WAVE feature, resembling a chimney effect, primarily centered close to, or on the equator that can extend for thousands of kilometers. CCKW consists of a WALL of rising AIR, which tends to tilt to the west with increasing altitude (primarily resulting from the earth's rotation). The rising air is accompanied by air masses that FLOW toward the earth's surface, generating a CIRCULATION which generates thunderstorms when reaching the air at lower altitude. These RAIN storms are reinforcing the effect combined with DIVERGENCE in air flow at higher altitude leading to cyclonic SPIN. Especially when two of these systems reach each other, for example, a westward rolling pattern across the Atlantic Ocean clashing with an eastward migrating CCKW can provide the mechanism for the generation of tropical cyclones. CCKW is also referred to as the active convective PHASE associated with the Madden–Julian OSCILLATION (named after the scientists and meteorologists from the United States: Roland A. Madden [1938–] and Paul Rowland Julian [1929–]; introduced in 1994). The El Niño—Southern Oscillation is a phenomenon that is related, only under a standing pattern. These waves generally travel with a PHASE VELOCITY of 10–17 m/s, which is a function of the geographic location. The time frame associated with Kelvin waves is exemplified by the fact that is takes approximately 45 days to 2 months for the CCKW to traverse the Pacific Ocean from west to east. However, the anomalies associated with the migration pattern (e.g., tropical storms) may arise as fast as weeks, over a DISTANCE in the order of several hundred kilometers (see Figure C.70).

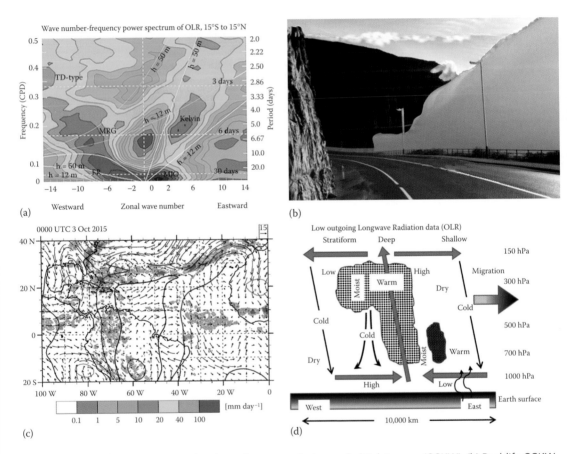

Figure C.70 (a) Schematic cross section through a convectively coupled Kelvin wave (CCKW). (b) Real-life CCKW phenomenon as observed on land in Dubai. (c) CCKW diagram at 40° and 70° latitude outlined by the "Frequency Zonal Wave number; frequency spectrum." (Courtesy of Michael Ventrice.) (d) OLR, outgoing longwave radiation data. (Courtesy of Paul Roundy.)

Cooper, Leon Neil (1930–)

[condensed matter, electronics, solid-state] Physicist from the United States. Leon Copper in collaboration with JOHN BARDEEN (1908–1991) and JOHN ROBERT SCHRIEFFER (1931–) developed a theoretical model for SUPERCONDUCTIVITY, referred to as the BCS theory of superconductivity. The Nobel Prize in PHYSICS was awarded to all three men in 1972 for their contributions.

Cooper pair

[condensed matter, electronics, solid-state] Pair of free electrons (conducting electrons) in a crystalline solid that behave as they are linked together by means of a long-range interaction. The Cooper pair provides a credible explanation of certain aspects associated with the phenomenon of SUPERCONDUCTIVITY. The first observation of superconductivity was in 1911 by the Dutch physicist HEIKE KAMERLINGH ONNES (1853–1926). The Cooper pair electrons will migrate through the LATTICE structure totally free of being impeded by the crystalline structure itself, nor any impurities, providing the conditions for zero RESISTANCE.

Copernicus, Nicolaus (1473–1543)

[astronomy, astrophysics, general] (alias of Niklas Koppernigk) German astrophysicist and MEDICINE student. His astrophysics work contributed (although delayed) to the adjustment of the calendar (Gregorian calendar 1581). Copernicus defended the HELIOCENTRIC planetary system (SUN as the center) with detailed scientific observations and theoretical proof in contrast to the accepted GEOCENTRIC system (EARTH as the center) postulated by CLAUDIUS PTOLEMAEUS (also PTOLEMY) (c. 90–168 AD). The Copernican heliocentric model also included the proof of a revolving Earth. Copernicus, however, still assumed the world to be a spherical object, which was challenged by WILLIAM GILBERT (1540–1603) in 1600, favoring an ellipsoidal model (see Figure C.71).

Figure C.71 Nicolaus Copernicus (1473–1543). (Courtesy of E. Scriven.)

Coriolis, Gaspard-Gustave de (1792–1843)

[fluid dynamics, general, mechanics] Physicist from France. In 1835 Coriolis recognized that the rotational MOTION affects the direction of the path of an object as well as the FLOW of liquids. Additionally, Coriolis introduced the displacement DISTANCE ($\vec{s}$) resulting from an applied force ($\vec{F}$) as work, $W = \vec{F} \cdot \vec{s}$, in 1829 (see Figure C.72).

Figure C.72 Gaspard-Gustave de Coriolis (1792–1843).

Coriolis effect

[computational, fluid dynamics, mechanics, thermodynamics] The principle that the direction of the path of an object as well as the FLOW of liquids is affected by rotational MOTION, introducing a *motion-dependent* acceleration and resulting redirection, while no apparent real external force is present. The acceleration of moving object under influence of a rotating reference frame was described by GASPARD-GUSTAVE DE CORIOLIS (1792–1843) in 1835. The phenomenon is however still also referred to as CORIOLIS FORCE. The Coriolis effect will for instance result in a higher water level on the east side of a river running south to NORTH on the northern hemisphere and on the western riverbank for a similar river on the southern hemisphere. The river itself on the northern hemisphere will also tend to starboard. The same will apply to a sail boat, but these effects are so small that they will be negligible, and may never be recognized due to other external forces, such as wind force. Alternatively, because of Coriolis effect the AIR flow on the northern hemisphere will be forced to move in counterclockwise direction, which is clearly visible when observing hurricanes ("Atlantic Ocean") and typhoons (cyclonic action in the northwest Pacific basin: western North Pacific ocean; between the 180° longitude and the 100° east longitude) as observed from a weather SATELLITE position (see Figure C.73).

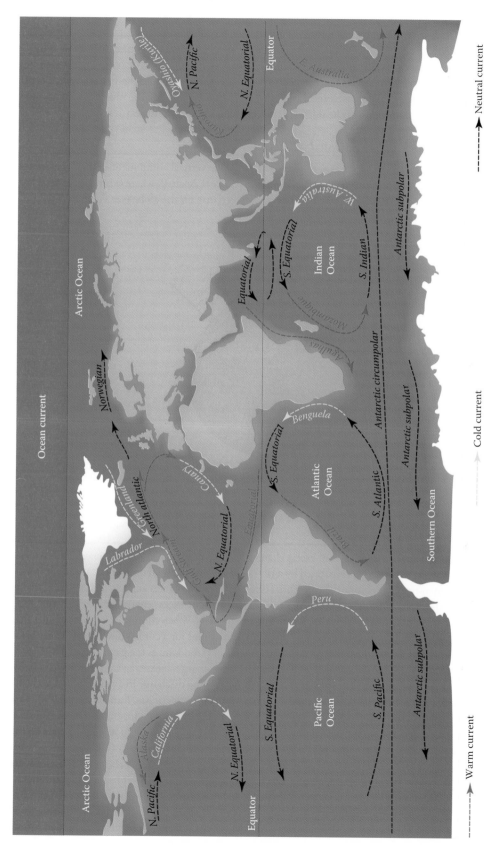

Figure C.73 The influence of the earth's rotation on the ocean currents under the Coriolis effect.

Coriolis force

[computational, fluid dynamics, mechanics, thermodynamics] Force exerted on an object with mass m by its movement with VELOCITY v with respect to a rotating system with ANGULAR VELOCITY ω expressed as $F = m * -2\vec{\omega} \times \vec{v}$, specifically identified by the acceleration aspect, $\vec{a} = -2\vec{\omega} \times \vec{v}$. The fictional force is primarily perpendicular to the MOTION of the object. This results, for instance, in having the water level on one side of a flowing river to be higher than the opposite side with respect to sea level due to the earth's rotation.

Coriolis frequency ($\nu_{coriolis}$)

[computational, fluid dynamics, mechanics, thermodynamics] The vertical component of the angular VELOCITY (ω), also referred to as the PLANETARY VORTICITY, defined as $\nu_{coriolis} = 2\omega \sin\theta$, where θ is the reciprocal ANGLE with respect to the normal to the earth's surface, that is, the LATITUDE on the globe (see Figure C.74).

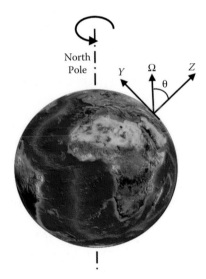

Figure C.74 The Coriolis frequency concept.

Corona (-discharge)

[electromagnetism, general] Electric FLUX of excess charges from a surface under the influence of a POTENTIAL DIFFERENCE V between two locations separated by a DISTANCE d. The potential difference may result from a variety of factors, man-made or natural. The potential difference thus generates an ELECTRIC FIELD E that is partially influenced by the radius of curvature (r_c) of the object with the excess charge (*also see* ELECTRIC FIELD IN CONDUCTOR as function of radius of curvature of the surface contour). When the radius is much smaller than the separation distance between the opposite charged points the conditions for a corona discharged arc are greater with exceeding ratio. The electric field resulting from the excess electron ionizes the gas(es) that fill the space between the charged points. The electric field and the discharge of charges creates a PLASMA in the GAS between the locations with opposite charge. One specific example is the glowing discharge in a neon light. Another example is lighting, which creates a PLASMA of the oxygen (O_2) in atmospheric AIR, forming ozone (O_3). Coronas are also used in copying MACHINES to charge nonconducting surfaces to hold ink for transfer of an IMAGE. Free ELECTRONS are drawn into the electric field in the CONDUCTOR, with a field distribution that depends on the radius of curvature, smaller radius higher field due to the higher surface charge density. The electrons (e) with electron charge are accelerated (a) by the electric field due to the ELECTRIC FORCE, $F = eE = ma$. The accelerated electrons collide with the resident neutral molecules at a high frequency of $\nu_{en} \sim 10^{12}$ Hz. At this point the ENERGY of the electrons is still limited by the INELASTIC COLLISIONS and remains below the PHOTOELECTRIC EFFECT energy. Only a limited number of electrons will escape from the surface of the conductor, contributing to the discharge arc. In standard air pressure with relative HUMIDITY 60%, a corona will form for a rounded surface with radius 200 μm under 5 kV electrical potential when placed at a distance of 10^{-2} m with a conductive plate. This discharge voltage increases to 10 kV when the diameter is 2 mm (*also see* ARC, LIGHTNING, SPARK, *and* ST. ELMO'S FIRE) (see Figure C.75).

Figure C.75 Corona discharge.

Cortical bone

[biomedical] Dense bone of the human skeleton, making up approximately 80% of the bone structure. The cortical bone is strong and can support large sheer force. The remaining bone structure is CANCELLOUS BONE. The main function of the cortical bone is to support and transfer force (e.g., femur, humerus, tibia, more generally [but not limited to] the bones in the arms and legs). Cortical bone forms the protective hard shell of all bone materials. The core material of skeletal bone still often consists of cancellous osseous tissue bones. It is also referred to as compact bone (see Figure C.76).

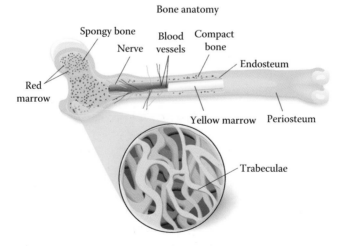

Figure C.76 Cortical (compact) bone layer of skeletal bone structure.

Cosmic force

[general] Set of laws applying to planetary and SATELLITE MOTION and gravitational interaction on galactic scale derived from the work by TYCHO BRAHE (1546–1601) and his successor JOHANNES KEPLER (1571–1630). The general concept of cosmic force is captured by the Kepler's THREE LAWS OF PLANETARY MOTION, next to the

description of satellite ORBITS (applying to comets and ASTEROIDS), and additionally, the description of the GRAVITATIONAL FIELD, including the hypothetical gravitational PARTICLE the "GRAVITON."

Cosmic rays

[astronomy/astrophysics, atomic] RADIATION consisting of both ELECTROMAGNETIC ENERGY as well as particles emerging from outer space as well as from the SUN. Cosmic rays consist of gamma rays, protons, ALPHA PARTICLES, neutrinos, and electrons next to a minor contribution from isotopes of heavy nuclei. The interaction of cosmic rays with atoms in the upper ATMOSPHERE of EARTH generate excitation and DECAY effects next to SCATTERING, which all result in (cascade of; depending on the incident energy) secondary radiation. The stream of PARTICLES and ELECTROMAGNETIC RADIATION reaching Earth from the SUN, the stars in our MILKY WAY as well as emission from other galaxies. The particles range in ENERGY from 10^6 eV to in excess of 10^{22} eV and are composed of PROTONS (~86%), HELIUM NUCLEI (alpha particles; 12.7%), heavy NUCLEI (1.3%) as well as ELECTRONS (~1%) and a TRACE of electromagnetic radiation (depending on the interpretation). At the lower end of the scale, GAMMA RAYS released in the production of π^0 eV MESONS have an energy of approximately 10^5 eV, higher energies are in electrons and positrons, increasing to ALPHA PARTICLES (He^{2+}). Additionally, NEUTRINOS and a mixture of nuclei are also detected using methods including BUBBLE–CHAMBER. At the highest range the energetic content is dominated by protons at increasing VELOCITY. The particle FLUX ("j") for PARTICLE in excess of a specific threshold energy (E) is expressed as $j(> E) = K_{cosmic} E^{-\delta}$, where K_{cosmic} and δ are positive constants in the energy regime described by the POWER LAW $E \geq 10\,\text{GeV/nucleon}$.

Cosmic string

[astrophysics, computational, nuclear, quantum, theoretical] In the relativistic QUANTUM theoretical description of symmetry breakdown there are hypothetical defects or DISCONTINUITIES in linear dimension that are referred to as cosmic strings. The theoretical concept is bound by ENERGY in excess of ~ TeV in very short dimensional PHYSICS. The cosmic string is part of STRING THEORY, in this case relating energy configurations of oscillating open string networks that extend to infinity as well as vibrating closed loops. The general concept of string theory is captured in SCHRIEFER UNIFIED THEORY.

Cosmonaut

[astronomy/astrophysics, general] Astronaut under the Union of Soviet Socialist Republics (USSR; now Russian Federation, "Russia") space flight program (see Figure C.77).

Figure C.77 Russian cosmonaut Yuri Gagarin.

Couette flow

[fluid dynamics] Laminar FLUID FLOW between plates with a gradual, and frequently linear, gradient in flow velocity. Couette flow obeys the simplified Navier–Stokes equation, $d^2u/dy^2 = 0$, where u represent the flow velocity parallel to the planes in the "X-direction" and y is the plate separation DISTANCE: the plates can move with respect to each other, the plates can be curved for the case of concentric cylinders, and the flow can describe the fluid MOTION between stationary plates. The flow velocity as a function of distance between two plates at separation h for a fluid with VISCOSITY η with respect to a PRESSURE (P) gradient is in the direction of flow x can be described under two conditions: (1) moving plate or (2) stationary plates. The moving (sliding) plates SOLUTION for relative velocity u_0 yields $u(y) = u_0(y/h) + (1/2\eta)(dP/dx)(y^2 - hy)$. The laminar Couette flow, with average flow velocity $u_{avg} = -(h^2/8\eta)(dP/dx)$ between stationary plates is described with respect to the center plane at $h/2$ in the form $u = 3/2u_{avg}[1 - (4y^2/h^2)]$, which has similarity to POISEUILLE FLOW and will develop into Poiseuille flow for larger flow velocity. Additional combination solutions are available, with a special case of plates moving in opposite direction. Couette flow generally describes shear stress-dependent flow, such as found in the connection between the piston rod and the crankshaft for an automobile ENGINE, however only below a critical REYNOLDS NUMBER. The flow velocity between two rotating concentric cylinders, with R_1 the inner cylinder radius, R_2 the outer cylinder radius, and ω the relative ANGULAR VELOCITY, as a function of radius (r) is $u_\theta(r) = (\omega R_1^2/r)[(R_2^2 - r^2)/(R_2^2 - R_1^2)]$. In Poiseuille flow, the flow velocity has a quadratic dependence to the location on the radius of the tube under an applied pressure gradient along the length of the tube. Similarly, in Couette flow the flow between two stationary plates has a quadratic dependence to the midpoint plane between the two plates. In comparison, the Poiseuille flow MAGNITUDE in a tube will have a dependency to the tube radius to the fourth power where Couette flow will only yield a maximum of quadratic dependence on plate separation. This type of flow is primarily found under low flow velocity, or alternatively low translational velocity of two plates with respect to each other, separated by a viscous fluid. Couette flow is not stratified and has constant vorticity. Under conditions of spanwise rotation, a plane turbulent Couette flow can develop where the velocity is SPIN in forward and backward flow with respect to the central line between the two moving object, again with only gradual gradient in flow velocity as a function of separation distance, and zero flow velocity on the central dividing plane (see Figure C.78).

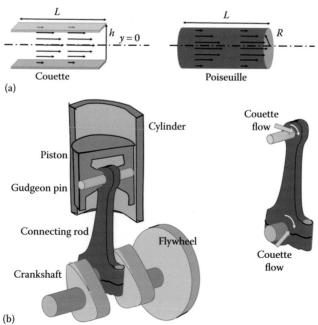

Figure C.78 Couette flow: (a) comparison between Couette and Poisseuille flow and (b) example of locations for potential Couette flow.

Coulomb, Charles-Augustin de (1736–1806)

[energy, general] French scientist whose publication of electric attraction in 1785 (*Histoire et Mémoires de l'Académie Royal des Sciences*) "Coulomb's Law" opened the doors for the BOHR ATOMIC MODEL as well as (see Figure C.79).

Figure C.79 Charles-Augustin de Coulomb (1736–1806).

Coulomb, unit

[chemical, electronics] ELECTRIC CHARGE unit, defined indirectly from the Lorentz force between two parallel wires conducting a current of 1 ampere, where $1A = 1C/s$ defines the FLOW of an electric charge of 1 coulomb across the cross-sectional area of the wire per SECOND.

Coulomb apparatus

[general] TORSION balance used by CHARLES-AUGUSTIN DE COULOMB (1736–1806) to derive the ELECTRO-STATIC FORCE between ELECTRIC CHARGES. A gold leaf-covered ball attached to a rod is suspended from a silver wire, counterbalanced by a loop with a paper disk. The suspended ball will be charged by a rod that has an electric charge applied, for instance, a GLASS rod that has been rubbed by silk. The suspended rod is in the same horizontal plane with a ball of identical material and size as the suspended ball is fixed in three-dimensional space by a nonconducting geometric device. When the two balls are in contact, and if the balls are touched by a charged rod, both will be loaded with the same electrostatic charge (both in sign and MAGNITUDE) and will REPEL. The balls will move away from each other to the point where the TORQUE and electrostatic force are balanced, based on NEWTON'S THIRD LAW, and the charge can be derived. Based on COULOMB'S LAW the electrostatic force will be equal to the force applied to the arm of the suspended rod, expressed as torque.

Coulomb potential

[atomic, condensed matter, energy] $V^{\text{Coulomb}} = e^2/r$, where $e = -1.6 \times 10^{-19}C$ is the electron charge, however, this also applies to protons as a POSITIVE CHARGE, and r the separation between the protons.

Coulomb's law, electrostatic charge

[chemical, electronics, mechanics] The force of attraction or repulsion (F_e) exerted between two electrostatic charges, Q_1 and Q_2, respectively, with a DISTANCE, s, separated by a medium of DIELECTRIC value, ε, is given by the equation $F_e = (1/4\pi\varepsilon)(Q_1 Q_2/s^2)$. Introduced by CHARLES-AUGUSTINE DE COULOMB (1736–1806).

CPT theorem

[computational, thermodynamics] Charge–space–time theorem. Invariance of the LAGRANGIAN THEORY under standard "proper" LORENTZ TRANSFORMATION, specifically pertaining to QUANTUM FIELD THEORY. Additionally the Lagrangian theory is invariant to TIME reversal (in either direction) (T), CHARGE conjugation (C), and SPACE INVERSION (P). In CHARGE CONJUGATION, all particles are transformed into their respective antiparticles. The independence to the space-time CONTINUUM requires the following condition to maintain invariance, $\Lambda_{\zeta\varkappa}\Lambda_{\zeta\mathcal{G}} = \delta_{\varkappa\mathcal{G}}$, yielding a KRONECKER DELTA FUNCTION for the MATRIX multiplication.

Crab Nebula, pulsar

[general] Remnants of a SUPERNOVA that presumably exploded in 1054 AD, based on documented observations by Chinese astronomers (keeping in mind the rudimentary tools available at the time and the DISTANCE). The Crab Nebula is at approximately 2000 parsecs from EARTH $\left(1 \text{ parsec} = 3.26 \text{ light-years} = 3.1 \times 10^{16}\,\text{m}\right)$. According to the documentation, the explosion was visible for two years, three weeks of this was also during the daytime. At the center of the nebula is a pulsating NEUTRON STAR, spinning in similar fashion to an old-fashioned lighthouse beam, with steady pulsating frequency of 30 Hz. The estimated size of this pulsating neutron star is merely 25 km, but has the mass of our SUN (see Figure C.80).

Figure C.80 Crab Nebula, part of the constellation Taurus. Remnant of a supernova that supposedly exploded in 1054 AD. The Crab Nebula has a strong pulsar neutron star in its core. (Courtesy of NASA/STScI, Space Telescope Science Institute, Baltimore, MD.)

Creep

[general, mechanics] Anelastic deformation of a solid medium. Anelastic creep is defined by a nonlinear elastic behavior described as a function of stress ($\sigma_{\text{stress}} = F/A$, the force F acting parallel to the area A) and strain ($\epsilon_{\text{strain}} = \Delta L/L_0$; L the length): $J_R\sigma + \tau J_u(\partial\sigma/\partial t) = \epsilon + \tau(\partial\epsilon/\partial t)$, where J_R, J_u, and τ are material constants and τ the RELAXATION TIME with respect to deformation. The material can be defined by a complex COMPLIANCE, $J^* = \epsilon_{\text{strain}}/\sigma_{\text{stress}} = J_1 + iJ_2$, where the real part of the compliance (J_1) is in PHASE with the applied stress. For a cyclically (ANGULAR FREQUENCY ω) applied stress the compliance is related to the constants as $J_1(\omega) = J_u + \left[(J_R - J_u)/(1+\omega^2\tau^2)\right]$ and $J_2(\omega) = (J_R - J_u)\omega\tau/(1+\omega^2\tau^2)$ (known as the DEBYE EQUATIONS). The relaxation process under elastic deformation adheres to $\tau = \tau_0 e^{E_A/k_b T}$, as a function of the activation ENERGY required for the process and the temperature T (Boltzmann coefficient: $k_b = 1.3806488 \times 10^{-23}\, \text{m}^2\text{kg/s}^2\text{K}$). Applying specific boundary conditions this will provide three specific types of creep: anelastic, secondary, and tertiary creeps. For instance, anelastic creep can be assumed under constant stress, which provides $\epsilon_{\text{strain}}(t)/\sigma_{\text{stress},0} = J_u + (J_R - J_u)[1 - e^{-(t/\tau)}]$. This process also applies to phonon propagation and "dislocated" LATTICE vibrations (vibrational damage). In biomedical context, this can relate to HEARING losses resulting from excessively loud music for extensive periods of time (see Figure C.81).

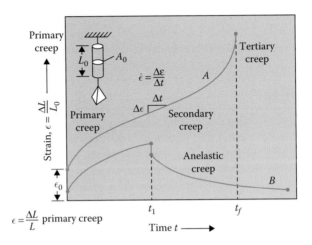

Figure C.81 The creep phenomenon.

Crick, Francis Harry Compton (1916–2004)

[biophysics, general] Molecular biologist, neuroscientist, and biophysicist from England, codiscoverer of the double helix structure of DNA with JAMES DEWEY WATSON (1928–), published in 1953. He dedicated his work to unraveling the mystery of the genetic code, and published the DNA–RNA–protein correlation as a one-way FLOW of genetic information (see Figure C.82).

Figure C.82 Francis Harry Compton Crick (1916–2004).

Critical angle (θ_c)

[general] ANGLE of incidence at an interface between two media of different index of REFRACTION at which total REFLECTION occurs. The critical angle is based on the refracted angle of 90° under SNELL'S LAW. The critical angle is of particular importance in FIBER-OPTIC transmission (see Figure C.83).

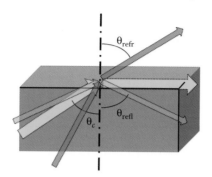

Figure C.83 Critical angle for refraction/reflection.

Critical point (c, subscript: "c")

[general, thermodynamics] The boiling point of liquids is directly correlated to the pressure. An increase in pressure will result in a proportional increase in boiling temperature. The difference in density between VAPOR and FLUID phase diminishes with increasing pressure and temperature until the critical point is reached and the density of the LIQUID and vapor phases become equal. This phenomenon is a SECOND-ORDER PHASE TRANSITION. Examples of media operating at the critical point are superconductors, ferro-magnets experiencing spontaneous magnetization, ferroelectrics exhibiting POLARIZATION, and binary fluids, which will separate out the two phases of the constituents (*see* CRITICAL TEMPERATURE [T_c] *and* CRITICAL PRESSURE [P_c]) (see Figure C.84).

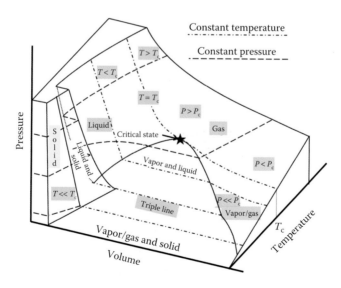

Figure C.84 Dependence of the conductivity of a medium with respect to the critical temperature.

Critical size

[nuclear] The minimum amount of a fissionable material that will support a CHAIN REACTION.

Critical temperature (T_c)

[general, thermodynamics] Temperature below which a material relinquishes its RESISTANCE and becomes superconductive (see Figure C.85).

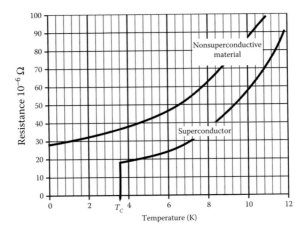

Figure C.85 Critical point in a phase diagram for a substance that contracts when melting, for instance, water. In contrast, carbon dioxide is a substance that will expand upon melting.

Crookes, William, Sir (1832–1919)

[atomic, electronics, nuclear] Physicist from Great Britain who introduced the investigational ELECTRODE tube, the CROOKES TUBE in 1870 (see Figure C.86).

Figure C.86 Sir William Crookes (1832–1919). (Courtesy of Leslie Ward, *Vanity Fair*, 1903.)

Crookes tube

[atomic, electronics, nuclear] VACUUM CRT introduced by SIR WILLIAM CROOKES (1832–1919), predecessor to the modern neon light signs. The Crookes tube was based on the earlier observations in a vacuum BAROMETER by the French physicist JEAN PICARD (1620–1682) (see Figure C.87).

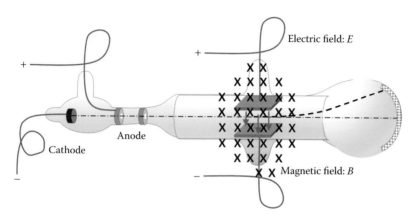

Figure C.87 Crookes tube and mechanism of operation.

Cross section

[atomic, general, mechanics, nuclear] The area subtended by an ATOM or molecule for the PROBABILITY of a reaction, that is, the reaction probability measured in units of area.

Cryodesiccation

[biomedical, energy, general] Freeze-drying, *also see* LYOPHILIZATION. Preservation mechanism for biological media used to slow down the chemical processes as well as facilitate the removal of water by means of deep-freeze quick cold storage.

Cryogenics

[atomic, biomedical, electronics, general, nuclear] Mechanism of producing extremely low temperature to observe the changes in physical properties of a material. The most known applications in cryogenics are in superconductivity. Most cryogenic processes take place at temperatures below 123K ~ −150°C. In biomedical applications the temperature requirements are not that extreme, for instance, the preservation of sperm for in vitro fertilization. In biological applications, generally, the use of frozen carbon dioxide (also known as "dry-ice") provide the mechanism for quick reduction of temperature by submersion. The MELTING POINT of the solid CO_2 is ~ −79°C. Attempts of cryogenics on large objects is primarily limited by the heat DIFFUSION and heat capacity of the object in question. FREEZING entire human bodies, for instance, will require PERFUSION with LIQUID CO_2, otherwise the core will remain too warm for too long a period and there will be risk for cracking of the solid outer shell due to the fact that most materials either expand or reduce in size with decreasing temperature. At lower temperatures, the molecular mobility will reduce (*see* TEMPERATURE) resulting in a reduction of the PROBABILITY for collision of migrating free electrons under the influence of an applied external electric field, that is, SUPERCONDUCTIVITY.

Crystalline lens

[biomedical, general, optics] ADAPTIVE OPTICS instrument in the HUMAN EYE that can be adjusted for FOCAL LENGTH by contraction and relaxation of the MUSCLE in the EYE attached to the LENS. When the muscles are relaxed the lens is most rounded with the IMAGE of the "near point" focused on the RETINA at approximately 24 mm from the lens. On contraction the lens flattens out, increasing the focal length, as defined by the LENS EQUATION, to the point of focusing "infinity" on the retina. The near point is at approximately 25 cm DISTANCE from the eye for the average adult, whereas children have a shorter near point (*see* ACCOMMODATION) (see Figure C.88).

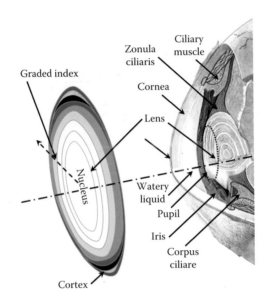

Figure C.88 Crystalline lens of the human eye forming an image on the retina for signal acquisition and processing by the brain.

Cubic amplifier

[computational, electromagnetism, general, theoretical] Nonlinearity in electric amplifier. Amplifier with a gain performance profile that can mathematically be described by a polynomial up to the third order (i.e., cubic) without significant loss in accuracy. The polynomial terms can incorporate the higher harmonics and desensitization of INTERFERENCE (i.e., band-block filter or "jammer") (*also see* VAN DER POL EQUATION).

C

Curie, Manya (Marie) Skłodowska (1867–1934)

[atomic, general, nuclear] Scientist and physicist, who left her homeland of Poland to start her doctoral studies at the French Sorbonne. Because of the exposure to RADIATION Marie Curie eventually died of cancer (see Figure C.89).

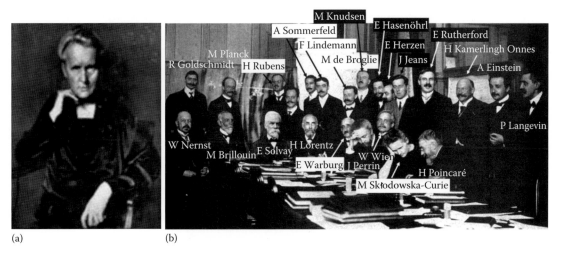

(a)   (b)

Figure C.89 (a) Manya (Marie) Skłodowska Curie (1867–1934) at the 1911 Solvay conference: "Conseil Solvay," organized by the industrialist Ernest Solvay by invitation only. Ernest Gaston Joseph Solvay (1838–1922) chemist from Belgium, shown in (b). The conference was held in Brussels in the latter part of 1911. The subject of the conference was a very popular topic of the day: "Radiation and the Quanta." The conference became open to the general scientific audience in 1912, upon peer review.

Curie, Paul-Jacques (1856–1941)

[atomic, general, nuclear] A scientist from France responsible for the discovery of the piezoelectric ACTIVITY of crystalline structures under deformation in 1880. Some of the crystalline structures under investigation were QUARTZ, tourmaline, and Rochelle SALT. Brother of PIERRE CURIE (1859–1906) (see Figure C.90).

Figure C.90 Paul-Jacques Curie (1856–1941).

Curie, Pierre (1859–1906)

[atomic, general, nuclear] Scientist from France mostly known for his work on RADIOACTIVITY with his wife MARIE CURIE (1867–1934). Exemplary contributions of Pierre Curie are the discovery of radium and polonium, next to his work on his work on crystals and MAGNETISM, specifically the formal description of piezoelectric behavior. The husband–wife research couple shared a Nobel Prize in PHYSICS with ANTOINE HENRI BECQUEREL (1852–1908) in 1903 for their work on RADIATION (see Figure C.91).

Figure C.91 Pierre Curie (1859–1906).

Curie, unit (Cu)

[atomic, nuclear] Unit of radioactive disintegrations, $1\,Cu = 3.70 \times 10^{10}$ disintegrations/s. The number of disintegrations closely resembles the radium emittance of ALPHA PARTICLES per SECOND from 1 g; however, the Curie is independent of the PARTICLE disintegration and particular particles emitted. The value of $1\,Cu$ indicates a very powerful RADIOISOTOPE: $7.4 \times 10^{4}\,[(\beta - \text{particle}/s)] = 2\mu Cu$.

Curie law

[atomic, general, nuclear] Behavior of paramagnetic devices/materials described by PIERRE CURIE (1859–1906). The Curie law relates the MAGNETIC DIPOLE MOMENT (M_B) to the applied MAGNETIC FIELD B as $M_B = C_c\,(B/T)$, where C_c is the material constant and T the ABSOLUTE TEMPERATURE (in Kelvin). The inverse proportionality with respect to temperature ties in with thermal agitation disrupting alignment.

Curie temperature

[atomic, general, nuclear] The temperature at which the magnetic structures (also called domains) in a PERMANENT MAGNET are disrupted by the kinetic MOTION and the magnet relinquishes its magnetic identity. This phenomenon was discovered by PIERRE CURIE (1859–1906) in 1894. The Curie temperature is material specific, for instance, for magnetite it is 575°C and for hematide 675°C, respectively.

Curium ($^{247}_{96}$Cm)

[atomic, nuclear] RADIOACTIVE ISOTOPE in the actinide series. The ELEMENT is named after Curie husband and wife research team: MARIE (MANYA) SKŁODOWSKA CURIE (1867–1934) and PIERRE CURIE (1859–1906).

Curl ($\nabla \times \overline{\mathbb{V}}$)

[computational] The mathematical vector operator describing "rotation" of a vector or function ($\overline{\mathbb{V}}$) within an area (A) at a location ($\vec{r}$); defining the rotation of the vector in infinitesimal steps, hence describing the projection of the vector upon all possible lines passing through the location in the three-dimensional space. The curl is expressed as $\nabla \times \overline{\mathbb{V}} \cdot \vec{n} = \lim_{A \to 0}\left(1/|A|\oint_C \overline{\mathbb{V}} \cdot d\vec{r}\right)$, observed as the closed integral over a loop (C), with direction defined by unit direction vector $\vec{n}$ which is the axis of vector rotation. It is also referred as cross-product. A vector field with curl = 0 is considered "irrotational."

Curl-free vector field

[computational, fluid dynamics] A fundamental component of HELMHOLTZ DECOMPOSITION in vector calculus. In three-dimensional space, the vector field composed of sufficiently smooth, rapidly decaying vectors can be decomposed into the sum of two "orthogonal" components, one consisting of an IRROTATIONAL (CURL-FREE) VECTOR FIELD and the second segment composed of rotational (divergence-free) or solenoidal vector field ($\vec{V}$). The formulation of this series is defined as a function of location ($\vec{r}$) within a volume in space (V) as $\mathbb{C} = (1/4\pi)\int_V^{\Omega} \nabla' \cdot \vec{V}\left(\vec{r}'\right)/|\vec{r} - \vec{r}'| dV' - (1/4\pi)\int_S^{\Omega} \vec{V}\left(\vec{r}'\right) \cdot dS'/|\vec{r} - \vec{r}'|$, the second term will have limit zero when the phenomenon comprises all of free space with the vector field rapidly decreasing.

Current (*I*)

[electromagnetism, general] Electron flow; more specifically the FLOW of "holes," since the current direction is in the opposite direction of the displacement of electrons. The current flow was based on the assumption that positive PARTICLE where involved, before the analysis of the atomic structure was performed or understood. The flow of electrons relies on a gradient in applied electrical potential (voltage) with an inherent electric field gradient.

Cyclone

[fluid dynamics, geophysics, meteorology] Low-pressure system, generating a FLUID volume rotating in the same direction as the EARTH. The pressure gradients associated with the cyclonic action generates winds that FLOW counterclockwise on the northern hemisphere and clockwise on the southern hemisphere. A meteorological cyclone is generally associated with thunderstorms and heavy rainfall (see Figure C.92).

Figure C.92 Image from satellite showing cyclone off the Florida coast.

Cyclotron

[atomic, nuclear] PARTICLE ACCELERATOR. Charged particles are released from the center of a circular accelerator and gain velocity with increasing radius (spiral path) as a result of an applied MAGNETIC FIELD. The cyclotron is used to investigate the structure of atomic nuclei and create new ELEMENTS. The cyclotron was invented by ERNEST ORLANDO LAWRENCE (1901–1958) in the 1920s, and was not feasible for practical use until a large enough magnetic field could be generated in the 1930s. Cyclotrons range in size from less than a meter to several meters. The largest cyclotron is the RIKEN in Japan, with a diameter of 19 m, 8 m high, and can operate with a maximum magnetic field strength of 3.8 T. The RIKEN cyclotron can reach 345 MeV of acceleration ENERGY per atomic mass unit. The cyclotron designed by Ernest Lawrence was built in 1939 has a diameter of 1.52 m. A block-wave periodic magnetic field is applied that matches the RESONANCE conditions to induce acceleration, based on the orbital period of the charge. The cyclotron frequency (angular velocity $\omega_{cyclotron}$; frequency $v_{cycloton} = \omega_{cyclotron}/2\pi$) is a function of the applied MAGNETIC field (B) and the MAGNITUDE of the CHARGE (q) on the MASS (m) as $\omega_{cyclotron} = qB/m$ (see Figure C.93).

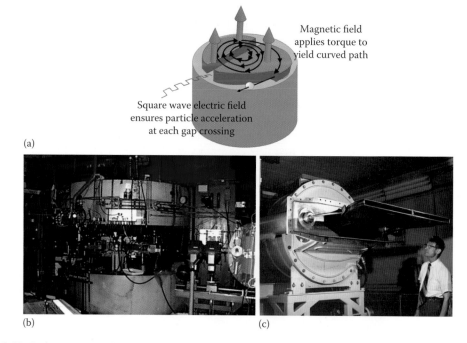

(a)

(b)   (c)

Figure C.93 Cyclotron: (a) mechanism of action for particle acceleration, (b) 1.52 m loop at Argonne National Lab, and (c) David Lind at 26 MeV cyclotron beam guide during the 1962 construction. (Courtesy of University of Colorado, Boulder, CO.)

Da Vinci, Leonardo; Leonardo di ser Piero da Vinci (1452–1519)

[biomedical, computational, fluid dynamics, general, mechanics] A scientist and "renaissance man" from Italy. Leonardo made a significant number of inventions and elaborate ENGINEERING drawings of devices not realized for many centuries (e.g., helicopter, submarine, and SCUBA GEAR). He was an initial proponent for the old VISION theory of EXTRAMISSION (approx. 1480), objects radiate out, but later changed his views after anatomical examination of the EYE and became a staunch supporter of the "refection" mode of observation. Additional anatomical observations made him conclude that the skeleton is an elaborate system of levers, powered by MUSCLE, hence introducing the concept of torque. Leonardo also made several devices and instruments for automation of certain tasks. He formed a rudimentary theoretical description of the movements of TECTONIC PLATES. He concluded that the surface on which an object is placed pushes back with the same force as what is resting on it (the weight). He also realized that FRICTION was proportional to the "weight" of an object, the normal force, but different depending on surface treatment or choice of materials. He also drew a seemingly insignificant but important conclusion that the forces between object and surface are independent of the contact area, hence without effect on the friction; placing a book flat or on its edge will not affect the friction. Leonardo's interpretation of FLUID waves also gave him the crucial realization that the WAVE does not transport fluid in the direction of the wave propagation. He also concluded that SOUND is propagated by means of waves; although ARISTOTLE (c. 384–322 BC) stated in 350 BC that sound travels to the EAR through the AIR medium, he did not specify the wave phenomenon as such. The principles of sound waves were further refined by GALILEO GALILEI (1564–1642) in 1600. Leonardo was an Italian civil engineer, physicist, architect, and artist (proverbial Renaissance man). As an artist, he introduced realism in the observation and depiction of the HUMAN BODY. His artistic studies of bird flight led to his conceptual drawings of equipment that would allow man to fly. His scientific work showed in-depth understanding of the human BLOOD CIRCULATION mechanism.

He also elaborated on human sight and, in the same line of thought, alluded to the use of a CAMERA to capture a still frame (see Figure D.1).

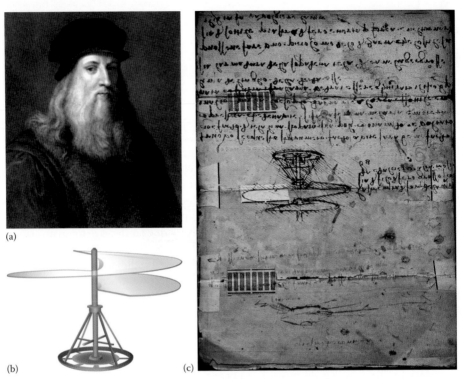

(a)

(b)   (c)

Figure D.1 (a) Leonardo di ser Piero da Vinci (1452–1519) the artist and scientist, with additional works representative of his insight into the geometry, anatomy, and science of people, objects, and devices, engraved by James Posselwhite; (b) design of a flight contraption, now called a plane, by Leonardo da Vinci; and (c) design of the helicopter concept based on the original ideas and drawings by Leonardo da Vinci.

Daguerre, Louis-Jacques-Mandé (1787–1851)

[optics] An artist and scientist from France. Louis Daguerre was instrumental in the invention and development of PHOTOGRAPHY (see Figure D.2).

Figure D.2 Louis-Jacques-Mandé Daguerre (1787–1851), photographed by Jean-Baptiste Sabatier-Blot.

Daguerreotype

[optics] A photographic IMAGE produced on a polished silver-plated sheet of copper, introduced in 1839 by LOUIS-JACQUES-MANDÉ DAGUERRE (1787–1851). The process was based on the CAMERA OBSCURA. The camera obscura was documented as created and first used by Mozi (c. 470–390 BC), a philosopher from China. Later in the sixth century, the Byzantine-Greek mathematician and architect Anthemius of Tralles (474–534 AD) describes the use of this mechanism for his work. The more established version was described by ROGER BACON (1214–1294), a scientist from Great Britain, during the observation of solar eclipses. The photosensitive material used was silver sensitized by iodine (see Figure D.3).

Figure D.3 Daguerreotype photograph of the 12th president of the United States: Zachary Taylor (1784–1850). (Courtesy of Mathew Brady.)

D'Alembert, Jean-Baptiste le Rond (1717–1783)

[computational] A mathematician from France. The work of d'Alembert was in the theoretical description of several PHYSICS phenomena, including FLUID DYNAMICS and equilibrium conditions, as well as in developing specific PARTIAL differential equations. Specific examples are the process of solving the one-dimensional WAVE EQUATION, inertia (potentially inspired by the laws of Newton, SIR ISAAC NEWTON [1642–1727], in the years 1743–1754), fluid dynamics (equilibrium conditions for fluid in MOTION, 1744), the motion of AIR ("wind," 1745), and the partial differential equation for the wave on a string (one-dimensional wave equation, 1747). Additional work of d'Alembert was in the description of celestial ORBITS, specifically the deviations and obliquity of the elliptic trajectory and perturbations, the PRECESSION of the equinoxes, and the general position of the EARTH in the SOLAR SYSTEM. The d'Alembert's paradox in fluid MECHANICS describes the fact that an object submerged in an inviscid, incompressible FLOW experiences no DRAG (see Figure D.4).

Figure D.4 Jean le Rond d'Alembert (1717–1783), engraved by W. Hopwood.

Dalton, John (1766–1844)

[atomic, chemical, energy, general, nuclear, quantum, thermodynamics] A scientist and school master from Great Britain. John Dalton was aware of the combination of ELEMENTS forming new structures (now known as MOLECULES), and with several elements known, he recognized that hydrogen was the lightest and assigned it the mass 1, which is still the accepted convention. He made attempts to construct a table of elements but was incorrect and incomplete. He did inspire the scientist DMITRI IVANOVICH MENDELEEV (1834–1907) from Russia to produce the PERIODIC TABLE OF ELEMENTS, which we currently use. Dalton was a contemporary of AMEDEO AVOGADRO, COUNT OF QUAREGNA (1776–1856), who in 1811 introduced the Avogadro number. In 1804, Dalton recognized that the relationship between the TEMPERATURE (T) and PRESSURE (P) for the boiling-point curve at the LIQUID-VAPOR equilibrium has a logarithmic dependency. This is clearly visible in the pressure versus temperature (P–T) PHASE DIAGRAM (see Figure D.5).

Figure D.5 John Dalton (1766–1844).

Dalton's law

[biomedical, fluid dynamics, thermodynamics] The sum of all the pressures (P) provided by the constituents of a GAS MIXTURE provides the total pressure; each component provides their respective partial pressure $P = \sum_{i=1}^{N} P_i$, where the partial pressure is the fractional component based on molar ratio (γ_m): $P_i = \gamma_{m,i} P$ (*also see* **DALTON'S LAW OF PARTIAL PRESSURES**).

Dalton's law of partial pressures

[thermodynamics] *See* **DALTON'S LAW**.

Damadian, Raymond Vahan (1936–)

[biomedical, computational, imaging] A scientist and mathematician from the United States, who also earned his medical degree in 1960. Damadian is the inventor of the nuclear MAGNETIC RESONANCE imaging modality in 1969. His research in tracking sodium and potassium during his appointment at SUNY (State University of New York) made him realize the potential of the imaging with magnetic fields. His research was partially based on the observations made during World War I where inanimate object exposed to strong alternating magnetic fields caused the nuclei to emit radiofrequency at characteristic wavelengths for the atomic structures involved. He discovered a change in RELAXATION TIME for radio-frequency emission from the nuclei of tissue cells resulting from pathological conditions, specifically cancer. It is now also a well-known fact that cancer cells have an enlarged NUCLEUS compared to the healthy equivalent.

Damage integral

[biomedical, energy] An estimation of the damage to material, chemical reaction, or biological tissue resulting from an insult. The PROBABILITY distribution is based on material properties and prior history of similar processes and materials. Generally, the insult involves the thermal response to exposure in excess of a verified threshold, or tiers or successive thresholds. When there are initially N_0 molecules of one species in a volume of medium (e.g., tissue), the number of surviving molecules is directly proportional to the native number, and the number of thermally altered molecules dN as a function of time can be shown to obey the following equation: $dN = k_r N dt$, where k_r is the reaction coefficient, specific for the molecules involved and generally a function of TEMPERATURE (T). The irreversible thermal changes to proteins are referred to as denaturation; boiling an egg provides the classic example, as is baking/grilling a steak or boiling milk. The number of surviving native molecules as a function of TIME (t) can be derived as $N(t) = N_0 e^{-\Omega(t)}$, where $\Omega(t)$ represents the damage integral. For both the egg and the steak, the thermal changes are clearly visible, as is the influence of time on the process. Another example is coagulation of collagenous tissue, which renders the collagen into gelatin, with the loss of collagen fiber bundle structure. The damage integral is defined by the ENTROPY (S) and ENTHALPY (H) of the "PHASE TRANSITION" or chemical transformation (e.g., denaturation). The damage integral is defined as $\Omega(t) = \int_0^t k_r(t')dt' = \int_0^t [k_B T(t')/h] e^{\Delta S/R} e^{-(\Delta H/RT(t))} dt' = A_* \int_0^t e^{-(E_*/RT(t))} dt'$, where R is the universal GAS constant (8.314×10^3 J/kmolK) and $k_B = 1.3806448 \times 10^{-23}$ m^{2}kg/s^{2}K is the BOLTZMANN COEFFICIENT. At any time during the thermal assault, the number of denatured molecules in the carbon-based medium, $N_{denatured}(t)$ [respectively chemically altered molecules: $N_{change}(t)$], equals $N_{change}(t) = N_0 - N(t)$. In general, the damage integral combines the average values for all molecular species involved in the denaturation process within a specific tissue volume. The rate constant k_r can have a dissimilar value for different transitions as a function of temperature and exposure time. Typical values of ΔS and ΔH for biological media as examples of thermal coagulation are listed in Table D.1. One common

Table D.1 Thermal denaturation parameters

TISSUE	ΔS (J/[MOL K])	H × 10³ (J/MOL)
Connective tissue	392	221
Muscle	510	259
Collagen	825	368

manner of setting the boundary conditions is to use the exposure time (t_Ω [s]) required to achieve $\Omega = 1$, which corresponds to $t_\Omega = 1/k_r$ as a function of the temperature. The more complex the molecular structures that are denatured, the sharper the rate constant that changes with temperature, because the change in entropy (ΔS) is greater as well as the number of bonds to be broken concurrently (ΔH) is larger in number. For the denaturation of a less-structured chemical matrix, the change in rate constant will proceed more gradually due to the fact that the changes in ΔS and ΔH do not respond as aggressive due to temperature. The choice of the change in enthalpy ΔH as the independent variable represents the concept of the ENERGY of all bonds that conjointly hold a molecular structure together. The entropy increase can be considered a dependent variable that describes the entropic energy change resulting from breaking the bonds.

Published data list the following average empirical values for SKIN tissue: $E_{*,skin} = 627\,kJ/mol$ and $A_{*,skin} = 3.10 \times 10^{98}\,s^{-1}$; for collagen, the published values are, respectively, $E_{*,collagen} = 89\,kJ/mol$ and $A_{*,collagen} = 1.0 \times 10^{130}\,s^{-1}$. The damage region $0 \le \Omega(t) \le 0.53$ is generally associated with no permanent damage or no damage at all to the tissue. The damage value in the range $0.53 < \Omega(t) < 5$ is associated with coagulation. Generally, the damage value $\Omega(t) < 1$ represents the creation of reversible tissue damage. The value $\Omega(t) = 1$ for biological media defines the point of onset of tissue coagulation, with a 63% probability of tissue denaturation. A value of $\Omega(t) = 4.6$ corresponds with a 99% probability of tissue denaturation. Tissue carbonization ("burning" the steak on the grill) can be considered to occur under the condition that the damage integral provides $\Omega(t) \ge 5$. Additionally, a second-degree burn can be indicated with $\Omega(t) > 10$, while a third-degree burn requires $\Omega(t) > 10,000$. Using these numbers to determine the time exposure required to obtain damage from skin exposure to 135°C resulting in a second-degree burn yields a value of approximately 150 s or 2.5 min, and a third-degree burn can be projected after an order of MAGNITUDE of 42 h (1 day and 18 h) total exposure. The damage integral is used to plan therapeutic procedures in clinical applications. One specific application is the removal of port-wine stain, while avoiding the formation of thermally generated scar tissue within the dermis. This can be accomplished by cooling the exterior, hence lowering the temperature and thus reducing the damage integral value for the dermis. For instance, laser heating of BLOOD vessels generates a high temperature transiently achieved subcutaneously from transdermal irradiation causing thermal damage to the vessel WALL and either shrinking of the vessel or eliciting clotting (thrombosis) within the vessel. The body's wound-healing response subsequently removes the clotted vessel, hence removing the port-wine stain lesion. Keeping in mind that irreversible damage to the skin tissue occurs at approximately 58°C or 60°C (short-term exposure). Comparatively, the instantaneous (i.e., less than a second) coagulation of blood does not occur until a temperature in the range 90°C to 100°C is achieved. This controversy makes avoiding unwanted thermal damage to the dermis difficult, even when selective wavelength choices that match the ABSORPTION SPECTRUM of blood are used (see Figure D.6).

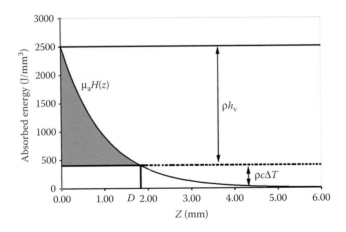

Figure D.6 Graphical illustration of the correlation between temperature and exposure to induce biological damage, represented under the damage integral.

Damage state

[biomedical, chemical, thermodynamics] The anticipated condition of the chemical reaction, and chemical and physical condition of a medium resulting from exposure to a certain temperature over an exposure time as described under the DAMAGE INTEGRAL. The damage state is generally determined as a discrete value, obtained over finite incremental steps in time. The damage calculation for time intervals of 0.1 s is, for instance, defined as $D_{damage} = \Omega(0.1) = 0.1A\text{th} \exp(-E\text{th}/RT)$, where $\Omega(t)$ is the damage integral, Ath and Eth are process constants (see DAMAGE INTEGRAL). On more general terms, the assumptions can be made Ath $= 3.1 \times 10^{98}$ s^{-1} is the rate constant, and Eth $= 6.28 \times 10^5$ J/mol the activation ENERGY for burn injury. The total cumulative damage follows from repeating the summation successively over the total time of thermal exposure, for instance, the use of radio-frequency or laser assisted heating. Keeping in mind that, for the summation process, the thermal response may change due to increase or decrease in temperature by cumulative heating, conduction, and convection processes. As a consequence, the damage integral will need to be adjusted within the summation process. The following damage conditions for biological tissues can generally be identified: reversible damage to tissue or no damage at all under the damage state $0 \leq D_{damage} \leq 0.53$, irreversible coagulation $0.53 < D_{damage} < 1$, and carbonization (starting out from the exterior with black boundaries) for $1 \leq D_{damage}$.

Damköhler number

[chemical] Reaction quotient for scaling the dominant factors in a chemical process, DIFFUSION or reaction or both. The chemical reactions are related by $N_{DA} =$ reaction rate/diffusion rate. The Damköhler number indicates the limiting factor; $N_{DA} \gg 1$ implies that the reaction is instantaneously at equilibrium, while $N_{DA} \ll 1$ means that the diffusion process will be at equilibrium well before the reaction process. Under condition of reaction rate proportional to the concentration of the bulk SOLUTION, the Damköhler number becomes $N_{DA} \sim k_R L^2/\alpha D_{AB}$, where k_R is the rate constant, L is the characteristic length, α is the representative of the geometry of the situation ($\alpha = 1$ in most cases), and D_{AB} is the DIFFUSIVITY. The relevance of these types of approximations and estimations is, for instance, applicable to transdermal drug delivery, which is diffusion dominated. Another aspect comes into play when a process has a periodic (temporal or spatially cyclic) delivery sequence, as found, for instance, in the developmental process of the embryo, with both spatially cyclic and temporarily oscillating patterns of chemical concentrations. The oscillations, for instance, can be long term (hormonal) or short term (vasodilation/vasoconstriction). The Damköhler number may also be found as Thiele modulus, for instance, with respect to catalytic reactions.

Damped harmonic motion

[general, mechanics] An undamped harmonic oscillation relies on an ELASTIC RESTORING FORCE (HOOKE'S LAW: $F = ku$, where k is the spring constant and u is the displacement) without loss. Generally, there will be a loss factor, FRICTION (surface or FLUID: VISCOSITY), or heat production in the spring (or, for instance, elastic band), next to gravitational influences. An undamped FORCED OSCILLATION is governed by the following equation: for a spring $\nabla^2 u(\vec{r}) = (1/v^2)[\partial^2 u(\vec{r})/\partial t^2] - F$ and for a flexing (torsional) structure $\nabla^4 u(\vec{r}) + (1/\alpha^2)[\partial^2 u(\vec{r})/\partial t^2] = F$, where $\nabla^4 = \nabla^2\nabla^2$ with $\nabla^2 = (d/dx^2) + (d/dy^2) + (d/dz^2)$ the Laplace operator, v is the speed of mechanical propagation in the structure, $u(\vec{r})$ is the displacement as a function of location, α is a parameter accounting for geometric and elastic influences (e.g., Young's modulus), and F is the applied force over time (t). Including dampening, for instance, the MOTION of a MASS (m) on a spring with the use of a DASHPOT with dampening factor b, in the spring system (e.g., a SHOCK ABSORBER) the WAVE EQUATION for the spring becomes a one-dimensional displacement $m(d^2u/dt^2) + b(du/dt) + ku = F_0 \sin \omega t$, where $\omega = 2\pi v$ is the oscillation ANGULAR VELOCITY and v is the frequency for the applied force (note that a steady-state force will only stretch the spring). A damped torsional VIBRATION (ANGLE of rotation: φ) for a homogeneous structure with uniform and constant density is described (in one-dimensional motion) resulting from an instantaneous torque providing an impulse: $\left(1/\alpha'^2\right)\left[\partial^2\varphi(\vec{r})/\partial t^2\right]-\sigma_1^2\left(\partial^2\varphi(\vec{r})/\partial x^2\right)+\gamma_1\left(\partial\varphi/\partial t\right)-I^* r_i\left[\partial^3\varphi(\vec{r})/\partial x^2\partial t\right]=\xi_{ab}K_1 e^{i\omega t}$, where α' is a ratio parameter accounting for geometric and elastic influences (e.g., YOUNG'S MODULUS) generally $\alpha'=1$, $\sigma_1^2=cI^*$ is a coefficient linked to the moment of INERTIA, $\gamma_1=\int_{x_1}^{x_2}\int_{t_1}^{t_2}r_e(x)/[\partial\varphi(x,t)/\partial t]\,dx\,dt$ is the dampening, where $r_e(x)$ is the coefficient of external DAMPING, r_i is the coefficient of internal damping, I^* is the moment of inertia per mass density, and K_1 defines the applied instantaneous torque. The total force is derived from the impulse Δp, where p is the momentum as $F=dp/dt$, with the impulse expressed as $p=\rho_\ell A_0\int_{x_1}^{x_2}\int_{t_1}^{t_1}\xi_{ab}(x)e^{i\omega t}dx\,dt$, where ξ_{ab} represents the force ELEMENT, ρ_ℓ is the density per unit length, and A_0 is indicative of the AMPLITUDE. Torsional OSCILLATION can be found in drive shaft, McPherson Struts, the bowl of a wine GLASS when stuck with a rod, tennis racket (off-axis) hitting a ball, cricket bat (off-axis), and PROPELLER blade (more complex mechanism-of-action, however), to name but a few. A third class of oscillations with inherent dampening properties is in flexural oscillations with respect to displacement $u(\vec{r})$. The time-dependent displacement can be derived based on the following expression: $(1/\alpha''^2)\left[\partial^2 u(\vec{r})/\partial t^2\right]-\sigma_1'^2\left[\partial^2 u(\vec{r})/\partial x^2\right]+\beta_1'^2\left[\partial^4 u(\vec{r})/\partial x^4\right]+r_e''\left[\partial^2 u(\vec{r})/\partial x\partial t\right]+r_{1e}'(\partial u/\partial t)-r_{it}'\left[\partial^3 u(\vec{r})/\partial x^2\partial t\right]-\gamma_i'''\left[\partial^4 u(\vec{r})/\partial x^2\partial t^2\right]+r_{i2}'\left[\partial^5 u(\vec{r})/\partial x^4\partial t\right]=\xi_{ab}'K_1' e^{i\omega t}$, where α'' is a parameter accounting for geometric and elastic influences (e.g., Young's modulus) as a fraction most often $\alpha''=1$, $\sigma_1'^2$ is the coefficient of shear (function of internal deformation), $\beta_1'^2$ is the coefficient of the ratio of inertia to force, r_e'' is the coefficient of external damping, r_{1e}' is the coefficient of external damping at the extreme, r_{it}' is the ratio of internal damping to tangential torque, γ_i''' is the coefficient of moment of inertia, r_{i2}' is the ratio of internal damping to inertia, ξ_{ab}' is the MAGNITUDE of the impulse, K_1' is the applied magnitude of instantaneous force, and ω is the ANGULAR VELOCITY. Since this describes a flexural oscillation, the applied force is generally an impulse, instantaneous, for example, hitting a tuning fork against a solid surface and jumping of a spring board at the swimming pool.

Damped motion

[general] A MOTION that is subject to restrictive external or internal forces, such as FRICTION, VISCOSITY, next to the ELASTICITY (YOUNG'S MODULUS) and COMPLIANCE of the materials used, as well as the interfaces between materials (see Figure D.7).

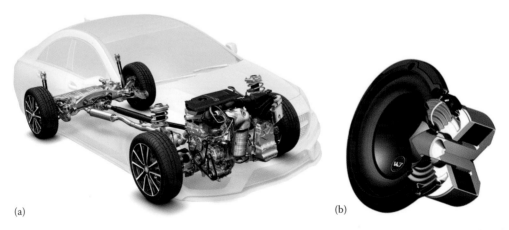

(a)   (b)

Figure D.7 Damped motion of the wheel on the axle of an automobile under influence of a pod: shock absorber, (a) automobile with shock absorbers at all four wheel attachments on the axles (Courtesy of Daimler AG) and (b) damped woofer. (Courtesy of Crutchfield, Charlottesville, VA.)

Damping

[general] A vibrational loss of AMPLITUDE due to conversion of ENERGY. An OSCILLATION can be underdamped with little loss of amplitude, critically damped yielding a pure exponential DECAY with oscillation, and overdamped providing a gradual decay in amplitude (see Figure D.8).

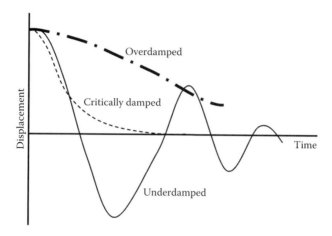

Figure D.8 Damping on harmonic motion.

Darcy's friction factor

[biomedical, fluid dynamics] In FLOW there are shear forces that result in LOSS OF HEAD (height that can be achieved by the flow, when flowing uphill), in addition to the loss resulting from discontinuities and shape changes such as found at fittings. The total HEAD LOSS in flow is the accumulated frictional loss $\left(h_L^{\text{friction}} = \sum K_L \left(v^2/2g\right)\right)$, where K_L is the "minor-loss" coefficient (a material property), v is the average flow velocity, and g is the GRAVITATIONAL ACCELERATION) plus the "shape" loss ($h_L^{\text{shape}} = \sum f_D(\ell/D)\left(v^2/2g\right)$, where f_D is the Darcy's friction factor, generally obtained from Moody plots, ℓ is the length of the tube, and D is the main tube diameter) expressed as $h_L^{\text{total}} = h_L^{\text{shape}} + h_L^{\text{friction}}$. The total head represents the height (h) that can be achieved under Bernoulli's flow when flow of FLUID with density ρ is forced into a PIPE leaning uphill under pressure P and with flow velocity v: $P + (1/2)\rho v^2 + \rho g h$.

Darcy–Weisbach equation

[fluid dynamics] A formula used to calculate pressure or HEAD LOSS due to FRICTION in ducts, pipes, and tubes. In PIPE friction, the head loss (h_ℓ) can be expressed as $h_\ell = f_{\text{Moody}}(L/D)\left(v_{\text{avg}}^2/2g\right) = \left(\Delta P/\rho\right)$, where f_{Moody} is the MOODY FRICTION FACTOR, L is the pipe length, D is the pipe diameter, v_{avg} is the average flow velocity, g is the GRAVITATIONAL ACCELERATION, ΔP is the pressure drop, and ρ is the density of the FLUID in FLOW.

Dark field

[acoustics, imaging, optics] In regular microscopy, the sample under investigation is illuminated by a filled cone of light. In dark-field microscopy, the illumination is performed by means of a hollow cone. The apex of the cone is designed to fall at the location of the sample. When no sample is placed, the light will exit from the plane of the sample holder again as a hollow cone. The hollow cone of light will encapsulate the MICROSCOPE objective, however, without entering the objective. Once a sample is placed in the illuminated plane at the apex of the cone, the light will scatter as a result of the particulate structure of the specimen. The IMAGE is made only from the rays of light that scatter from the specimen and enter the objective LENS. The dark image without specimen construes the "dark field."

Dark matter

[astronomy/astrophysics, atomic, nuclear, quantum] A hypothetical cosmic substance that has been introduced to account for gravitational phenomena that require the presence of mass where no mass has been identified on a galactic scale. In practicality, the elusive dark matter does not interact with ELECTROMAGNETIC RADIATION (no light absorption or SCATTERING). The dark matter concept has overlapped with the concept of the "GRAVITON," the gravitational particle. The principle of dark matter was introduced in 1932 by the Dutch astronomer JAN HENDRIK OORT (1900–1992). The dark matter is a substance that can account for the deviations in planetary and stellar ORBITS in the MILKY WAY that are apparently dependent on the presence of other planets; however, there is no observable evidence of those celestial objects. Some of the prevailing evidence supporting the existence of dark matter is the observed changes in galactic movement (the EXPANDING UNIVERSE), specifically by the Hubble TELESCOPE. The UNIVERSE has indication that is expanded more slowly some time back than it is today, the universe is accelerating. So far, the support for the existence of dark matter is purely based on theoretical analysis and still hypothetical. Based on these theoretical processes, more than 68% of the universe would consist of dark matter and only 5% is observable MATTER.

Dashpot

[general] A mechanical resistive device, for instance, by means of viscous FLOW through a porous structure under the influence of a moving piston or resulting from contact FRICTION. The equivalent in general MECHANICS is the use of a crumple zone in cars to absorb the ENERGY involved in a collision. The mechanical deformation provides the mechanism of conversion of kinetic energy in order to prevent exposure of the driver and passengers in the car to the full impulse (see Figure D.9).

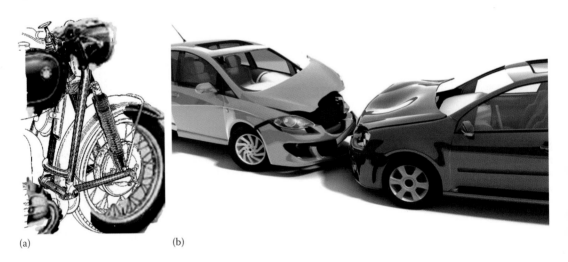

(a)   (b)

Figure D.9 (a) Dashpot dampener and (b) crumple zone of an automobile.

Daughter nucleus

[general, nuclear] A nuclide configuration resulting from atomic DECAY process. The parent–daughter process will emit a PARTICLE (e.g., helium $_2^4 He$) and will release dissociation or disintegration ENERGY, defined by the rest MASS BALANCE ($Q_{diss} = m_X c^2 - m_Y c^2 - m_{He} c^2$, where $c = 2.99792458 \times 10^8$ m/s is the speed of light): $_Z^A X \rightarrow _{Z-2}^{A-4} Y + _2^4 He + Q_{diss}$.

Davis, Raymond Jr (1914–2006)

[nuclear] A physicist and chemist from the United States. The work of Raymond Davis on neutrinos resulted in the discovery of the antineutrino, which brought him the Nobel Prize in Physics in 2002.

Davisson, Clinton Joseph (1881–1958)

[computational] A physicist from the United States. In collaboration with LESTER HALBERT GERMER (1896–1971), Clinton provided conclusive evidence supporting de Broglie's hypothesis of a WAVE phenomenon associated with moving MATTER in 1927 (see Figure D.10).

Figure D.10 Clinton Joseph Davisson (1881–1958). (Courtesy of Nobel Foundation, Stockholm, Sweden, 1937.)

Davisson–Germer experiment

[atomic, nuclear, quantum] An experimental observation made by CLINTON JOSEPH DAVISSON (1881–1958) and LESTER HALBERT GERMER (1896–1971) in 1925 verifying the postulated WAVE characteristics of any moving object. The observation was made somewhat accidental after the experimental setup had been modified due to a fire in the lab, in which the polycrystalline nickel target had been heated into a reformatted crystalline structure, at first unbeknownst to them. The initial experiment, prior to the explosion, had failed to show any conclusive evidence. During irradiation of the nickel crystal (new crystal) with an electron beam, a diffraction pattern similar to X-ray ELECTROMAGNETIC RADIATION was observed, hence verifying de Broglie's hypothesis.

Davy, Humphry (1778–1829)

[general] A scientist who discovered attraction between a PERMANENT MAGNET and a current-carrying wire. The observation of Humphry Davy is tied in with the work of his contemporary ANDRÉ MARIE AMPÈRE (1775–1836) (see Figure D.11).

Figure D.11 Sir Humphry Davy (1778–1829). Portrait by Thomas Phillips; Copyright National Portrait Gallery, London: NPG 2546.

DC (direct current)

[general] Steady-state electrical phenomenon of moving charges.

De Benedetti, Sergio (1912–1994)

[nuclear] A physicist from Italy. The work of Sergio De Benedetti involved the description of the nuclear structure and ENERGY configuration, specifically the positron lifetime as well as the discovery of the POSITRONIUM.

De Broglie, Prince Louis Victor Pierre Raymond duc (1892–1987)

[computational, nuclear] A French mathematician and aristocrat. De Broglie gave an in-depth theoretical explanation of the WAVE properties of particles in 1923, electrons in particular, but also applicable to the nuclei: protons and NEUTRON (see Figure D.12).

Figure D.12 Prince Louis Victor Pierre Raymond duc de Broglie (1892–1987) in 1929.

De Broglie wave

[nuclear physics] A realization by Louis de Broglie (1892–1987) in 1923 that all moving objects have a wave function associated with the movement with a wavelength (λ) that is the Plank's constant $\left(h = 6.62606957 \times 10^{-34} \text{ m}^2\text{kg/s} \right)$ divided by the momentum ($p = mv$, m the object's mass and v its velocity) as $\lambda = (h/p) = (h/mv)$. The wavelength of any energy form (particle or pure energy such as photons). For photons: $p = E/c$, where the energy $E = h\nu$, is proportional to the frequency (ν) of the electromagnetic radiation. This concept ties in with the Schrödinger equation for electrons, neutrons, and protons, as well as explains the movements of large objects. One specific application of the "de Broglie wave" theorem is in electron microscopy, providing a resolution associated with the electron bombardment that fits the Rayleigh criterion.

De Fermat, Pierre (1601–1665)

[computational, general, optics] A French mathematician, *see* **Fermat, Pierre de** (see Figure D.13).

Figure D.13 Pierre de Fermat (1601–1665).

De Forest, Lee (1873–1961)

[atomic, electronics, nuclear] An inventor and scientist from the United States. In 1906, Lee de Forest invented the triode VACUUM tube and realized the opportunities for SIGNAL amplification with this device. He perfected the vacuum tube concept as invented by SIR JOHN FLEMING (1849–1945). His work proceeded to move him into RADIO transmissions, specifically establishing radio broadcast stations. He is also credited for adding SOUND to movies, ending the "silent movie" period (see Figure D.14).

Figure D.14 Lee De Forest (1873–1961) as published in the 1904 issue of "The Electrical Age."

De la Tour, Charles Cagniard, Baron (1777–1859)

[general] An inventor from France who introduced the "siren" in 1819, based on his analogy between the legend of the sirens and the initial invention of the Scottish mathematician and scientist JOHN ROBINSON (1739–1805) in 1801. Robinson's invention produced puffs of AIR that, when blown through holes on a rotating disk, would create a SOUND that he equated to the VOICE of a woman singing. De la Tour managed

to repeat the process with a similar device under water, and hence, the analogy with the Greek mythology sea-nymphs was made (see Figure D.15).

Figure D.15 Baron Charles Cagniard de la Tour (1777–1859).

De la Vallée-Poussin, Charles-Jean Étienne Gustave Nicolas, Baron (1866–1962)

[computational] A mathematician and physicist from Belgium. De la Vallée-Poussin is known for proving the prime-number theorem in 1896. His other work was in approximation theory.

De Laval, Karl Gustaf Patrik (1845–1913)

[fluid dynamics] A scientist and inventor from Sweden. De Laval made significant contributions to the design and operations of steam turbine MACHINES as well as several devices used in the dairy industry. One specific device is the milk separator, separating cream from milk under high volume processing. Other contributions include improvements to milking machines (commercial introduction in 1918) and injection nozzles (see Figure D.16).

Figure D.16 Karl Gustafs Patrik de Laval (1845–1913), 1875.

De Laval nozzle

[fluid dynamics] A NOZZLE introduced based on the fluid-dynamic analysis by GUSTAF DE LAVAL (1845–1913) used in rocket engines. The specific design of the De Laval nozzle guides pressurized GAS from subsonic to SUPERSONIC FLOW velocity resulting in optimal ENERGY conversion from the heated entry gas to kinetic flow velocity using a choked flow condition. The De Laval nozzle was initially designed for steam turbines and later found perfect applications in super-sonic JET engines (see Figure D.17).

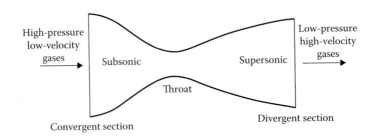

Figure D.17 De Laval nozzle.

De Maricourt, Pierre (1220–1290)

[electromagnetism, solid-state] A French engineer that described the use of metallic needles that were placed on the surface of a spherically polished LODESTONE (naturally magnetic material) the METAL silvers aligned themselves with the MAGNETIC FIELD lines running from pole to pole, while at the pole they would stand erect. In record books, de Maricourt is also named as Petri Peregrinus (*also* Peter de Pilgrim). The meridian circles formed by connecting the positions with equal magnetic force would converge at the respective opposing POLES, where the magnetic field is the strongest (primarily because the field-line density is the highest here). While MAGNETISM was supposedly known since the thirteenth century BC, as was the interaction between magnetic particles and metals, no formal description of the magnetic path was described in detail.

De Medicina

[biomedical] The medical encyclopedia assembled by the Egyptian scholar CELSUS. (second century AD). De Medicina is the most complete known work describing the medical knowledge acquired during the Alexandrian Empire.

De Mondeville, Henri (1260–1320)

[biomedical, chemical, optics] A French physician who reportedly used RED light in the earlier recorded attempts of using light to treat medical conditions, in this case the treatment of smallpox. The Danish physician NIELS RYBERG FINSEN (1860–1904) built on this initial endeavor in the early

1900s when he received the Nobel Prize for his work on the treatment of smallpox with light in 1903 (see Figure D.18).

CHIRURGIE

DE

MAITRE HENRI DE MONDEVILLE

CHIRURGIEN DE PHILIPPE LE BEL, ROI DE FRANCE

COMPOSÉE DE 1306 A 1320

TRADUCTION FRANÇAISE

AVEC DES NOTES, UNE INTRODUCTION ET UNE BIOGRAPHIE

Publiée sous les auspices du Ministère de l'Instruction publique

PAR

E. NICAISE

PROFESSEUR AGRÉGÉ A LA FACULTÉ DE MÉDECINE DE PARIS
CHIRURGIEN DE L'HÔPITAL LAËNNEC
ANCIEN PRÉSIDENT DE LA SOCIÉTÉ DE CHIRURGIE DE PARIS
ANCIEN MEMBRE DU CONSEIL DE SURVEILLANCE DE L'ASSISTANCE PUBLIQUE

AVEC LA COLLABORATION

DE Dr SAINT-LAGER ET DE F. CHAVANNES

Ce monument de la Chirurgie française méritait de trouver sa place parmi ceux des prédécesseurs de Guy de Chauliac.
LITTRÉ, *Hist.*... t. XXVIII, p. 349.

PARIS

ANCIENNE LIBRAIRIE GERMER BAILLIÈRE ET Cie

FÉLIX ALCAN, ÉDITEUR

108, BOULEVARD SAINT-GERMAIN, 108

1893

Figure D.18 Title page of the manuscript written by Henri de Mondeville in 1306. (Courtesy of Bibliothèque de France, Paris, France.)

De V. Weir, John B.

(surname potentially: Weir de Vere (de Vere), Weir; twentieth century scientist; no additional personal information available) [biomedical, thermodynamics]: an engineer and physiologist from Scotland, Great Britain, who derived the relationship between the conversion of oxygen to carbon dioxide (CO_2) and the generated heat and CHEMICAL ENERGY, specifically applied to the METABOLISM of the HUMAN BODY, published in 1948. The total ENERGY (E_{Ox}) released per VOLUME (V) oxygen ($[O_2]$) was empirically found to obey $E_{Ox} = 3.941 + 1.106 \log(V_{CO_2}/V_{O_2})$, expressed in kcal/liter O_2. This equation was later corrected to incorporate the influence of proteins, measured by means of the production of urinary nitrogen ($[N]$, in urinary solution U_N): $E_{Ox} = 3.941 + 1.10 E_{Ox} = 3.941 + 1.106 \log(V_{CO_2}/V_{O_2}) - 2.170(U_N/V_{O_2})$. The WEIR EQUATION is a standard in biomedical CALORIMETRY.

Dead state

[thermodynamics] A description of the environmental conditions pertaining to a system within a reservoir where no ENERGY is exchanged, nor added. This situation will never support work or chemical reactions, no useful efforts (*also* PASSIVE STATE).

Dead zone

[acoustics] A region in underwater acoustic imaging where there is no discrimination possible between objects and life forms with respect to the sea bottom. The dead zone height (h_{dz}; DISTANCE from the bottom that cannot be resolved acoustically) calculations can apply, for instance, to the location of concentrations of fish. The imaging constraints are a function of the operational pulse duration (τ_p), the depth at which the bottom of the body of water resides (d_B), the viewing ANGLE (θ) for determination of the school of fish with respect to the acoustic axis, and the speed of SOUND of the WATER (v_s), expressed as $h_{dz} = d_B(1 - \cos\theta) + (v_s \tau_p/2)$.

Dean, William Reginald (1896–1973)

[fluid dynamics, mechanics] A scientist from Great Britain. The work of William Dean involved FLOW at low REYNOLDS NUMBER. Specific research provided details about the perturbations in an otherwise LAMINAR flow process resulting from a gap in the WALL of a PIPE. His main contributions to the understanding of flow were with respect to the turbulent effect in curved tubes.

Dean equations

[biomedical, fluid dynamics] FLUID flow equations under low REYNOLDS NUMBER defined by WILLIAM REGINALD DEAN (1896–1973) in 1928. The FLOW equations for a tube with slight curvature R and radius a were subjected to PRESSURE (P) gradient $P_G = -(1/r)(\partial P/\partial \phi)$, expressed in cylinder coordinates. The use of cylinder symmetry of the tube makes the cylinder coordinates useful (r, ϕ, z). The system is defined by flow velocity $\vec{u} = (u_r, u_\phi, u_z)$ with MAGNITUDE ("reasonable value") $U_f \cong P_G a^2/\eta_{dyn}$, where η_{dyn} is the dynamic VISCOSITY. Consider only the torus portion of the cylinder $(r - b)^2 + z^2 = a^2$, where $b \gg a$, scaling $r = b + ax^*$. Under entry LAMINAR FLOW, the down-pipe flow velocity distribution is $v(r) = 2v_{max}[1 - (r^2/a^2)]$. The flow process can be described as follows: $Dn[(Du_x^*/Dt) - u_\phi^{*2}] = -Dn(dP/dx^*) + \nabla^2 u_x^*$, where $Dn = Re\sqrt{(2a/2R)} = (L\rho v/\eta_{dyn}g)\sqrt{(2a/2R)}$ the DEAN NUMBER, with L the PIPE length, and the convective derivative $(D/Dt) = (\partial/\partial t) + u_r(\partial/\partial r) + u_z(\partial/\partial z)$, $u_{r,z}^* = v_{r,z}/U_f\sqrt{(a/R)}$, $u_\phi^* = u_\phi/U_f$, and $u_z^* = u_z/U_f$ the axial velocity. Additional expressions in the Dean equations are $Dn(Du_\phi^*/Dt) = 1 + \nabla^2 u_\phi^*$ and $Dn(Du_z^*/Dt) = -Dn(\partial P/\partial z) + \nabla^2 u_z^*$, under boundary condition $(\partial/\partial x)u_x^* + (\partial/\partial y)u_z^* = 0$. The Dean equations have close similarity to the NAVIER–STOKES EQUATIONS, in this case specifically applied to a NEWTONIAN FLUID flow in a toroidal pipe with a steady axially uniform flow profile. For pipe configuration with small radius of curvature in the flow, the Dean number can be expanded in a series ($Dn < 956$, above this number the solutions tend to become unstable), reducing the mathematical complexity.

Dean number ($Dn = Re \times \sqrt{(D/2R)}$)

[energy, fluid dynamics] An indication of the influence of the CENTRIPETAL FORCE on the FLOW pattern for flow in a curved pipe, where Re is the REYNOLDS NUMBER, D is the diameter of the PIPE, and R is the radius of curvature of the curved pipe. It was introduced by WILLIAM REGINALD DEAN (1896–1973) in 1927. The phenomenon has similarity with the CORIOLIS EFFECT.

Dearnaley, Geoffrey (1930–2009)

[electronics, nuclear, solid-state, thermodynamics] A physicist from Great Britain. Geoffrey Dearnaley provided significant insight into low-energy nuclear PHYSICS, next to ION channeling processes. His additional efforts were in semiconductor impregnation process development and semiconductor construction (see Figure D.19).

Figure D.19 Geoffrey Dearnaley (1930–2009). (Courtesy of Southwest Research Institute, San Antonio, TX. Copyright SwRI.)

Debakey, Michael Ellis; born "Dabaghi" (1908–2008)

[biomedical, fluid dynamics, mechanics] A surgeon and scientist from the United States who significantly improved on the operations and functionality of the roller PUMP. The roller pump is a machine that circulates BLOOD and other fluids without significant risk to the integrity of the constituents, specifically not damaging RED blood cells. The mechanism of action is a PERISTALTIC MOTION. The FLUID is transported through a tube using the Bernoulli principle and the mechanical advancement of an enclosed segment of LIQUID in a tube by means of a set of rollers that are critically spaced to continuously keep an amount of liquid trapped for transport, which is based on the tube diameter. The alternatives are piston pump or diaphragm pump. Since the tubing can be removed and constitutes an ISOLATED SYSTEM, the peristaltic roller pump has all the conditions for sterility and safety in place. The roller pump is an essential tool in cardiac bypass surgery, known as the HEART–lung machine. Michael DeBakey is also known for his pioneering efforts in bypass replacement surgery, implanting Dacron® grafts to replace occluded coronary vessels. He was also instrumental in the development of the first artificial heart and subsequent device design improvements and surgical implant procedures (see Figure D.20).

Figure D.20 Michael DeBakey (1908–2008).

Deborah number ($De = t_f |\vec{v}|/L$)

[fluid dynamics] A dimensionless number used in VISCOELASTIC (i.e., non-Newtonian) FLOW to scale the deformation rate for nonconstant stretch, in which case ENERGY is stored and released in the elastic format. The Deborah number is used to describe the ratio of the RELAXATION TIME to the flow timescale $De \sim t_r/t_f$, where $t_f = \sqrt{(\mathrm{tr}(\dot{\varepsilon}))^2/2} = \mathrm{tr}(\dot{\varepsilon})/\sqrt{2}$ describes the flow timescale, tr denotes TRACE, $\vec{v}$ is the flow velocity, and L is the PIPE length or characteristic size of the phenomenon (over which the elastic energy is stored or released). The flow timescale is dependent on the strain rate $\dot{\varepsilon}$ (strain: ε), which is defined by the flow velocity as $\dot{\varepsilon} = \nabla\vec{v} + (\nabla\vec{v})^T$, where T is the transposed matrix. The Deborah number becomes important when the flow timescale is less than the relaxation time (τ_r) and elastic effects become dominant. For a steady low REYNOLDS NUMBER shear flow, the flow time factor is a function of the shear rate (γ): $t_f = 1/\gamma$, whereas the flow timescale for extensional flow is $t_f = 1/\varepsilon$, where ε is the extension rate or strain of the flow. The WEISSENBERG NUMBER is closely related to the Deborah number ($De = t_f |\vec{v}|/L = \mathrm{Wi}\,(D/L)$), where D is the diameter of the phenomenon or tube diameter).

Debye, Petrus (Peter) Josephus Wilhelmus (1884–1966)

[general, thermodynamics] A electrical engineer, physicist as well as physical chemist from the Netherlands. Debye's work on heat capacity of solids provided him with the Nobel Prize in Chemistry in 1936, and he received various other awards (e.g., Max Planck Medal in 1950; Priestley Medal of the American Chemical Society in 1963, and Kommandeur des Ordens Leopold II in 1956). Professor Debye moved to the United States in 1940 after working at many renowned European universities (see Figure D.21).

(a)

(b)

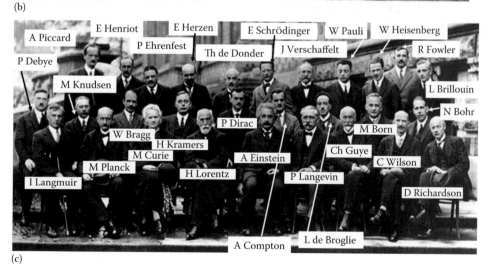

(c)

Figure D.21 (a) Petrus (Peter) Josephus Wilhelmus Debye (1884–1966). (Courtesy of Museum Boerhaven, Leiden, the Netherlands.) (b) Courtesy of Solvay conference 1924 "Electrical Conductivity of Metals and Related Problems." (c) Courtesy of Solvay conference 1927: "Electrons and Photons."

Debye (*D*)

[electromagnetism, general] A unit for DIPOLE moment attributed to and named in honor of PETER DEBYE (1884–1966), expressed in a derivative of the SI UNITS, in CENTIMETER gram seconds (CGS) units. The unit transforms as $1D = (1/299792458) \times 10^{-21}$ cm.

Debye equations

[mechanics] The relationship between force and static stress (CREEP) that determines the region of ANELASTICITY (i.e., NONELASTIC behavior) for a medium, expressed as a function of the complex COMPLIANCE $J^* = \text{strain/stress} = \epsilon_s/\sigma_s = J_1 - iJ_2$. For a cyclic LOAD with ANGULAR FREQUENCY ω, this provides the Debye equations: $J_1 = J_U + [(J_R + J_U)/(1 + \omega^2\tau^2)]$, where τ is the relaxation time, and $J_1 = (J_R - J_U)[\omega\tau/(1 + \omega^2\tau^2)]$. This is connected to the material conditions as $J_R\sigma_s + \tau J_U(d\sigma_s/dt) = \epsilon_s + \tau(d\epsilon_s/dt)$. The imaginary part of the compliance (J_2) lags the STRESS (σ_s) by $\pi/2$. The anelastic behavior is described by the creep defined by $\epsilon_s(t)/\sigma_0 = V + (J_R - J_U)[1 - \exp(-t/\tau)]$, where σ_0 is a static stress (creep). Plotting this expression as a function of time will illustrate a discontinuity at the moment the external force is removed. The Debye relations illustrate a form of DAMPING. When plotting the imaginary part (J_1) of the Debye functions against $\log \omega t$ will yield a graph that is symmetric around a point, referred to as the Debye peak. Under thermally activated processes, the Debye peak associated with the anelastic behavior will shift over time with TEMPERATURE (T) according to $\tau = \tau_0 e^{Q_{ch}/k_bT}$, where τ_0 is the steady-state SOLUTION for the RELAXATION TIME, Q_{ch} is the chemical activation ENERGY, and $k_B = 1.3806488 \times 10^{-23}$ m^{2}kg/s^{2}K is the Boltzmann coefficient.

Debye expression

[quantum] A definition of the SPECIFIC HEAT of a solid based on QUANTUM THEORY, defined as $C_P(T) = 3RD(\xi)(T/\theta_D)$, where $C_P(T)$ is the specific heat at constant pressure, θ_D is the "Debye temperature" with $D(\xi) = 12\xi\int_0^{1/\xi}[x^2dx/(e^x - 1)] - [(3/\xi)/(e^{1/\xi} - 1)]$, and R is the universal GAS constant.

Debye length (λ_D)

[thermodynamics] The confinement requirement for unneutralized plasmas $\lambda_D = \sqrt{(k_BT/4\pi e^2 n)} = \sqrt{(\varepsilon_{0k_b}/e^2)/[(n_e/T_e) + \sum_{ij} j^2 n_{ij}/T_i]}$, where $k_B = 1.3806488 \times 10^{-23}$ m^{2}kg/s^{2}K is the Boltzmann coefficient, T is the PLASMA temperature in KELVIN, respectively, T_i is the ION temperature and T_e is the electron temperature, $e = 1.60217657 \times 10^{-19}$ C is the electron charge, and n is the number of components (electrons and ions), respectively, n_e is the plasma electron density and n_{ij} is the density of all the various ions with charge je, and the permittivity of free space: $\varepsilon_0 = 8.85419 \times 10^{-12}$ C^2/Nm^2. The pressure requirement for separation at the Debye length is standard thermal pressure $P = nk_bT$. Also found as $\lambda_D = 69\sqrt{(KE/n_e)}$, where KE is the KINETIC ENERGY of the ELECTRON, and n_e is the electron density.

Debye sphere

[nuclear, thermodynamics] The confinement for plasma PHYSICS, describing the interaction between ions and electrons (SCATTERING), outlined by the DEBYE LENGTH $\lambda_D = \sqrt{(k_bT/4\pi e^2 n)}$, where $k_B = 1.3806488 \times 10^{-23}$ m^{2}kg/s^{2}K is the Boltzmann coefficient, T is the temperature in KELVIN, $e = 1.60217657 \times 10^{-19}$C is the electron charge, and n is the number of components (electrons and ions). The collision time (τ_c) for the large PARTICLE population within the DEBYE SPHERE (N), at temperature T, under particle density n, for ions with mass m with charge Ze (electron charge $e = 1.60217657 \times 10^{-19}$C) is $\tau_c = m^{1/2}(3k_bT)^{3/2}/8\pi n Z^4 e^4 \ln \Lambda$, where $\ln \Lambda \sim \ln N$ represents the collision PROBABILITY.

Debye streamers

[nuclear, thermodynamics] Corona channels of ionized material (electrons and ions) traveling in a confined current density that acts as a screen to external electric and magnetic fields, as sheaths (i.e., Coulomb shielding). The sheaths have a "thickness" in the order of the DEBYE LENGTH ($\lambda_D = 69\sqrt{T_e/n_e}$, where T_e represents the electron temperature of the PLASMA and n_e is the electron density. These streamers form in the time frame defined by the PLASMA FREQUENCY ($\nu = \omega/2\pi = 8.97\sqrt{n_e}$). For a VACUUM tube filament discharge corona, this translates in an electron density range $10^{18} < n_e < 10^{21}\,\text{m}^{-3}$, with associated applied potential resulting in $T_e = 1.0\,\text{eV} \sim$ (INFRARED light), and this will yield $7.4 \times 10^{-6} < \lambda_D < 2.0 \times 10^{-7}\,\text{m}$ over a discharge timescale of $1.1 \times 10^{-10} < 2\pi/\omega < 3.5 \times 10^{-12}\,\text{s}$. For a thunderstorm with LIGHTNING, this may be in the range of $T_e = 4 \times 10^4\,\text{K} = 3.6\,\text{eV}$ (visible light), with currents in excess of $30\,\text{kA}$, with the discharge lasting in the order of $1 \times 10^{-4}\,\text{s}$. During discharge, the electrons travel at several thousand times the DRIFT VELOCITY ($v_{\text{drift}} \sim 10^{-6}\,\text{m/s}$; $v_{e,\text{discharge}} \sim 10^3\,\text{m/s}$), as do the ions, reaching in excess of $v_{\text{ion,discharge}} \sim 10^1\,\text{m/s}$. Next to lightning, the phenomenon of "dry discharge" under St. Elmo's fire falls in the same category but the conditions are different due to the static buildup of charge and the narrow spike (e.g., ship's mast) that forms the wide-spread spray of discharges with limited confinement (see Figure D.22).

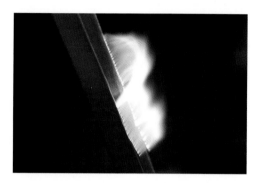

Figure D.22 Example of Debye streamers; plasma Jacob's ladder. (Courtesy of "Chocolateoak.")

Debye temperature (Θ_D)

[thermodynamics] A thermal coefficient that defines the TEMPERATURE (T) dependence of the SPECIFIC heat under constant volume (c_V) and provides the approximation under condition $T \gg \Theta_D$: $c_V \approx 3R\left[1 + (1/20)\left(\Theta_D/T\right)^2\right]$, where $R = 8.3144621(75)\,\text{J/Kmol}$ is the universal GAS constant, named after PETRUS DEBYE (1884–1966). The Debye temperature has been tabulated for a range of media.

Debye–Scherrer equation

[atomic, imaging, nuclear] See SCHERRER EQUATION. It was introduced by PAUL SCHERRER (1890–1969), and experimental verification was supported by PETER DEBYE (1884–1966).

Debye–Scherrer method

[atomic, imaging, nuclear] An X-RAY crystallographic study where the medium is GROUND up for analysis, in contrast to the Bragg and Laue method that uses the single crystal. Debye–Scherrer method was named after PETER DEBYE (1884–1966) and PAUL SCHERRER (1890–1969) and was proposed in 1916. The analysis applies a narrow parallel beam of monochromatic X-ray. Interaction with the crystalline powder produces a series of coaxial X-ray diffraction CONES, with the incident BEAM FORMING the central axis. The apex ANGLES of the respective cones are determined from the Bragg diffraction rule: $n\lambda = 2d\sin\theta$ (where n is a positive integer, λ is the wavelength of the X-rays, d is the DISTANCE between the parallel planes [interplanar spacing] of the points of the space crystal LATTICE, and θ is the angle between the reflecting plane and the incident beam). Under these conditions, the REFLECTION angle (ϕ) is one-fourth of the cone's apex angle. The diffraction pattern is captured on a PHOTOGRAPHIC PLATE, or scintillographic digital recording, providing the tools for derivation of the distance of the interplanar distance.

Decay

[nuclear] The disintegration of the NUCLEUS of an unstable ELEMENT by the spontaneous emission of charged particles and/or photons (*see* RADIOACTIVE DECAY *and* COMPOUND NUCLEUS).

Decay constant ($\lambda_{1/2}$)

[atomic, biomedical, computational, general, nuclear] The change in the number of isotopes of reactive atoms N over TIME (t) with respect to the original quantity by half $\lambda_{1/2} = (\Delta N/\Delta t)/N$, units $1/s$ or s^{-1}. This is a fixed value for any specific ELEMENT or ISOTOPE. This rewrites as $N = N_0 e^{-\lambda_{1/2}t}$, where N_0 is the base quantity.

Decay rate (R)

[general] $R = \left(1/2\right)^n R_0$, where $R_0 = \lambda_{1/2} N_0$, $\lambda_{1/2}$ is the DECAY constant, N_0 is the original quantity of radioactive nuclides, and n is the number of half-lives: $\tau_{1/2} = 1/\lambda_{1/2}$, or the total disintegration time period.

Decibel

[acoustics, biomedical, electronics, general, imaging, optics] The log of intensity (I), primarily used for quantifying relative signal MAGNITUDE, used in ELECTRONICS, ACOUSTICS, OPTICS (specifically as a material property [specification] for fiber optics), and IMAGE processing. DECIBEL was named after the Scottish inventor and scientist ALEXANDER GRAHAM BELL (1847–1922) and was introduced at the Bell laboratories. The decibel scale has two separate definitions: $dB = 10\log I/I_0$, where I_0 is a reference intensity. Under these conditions, double the intensity equates to a 3 dB increase. Alternatively, in acoustics, the magnitude may be measured in PRESSURE (P) $dB = 20\log\left(P_{RMS}/P_0\right)$ (note that intensity is equal to the square value of the magnitude: $I \sim P^2$), where in AIR $P_0 = 2.0 \times 10^{-5}$ Pa and in water $P_0 = 1 \times 10^{-6}$ Pa, respectively. In biomedical acoustics (HEARING), the reference level is the threshold for hearing $I_0 = 1 \times 10^{-12}$ W/m^2. The threshold for pain with respect to human hearing is in the order of 120 dB across the entire audible FREQUENCY SPECTRUM, whereas the hearing limit is at its lowest threshold for 3000 Hz. The human hearing, as well as that for other mammals, has an enormous dynamic range, in the order of 10^{14}. The decibel is used in quality control as a threshold for acceptance or in quantifying degradation of SIGNAL strength and MAGNITUDE in general.

Defibrillator

[biomedical] An electronic device used to apply a short-duration, high-current, single-pulse electrical discharge to the HUMAN BODY in order to depolarize the HEART MUSCLE and attempt to reinitiate the DEPOLARIZATION sequence in normal order. The heart is composed of the atria and ventricles, which can, as a result of medical and chemical (e.g., cocaine) conditions, reach a high rate of depolarization. The high depolarization rate will force the heart to contract at rates that are not conducive to the COMPLIANCE of the vasculature, resulting in a "no-flow" condition. Arrhythmias may result in death when not properly attended to. Ventricular fibrillation or ventricular tachycardia (the high rate of depolarization; arrhythmia) is more directly linked to the whole-body FLOW, whereas atrial fibrillation will result in poor ventricular filling, ensuing poor ventricular outflow (see Figure D.23).

Figure D.23 Image of cardiac defibrillator.

Deflagration

[chemical, energy, fluid dynamics] A subsonic combustion propagating by THERMAL CONDUCTIVITY. The addition of water to specifically OIL that is being incinerated will result in the formation of water VAPOR, which adheres oil droplets that are ejected into the flames, causing the flames to erupt in a pyrogenic fireworks display.

Degeneracy

[atomic, computational, nuclear, quantum, thermodynamics] In QUANTUM MECHANICS, the degeneracy of a state outlines the entropy of the GROUND STATE, defined as $S_g = k_B \ln D_g(n, \beta)$, where $D_g(n, \beta)$ is the degeneracy that ties in to the SECOND and THIRD LAW OF THERMODYNAMICS, signifying that a state is nondegenerate when $D_g(n, \beta) = 1$ for all values of the number of constituents (n) and a range of parameters (β) (including but not limited to volume, pressure, and temperature), and $k_B = 1.3806488 \times 10^{-23}$ m^{2}kg/s^{2}K is the Boltzmann coefficient. The nondegenerate state implies that there is only QUANTUM STATE for one particular ENERGY value. In contrast, an energy value is degenerate if it is shared between multiple quantum states. The computational equivalence is that a nondegenerate state only has one EIGENFUNCTION as SOLUTION to the Schrödinger equations.

Degenerate state

[atomic, computational, nuclear, thermodynamics, quantum] the computational principle of a degenerate state is that more than one EIGENVALUE exists for the Hamiltonian at one condition. In quantum PHYSICS, this translates into a condition where one ENERGY state has more than one QUANTUM STATE, captured by multiple solutions for the eigenfunctions of the Schrödinger equations.

Degrees of freedom

[atomic, general, nuclear, solid-state, thermodynamics] The potential MOTION capabilities of a system, rotation of an object on its own axis, rotation around a centralized object, translation, acceleration, as well as PRECESSION, and the fact that an elliptical orbit can "swing" around one of the focal positions (found to apply to nuclear ORBITS in QUANTUM PHYSICS). The latter is captured by the "FINE-STRUCTURE CONSTANT," for instance, applied to the ZEEMAN EFFECT and other spectral phenomena.

Delayed neutron

[atomic, nuclear] In a NUCLEAR REACTOR, the neutrons are generally emitted within 10^{-14} s after the FISSION process with the incident NEUTRON. For ^{235}U, the half-life involving neutron DECAY ranges from 2.3×10^{-1} s to 56 s. During the fission process, other radioactive ELEMENTS are formed, some of which are neutron rich but these will decay with rates that are up to three orders of MAGNITUDE slower, referred to as the delayed neutrons. The delayed neutrons become an integral part of the reactor process control function when the reactor becomes "critical."

Delta ray

[nuclear] A recoil PARTICLE that generates secondary IONIZATION produced during atomic interactions with ALPHA PARTICLES (helium ions). The term "delta ray" was introduced by JOSEPH JOHN THOMSON (1856–1940), in the early nineteen hundreds. Later, this term was refined to describe electrons that were released from an ATOM and have enough kinetic ENERGY to move away from the incident primary particle beam and subsequently ionize neighboring atoms. Additionally, in PARTICLE ACCELERATOR, the delta ray indicates electrons with significantly less kinetic energy, which will exhibit a spiraling DECAY process in a BUBBLE CHAMBER, decelerating at a characteristic rate.

Demodulator

[electronics, theoretical] An electronic device that extracts the encoded information from a CARRIER WAVE, demodulation. With respect to digital encoding, the demodulator is a software program (application process). One specific application of the demodulator is the PHASE-LOCKED LOOP and specifically RADIO (FM) tuning.

Dempster, Arthur Jeffrey (1886–1950)

[atomic, nuclear] A physicist from Canada. Arthur Dempster was an integral collaborator on the MANHATTAN PROJECT. In 1918, he developed the concept of the MASS SPECTROMETER and created a functional device. He also provided the theoretical foundation of MASS SPECTROSCOPY for today's applications (see Figure D.24).

Figure D.24 Arthur Jeffrey Dempster (1886–1950). (Courtesy of Emilio Segrè Visual Archives c. 1925.)

Dempster's mass spectrograph

[atomic, nuclear] The first known MASS SPECTROMETER was designed by ARTHUR JEFFREY DEMPSTER (1886–1950) in 1918. The mass spectrometer separates particles a function of their mass and provides the scientific details to reconstruct the ISOTOPE configuration with respect to an element (see Figure D.25).

Figure D.25 Components of the Dempster mass spectrometer.

Density (ρ)

[general] Mass (m), respectively, charge ($Q = Ze$, $e = 1.60217657 \times 10^{-19}$ C the electron charge) per unit volume (V). Primarily, the mass density is $\rho = m/V$. Alternatively, the charge density is $\rho_e = Q/V$. In general, any parameter can be expressed as a density commodity, per unit volume.

Depletion layer

[electronics, general] In a semiconductor device construction, n-type and p-type materials are placed side by side with a transition layer forming in between to allow for a transition between the respective ENERGY levels of the two media. The depletion layer is virtually devoid of charge carriers, acting as an INSULATOR. The charge carriers are drawn to the p-type leaving the n-type side with holes (positively charged) due to the electrical potential difference generated by the close proximity placement forming a CAPACITOR across the depletion layer. Application of an external voltage will reduce the "width" of the depletion layer, inducing conduction as a function of the applied voltage, depending on the respective semiconductor materials: >300 mV for germanium and >650 mV for silicon. During reverse-bias situation, the electric field is in the opposite direction, broadening the depletion layer, attracting electrons in the n-type material to the positive pole and holes in the p-type material to the negative external pole.

Depolarization

[biomedical, chemical, electronics] A change with respect to electric equilibrium for a solid object or a cellular MEMBRANE. A polarized layer has an electrical discontinuity based on either differences in the concentrations of chemical constituents or electron displacement resulting from an applied electric field. A dry-cell BATTERY uses a depolarization layer to enhance the emf resulting from the buildup of ions formed in the generation of the emf. In biomedical applications, the depolarization is the process where the steady-state ionic differences between the intra- and extracellular LIQUID is disturbed under the influence of a chemical catalyst effect that changes the ION flow through specialized ports in the CELL MEMBRANE. The result is the formation of an ACTION POTENTIAL. The polarized state of the cell forms an electrical potential that obeys the NERNST EQUATION, where the depolarization process follows the process described by DAVID GOLDMAN (1910–1998), and in particular SIR ALLAN HODGKIN (1914–1998) and SIR ANDREW HUXLEY (1917–2012), the Hodgkin–Huxley equation.

Descartes, René (1596–1650)

[general, optics] A French scientist and mathematician. Descartes wrote a theoretical verification of the corpuscular properties of light in 1637: La Dioptrique, adopting the philosophies of the ancient Greek. With the corpuscular description, he was able to formulate a description of REFRACTION, and hence in France, the LAW OF REFRACTION (i.e., SNELL'S LAW) is often referred to as Descartes' law. He also formulated a thorough foundation of geometry in his book *Géométrie* in 1637. His influence in geometry includes the introduction of the mathematical notation of equations with known constants in an equation as the first letters of the alphabet: a, b, c, …, and the variables or values on the axis denoted with the last letters: x, y, z. This lead to the formulation of the Cartesian coordinate system, named after Descartes. In his commitment to the newly adopted integral and differential formulation, he describes a cycloid with the equation $x^3 + y^3 = 3axy$, which is centered around the origin in this formulation, and has an enclosed segment with area $3/2\,a^2$. Until SIR ISAAC NEWTON (1642–1727) introduced his gravitational theory, the scientific world was in favor of the VORTEX model of Descartes with regard to the planetary COHESION around the SUN, in this model the revolving Sun generates an AETHER vortex that pulls the planets in tow and holds them confined in "orbit." In 1644, Descartes also postulated that the total amount of momentum in the GALAXY is constant and "can only be changed by god." The statement by Descartes provided the basis of one of the first

D

CONSERVATION LAWS, followed by the CONSERVATION OF MASS law introduced by ANTOINE-LAURENT DE LAVOISIER (1743–1794) in 1789. This principle law of PHYSICS was followed in 1853 by the proclamation of the CONSERVATION OF ENERGY by WILLIAM JOHN MACQUORN RANKINE (1820–1872) (see Figure D.26).

Figure D.26 René Descartes (1596–1650), engraved by W. Holl.

Deslandres, Henri Alexandre (1853–1948)

[astronomy, astrophysics, optics] A French astronomer and physicist, contemporary of GEORGE ELLERY HALE (1868–1938). Deslandres developed a spectral PHOTOGRAPHY mechanism ("spectroheliograph") that allowed for spectral analysis of solar emissions as well as comets independent of the construction by Hale. In addition, his "spectro-enregistreur des vitesses" presumably allowed for time-resolved detection of chromospherical cloud emission (solar flares) that yielded the radial velocity of DRIFT of the flare (see Figure D.27).

Figure D.27 Henri Alexandre Deslandres (1853–1948). (Courtesy of Bulletin de la société astronomique de France, 1913.)

Deuterium (^{2}H)

[atom, nuclear] referred as "heavy hydrogen." Symbol: D or 1H2. The NUCLEUS of a deuterium ATOM contains 1 proton and 1 neutron. In atom count, deuterium accounts for approximately 0.0156 of all "hydrogen" and related isotopes in the earth's water.

Deuteron

[atom, nuclear] NUCLEUS of deuterium, referred to as "heavy hydrogen." Deuterium is an ISOTOPE of hydrogen: the nucleus of a deuterium ATOM contains 1 proton and 1 neutron. The other hydrogen isotope: protium ($^1_1$H) contains no NEUTRON in the nucleus and accounts for 99.98% of all hydrogen on the PLANET and is as known so far the most abundant ELEMENT in the UNIVERSE.

Deutsches Institut für Normung

[biomedical] Translation: German Institute of Standards, an organization responsible for standards in Germany and adopted by most of the world as DIN-standard (Deutsche Industry Norm). The equivalence is in the following standardization organizations: ISO (International Organization for Standardization), ANSI (developed by the American Society for Testing and Materials: ASTM), and IEEE (Institute of Electrical and Electronics Engineers), each providing standards on quality, communication, material choices, and safety (see Figure D.28).

Figure D.28 The DIN building in Berlin.

Dew

[geophysics, thermodynamics] Water VAPOR condensation, particularly observed in the morning. Dew forms from saturated AIR condensation on cooling surfaces, resulting from heat release to the ATMOSPHERE. Dew form on grass, plants, and objects, not on soil due to its respective conductive heating (see Figure D.29).

Figure D.29 Dew on spider web and brush.

Dew line

[thermodynamics] The thermodynamic description of a DISTILLATION process for a two-phase system (f, LIQUID and g, VAPOR), where the pressure per constituent, at temperature T, is defined as the fraction of the component (y_{if}) multiplied with the SATURATION pressure of that component ($P_{sat,i}$), under FLUID, and the remaining components with their respective contributions (Dalton pressure law) as $P = y_1 P_{sat,1f}(T) + \sum_{i=2}^{n}(1 - y_{if}) P_{sat,ii}(T)$, with P_{ii} the partial pressure of constituent i in unique SOLUTION. RAOULT'S LAW defines the gas-phase pressure ($P_{sat,ig}$) of constituent i in relation to the pressure as dissolved in liquid ($P_{sat,if}$) as $y_{ig}P = y_{ig}P_{sat,ii}(T)$, and the CHEMICAL POTENTIAL (μ_{1f}) for constituent 1 in solution in reference to the unique substance μ_{11} is defined by $\mu_{1f} = \mu_{11}[T, P_{sat,11}(T)] + V_{11f}[P - P_{sat,11}(T)] + RT\ln y_{1f}$, where is the fluid volume for solution 1 and $R = 8.3144621(75)$ J/Kmol is the universal GAS constant. The fractional composition between liquid and vapor is linked through HENRY'S LAW as $y_{1g}P = y_{1f}\mathcal{H}_{1f}(T,P)$, where $\mathcal{H}_{1f}(T,P)$ represents the HENRY'S CONSTANT for SOLUTE 1 in liquid-phase solution. The full definition for the Henry constant is $\mathcal{H}_{1f}(T,P) = P_{sat,11}(T)\exp[[c_{1f}(T,P) - \mu_{11f}(T, P_{sat,11}(T))]/RT]$, where $c_{1f}(T,P)$ is a NORMALIZATION constant, and for an IDEAL SOLUTION $\mathcal{H}_{1f}(T,P) = P_{sat,11}(T)$. The dew line for constituent 1 in the distillation process for a two-component solution is now defined by $(1/P) = [y_{1g}/P_{sat,11}(T)] + [1 - y_{1g}/P_{sat,22}(T)]$, where now y_{1g} constitutes the solubility of liquid component 1 in gaseous SOLVENT constituent 2 and y_{2f} represents the solubility of gaseous component 2 in liquid solvent constituent 1. Note that the slope of the dew line changes with distillation concentration (compare the slope at $y_1 = 0$ to that at $y_1 = 1$). The dew line can be plotted in pressure versus fraction portion (P vs. y_1) or as temperature versus constituent portion (T vs. y_1).

Dew point

[thermodynamics] A temperature point where water VAPOR condenses to form LIQUID water. The temperature where the dew point is established is the DEW POINT TEMPERATURE.

Dew point temperature

[thermodynamics] The temperature where the water VAPOR has a partial pressure that has reached the SATURATION pressure and the water vapor will condensate and form LIQUID water.

Diamagnetism

[atomic, general, nuclear] The material repercussions in the magnetic orientation of an object caused by the internal EDDY CURRENTS in result of the application of an external MAGNETIC FIELD. Two kinds of orientation of the internal magnetic field can be recognized, parallel and anti-parallel. In case the internal magnetic field is in the opposing direction from the external filed, the magnetization is referred to as diamagnetic, whereas the congruent lineup of magnetic field lines is called PARAMAGNETISM. The object is hence diamagnetic or paramagnetic. This phenomenon is closely related to FERROMAGNETISM. Under diamagnetism, the external magnetic field is almost canceled in its entirety by the material's induced internal magnetic field. The magnetic permeability under diamagnetism hence has a value less than VACUUM (vacuum: $\mu_m^0 = 4\pi \times 10^{-7} \, Tm/A$), however slightly only. The magnetic behavior of diamagnetic materials is opposite to that of ferromagnetic materials or paramagnetic materials. The latter two are attracted to the MAGNETIC POLES, whereas the diamagnetic materials are repelled. Even though magnetization can be ferromagnetic or paramagnetic, virtually all materials have a tendency for diamagnetism, however in most cases understated. This phenomenon was first documented by MICHAEL FARADAY (1791–1867) in 1845. Faraday observed that certain materials (e.g., wood, GLASS, and gasses) exposed to a strong magnetic field are consistently repelled.

Diastolic pressure

[biomedical, general] Arterial pressure in mammals during the relaxed state of the HEART MUSCLE (fully distended heart muscle), in contrast to the SYSTOLIC PRESSURE. The systolic pressure constitutes the maximum pressure obtained at the peak of the contraction point by the heart muscle. The diastolic pressure is a function of the vascular COMPLIANCE and BERNOULLI'S LAW (see Figure D.30).

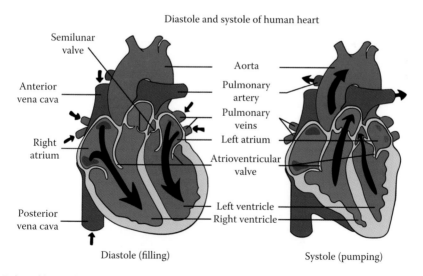

Figure D.30 Relaxed heart during diastolic pressure condition.

D

Diathermy

[general, thermodynamics] A method employed to add heat to the HUMAN BODY. One specific mechanism of action is the controlled application of high frequency current that is converted into heat locally and gradually (see Figure D.31).

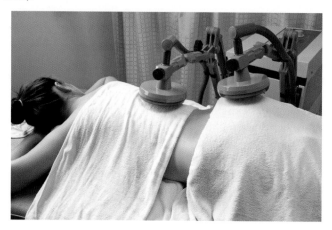

Figure D.31 Woman receiving short-wave diathermy.

Dicrotic notch

[biomedical, fluid dynamics, mechanics, theoretical] A sharp rise and fall in ARTERIAL pressure directly behind the AORTIC VALVES in the AORTA immediately following SYSTOLE resulting from the relaxation of the VENTRICLE of the HEART during DIASTOLE, which causes the aortic valve LEAFLETS to close and in fact become pulled into the ventricle under the apparent relative VACUUM. The cupping of the aortic leaflets results briefly in a backflow from the aorta toward the ventricle, immediately followed by a cessation of FLOW, before the left ventricle prepares from systolic CONTRACTION and expulsion.

Diderot, Denis (1713–1784)

[general] A philosopher and scientist from France. Denis Diderot published a scientific encyclopedia (Encyclopédie) with JEAN LE ROND D'ALEMBERT (1717–1783). Denis Diderot expressed certain philosophies on the dimensions of space and time that were based on the work by FRANÇOIS-MARIE AROUET (*also* VOLTAIRE) (1694–1778) (see Figure D.32).

Figure D.32 Denis Diderot (1713–1784), painted by Louis-Michel van Loo in 1767.

Dielectric

[electromagnetism, general, thermodynamics] an equivalent for INSULATOR. Charges are refrained from moving freely. Under the influence of an external electric field, the dielectric medium will become polarized or made to retain a static charge. In contrast, a CONDUCTOR has free migration of charges, which provides for the phenomenon of current conduction. A conductor will not retain a static charge, even though an external electric field will separate the charges according to the electric field direction for the duration of the applied field. The conductor may remain charged if the free-moving electrons are siphoned off by toughing the polarized conductor with another conductor. Dielectrics are used in capacitors to maintain a charge separation between conducting plates. A material used to perform a specific electronic function with respect to shielding or insulation, dielectric material. A dielectric generally refers to a type of insulator, material with limited conductivity. The dielectric can be polarized to hold charge in most cases. A range of dielectric materials is used in the construction of capacitors to provide a specific value due to the relative dielectric constant ε of the medium, where the PARALLEL-PLATE CAPACITOR value (C) is defined as $C = \varepsilon A/d$, where d is the plate spacing and A is the area of the capacitor plates. The documented specific conductivity of certain materials was first investigated with some rigor by the British amateur astronomer STEPHEN GRAY (1666–1736) in 1729.

Dielectric breakdown

[general] An electric field on the surface of a CONDUCTOR exceeding $\sim 3 \times 10^6$ V/m will break the electronic bond and results in IONIZATION of dry AIR, creating sparks. The static electric sparks resulting from rubbing polyester and NYLON mixtures together (e.g., socks on carpet) under dry air conditions can generate sparks under a static electrical potential in the order of $\sim 1 \times 10^4$ V. The electrical potential associated with an ELECTRIC CHARGE was described in detail by CHARLES DE COULOMB (1736–1806).

Dielectric constant (K_e)

[electronics, general, thermodynamics] The ratio of the DIELECTRIC permittivity of VACUUM (ε_0) to the dielectric permittivity of a medium (ε): $K_e = \varepsilon/\varepsilon_0$. The dielectric constant for water is $K_e = 80$, for Teflon $K_e = 2.1$, GLASS ranges from $K_e : 5 - 10$, and for NYLON $K_e = 3.5$ to name but a few.

Diesel cycle

[thermodynamics] A thermal cycle in ENERGY exchange, without the introduction of external energy to achieve an EXPANSION (resulting from combustion; the OTTO CYCLE, used to describe gasoline engines, requires a spark plug). The diesel cycle for a diesel ENGINE compresses the fossil fuel to a density/pressure that exceeds the ignition point, resulting in spontaneous combustion. The diesel cycle is comprised of the following four stages: (1) isobaric heating and expansion; (2) ISENTROPIC expansion; (3) isochoric cooling; and (4) isentropic compression. The diesel cycle can be executed by a reciprocating piston-cylinder engine. The work ($w_{\text{net}}^{\rightarrow}$) performed by the diesel engine, assuming a PERFECT GAS is defined by the heat produced per unit gas ($q_{12}^{\leftarrow}$, function of the "quality" of the fossil fuel) with respect to the respective temperature profiles in the various stages of the cycle (T_i) as $w_{\text{net}}^{\rightarrow} = q_{12}^{\leftarrow} \left\{ 1 - (1/\gamma_c) \left[(T_3/T_4) - 1 \right] T_4 / \left[(T_2/T_1) - 1 \right] T_1 \right\}$, where $\gamma_c = c_p/c_v$ is the ratio of the SPECIFIC HEAT under constant pressure over the specific heat at constant volume. The efficiency of the diesel cycle is the ratio of the work over the heat production $\gamma_{\text{eff}} = w_{\text{net}}^{\rightarrow}/q_{12}^{\leftarrow}$. The compression ratio (r_{comp}) is the relative VOLUME (V) change between stage 4 and stage 1, defined as $r_{\text{comp}} = V_4/V_1$, which ties the temperature difference between the two stages as $T_4/T_1 = r_{\text{comp}}^{-(\gamma_c - 1)}$.

Dieterici, Conrad Heinrich (1858–1929)

[thermodynamics] A physicist from Germany (Prussian Empire). Conrad Dieterici focused his efforts on the explanation of thermodynamic processes, specifically the theoretical evaluation and definition describing specific circumstances (see Figure D.33).

Figure D.33 Conrad Heinrich Dieterici (1858–1929) c. 1907. (Courtesy of Universität Rostock, Rostock, Germany.)

Dieterici equation of state

[thermodynamics] Introduced as a correction to the VAN DER WAALS EQUATION, and actually more accurate, defined as $P(V - b_d) = \mathrm{RT}e^{a_d/\mathrm{VRT}}$, where $a_d = 4R^2 T_c^2 / e^2 P_c$ is a factor describing the molecular interaction, $b_d = (1/2)V_c = (\mathrm{RT}_c / e^2 P_c)$ accounts for the molecular size of the Dieterici constant (different from the van der Waals constants), V_c is the critical volume, T_c is the CRITICAL TEMPERATURE, P_c is the critical pressure (all derived experimentally), $R = 8.3144621(75)$ J/Kmol is the universal GAS constant, V represents volume, P represents pressure, T represents the temperature of the gas, with the number $e = 2.7182(8) = \sum_{n=0}^{\infty}(1/n!)$. Generally, the constants are listed in look-up tables for general use. The Dieterici equation of state was introduced by CONRAD HEINRICH DIETERICI (1858–1929) in 1899. The Van der Waals equations were introduced by JOHANNES VAN DER WAALS (1837–1923) in 1873.

Differential entropy

[imaging, theoretical] In IMAGE processing, specifically image registration, the comparison between pixels of image pairs can be classified under two categories: parametric and nonparametric. The point distribution of the image can be captured by the entropy outlining the PROBABILITY for the MEASUREMENT of the value μ_m with respect to the reference value v_m, defined under integral entropy $H_{f,v_m}(\mu_m) = -\int f(d\mu_m/dv_m)dv_m$, where f is a continuous function. Alternatively, the entropy of a random variable X can be defined by the probability of this value occurring in the data set $(p(x))$, given by the differential entropy or SHANNON ENTROPY $H_S(X) = -\int_\Omega p(x)\log(p(x))dx$. Under image registration, the data are compared against a standard with value Y, giving the joint (differential) entropy as outlining the entropy of finding similarities between X and Y, as the "conditional (differential) entropy" $H_S(X|Y) = -\int_\Omega p(x|y)\log(p(x|y))dxdy$ (*also see* **SHANNON ENTROPY**).

Diffraction, far field

[general, optics] DIVERGENCE and INTERFERENCE phenomena associated with the WAVE mechanisms due to edge phenomena. The redirection of mechanical waves due to obstructions was described by the Greek philosophers and for light by the Jesuit priest FRANCESCO GRIMALDI (1618–1663) from Italy in 1660. Furthermore, documented observations in 1723 by GIACOMO FILIPPO (JACQUES PHILIPPE) MARALDI (1665–1729) from Italy described the wave concept of light; however, this was not accepted by the scientific community. Without diffraction, an APERTURE or obstruction would project an IMAGE that has a uniform radiance across the entire imaging field, a single uniform bright spot from the aperture. In reality, the aperture generates a centralized bright spot with a less-bright (potentially totally void of RADIATION) minimum surrounding this central maximum, followed by additional, progressively less-bright maxima and intermitting minima in the radial direction of the imaging plane. The full-wave concept for light (ELECTROMAGNETIC RADIATION) was not fully verified until the experimental and theoretical work of THOMAS YOUNG (1773–1829) in 1801. Thomas Young illustrated the interference of waves from double slit with a single-LIGHT SOURCE. These observations were further validated by Agustin Fresnel (1788–1828) in 1818. In comparison, the basic principles of optical and mechanical REFRACTION at the interface of two media were defined through Snell law introduced by WILLEBRORD SNEL (SNELL) (WILLEBRORD SNEL VAN ROYEN [1591–1626]) in 1621 followed by the work on diffraction by AUGUSTIN JEAN FRESNEL (1788–1827). The "bending" of waves under the influence of the edge of on object in the path of electromagnetic radiation as well as mechanical waves (e.g., water waves) is described as refraction. When a WAVEFRONT with dimensions greater than the obstructions, greater than an opening (i.e., aperture), respectively, outer dimensions (e.g., the MOON between the SUN and the EARTH during a total eclipse), the wave aspect provides significant theoretical and practical implications. Under the Huygens–Fresnel principle (based on the work of CHRISTIAAN HUYGENS [1629–1695] and AUGUSTIN-JEAN FRESNEL [1788–1827]), each individual location in space in the path of a wave forms a secondary source emitting a SPHERICAL WAVE (depending on the geometry of the medium) in direction r. Hence, any wavefront is constructed of an infinite number of spherical waves, in particular for electromagnetic radiation. The secondary source has the same WAVELENGTH (λ, associated frequency $\nu = c/\lambda$, ANGULAR FREQUENCY $\omega = \nu 2\pi$, and c the speed of light) and PHASE as the primary source. The total wavefront is formed by means of superposition of all the secondary spherical wavefronts. In free space, the superposition provides a continuing PLANE WAVE. In case of an aperture, or LENS, the cross-sectional area is filled with secondary spherical sources. These individual sources are defined by the WAVE EQUATION with SOLUTION providing a MAGNITUDE as a function of DISTANCE as $E = \left(\xi_0/r\right)e^{i(kr-\omega t)}$, where ξ_0 is the generic source strength of each respective coherent source point and $k = 2\pi/\lambda$ is the wave number. The interference pattern from the infinitesimal sources in the aperture is observed at a distance $\vec{R}$ in the imaging plane. A detailed description of far-field diffraction was presented by JOSEPH VON FRAUNHOFER (1787–1826), that is, FRAUNHOFER DIFFRACTION. In imaging and general VISION, the ability to distinguish objects is the result of the diffraction limit. One application of interest is microscopy, where the investigation of the details is limited by the wavelength of the probing device (light, electron beam) as well as the design of the instrumentation used for delivery and collection of the ray of waves. Considering the collection of infinitesimal sources with radiance $\partial E = \left(\xi_a/r\right)e^{i(kr-\omega t)}$, emitted by infinitesimally small "surface areas" ds, located at distance $\vec{r'}$ from the origin, assuming a centrally located reference frame. The "projection" from each source is defined by the vector $\vec{r} = \vec{R} - \vec{r'}$ with magnitude $r = \|\vec{r}\| = [(\vec{R}-\vec{r'})\cdot(\vec{R}-\vec{r'})] = \sqrt{(R^2 + r'^2 - 2\vec{R}\cdot\vec{r'})}$, providing $\partial E = \left[\xi_a/(R^2 - r'^2 - 2\vec{R}\cdot\vec{r'})^{1/2}\right]e^{i\left[k(R^2+r'^2-2\vec{R}\cdot\vec{r'})^{1/2}-\omega t\right]}$. In the far-field case $\vec{R} \gg \vec{r'}$, allowing for a series approximation of $r_{\text{denominator}} = \left(1/R^2\right)\{1 - [(2\vec{R}\cdot\vec{r'} + r'^2)/R^2]\}^{1/2} = R\{1 + (1/2)[(2\vec{R}\cdot\vec{r'} + r'^2)/R^2]\} + (3/8)[(2\vec{R}\cdot\vec{r'} + r'^2)/R^2]^2 + \cdots$, next to the exponential expression of the location: $r_{\text{exponent}} = R\{1 + [(r'^2 - 2\vec{R}\cdot\vec{r'})/R^2]\}^{1/2} = R\{1 + (1/2)[(r'^2 - 2\vec{R}\cdot\vec{r'})/R^2]\}^{1/2} + (3/8)[(r'^2 - 2\vec{R}\cdot\vec{r'})/R^2]^2 + \cdots$. Truncating the binomial series $\left((r'^2/R) \rightarrow 0\right)$ after the first term, this yields $\partial E = (\xi_a/R)e^{i(kR-\omega t)}e^{i[k(\vec{R}\cdot\vec{r'}/R)]}$. This term can be integrated over the area ds (total area: A) and can be expressed in Bessel function (J_1) as $E = (\xi_a/R)e^{i(kR-\omega t)}(2\pi/kr_i)RR_1 J_1(kr_iR_1/R)$, where r_i is the distance from the center point in the imaging plane and R_1 is the radius of the aperture. The intensity of the electric field is the complex conjugate multiplied by the regular function providing an expression that is known as the AIRY DISK. The central maximum is identified for $J_1(kr_iR_1/R) = 0$, which is valid for $(kr_iR_1/R) = 3.83$. From this, the radial distance to the first minimum is $r_i = 1.22\lambda R/2R_1$, which translates to the ABBE CONDITION for an image in the FOCAL POINT (f) of lens with diameter $D = 2R_1$, at

D

distance $R = f : \theta \cong \sin\theta = r_i / f = 1.22\lambda / D$, named after ERNST ABBE (1840–1905). Two maxima from two separate sources can hence be identified when one maximum falls in the minimum of the diffraction pattern from the other source. Alternatively, the RAYLEIGH CRITERION expresses this in a different manner, with respect to the NUMERICAL APERTURE of the lens $NA = n\sin\alpha$, giving the resolving power as RESOLVING POWER $= 0.61\lambda / NA$, where α represents half the acceptance ANGLE for the optical device (e.g., lens and fiber optic). A final remark is with respect to the binomial series for the far-field approximation; in the near-field approximation, the higher-order terms do conceal relevant information about the MICROSCOPIC features of the object being imaged. This latter becomes important in near-field scanning optical microscopy (NSOM) (see Figure D.34).

Figure D.34 Far-field diffraction, color pattern cast in the sky by the Sun diffracting from clouds.

Diffraction grating

[atomic, general, nuclear, optics] A series of parallel grooves on a surface of a transparent plate or a highly regular crystal structure used for WAVELENGTH (λ) analysis as well as material diagnostics. The diffraction grating can be considered a type of INTERFEROMETER. The first documented use of a diffraction GRATING was by JOSEPH VON FRAUNHOFER (1787–1826) in 1823. A diffraction grit with n slits, separated by DISTANCE d_{line} and width b (diffraction constant), produces a radiant intensity profile (i.e., radiance measured electronically: I) as a function of the ANGLE with the normal (θ) with the diffraction plate expressed as $I = I_0 \left(\sin n\alpha' / \sin\alpha' \right) \left(\sin\beta' / \beta' \right)$, where I_0 is the incident MAGNITUDE of radiance, $\alpha' = \pi d \sin\theta / \lambda$ and $\beta' = \pi b \sin\theta / \lambda$. The respective ($m = 0, 1, 2, \ldots$) principal maxima will fall in the following directions: $d_{\text{line}} \sin\theta_m = m\lambda$ (i.e., GRATING EQUATION). For non normal incidence (at angle θ_i), this becomes $m\lambda = d_{\text{line}} \left(\sin\theta_i + \sin\theta_m \right)$.

Diffusion

[biomedical, chemical, computational, electronics, thermodynamics] The movement of small particles, in most cases the constituents in a SOLUTION. The diffusion process is driven by several factors; thermal agitation, potential gradient, and concentration gradient are the dominant factors. One specific example of electric-field-mediated diffusion is the transport of electrical current over a wire. Additionally, photons may also diffuse in a SCATTERING process, "gradually" making their ways in randomly distributed directions. Diffuse light emitted from a milky-WHITE LIGHT bulb, or frosted GLASS, involves the scattering process that randomly changes the propagation direction, classifying the PHOTON migration as diffusion. The process of facilitated (chemical) diffusion relies on external factor that provides "permission" for atoms and molecules to pass through a BARRIER, such as a SEMIPERMEABLE MEMBRANE. In biological systems, the transport of chemical through the cellular membrane is mediated through LIGAND-gated channels, which assist in the selective passage of chemicals under the influence of a catalyst chemical in the proximity of the gated channel, thus activating the chemical mechanism of action for the transportation, or simply widening the ORIFICE supporting a higher transfer rate. One organ in particular is highly organized in the regulation of the diffusion, the KIDNEY. Ligand-gated membranes regulate processes such as (but not limited to) sleep/WAKE, pain, memory, and anxiety, which can be altered under the influence of chemical stimulants (legal and illegal drugs). In manufacturing, diffusion is applied to form crystal growth, next to marinating food for consumption as well as adding fungicides for preservation. Dialysis MACHINES use selective diffusion for the filtration process of BLOOD, replicating the functional processes of the KIDNEY. The diffusion process can be described by the DIFFUSION EQUATION, which has several forms. The most well-known DIFFUSION EQUATION is FICK's LAW: $J = -D_{chem} \left(\partial [C]/\partial z \right)$, where $[C_i]$ is the concentration of SOLUTE i, with gradient in direction z, $D_{chem} = D_0 e^{-Q_{ac}/RT}$ is the diffusion coefficient (D_0 is a material constant, Q_{ac} is the activation enthalpy, $R = 8.3144621(75) \, J/Kmol$ is the universal GAS constant, and T is the temperature in KELVIN), and J is the particulate FLUX density, and as a function of TIME (t) provides Fick's SECOND LAW: $\partial [C]/\partial t = D_{chem} \left(\partial^2 [C]/\partial z^2 \right)$. Alternatively, Onsager's equation treats the process from THERMODYNAMICS principles: $J = -L_{diff} \left(\partial \mu_c/\partial z \right)$, where μ_c is the chemical potential and L_{diff} is the thermodynamic diffusion coefficient. Fick's first law can be expanded to include convection as $J = -D_{chem} \left(\partial [C]/\partial z \right) + [C]U_z$, where U_z is the convective migration velocity. For electrons, the diffusion is generally referenced as electron DRIFT with a current density J_e defined as $J_e = en\mu_e E + eD_e \left(dn/dz \right)$, where $e = 1.60217657 \times 10^{-19} \, C$ is the electron charge, n is the number of electrons/"holes" per volume (charge density), μ_e is the electron mobility (material property), and $D_e = \mu_e v_t = \mu_e \left(k_b T/e \right)$ is the electron diffusion coefficient ($v_t = k_b T/e$ is the Einstein's relationship for electrical mobility and $k_B = 1.3806488 \times 10^{-23} \, m^2 kg/s^2 K$ is the Boltzmann constant). The total transport of particles in the diffusion process was introduced by Stefan Boltzmann in 1872 as a PROBABILITY function, using a functional correlation between the velocity of a PARTICLE (finite range of MAGNITUDE and direction) and its respective location as a function (with uncertainty) of time next to the net result of loss and gain from a unit volume as the TRANSPORT EQUATION. The statistically anticipated number of particles with velocity v with bandwidth dv in location r with spread dr is given for constituent i at time t as $f_i \left(\vec{r}, \vec{v_t}, t \right)$, which is linked in transport to the interparticle collision, defined by the collision CROSS SECTION σ_{ij} and the respective Coulomb interaction $g_{ij} \left(r \right)$ to yield the loss and gain balance for a volume of space (Ω), summarized over all components: $\sum_{j=1}^{n} \iiint \left(f_i' \left(\vec{r}, \vec{v_t}, t \right) f_j' \left(\vec{r}, \vec{v_t}, t \right) - f_i \left(\vec{r}, \vec{v_t}, t \right) f_j \left(\vec{r}, \vec{v_t}, t \right) \right) \sigma_{ij} g_{ij} \left(\vec{r} \right) \left(g_{ij} \left(\vec{r} \right), \vec{r} \right) d\vec{v_t} d\Omega$. The net transport need to match the diffusion for six DEGREES OF FREEDOM with respect to the N number of particles, and resulting $6N$ first-order differential equations, which is generally contracted to an equivalent single-particle distribution following: $\left(\partial f_i \left(\vec{r}, \vec{v_t}, t \right)/\partial t \right) + \vec{v_i} \cdot \left(\partial f_i \left(\vec{r}, \vec{v_t}, t \right)/\partial \vec{r} \right) + \vec{D} \cdot \left(\partial f_i \left(\vec{r}, \vec{v_t}, t \right)/\partial \vec{v_i} \right) =$ $\sum_{j=1}^{n} \iiint \left(f_i' \left(\vec{r}, \vec{v_t}, t \right) f_j' \left(\vec{r}, \vec{v_t}, t \right) - f_i' \left(\vec{r}, \vec{v_t}, t \right) f_j' \left(\vec{r}, \vec{v_t}, t \right) \right) \sigma_{ij} g_{ij} \left(\vec{r} \right) \left(g_{ij} \left(\vec{r} \right), \vec{r} \right) d\vec{v_t} d\Omega + S \left(\vec{r}, \vec{v_t}, t \right)$, where $S \left(\vec{r}, \vec{v_t}, t \right)$ is a particle source, which may be IONIZATION (electron/hole diffusion) or a chemical reaction or a reactive DECAY process; the source may be absent for general diffusion processes. An equivalent description is applied to photon transport described under the radiative transport equation (also the equation of radiative transport).

One elementary example is the process of infusing food with special flavors while marinating, or the natural preservation of food by submersion in vinegar or other solution (see Figure D.35).

Figure D.35 Marinating peppers.

Diffusion equation

[biomedical, chemical, computational] A description of the migration of particles, specifically small molecules across a BARRIER that can be either chemical, mechanical, or electrical (POTENTIAL BARRIER), also known as FICK'S EQUATION. DIFFUSION is associated with the transport of gases (e.g., O_2, CO_2, and NO), as well as small molecules, and molecules soluble in polar solvents migrating through the CELL MEMBRANE. The diffusion process is driven by concentration gradients and will continue until an equilibrium is reached, both based on OSMOTIC PRESSURE as well as on chemical concentration ([Chem]). The diffusion rate (J_{Chem}) is directly proportional to the cross-sectional area (A) of the partition (e.g., MEMBRANE), the concentration gradient $\Delta[\text{Chem}]$, the diffusion coefficient for the chemical substance in question (D_{Chem}), and inversely proportional to the DISTANCE (Δx) over which the chemical needs to traverse $J_{Chem} = -D_{Chem} A (\Delta[\text{Chem}]/\Delta x)$, known as Fick's law. The associated diffusion time is a function of the diffusion distance and the diffusion coefficient, also in reference to the permeability coefficient (based on the Einstein relation) $t_{diff} = (\Delta x)^2/2D_{Chem}$. The diffusion coefficient for small molecules is inversely proportional to the MOLECULAR WEIGHT (in Dalton). For large "spherical" molecules, the diffusion coefficient can be derived from the Stokes–Einstein relationship, based on the radius of the dissolved molecule ($r_{molecule}$) and yields $D_{Chem} = RT/6N_a \pi r_{molecule} \eta$, where $R = 8.3144621(75)$ J/Kmol is the GAS constant, T is the temperature, η is the VISCOSITY of the LIQUID, and $N_a = 6.022137 \times 10^{-23}$ J/K is Avogadro's number. The molecular radius is the effective Stokes radius, expressed as $r_{solute} = (3MW/4\pi\rho N_a)$, where ρ is the SOLUTE density and MW is the molecular weight in Daltons ("dimensionless"; 1/12 the mass of one single ^{12}C ATOM).

Diffusion length

[atomic, chemical, electronics, fluid dynamics, nuclear] In electronic design, the diffusion length applies to semiconductor design, specifically p–n and np junctions. The diffusion length for SEMICONDUCTORS relates the average DISTANCE traveled by electrons to initiate recombination. The recombination process is the lifetime of the charge CARRIER and the materials used in the formation of the semiconductor, the number of defects created in the materials as well as the DOPING medium. In heavily doped silicon, the Auger recombination will be the prevailing process in recombination. The diffusion length (L_{diff}) for the semiconductor is defined as the square root of the product of the lifetime (τ_{recom}) and the DIFFUSIVITY (D_{diff}): $L_{\text{diff},e} = \sqrt{D_{\text{diff}}\tau_{\text{recom}}}$, where the lifetime can be as long as 1ms, and the diffusivity (equivalent to diffusion coefficient) is the electron diffusion coefficient $D_{\text{diff},e} = \mu_e\left(k_BT/e\right)$ ($k_B = 1.3806488\times10^{-23}$ m²kg/s²K is the Boltzmann constant, $e = 1.60217657\times10^{-19}$C is the electron charge, μ_e is the electron mobility [material property]). For the hole diffusion coefficient: $D_{\text{diff},b} = \mu_b\left(k_bT/e\right)$ (μ_b the hole mobility [material property]). For solar cells, the typical diffusion length is in the order of $100-300\,\mu$m. In chemical diffusion, the diffusion length is the square root of the product of the diffusion coefficient (D_{chem}) and the diffusion process timeframe (t): $L_{\text{diff,chem}} = \sqrt{D_{\text{chem}}t} = 2\sqrt{\left(\int_{t'}D_{\text{chem}}\left(t'\right)dt'\right)}$.

Diffusion-limited aggregation

[chemical, computational] A reactant process limited in its MAGNITUDE by the DIFFUSION process. Next to diffusion, transport can take place based on convection or chemical reaction itself, also known as diffusion limited process (*also see* DAMKÖHLER NUMBER).

Diffusivity

[biomedical, chemical, fluid dynamics, general, mechanics] For electromagnetic induction on fluids, the FLUID motion is a factor that needs to be taken into account for the influences of ELECTROMAGNETIC RADIATION on the FLOW patterns of conducting fluids. This mechanism of action falls under MAGNETOHYDRODYNAMICS. The interaction time for the MAGNETIC FIELD with the fluid is captured by the electromagnetic DIFFUSION time: $\tau_\eta = L_{\text{EM}}^2/\eta_m$, where L_{EM} is the characteristic electromagnetic DIFFUSION LENGTH (which can potentially be excessively long, specifically when considering stars, this may exceed the lifetime of the STAR). The magnetic diffusivity is defined as $\eta_m = 1/\mu_e\sigma_e$, where $\mu_e = \mu_r\mu_{e0}$ is the permeability of the medium, $\mu_{e0} = 4\pi\times10^{-7}$ H/m is the permeability of VACUUM, $\sigma_e = 1/\rho_e$ ($\rho_e = R(A/\ell)$ is the resistivity, R is the RESISTANCE, A is the cross-sectional area of charge flow, and ℓ is the characteristic length) is the electrical conductivity. The fluid can be a watery mix, but also may be PLASMA or magma. With respect to the magnetic induction, the MAGNETIC REYNOLDS NUMBER can be introduced $Re_m = vL_{\text{EM}}/\eta_m$, where v is the fluid VELOCITY (*also see* ALFVÉN THEOREM).

Diffusivity constant

[biomedical] The scale of migration of SOLUTE a through SOLVENT b: $D_{ab}\left(T,\eta\right)$ (function of the solvent's chemical and material properties, TEMPERATURE T, and VISCOSITY η), as applicable to Fick's first law (*also see* DIFFUSION COEFFICIENT).

Dilatant fluids

[biomedical, fluid dynamics] SHEAR-THICKENING FLUID, fluid that increases in SHEAR FORCE with increasing FLOW velocity. The associated viscous phenomena also increase with a rise in VISCOSITY per se, in contrast with a NEWTONIAN FLUID or PSEUDOPLAST FLUID.

Dilation of time

[atomic, quantum] Considering two independent systems (system 1 and 2, respectively, denoted by the primary subscript) traveling with respect to each other while two sequential events takes place within one of the system that is observed in both systems. Under classical MECHANICS, the transfer of information would take place at phenomenal velocity, making the observation unquestionably simultaneous for both systems.

However, under the acceptance that the speed of light is the fastest known velocity for communication of data, the SIMULTANEITY between the two systems becomes questionable when either or both system travel at velocities approaching the speed of light, in particular that both systems are moving with respect to each other at relative velocity v_r. In case the two systems are traveling on the same axis, the concept of dilation of time can easily be illustrated by using the respective origins as the reference points. As two sequential events occur in system one (1), respectively, at location $x_{1,1}$ at time $t_{1,1}$ followed by the second event at location $x_{1,2}$ at time $t_{1,2}$, the Lorentz–Einstein relativistic time transformation yields for the two observers traveling in the two separate systems: location $x_{2,1} = \{1/\sqrt{[1-(v_r^2/c^2)]}\}(x-v_rt)$ $(c = \sqrt{\mu_0\varepsilon_0}^{-1} = (\mu_0\varepsilon_0)^{-1/2} = 2.99792458\times10^8$ m/s is the speed of light in VACUUM), while the time transformation yields $t_{2,1} = \{1/\sqrt{[1-(v_r^2/c^2)]}\}[t_{1,1} - (v_rx_{1,1}/c^2)]$ and $t_{2,2} = \{1/\sqrt{[1-(v_r^2/c^2)]}\}[t_{1,2} - (v_rx_{1,2}/c^2)]$, respectively, signifying that there is a difference between the time observed in reference frame 2 with respect to reference frame 1, expressed as $t_{2,2} - t_{2,1} = \{1/\sqrt{[1-(v_r^2/c^2)]}\}[(t_{1,2} - t_{1,1}) - (v_r/c^2)(x_{1,2} - x_{1,1})]$. Setting the frame 1 as the standard, reflecting "proper time" (local time), and syncing both events, the time difference between the two system is now $(t_{2,2} - t_{2,1}) = \{1/\sqrt{[1-(v_r^2/c^2)]}\}(t_{1,2} - t_{1,1})$, which also applies vice versa and validates simultaneity between the two systems. However, when the two events in system 1 do not occur at the same location $(x_{1,1} \neq x_{1,2})$, the two events can be observed in system two in three different situations: (1) the sum in the square brackets can still yield zero in which case the events are still observed simultaneously in system 2; (2) the sum yields a positive value while $t_{1,2} > t_{1,1}$ the events are observed in the same order in system 2; and (3) the sum in the square brackets yields a negative value while still $t_{1,2} > t_{1,1}$ the events are observed in the reverse order in system 2. This can be observed theoretically after a rewrite: $t_{2,2} - t_{2,1} = \{(t_{1,2} - t_{1,1})/\sqrt{[1-(v_r^2/c^2)]}\}\{1-(v_r/c)[(x_{1,2} - x_{1,1})/c(t_{1,2} - t_{1,1})]\}$, illustrating condition "3" occurs only when $[(x_{1,2} - x_{1,1})/c(t_{1,2} - t_{1,1})] > (c/v_r)$. This type of inequality will only be realized when the DISTANCE between the location of the two events $(x_{1,2} - x_{1,1})$ spans a greater distance than the distance traversed by the light within the time frame $(t_{1,2} - t_{1,1})$. Keep in mind that in case the spatial separation would be large enough for the latter reversal to occur the light leaving event 1 could never reach event 2 prior to its manifestation (III). This condition rules out that a causal sequence of event can never be reversed, even observed in system 2. One notable consequence of time dilation is often referenced as the twin paradox (*also* TIME DILATION and LENGTH CONTRACTION) (see Figure D.36).

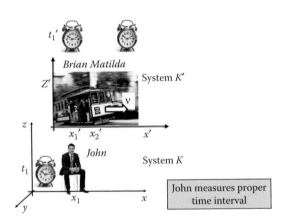

Figure D.36 Dilation of time for a reference-frame moving with respect to a stationary reference frame.

Dilution law of Ostwald

[atomic, biomedical, chemical, general, quantum] Correlation between the dissociation constant $K_d = \{[C^1]^a \times [C^2]^b\}/[C^1C^2]$ for a chemical reaction $(C^1_aC^2_b \rightleftarrows aC^1 + bC^2$, with chemicals C^1 and C^2 in concentrations $[C^1]$ and $[C^2]$, respectively) to the degree of dissociation with respect to a weak electrolytic SOLUTION.

The dissociation is the process of ionic compounds splitting in their constituents (e.g., salts). The degree of dissociation is defined as the fraction of original molecules that have split-off in their ionic components. The degree of dissociation is linked to the VAN'T HOFF FACTOR (i_H) for a SOLUTE with a maximum of n potential ions as $i_H = 1 + \alpha_d (n-1)$. The dilution law of Ostwald is now $K_d = (\alpha_d^2/1 - \alpha_d)[C_0^e]$, where $[C_0^e]$ is the original concentration of ELECTROLYTE. This was introduced by the chemist FRIEDRICH WILHELM OSTWALD (1853–1932) (from what is now known as Latvia) in 1902. The IONIZATION constant in the ARRHENIUS EQUATION for ELECTROLYTIC DISSOCIATION is independent of the concentration of electrolytes. The Ostwald law will also apply to the various stages of QUANTUM dot formation (also referred to as "ripening") and dissociation.

Dimensionless numbers

[computational, general] There are numerous scaling parameters that are used as either calibration factors or as an indication of which computational regiment a specific phenomenon is grouped under, such as turbulent versus LAMINAR FLOW. The following dimensionless numbers are identified, please check the respective entries for a full explanation: ALFVÉN NUMBER $A_L = ev_{\text{fluid}}/v_a$); ARCHIMEDES NUMBER ($Ar = g\rho L^3/\mu^2$); ARRHENIUS NUMBER ($\alpha^* = E_0/RT$); Bernoulli number ($x/(e^x - 1) = \sum_{n=0}^{\infty} B_n x^n/n$); Bernoulli number of the second kind ($b_n = \int_0^1 (x)_n \, dx$), BINGHAM NUMBER ($Bm = \tau_\gamma L/\eta v$); BIOT NUMBER ($Bi = hL/k$); BIOT NUMBER, MASS TRANSFER ($Bi_m = h_m L/D_{AB}$); BLAKE NUMBER ($B\ell = v\rho/\eta(1-\varepsilon) A_s$); BODENSTEIN NUMBER ($Bs = vL/D_{v,a}$); BOND NUMBER ($Bo = [g(\rho_\ell - \rho_v)L^2]/\gamma_s$); BRINKMAN NUMBER ($Br = \eta v^2/\kappa_{\text{cond}}(T - T_0)$); Brinkman number, modified ($Br^* = (T - T_0)_{\text{iso}}/(T - T_0)$), CAPILLARY NUMBER ($Ca = \eta v/\gamma_s$); CAUCHY NUMBER ($C = \rho v^2/E$); CAVITATION NUMBER ($\sigma_c = 2(P - P_v)/\rho v^2$); COEFFICIENT OF FRICTION ($C_f = \tau/(1/2)\rho v^2$); Colburn j-factor, heat transport ($j_H = St Pr^{2/3}$); Colburn j-factor, mass transport ($j_m = St_m Sc^{2/3}$); CONDENSATION NUMBER ($Co = g\rho^2 \Delta H_{\text{vap}} L^3/k\eta\Delta T$); DEAN NUMBER ($Dn = Re \times \sqrt{(D/2R)}$); DEBORAH NUMBER ($De = t_f |\bar{v}|/L$); DRAG COEFFICIENT ($C_d = g(\rho - \rho_f)L/\rho v^2$); drag coefficient ($C_D = F_D/[(1/2)\rho v^2](\pi R^2)$); Eckert number ($Ec = v^2/c_p(T_s - T_\infty)$); ELASTICITY NUMBER ($E\ell = \theta_r\eta/\rho R^2$); Eötvös number ($Eo = (\rho - \rho_f)L^2/\sigma_{\text{surf}}$); EULER NUMBER ($Eu = \Delta P/\rho v^2$); Euler number ($(\cosh t)^{-1} = 2/(e^t + e^{-t}) = \sum_{n=0}^{\infty} (E_n/n!) \times t$); FINE-STRUCTURE CONSTANT ($\alpha_{\text{fine}} = e^2/4\pi\hbar c$); FLUIDIZATION NUMBER ($N = \rho s3d \, 4g2/\mu \, f \, 2 \, E \, M$) [http://www.scribd.com/doc/88982439/Fluidization-Gupta, p20], FOURIER NUMBER ($Fo = \kappa t/c\rho L^2$); FOURIER NUMBER, MASS TRANSFER ($Fo_m = D_{AB}t/L^2$); FRESNEL NUMBER ($N_F = a^2/\lambda z$); FRICTION FACTOR ($f = \Delta P/(L/D)[(1/2)\rho v^2]$); FROUDE NUMBER ($Fr = v^2/gL$); GALILEO NUMBER ($Ga = gD^3\rho^2/\eta_{\text{kin}}^2$); GRASHOF NUMBER ($Gr = g\beta\rho^2(T_s - T_\infty)L^3/\eta_{\text{kin}}^2$); GRASHOF NUMBER, MASS TRANSFER ($Gr_m = (1/\rho)(\partial\rho/\partial[a])_{T,P} g([a]_s - [a]_a)L^3/\eta_{\text{kin}}^2$); GRÄTZ NUMBER ($Gz = \dot{m}c_p/kL$); HODGSON NUMBER ($H = fV\Delta P/Q\bar{P}$); JAKOB NUMBER ($Ja = c_p(T_s - T_{\text{sat}})/\Delta H_{\text{vap}}$); KNUDSEN NUMBER ($Kn = \lambda/L$); LEWIS NUMBER ($Le = \kappa/D_{AB}$); LOCKHART–MARTINELLI NUMBER ($\chi = \dot{m}_\ell/\dot{m}_g \sqrt{\rho_g/\rho_\ell}$); Lorentz number ($L_{t-e} = k_{\text{cond}}/\sigma_e T = \pi^2 k_{\text{cond}}^2/3e^2$); LUNDQUIST NUMBER ($Lu = v_a L/\eta$); MACH NUMBER ($Ma = v/v_{\text{sonic}} = \sqrt{Re \times Wi}$); MIXING TIME ($\theta_m = t_m/t_C$); MORTON NUMBER ($Mo_{\text{Liquid}} = g\eta_f^4\Delta\rho/\rho_f^2\sigma^3$); Newton number ($Ne = F_{\text{drag}}/\rho_f v^2 \ell^2$); NUSSELT NUMBER ($Nu = h_t L/\kappa$); OHNESORGE NUMBER ($Z = \eta/\sqrt{\rho L\sigma}$); PÉCLET NUMBER ($Pe = vL/\alpha$); Péclet number, mass transfer ($Pe_m = vL/D_{\text{dif}}$); PIPELINE PARAMETER ($\rho'' = v_{\text{gr}}v_0/2gH$); POISEUILLE NUMBER ($Ps = v_{\text{ph}}\eta/gd_p^2(\rho_s - \rho_f)$); POWER NUMBER ($N_P = c_{Np}(P_u/\omega_N^3\rho L^5)$); PRANDTL NUMBER ($Pr = v/\alpha$); PRESSURE COEFFICIENT ($Pt = (P - P_\infty)/(1/2)\rho_\infty v_\infty^2$), RAYLEIGH NUMBER ($Ra = L^3\rho^2 g\alpha_{\text{exp}}\Delta Tc_p/\eta\kappa = GrPr$); REYNOLDS NUMBER ($Re = vL\rho/\eta$); REYNOLDS NUMBER, MAGNETIC ($Re_m = \mu\sigma v_{\text{fl}}L$); SABERIAN NUMBER ($Sa =$); SCHMIDT NUMBER ($Sc = \eta/\rho D_{AB}$); SHERWOOD NUMBER ($Sh = h_m L/D_{AB}$); STANTON NUMBER ($St = h/\rho vc_p = Nu_L/Re_L Pr$); Stanton number, mass transfer ($St_m = h_m/v = Sh_L/Re_L Sc$); STROUHAL NUMBER ($Sr = v_s L/v$); WALL-FACTOR 1 ($K_2 = 2(d_{\text{PV}}/D_C)$); WALL-FACTOR 2 ($K_3 = (d_{\text{PV}}/D_C)(Z/D_C)$); WEBER NUMBER ($We = \text{Const}(\rho v^2 L/\sigma)$); Wiedermann–Franz ratio ($L_{t-e} = (k_{\text{cond}}/\sigma_e T) = \pi^2 k_{\text{cond}}^2/3e^5$); Weissenberg number ($Wi = t_f |\bar{v}|/D$); WOMERSLEY NUMBER ($\alpha_{Wo} = r\sqrt{\rho\omega/\eta}$).

Diode

[electronics, general] A VACUUM-tube configuration (inception: Thomas Alva Edison [1847–1931] 1883) in its primary form as well as semiconductor with a single charge BARRIER constructed with charge DOPING of silicon or germanium LATTICE structures. A junction with on one side a medium that is doped with acceptors and produces "positive charges" (i.e., holes), the p-type material, mated with a NEGATIVE CHARGE emitter doping (donor type; n-type), separated by a DEPLETION LAYER is referred to a p–n junction. The characteristic electronic property of the diode is that it will only allow current to transfer in one direction (for the p–n junction, from p-type side to n-type side) (*also see* SEMICONDUCTOR, P-N JUNCTION, *and* N-P JUNCTION) (see Figure D.37).

Figure D.37 Picture of representative diodes.

Diode laser

[general, optics, quantum, solid-state] Light amplification by STIMULATED EMISSION of RADIATION with respect to the use of a DIODE LIGHT SOURCE. The p–n junction offers a highly effective means for ELECTROLU-MINESCENCE, initially used to produce LED's (light-emitting diodes), introduced in 1962 by General Electric® led by ROBERT N. HALL (1919–). The discovery of electroluminescence dates back to Henry JOSEPH ROUND (1881–1966) at the Marconi Laboratories in 1907. The choice of materials used for the p–n junction sets the scale for the emitted wavelength base on the PHOTOELECTRIC EFFECT $v = E/h$, where $v = \lambda/c$ is the frequency, λ is the wavelength, $c = 2.9979245(8) \times 10^8$ m/s is the speed of light, E is the ENERGY in reference to the work function (ϕ_w; $hv = KE - \phi_w$, KE is the additional kinetic energy of the electron in the VALENCE BAND) of the material and $h = 6.62606957 \times 10^{-34}$ m^{2}kg/s is the Planck's constant. The combination of the work function of a material and the maximum kinetic energy will provide the EMISSION SPECTRUM limits of the diode. Placing the LED material inside a cavity constructed of two opposing mirrors can under specific conditions create the platform for laser excitation and laser emission. Specific configurations of diode laser are horizontal emitting laser (HCSEL: horizontal cavity, surface-emitting laser) and vertical emitting diode (VECSEL: vertical-external-cavity surface-emitting laser). Over the years, the emission wavelength has gradually diminished, reaching into the ultraviolet-C (UV-C) ($\lambda \sim 260$ nm) currently, while during its introduction only INFRARED emission (~ 870 nm) was possible. Ever since then the upper and lower limits are continuously challenged, reaching in the mid-infra-red 3330 nm, GaInAsSb. The significant advantage of diode laser is their potentially small size (see Figure D.38).

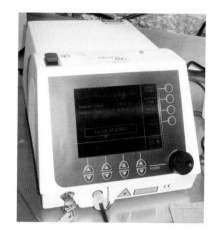

Figure D.38 Picture of a diode laser, this one in particular operates at 830 nm.

Diopters (D)

[general, optics] The power of a LENS, captured as the reciprocal value of the focal length. Due to its definitions, the lens power in diopters is additive. For VISION correction, a far-sighted person needs to gain close DISTANCE vision by means of a lens system with a positive value, whereas near-sightedness can be corrected by a negative diopter lens. A positive diopter lens is a magnifier, whereas a negative diopter will result in a virtual IMAGE.

Dipole

[atomic, biomedical, chemical, condensed matter, electromagnetism, electronics, general, solid-state] A charge configuration that results in a respective positive and negative net-charge concentration in three-dimensional space of equal MAGNITUDE that is separated by a fixed or variable DISTANCE. Certain animals produce a DIPOLE FIELD to aid in orientation of movement and target location for hunting. Examples of dipole-equipped animals are the Platypus and the elephantnose fish Gnathonemus (African freshwater fish). A dipole pruces a well-organized electric field configuration. RADIO waves emitted from an ANTENNA use the antenna as the dipole in which the magnitude of the respective charge fluctuates over time with the driving modulation frequency, either in AMPLITUDE (magnitude of SIGNAL strength, meaning magnitude of charge; i.e., A.M. modulation) or in OSCILLATION frequency (high-frequency CARRIER WAVE [kHz or predominantly Mhz] with super-imposed modulation, usually in the audible frequency range) (see Figure D.39).

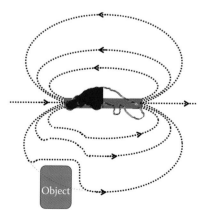

Object

Figure D.39 Platypus as an electric dipole.

This phenomenon applies both to ELECTRIC CHARGES and MAGNETIC FIELD. In electric charge distribution, an object that has a cluster of POSITIVE CHARGE separated forms an equal amount of NEGATIVE CHARGE, where the virtual charge centers are at a distance that can be determined. An oscillating variable distance produces an antenna, emitting ELECTROMAGNETIC RADIATION. The electric field lines created by the ELECTRIC DIPOLE are very uniform and form a perfect mirror IMAGE from the positive to the negative pole with respect to the central plane between the charge centers. Due to the split positive and negative charge, the dipole creates a dipole moment. A MAGNET that by definition has a NORTH and south pole is a MAGNETIC DIPOLE, specifically since the single object will by definition have equal field lines coming in as there are emanating, essentially due to the fact that magnetic field lines are closed (*see* **GAUSS'S LAW**).

D

Dipole field

[electromagnetism, general] The electric field lines connecting the positive and negative charges of a DIPOLE. Due to the equal MAGNITUDE of the net charges of a dipole, the electric field configuration on a MACROSCOPIC scale is highly uniform and symmetric.

Dipole layer

[biomedical, chemical, electronics, general, solid-state] A layer that has two opposite sign charges on either side. The DIPOLE moment for a point charge distribution with MAGNITUDE q (equal magnitude but opposite charges) is $\overrightarrow{p_{\text{dipole}}} = q\delta$. The electrical potential (V_p) resulting from a dipole charge distribution in point P at great DISTANCE from the dipole ($r \gg \delta$; r the distance from the center of the dipole to the point P, δ the dipole distance) is defined as $V_p = K\left(\overrightarrow{p_{\text{dipole}}} \cdot \vec{r}/r^3\right) = K\left(p_{\text{dipole}}\cos\theta/r^2\right)$, where θ represents the ANGLE with the normal to the axis of the dipole and K represents a constant. Replacing the POINT CHARGES with layers of equal charge results in reconfiguring the dipole moment in a dipole moment per surface area (A): $m = q\delta/A = \sigma_{\text{charge}}\delta$, now δ is the plate separation. In this situation, the electrical potential at a distant point P is defined as a function of location with respect to the entire plate with charges, in reference to the central location on the plate, as $dV_p = K\left(\overrightarrow{m_{\text{dipole}}} \cdot \overrightarrow{dA}/r^3\right) = K\left(m_{\text{dipole}}\cos\theta\,dA/r^2\right) = Km_{\text{dipole}}d\Omega$, where Ω is the SOLID ANGLE. Considering the entire surface contribution yields for the potential $V_p = \int_{\text{surface-angle}} Km_{\text{dipole}}d\Omega = Km_{\text{dipole}}\Omega$, point P observes the entire surface with the solid angle Ω.

Dipole moment (p_{dipole})

[chemical, condensed matter, electronics, general, solid-state] Generally referring to an ELECTRIC DIPOLE, the dipole associate with a MAGNETIC FIELD distribution is generally specifically named the "MAGNETIC DIPOLE" with a "magnetic moment." The charge "pivot" (two equal but opposite charges [q^+; q^-] separated by a DISTANCE d) expressed at a distance $\vec{r}$ is defined as $\overrightarrow{p_{\text{dipole}}} = q\vec{r}$. POLARIZATION is a special case of dipole moment, signifying the ELECTRIC DIPOLE MOMENT per unit volume. Placing a dipole in an external electric field ($\vec{E}$) will generate a torque with MAGNITUDE and the direction perpendicular to the field $\overrightarrow{\tau_E} = \vec{E} \times \overrightarrow{p_{\text{dipole}}}$. Note, compare this to the torque from a magnetic field ($\vec{B}$) on the magnetic moment $\tau_B = \vec{B} \times \mu_B$, where the magnetic moment is defined as $\overrightarrow{\mu_B} = \overrightarrow{m_B} = (1/2)q\vec{r} \times \vec{v}$ for an ELECTRIC CHARGE moving in a circle with radius r at velocity v (creating a current), or expressed with respect to a current loop (I) with vector field area $\overrightarrow{A_S}$: $\overrightarrow{\mu_B} = I\overrightarrow{A_S}$. POLARIZATION ($P_{\text{pol}}$) in this case refers to the average dipole moment of a group of N elementary dipoles (atomic or molecular) defined as $P_{\text{pol}} = N\left(p_{\text{dipole}}\right) = N\alpha_{\text{pol}}E = \chi(E/4\pi)$, where E is the electric field strength, α_{pol} is a material property referred to as the polarizability, and χ is the susceptibility of the electric system (see Figure D.40).

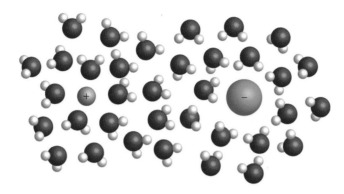

Figure D.40 Diagram of interaction between ion and dipole.

Dipole radiation

[electromagnetism, general] The steady-state charge configuration of a DIPOLE constitutes a constant electric field configuration, both in time and in space. Moving charges (i.e., current) provide a changing electric field that can be periodic in time, oscillating. The moving charges and associated current induces an alternating MAGNETIC FIELD, which combined with the alternating synchronized electric field generates electromagnetic (EM) RADIATION, that is, RADIO and television waves. The electric and magnetic field lines generated by the oscillating charges obey the standard MAXWELL EQUATIONS and are perpendicular to each other at all times. Naturally occurring EM radiation originates, for instance, from the partial displacement of the center of the positive and NEGATIVE CHARGE "cloud" on atomic and molecular scale. Assuming a stable ELEMENT, the charges are distributed with the POSITIVE CHARGE in the NUCLEUS and the negative (electron) charge in stable orbit around the nucleus. Disrupting the charge balance by removing an electron from orbit, specifically converting the "free" electron to another element (i.e., ION) will temporarily or permanently form a dipole. The temporary dipole creates the platform for atomic dipole EM emission (see Figure D.41).

Figure D.41 Dipole radiation emitted from analog radio antenna.

Dipole–dipole interaction

[chemical, nuclear, solid-state] The forces between two polar molecules, respectively, the positive and negative side of either. These forces are Coulomb forces and are attractive and strong, ranging from 5 kJ/mol to

20 kJ/mol. The potential ENERGY in the interaction is captured by the COULOMB POTENTIAL between the two net charges of the respective POLES of the individual dipoles. WATER (H_2O) provides an excellent example, as does hydrogen chloride (HCl).

Dirac, Paul Adrien Maurice (1902–1984)

[atomic, computational, mechanics, nuclear, quantum, relativistic, thermodynamics] A theoretical physicist from Great Britain. The work of Paul Dirac combined with the efforts of ERWIN SCHRÖDINGER (1887–1961) resulted in them sharing the Nobel Prize for Physics in 1933. Paul Dirac combined the knowledge of JAMES CLERK MAXWELL (1831–1879) on QUANTUM THEORY with special relativity to form the concept of QUANTUM ELECTRODYNAMICS (see Figure D.42).

Figure D.42 Paul Adrien Maurice Dirac (1902–1984) at the 1927 Solvay conference on "Quantum Mechanics."

Dirac delta function

[computational] A generalized function that is zero everywhere except for the (transformed) coordinate zero $\delta\left(\vec{r} - \vec{r'}\right) = (1/2\pi)\int_{-\infty}^{\infty} e^{ip\left(\vec{r} - \vec{r'}\right)} dp$, where $\delta\left(\vec{r} - \vec{r'}\right) = 0$ for $\vec{r} - \vec{r'} \neq 0$ and $\delta\left(\vec{r} - \vec{r'}\right) = +\infty$ for $\vec{r} - \vec{r'} = 0$ (see Figure D.43).

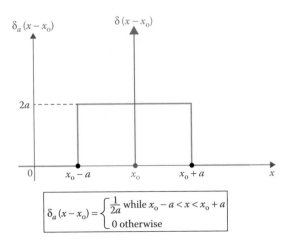

Figure D.43 Dirac delta function.

Dirac equation

[general, nuclear, quantum] Combining the concepts of the SCHRÖDINGER EQUATION with QUANTUM THEORY and special relativity yields a relativistic WAVE EQUATION that provides a DENSITY FUNCTION ($\Psi(\vec{x},t)$), as function of space $\vec{x}=(x,y,z)$ and time t) of the PROBABILITY of the location of a PARTICLE in MOTION (predominantly moving at extremely high velocities on the order of the speed of light). The wave function specifically applies to the electron orbital motion. The concept was derived from the ENERGY equation $E^2 = c^2 p^2 + m^2 c^4$ with speed of light $c = 2.99792458\times10^8 \ m/s$, $p=\hbar k$, $\hbar = h/2\pi$, $h = 6.62606957\times10^{-34}\,\mathrm{m}^2\mathrm{kg/s}$ the Planck's constant, p the momentum of a particle in motion (*Note*: this function also identifies the momentum operator in the Schrödinger equation), $k = 2\pi/\lambda$ the WAVE NUMBER (also referred to as wave vector) for a wave function with wavelength λ, m the particle REST MASS. The energy equation theoretically yields two solutions, positive and negative. The concept of negative energy is not permitted in classical MECHANICS, while in QUANTUM MECHANICS (quantum ELECTRODYNAMICS), this negative energy (i.e., "holes" as devised at the time) occupied virtual space that was ideologically converted into POSITIVE CHARGE, the antielectron later confirmed as the positron by CARL DAVID ANDERSON (1905–1991). The Dirac equation for a particle with spin 1/2 (such as the ELECTRON and the QUARKS) is expressed as $\left(\beta mc^2 + \sum_{l=1}^{3}\alpha_l p_l c\right)\Psi(\vec{x},t) = i\hbar\left(\partial\Psi(\vec{x},t)/\partial t\right)$, which provides a 4×4 matrix, summarized over the three DEGREES OF FREEDOM (l), with

$$\beta = \begin{bmatrix} 1 & 0 \\ 0 & -1 \end{bmatrix}$$

and α_l coefficients that conform to the boundary conditions:

$$\alpha_1 = \begin{bmatrix} 0 & \sigma_x \\ \sigma_x & 0 \end{bmatrix}, \alpha_2 = \begin{bmatrix} 0 & \sigma_y \\ \sigma_y & 0 \end{bmatrix}, \text{ and } \alpha_3 = \begin{bmatrix} 0 & \sigma_z \\ \sigma_z & 0 \end{bmatrix}$$

Pauli spin matrices with SPIN $\sigma_{l'}$.

Dirichlet, Peter Gustav Lejeune (1805–1859)

[computational] A mathematician and scientist from Germany (French Empire/Prussian Empire) dedicated to analytical number theory, also referred to as arithmetic. Arithmetic deals with whole numbers and has traceable roots dating back to documentation (clay tables) from approximately 1800 BC (see Figure D.44).

Figure D.44 Drawing of Peter Gustav Lejeune Dirichlet (1805–1859).

Dirichlet boundary condition

[computational, electromagnetism, fluid dynamics, mechanics, thermodynamics] Boundary conditions named after the work by PETER GUSTAV LEJEUNE DIRICHLET (1805–1859) that define the limitations to an ordinary or partial differential equation at the edges of the domain for the boundary of the solutions. The Dirichlet boundary conditions apply to cases where the following conditions determine to process, for instance, a surface maintained at a fixed temperature; an electric junction at a fixed voltage; in FLUID FLOW the boundary will have a no-slip condition, with ensuing zero velocity at the boundary; and a geometric construction that is fixed at one end. The Dirichlet boundary condition is captured by $\nabla^2 f(\vec{r}) + f(\vec{r}) = 0$, which for the ordinary form is written as $f''(\vec{r}) + f(\vec{r}) = 0$ and ∇^2 is the Laplacian differential operator.

Dirichlet distribution

[computational] A discontinuous collection of multivariate PROBABILITY distributions, frequently denoted as $Dir(\vec{\alpha})$, where $\vec{\alpha}$ are positive real number vectors $(\vec{\alpha} = (\alpha_1, \alpha_2, \ldots, \alpha_i, \ldots \alpha_K))$. The probability function is described as $\mathrm{Dir}(\vec{\alpha}) = \left[\prod_{i=1}^{K}\Gamma(\alpha_i)/\Gamma\sum_{i=1}^{K}\alpha_i\right]^{-1}\prod_{i=1}^{K}x_i^{(\alpha_i - 1)}$, with K the number of categories ($K \geq 2$), $\sum x_i = 1$, $\Gamma(x) = (n-1)!$ with n a positive integer or for complex numbers $\Gamma(x^*) = \int_0^\infty t^{*x^*-1}e^{-t^*}dt^*$, the gamma function, describing the probability components, and α_i concentration or weight parameters. The "mode" of the distribution function is defined by $x_i = (\alpha_i - 1)/\sum_{i=1}^{K}\alpha_i - K$, setting $\alpha_i > 1$. The entropy of this distribution is $H(X) = \log B(\vec{\alpha}) + (\alpha_0 - K)\psi(\alpha_0) - \sum_{j=1}^{K}[(\alpha_j - 1)\psi(\alpha_j)]$, where $\psi(\alpha_j)$ is the digamma function $(\psi(x) = (d/dx)\ln[\Gamma(x)] = [\Gamma'(x)/\Gamma(x)])$ and $B(\vec{\alpha}) = \prod_{i=1}^{K}\Gamma(\alpha_i)/\Gamma\sum_{i=1}^{K}\alpha_i$. The Dirichlet probability function is also found, defined as $\mathrm{Dir}(\vec{\alpha}) = \lim_{m\to\infty}\lim_{n\to\infty}[\cos^{2n}(m!\pi\vec{\alpha})]$, plotted as a discrete "AMPLITUDE" for probability as a function of location $\vec{\alpha}$ (see Figure D.45).

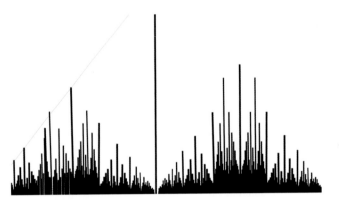

Figure D.45 Discrete Dirichlet distribution.

Discrete vortex

[computational, fluid dynamics] The theoretical approach to the turbulent FLOW across a vertical plane with respect to the surface of the WING of a plane, however at a significant DISTANCE behind the wing (i.e., Trefftz plane). The approximations for solving the flow pattern are that each aerodynamic surface is outlined by vortices in the shape of discrete horseshoe geometry. The collection of discrete vortices is defined by the Kutta–Joukowski theorem and obeys the BIOT–SAVART law. The Kutta–Joukowski theorem describes the CIRCULATION and is generally solved numerically (see Figure D.46).

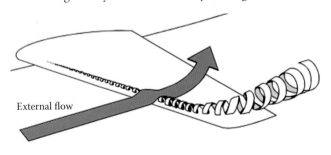

External flow

Figure D.46 Discrete vortex formed over a wing.

Dispersion

[computational, general, optics] Waves are identified by both a PHASE VELOCITY and a group velocity. Under conditions of inhomogeneity and anisotropy, the group VELOCITY (v_g) will be a function of the WAVELENGTH (λ): $v_g = d\omega/dk = v + (dv/dk)$, where $\omega = (2\pi\lambda/c) = 2\pi\nu$, $k = 2\pi/\lambda = \omega/v$, ν is the frequency, and $c = \sqrt{\mu_0\varepsilon_0}^{-1} = (\mu_0\varepsilon_0)^{-1/2} = 2.99792458 \times 10^8$ m/s is the speed of light. Respectively, the phase velocity $v_g = \omega/k$. The dispersion thus results in speed of propagation differences as a function of wavelength, specifically applied to a waveform that is constructed from superposition of multiple waves with different wavelengths (e.g., WHITE LIGHT). In 1875, the Scottish physicist JOHN KERR (1824–1907) demonstrated the induction of anisotropy in GLASS with index of REFRACTION $n = n(\lambda)$ under the influence of an external electric field (KERR EFFECT, *also see* **KERR CELL**). The index of refraction for most transparent materials is a function of wavelength, providing a mechanism resulting in spectral spread for white light, or general broad band source. A specific example of anisotropy is found for the material calcite. For a PRISM, the deviation ANGLE (δ) for the ray of light at a particular WAVELENGTH (λ) exiting from the prism with respect to the vector of the incident ray provides the dispersion: $D_{prism} = d\delta/d\lambda = (d\delta/dn)(dn/d\lambda)$. The latter condition introduces the condition of Cauchy for the index of refraction as a function of wavelength as $n = C + (B/\lambda^2)$, with B and C material-specific constants. Dispersion in imaging introduces the concept of CHROMATIC ABERRATION. In case the phase velocity decreases with increasing frequency, the dispersion is referred to as normal dispersion, whereas an increase is defined as anomalous dispersion (*also see* **WAVE DISPERSION** *and* **KRAMERS–KRONIG RELATIONS**) (see Figure D.47).

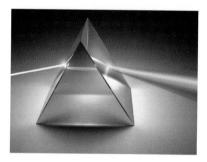

Figure D.47 Wavelength-dependent propagation leading to dispersion in a prism.

Distance

[general] The length of separation between two objects, unit meter. The meter was originally defined in 1799 as one-ten millionth of the distance from the NORTH Pole to the equator along the meridian that runs through Paris. In 1960, the meter was redefined under the Système International (SI) unit scheme as the length of a stick of alloy composed of platinum–iridium under normal conditions (20°C; 1 atm; 60% HUMIDITY), also on file in Paris. Later that decade, the krypton-86 ISOTOPE DECAY emission WAVELENGTH (orange–red: 605.78 nm) was used to define the unit length, and currently, the distance traveled by ELECTROMAGNETIC RADIATION sets the standard for a time frame of 1/299792458 s. Early definitions of length involve the distance between the tip of the middle finger of the stretched arm of King Henry I of England to the tip of his nose, as documented in 1120, this being the yard. The early Greek units of measure were dimensions related to body parts, such as the digit of a finger (approx. 19 mm), the length of

a foot (which is 1% longer than the foot in the English system of measurements), and the fathom (approx. 180 cm) described in 450 BC (see Figure D.48).

Figure D.48 Sign post with distances to remote locations.

APPROXIMATE DISTANCE OR SIZE FOR OBJECT	VALUE [M]
Earth to known quasar at greatest distance	1.4×10^{26}
Earth to known most remote galaxy	9×10^{25}
Earth to Andromeda (closest galaxy; M31)	2×10^{22}
Sun from nearest star (Proxima Centauri)	4×10^{16}
Earth's orbit radius	1.5×10^{11}
Earth mean radius	6.3×10^{6}
Average human	1.6
Sparrow	1.5×10^{-2}
Average cell diameter	1×10^{-5}
Blue light wavelength	3.30×10^{-7}
Hydrogen atom diameter	1×10^{-10}
Hydrogen nucleus	1×10^{-14}
Neutron	1×10^{-15}

Distillation

[thermodynamics] A selective condensation of LIQUID constituents from a mixture (e.g., separate ALCOHOL from water, OIL refinement for separation into fossil fuels, and raw ingredients for polymerization). In a distillation column, a mixture of fluids is introduced in a column that has several stages arranged in horizontal planes at increasing altitude that operate in mutual equilibrium between the liquid and the vapor PHASE. The more volatile constituent (lower vapor point) is evaporated as BUBBLES upward to the next higher stage, eventually making its way to the top of the column, while the condensate that has been depleted of the more volatile components will drain back down to the bottom of the column, where the

"SOLVENT" can be drained off. The evaporated more volatile constituent can be vented and collected at the top of the column (*also see* DEW LINE) (see Figure D.49).

(a)   (b)

Figure D.49 (a) Picture of an old-fashioned "moon-shine" distillery (Courtesy of Southwestern Distillery, Cornwall, United Kingdom.) and (b) modern large volume copper distillery.

Diurnal

[biomedical, energy, geophysics] Daily (diurnal cycle) or related to ACTIVITY occurring primarily in the daytime (e.g., flowers closing at night) (*also see* CIRCADIAN RHYTHM) (see Figure D.50).

Figure D.50 Diurnal events: sunrise at Mount Fuji, Japan.

Diurnal arc

[biomedical, energy, geophysics] The daily time frame that a celestial body is above the horizon for the observer (e.g., the daily cycle of the SUN has a diurnal arc lasting from sunrise to sunset).

Divergence

[biomedical, computational, fluid dynamics, optics] The vector spread due to a computational algorithm or due to a device. A LENS with a negative focal length will provide divergence of the optical rays, forming a virtual IMAGE. An EXPANSION with gradual or discrete increase in diameter with increasing DISTANCE from a reference point following a cylindrical flow will result in diverging flow or general mass movement. Divergence represents the vector density (e.g., flow density) as a function of volume with respect to a vector field initiating from an infinitesimal volume. Computationally, this is captured by the vector operator gauging the MAGNITUDE of the source or sink for a vector field ($\vec{F}_v$), expressed by $\operatorname{div}\vec{F}_v(\vec{p}) = \nabla \cdot \vec{F}_v(\vec{p}) = \lim_{V \to \{p\}} \iint_{S(V)}(\vec{F}_v \cdot \vec{n}/|V|)dS \cong \lim_{\Delta V \to 0}\{(1/\Delta V)\oint_S \vec{F}_v \cdot d\vec{S}\}$, where $\nabla = (d\vec{F}/dx) + (d\vec{F}/dy) + (d\vec{F}/dz)$, $m = \rho V$ is the mass, ρ is the density (respectively, line density [vector field] or mass density [flow]), V is the local volume, with respect to points in space p that identifies the origin of the vector field (e.g., "source" of flow), note that $\Delta V \to 0$ implies convergence on the "origin," S is the surface of the boundary of the volume V, with $\vec{n}$ the normal to this surface. In case the flux-density vector $\vec{F}_v$ describes the FLOW of charge (current) or mass (FLUID), the closed integral signifies the deviation from the CONSERVATION OF MASS with respect to the volume enclosed. The flow of mass can be represented by a velocity field $v(x,y,z,t) \sim v(r,\theta,\phi,t)$ leading to $\oint_S \rho\vec{v} \cdot d\vec{S} = -\int_{\Delta V}(\partial\rho/\partial t)dV$, which yields the EQUATION OF CONTINUITY under Gauss's theorem $\nabla \cdot (\rho\vec{v}) = -(\partial\rho/\partial t)$, where ρ is a function of location. Applying the Stokes theorem, the convergence also implies $\int_{\Delta S}\nabla \times \vec{F}_v \cdot dS = \oint_L \vec{F}_v \cdot d\vec{\ell}$, with L the circumference of the closed loop, and $\vec{F}_v \cdot d\vec{\ell}$ may need to be developed into a series $\vec{F}_v \cdot d\vec{\ell} = (\vec{F}_v + (d\vec{F}_v/dx)\Delta x + \cdots)\Delta y$ for approximate analog solving for the divergence (see Figure D.51).

Figure D.51 Divergence.

Divergence-free vector field

[computational, fluid dynamics] A fundamental component of HELMHOLTZ DECOMPOSITION in vector calculus. In a three-dimensional space, the vector field ($\vec{V}$) composed of sufficiently smooth, rapidly decaying vectors can be decomposed into the sum of two "orthogonal" components, one consisting of an IRROTATIONAL (CURL-FREE) VECTOR FIELD and the second segment composed of rotational (divergence-free) or solenoidal vector field. The formulation of this series is defined as a function of location ($\vec{r}$) within a volume in space (V) as $\mathbb{D} = (1/4\pi)\int_V \nabla' \times \vec{V}(\vec{r}')/|\vec{r} - \vec{r}'|dV' + (1/4\pi)\int_S \vec{V}(\vec{r}') \times dS'/|\vec{r} - \vec{r}'|$. The second term will have limit zero when the phenomenon comprises all of free space with the vector field rapidly decreasing.

Divergenceless

[computational] A vector field that implies charge conservation $(\partial/\partial t)\vec{J} + \nabla \cdot \vec{J} = 0$, where $\vec{J}$ represents the charge FLOW density. For a general steady-state vector field $(\overline{F_v})$, the concept of divergenceless entails $\nabla \cdot \overline{F_v} = 0$, which means that there exists a function $\vec{G}$ that provides $\overline{F_v} = \nabla \times \vec{G}$, also referred to as a SOLENOID field (as introduced by SIR HORACE LAMB [1849–1934] in 1932) (*also see* DIVERGENCE-FREE VECTOR FIELD *and* EARNSHAW'S THEOREM).

Divergent flow

[fluid dynamics] A FLOW that proceeds with a diminishing vector density at greater DISTANCE from the exit point (*see* DIVERGENCE).

Diverging lens

[general, optics] A LENS that will form a virtual IMAGE, diverging rays of light emanating from the lens under illumination by a directional source. A parallel ray entering the lens will emerge with a positive ANGLE with respect to the optical axis, also referenced as a lens with a negative focal length. This applies to optical lenses, magnetic, and electronic lenses (*see* LENS).

Diverging waves

[fluid dynamics] Free SURFACE WAVES can be regarded as diverging waves in a circular path (of uniform depth). As such, TIDAL WAVES behave as long waves on a plane sheet of water, which can be approached as a steady-state WAVE propagating in the r-direction. This problem is presented as $(d^2\phi_w/dr^2) + (1/z)(d\phi_w/dr) + \phi_w = 0$, where ϕ_w represents the VELOCITY POTENTIAL of the surface. In an asymptotic SOLUTION, the diverging wave solution incorporates a zero-order Bessel function of the first kind $J_0(r) = (2/\pi)\int_0^\infty \sin(z\cosh\xi)d\xi$, where ξ is the ANGLE in the complex plane (representing DIVERGENCE) expressed in series approximation as $\phi_w^0(\chi r) = (1/4)\{(2/\pi)(\log(\chi r/2) + \gamma_w + (1/2)i\pi)J_0(\chi r) - (2/\pi)[(\chi^2 r^2/2^2) - s_2'(\chi^4 r^4/2^2 2^4) + s_3'(\chi^6 r^6/2^2 2^4 2^6) - \cdots]\}e^{i\omega t}$, where $\chi = \omega/\sqrt{gh}$ is the WAVE NUMBER (wave propagation velocity: $v = \sqrt{gh}$, g is the GRAVITATIONAL ACCELERATION, h is the height of the water [depth], $\omega = 2\pi v$ is the ANGULAR FREQUENCY, v is the OSCILLATION frequency [applied external "pressure" oscillation]), γ_ω is the integration constant for the real part [derived from $\int_{r/2}^\infty (e^{-\varsigma}/\varsigma)d\varsigma = -\gamma_w - \log(r/2) + \text{Rest}$; Rest represents a rest term], and $s_i' = 1 + \{(1/2) + (1/3)\} + \cdots + (1/i)$ is a NORMALIZATION constant respectively air/water where ϕ_w represents the wave phenomenon. On the other hand, a diverging wave in AIR will be a three-dimensional process, expressed as a SPHERICAL WAVE obeying (in spherical coordinates) $(\partial^2\phi_a/\partial t^2) - 2(\partial\phi_a/\partial r)(\partial^2\phi_a/\partial r\partial t) + (\partial\phi_a/\partial r)^2(d^2\phi_a/dr^2) = v_s^2[(\partial^2\phi_a/\partial r^2) + (2/r)(d\phi_a/dr)]$, where ϕ_a represents the velocity potential of the air volume, $v_s \sim \sqrt{P_0\rho_0}$ is the speed of SOUND, P_0 is the pressure, ρ_0 is the density, derived from the velocity potential (ϕ_a) through $v_s^2 c_s = (\partial\phi_a/\partial t)$, with c_s the "condensation," which is an indication of the AMPLITUDE of the pressure. For a simple HARMONIC OSCILLATION with frequency $v = v_s/\lambda$ (λ is the wavelength), the solution is rather conveniently written as $\phi_a = e^{-ikr}/4\pi k$, where $k = 2\pi/\lambda$ is the wave number. A rather complex solution for a complex disturbance is described in detail dedicating book volumes to this problem (see Figure D.52).

Figure D.52 Divergent wave separating tectonic plates during earthquake.

DNA

[biomedical, chemical, general, theoretical] Deoxyribonucleic ACID (genetic molecule). DNA contains the information about the species and the construction details for functional components (i.e., organs), the hereditary information. DNA is contained in the cellular chromosome. DNA was identified and isolated by the Swiss physician FRIEDRICH MIESCHER (1844–1895) in 1868. DNA is constructed of four elementary molecular chains: guanine (G), adenine (A), thymine (T), and cytosine (C). The molecular structure of DNA was discovered by ROSALIND FRANKLIN (1920–1958) in 1951. In general, the DNA molecules are comprised of two biopolymer strands. The strands are coiled around each other to form what is known as the double helix. The double-helix structure of DNA was unraveled based on the efforts from Rosalind Franklin by the American molecular biologist JAMES D. WATSON (1928–) and the English molecular biologist and biophysicist FRANCIS HARRY COMPTON CRICK (1916–2004) in 1953. DNA is contained in every living CELL of all organisms and viruses. The DNA is apparently involved in a communication system (chemical, electronic, and potentially optically) that allows individual cells to split, congregate (cell proliferation), and form a specific structure, based on their location with respect to neighboring cells. DNA performs a process of chemical encoding, decoding, and regulation. With each cell division, all material in the cell is duplicated. The cell duplication process is initiated by a DNA synthesis process, the DNA coils split longways, and a messenger RNA (ribonucleic acid) is formed that reproduces the identical DNA strand. The nucleic acids DNA and RNA in combination with proteins form the essential constituents and foundation of life (see Figure D.53).

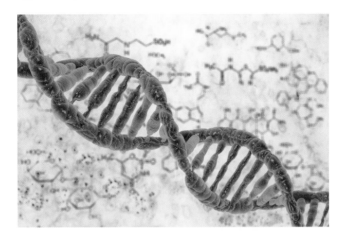

Figure D.53 The complex structure of the DNA molecular chain.

Dolphin

[general] A sea mammal that relies on ultrasonic detection for orientation and guidance. The DOLPHIN ultrasound is comprised of short chirp WAVE trains with a wavelength of approximately 1.4×10^{-2} m (see Figure D.54).

Figure D.54 Dolphins.

Donnan, Frederick George (1870–1956)

[biophysics, chemical, electromagnetism, energy] A physical chemist from Ireland. Frederick Donnan is best known for his work on the chemical equilibrium at the cellular MEMBRANE in biological media. The DONNAN EQUILIBRIUM describes the ION transportation across the CELL MEMBRANE (see Figure D.55).

Figure D.55 Frederick George Donnan (1870–1956). (Courtesy of *Journal of Chemical Education*, 4[7], 1927.)

Donnan equilibrium

[biophysics, chemical, electromagnetism, energy] A chemical equilibrium across the MEMBRANE of biological tissues as derived by FREDERICK DONNAN (1870–1956).

Doping

[atomic, electronics, energy, general] Introduction of a foreign ATOM in the atomic structure of a semiconductor throwing the electron balance off, which increases the PROBABILITY of an electron to move from the VALENCE BAND to the CONDUCTION BAND (electron surplus: N-TYPE) or vice versa (electron shortage or "electron–hole" [hole] surplus: P-TYPE).

Doppler, Christian Johann (1803–1853)

[acoustics, mechanics, optics] A physicist from Austria who explained the changes on observed SOUND frequency resulting from MOTION in 1842. For instance, when waiting at a railroad crossing while a train is approaching that blows its whistle, the sound will be higher on approach than on departure with respect to the location of the observer. In actuality, the sound generated by the train while increasing its DISTANCE is lower than the sound acquired when the train and the observer are fully aligned (see Figure D.56).

Figure D.56 Christian Doppler (1803–1853).

Doppler

[acoustics, biomedical, biophysics, electromagnetism, engineering, imaging, instrumentation, theoretical]
A FREQUENCY (f or v) shift resulting from relative MOTION between source and observer. This frequency
phenomenon was described by the Austrian physicist CHRISTIAN JOHANN DOPPLER (1803–1853) in 1842.
The Doppler effect holds true for all types of waves. When light interacts with moving objects or if the LIGHT
SOURCE itself is moving, there will be a frequency shift. This frequency shift has been described in various
textbooks for both SOUND and light. The underlying principle to the frequency shift with regard to a moving
source or observer is the fact that the perceived wavelength will be caught at progressively earlier (respectively
later) PHASE intervals when source and observer are moving toward (respectively away from) each other. We
leave the derivation of the expression to the basic PHYSICS textbooks and will only show the final statement
for a frequency shift resulting from both moving source and observer. The perceived frequency by the
observer in relative motion to the source is given by $f' = \left[\left(1\pm\frac{u_o\cos(\theta)}{c'}\right)\Big/\left(1\mp\frac{u_s\cos(\theta)}{c'}\right)\right]f = f + \Delta f$, where f is
the frequency of light, u_o is observer velocity, or the speed of the modulation device in the detector arm, u_s is the
source velocity, or the speed of the modulation device in a reference arm of the detector arrangement, c' is the
speed of light in the medium. The signs on top, for the numerator: $\pm$, and for the denominator: $\mp$ respectively,
are for the situation where the source is moving toward the detector and the signs on the bottom means that the
source is moving away from the observer. One will try to ensure that multiple modulations will synchronize
and reverse direction simultaneously within in one period. The term $\cos(\theta)$ represents the component normal
to the source (see Figure D.57).

Figure D.57 Doppler effect.

Double-slit experiment

[general, optics] *See* THOMAS YOUNG DOUBLE-SLIT EXPERIMENT.

Doublet

[fluid dynamics] FLOW velocity field potential (ϕ) that has a clockwise rotation in the upper quadrant versus a counterclockwise rotation in the bottom quadrant. In this situation, the streamlines (ψ) are represented as $\psi = -(A/r)\sin\theta$, where A represents the AMPLITUDE, r is the radial location, and θ the angular orientation.

Doublet state

[atomic, chemical, nuclear] Quantum-MECHANICAL ENERGY configuration that describes a system with a SPIN of ½, consequently providing two spin components, respectively: $-½$ and $+½$. One example of the influence of the quantum state doublet is found for sodium, specifically the transition from 3p (QUANTUM NUMBER $\ell = 1$) to the 3s level (quantum number $\ell = 0$), where the 3p state has two states: $j = 3/2$ and $j = 1/2$ with ensuing spin-orbit split yielding two spectral lines 589.0 nm and 589.6 nm, respectively. The difference in energy (ΔE) for the doublet is defined by the BOHR MAGNETON (μ_B), the ELECTRON SPIN factor g^*, and the MAGNETIC FIELD (B) required to produce this split: $\Delta E = \mu_B g^* B$. Based on the emission line energy difference, the magnetic field strength can be determined (*also see* ZEEMAN EFFECT).

Down quarks

[general] A QUANTUM condition of elementary PARTICLE. Quarks can have six (6) "flavors": up, down, strange, charmed, bottom, and top (*see* QUARK).

Drag

[fluid dynamics, mechanics, thermodynamics] *Also* DRAG FORCE: a FRICTION force or resistive force, primarily velocity dependent $F_d = F_F = -b_F v$, where b_F is a geometric coefficient relating the shape and size of the body in MOTION to the motion as well as the characteristics of the media in contact, and v is the relative velocity of the body in motion with respect to the surroundings, primarily FLUIDS. It is also defined as the resistive force that is proportional to the area of the object (A) in relative motion with respect to a FLUID medium with relative VELOCITY v, multiplied by the DRAG COEFFICIENT, which is also dependent on the velocity, the DENSITY of the object (ρ) and fluid, respectively, as well as size and shape of the object. The drag force is in the opposing direction of the relative FLOW velocity: $\vec{F_d} = -\left(\rho v^2 C_d A/2\right)\vec{e_v}$, where $\vec{e_v}$ represents the unit vector for the flow direction (*also see* DRAG COEFFICIENT [C_d]). An object exposed to drag in free fall will eventually reach TERMINAL VELOCITY under balancing forces, instead of continuously accelerating under the influence of the governing acceleration phenomenon (primarily GRAVITATION, e.g., dropping a ball from the Eiffel Tower) (see Figure D.58).

Figure D.58 The concept of drag visually interpreted.

Drag coefficient $\left(C_D = F_D / [((1/2)\rho v^2)(\pi R^2)]\right)$

[fluid dynamics, mechanics] A dimensionless number expressed as the quotient of the DRAG force with respect to the kinetic ENERGY multiplied by the cross-sectional area of the object exposed to FLOW. When considering a sphere with radius R subjected to a flow with velocity v in a FLUID with density ρ while subjected to a measured drag force F_D. Certain objects have their own drag coefficient, primarily due to the frequent computational efforts on the subject. To list a few specific drag coefficients, the following are representative: the drag coefficient of a cone (C_D), drag hemisphere, cylinder, oblong board, and passenger car. One of the lowest drag coefficient mass-produced automobiles is the Mercedes CLA (see Figure D.59).

Figure D.59 Mercedes CLA with $C_D = 0.23$. (Courtesy of Daimler AG.)

Drag coefficient $\left(C_d = g(\rho - \rho_f)L / \rho v^2\right)$

[mechanics] The quotient of gravitational force over inertial force (-density) applied during resistive FLOW and settling velocities due to DRAG, where g is the GRAVITATIONAL ACCELERATION, L is characteristic length, ρ is the object density, ρ_f is the FLUID density, and v is the velocity.

Dreyer, John Louis Emil (1852–1926)

[astronomy, astrophysics, general] An astronomer from Denmark. John Dreyer is known for his validation work on the observations made by TYCHO BRAHE (1546–1601). He also made several distinguished observations about the surface structure on the MOON and has one crater named after him (see Figure D.60).

Figure D.60 John Louis Emil Dreyer (1852–1926).

Drift

[biomedical, mechanics] Small eye-movements in the classification of fixation on a point. Other classifications are micro-SACCADE (jerky) and TREMOR.

Drift velocity

[electromagnetism, general] Average VELOCITY (v_d) of charged particles, specifically electrons, under the influence of an external ELECTRIC FIELD (E): $v_d = \mu_{ch} E = M_{conductor} V / \rho_c N_A \ell e n_f \rho_0' (1 + \alpha_0 T)$, where μ_{ch} represents the "mobility" of the charged PARTICLE, $M_{conductor}$ is the MOLAR MASS of the CONDUCTOR, ρ_c is the density of the conductor, V is the applied electrical potential, N_A is the Avogadro's number, e is the fundamental electron charge, ℓ is the length of the conductor, n_f is the atomic number of VALENCE electrons that can be released as free electrons, T is the temperature of the conductor (METAL), and α_0 is the temperature coefficient for the SPECIFIC RESISTANCE of the conductor. The associated current density (J_e) is expressed as $J_e = \rho v_d$, where ρ is the charge density.

Driven damped harmonic oscillator

[acoustics, atomic, electromagnetism, fluid dynamics, general, mechanics, nuclear] A harmonic OSCILLATOR that is under continuous periodic externally applied excitation force with additional dampening force applied. The OSCILLATION can be electronic in an electronic LRC (L: INDUCTOR; R: RESISTOR; C: CAPACITOR) circuit or mechanical. The mechanical force may be the AIR stream blowing over a reed in a clarinet. Any driven damped harmonic oscillator can be described by an ordinary differential equation with respect to the ANGLE of rotation (ϕ) as $\left(\partial^2 \varphi / \partial t^2 \right) + 2\gamma_d \left(\partial \varphi / \partial t \right) + \omega_0^2 \varphi = A_0 \cos \left(\omega_d t \right)$ for a driven oscillation with frequency ω_d, AMPLITUDE A_0, with dissipation γ_d, for a system with NATURAL FREQUENCY $\omega_0 = 2\pi\nu$, ν the frequency as a function of time t. Generally, the amplitude of the damped oscillation is small and $\varphi(t) \cong \sin \varphi$. The SOLUTION to the WAVE EQUATION consists of two components: a homogeneous solution as well as a particular solution. In case the DAMPING is non-vanishing $\gamma_d > 0$, only the particular solution remains over time. The initial conditions will include the homogenous solution however, where $\partial \varphi / \partial t |_{t=0} = v_0$ provides the initial velocity, while the object is initially at rest $\varphi(0) = 0$. The transient period, leading to the particular solution is defined by the time constant $t \gg \gamma_d^{-1}$. During the transient process of the oscillation, there may be a PHASE shift θ, leading to the "proposed solution" for the amplitude $\chi(t)$: $\chi(t) = \chi_{\omega_d} \exp\{i(\omega_d t - \theta)\}$, with associated oscillation force $F_{osc} = A_0 e^{-i\omega_d t}$, providing $|\chi_{\omega_d}| = A_0 / \sqrt{\{(\omega_d^2 - \omega^2)^2 + 4\gamma_d^2 \omega_d^2\}}$. The persistence of the oscillation can be defined by the QUALITY FACTOR $Q = \omega_0 / 2\gamma_d$ (*also see* OSCILLATION) (see Figure D.61).

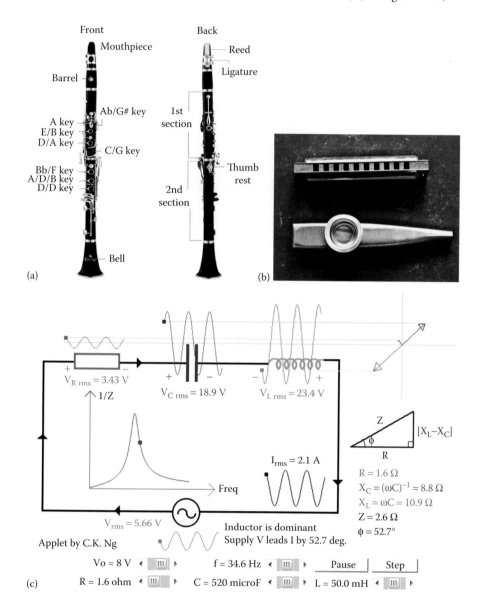

(a)

(b)

(c)

Figure D.61 Driven harmonic oscillator examples: (a) Clarinet and (b) Harmonica and Kazoo. (Courtesy of Chiu-king Ng, Hong Kong, China; http://ngsir.netfirms.com.) (c) RLC-circuit. (Courtesy of Chiu-king Ng, Hong Kong, China; http://ngsir.netfirms.com.)

Du Fay, Charles François de Cisternay (1698–1739)

[atomic, nuclear] A French botanist and chemist. His work contributed to the basic understanding of the atomic structure by describing the phenomenon of ELECTRIC CHARGE. Specifically his description of the charges with associated repulsion and attraction when rubbing a GLASS rod with silk and AMBER rubbed with fur between like materials and unlike materials, respectively, published around 1734 (see Figure D.62).

Figure D.62 Charles François de Cisternay du Fay (1698–1739).

Dubridge, Lee Alvin (1901–1994)

[atomic, nuclear] A physicist from the United States. Alvin DuBridge was involved in pioneering efforts on the PHOTOELECTRIC EFFECT.

Duhem, Pierre Maurice Marie (1861–1916)

[chemical, thermodynamics] A physicist from France. Pierre Duhem was an avid supporter of the concept that THERMODYNAMICS provided the foundations to all aspects of PHYSICS and ENGINEERING, "energetics." Duhem's work in chemistry and thermodynamics provides theoretical support for reversible processes (*see* **Duhem–Margules relation**). Additional work was in collaboration with Willard Gibbs (1839–1903), where Duhem provided proof for the Gibbs phase rule (see Figure D.63).

Figure D.63 Pierre Maurice Marie Duhem (1861–1916), 1913.

Duhem–Margules relation

[thermodynamics] An equation describing the thermodynamic equilibrium between two constituents of a FLUID in comparison to the VAPOR mixture as proposed by Pierre Duhem (1861–1916) and Max Margules (1856–1920).

Dull, Charles Elwood (1878–1947)

[general] A physicist from the United States. Charles Dull provided an elaborate description of "modern physics" in 1922.

Dulong, Pierre Louis (1785–1838)

[thermodynamics] A French physicist. Pierre Louis Dulong (1785–1838) is specifically known for his contributions to the THERMODYNAMICS of SPECIFIC HEAT and the LAW OF DULONG AND PETIT (see Figure D.64).

Figure D.64 Pierre Louis Dulong (1785–1838). (Courtesy of Wilhelm Ostwald; Elektrochemie—Ihre Geschichte und Lehre; Veit & comp., Leipzig, Germany 1896.)

Dulong–Petit law

[atomic, thermodynamics] The relation for the SPECIFIC HEAT under constant volume ($c_v = du/dT = T(dS/dT)$) asymptotically approaches the value of the universal GAS CONSTANT (R) (or Boltzmann constant for a molecular or atomic base, expressed per ATOM/molecule), however not limited to gaseous form. When the temperature (T) of the medium is much larger than the DEBYE TEMPERATURE (Θ_D), the specific heat reduces as $c_v(T) \cong 3R\{1-(1/20)[\Theta_D/T]^2\} \rightarrow 3R$, established by PIERRE LOUIS DULONG (1785–1838) and ALEXIS PETIT (1791–1820) in 1819.

Dunning, John Ray (1907–1975)

[atomic, nuclear] A physicist from the United States who was instrumental in the development of the ATOMIC BOMB. Additional work by Dunning involved the development of the process for ISOTOPE separation by means of GAS DIFFUSION (see Figure D.65).

Figure D.65 John Ray Dunning (1907–1975) on the right along with Hubert Thelen taken in 1957.

Dushman, Saul (1883–1954)

[atomic, nuclear] A scientist from Russia. Dushman was a contemporary of WALTER SCHOTTKY (1886–1976), and he used his work in ELECTRON TUBE functionality formulation (*also see* EDISON EFFECT).

Dussik, Karl Theodore (1908–1968)

[acoustics, biomedical, imaging] A physician from Austria with a specialization in neurology. Dussik published the first use of high-frequency ACOUSTICS (now known as ULTRASOUND imaging) for imaging purposes in 1942. The work of Karl Dussik was inspired by his professional interest in the location of brain tumors (see Figure D.66).

Figure D.66 Karl Theodore Dussik (1908–1968).

Dynamics

[general, mechanics, thermodynamics] An investigation of the cause of MOTION (in contrast to KINEMATICS). The study of dynamics, for instance, includes the concepts of FORCE and MASS, NEWTON'S LAWS (incorporating for instance COLLISION), in addition to the characteristics of GRAVITATION. In greater detail, the phenomena of FRICTION and KEPLER'S LAW are part of this field of MECHANICS.

Dyne

[general] A unit of force that, when acting upon a mass of $1\,g$, will produce an acceleration of $1 \times 10^{-2}\,\mathrm{m/s^2}$.

Dynode

[electronics] An ELECTRODE that produces secondary electron emission in a VACUUM tube. Dynodes in a PHOTOMULTIPLIER TUBE are arranged at incremental electrical potential in order to provide an avalanche effect (see Figure D.67).

Figure D.67 Dynode vacuum tube.

e

[electromagnetism, energy, general, thermodynamics] Electron, the fundamental nuclear PARTICLE. The existence of charges (disparity in charge to be exact) was first described in 1734 by the French botanist and chemist CHARLES FRANÇOIS DE CISTERNAY DU FAY (1698–1739), and later by BENJAMIN FRANKLIN (1706/1705?–1790) in 1760. The electron is on the outer shell of the ATOM with the PROTON and NEUTRON in the NUCLEUS of the atomic structure, as aptly described by the ATOM MODEL of NIELS HENRIK DAVID BOHR (1885–1962). The electron was described in detail with the associated effects by CHARLES-AUGUSTIN DE COULOMB (1736–1806) in the 1785 publication *Histoire et Mémoires de l'Académie Royale des Sciences*. The revelation of the electron charge formed the SOLIDIFICATION that charged object either attract or repel, depending on the respective charge configuration ("opposites attract"). The MAGNITUDE of the electron charge is expressed in COULOMB UNITS.

Eadie–Hofstee plot

[biomedical, chemical, computational] Reaction analysis for enzyme-catalyzed LIGAND binding processes. Graphical representation of the enzyme kinetics with respect to the ratio of the substrate dissipation rate (V_{max}) and the substrate concentration ($[S]$) illustrating the reaction rate ($v_{rèact}$), as $v_{rèact} = -K_M \left(v_{rèact}/[S] \right) + V_{max}$, where K_M is the Michaelis–Menten constant, as derived from the MICHAELIS–MENTEN EQUATION. The plot can be used to identify the Michaelis–Menten constant (slope) (see Figure E.1).

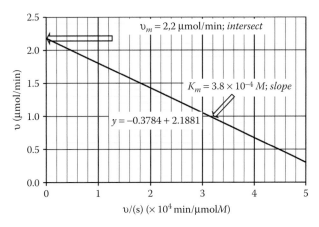

Figure E.1 Graphical representation of Eadie–Hofstee plot.

Ear

[acoustics, biomedical, general] Biological sensing device that can acquire PRESSURE WAVES (i.e., SOUND). The EAR has several structural components as well as various functional components. Structurally, the ear is divided in three segments: the OUTER EAR, the MIDDLE EAR, and the INNER EAR. The outer ear is the exterior shell (pinna) with magnified areal size in comparison to the AUDITORY MEATUS or EAR CANAL to ensure effective and high PROBABILITY of collecting large MAGNITUDE pressure waves that captures the pressure waves and guides them into the AUDITORY CANAL leading to the middle ear. At the middle ear, the pressure variations are transferred into MECHANICAL VIBRATIONS of a solid. The middle ear is separated from the auditory canal by the TYMPANIC MEMBRANE or EAR DRUM the tympanic MEMBRANE is the percussion device that converts pressure variations into mechanical vibrations, facilitating further transfer to a LIQUID-filled sensory compartment (inner ear). Inside the middle ear, the tympanic membrane pushes against a LEVER system of three interlocking bones (the OSSICLES) to transfer the mechanical vibrations to the OVAL WINDOW of the inner ear. The bone structure components are (moving inward, respectively) the MALLEUS (hammer), connected to the INCUS (anvil) connected to the STAPES (stirrup), which is connected to the oval window. The middle ear is generally pressure stabilized by maintaining contact with the external pressure through the EUSTACHIAN TUBE, which connects to the nasal cavity. The SOUND intensity transferred in the middle ear is capable of being regulated by various means. The intrinsic regulation of energy transfer is in the aspect ratio of the components. Primarily, the ELASTICITY of the tympanic membrane can be controlled by pressure variations of the middle ear mediated by the Eustachian tube. The VIBRATION magnitude is controlled by spacing the three bones through MUSCULAR interaction (for instance, the malleus is attached to the TENSOR tympani MUSCLE), providing a means to control the LOUDNESS of the sound as well as offering protection from mechanical damage to the inner ear. The oval window is the passage to the inner ear, which consists of the COCHLEA. The cochlea doubles back to the final window, the round window, close to the oval window in the middle ear. The inner workings of the cochlea are described in detail in the entry for cochlea. The intricate mechanism of conversion of mechanical vibrations to electrical impulses is transmitted over the cochlear branch of the VESTIBULOCOCHLEAR NERVE for DATA ACQUISITION and SIGNAL PROCESSING by the BRAIN (*also see* HEARING). In addition to HEARING, the inner ear also has a construction attached that functions as a means to detect and preserve equilibrium (see Figure E.2).

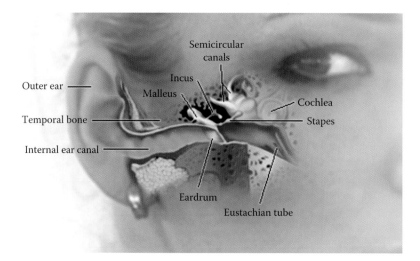

Figure E.2 Anatomy of the ear. Detailed illustration of working components of the ear: outer ear, middle ear, inner ear, and equilibrium system.

Earnshaw, Samuel (1805–1888)

[computational, electromagnetism, general] An English clergyman, mathematician, and mathematical physicist. Earnshaw made significant contributions to the theoretical description of passive MAGNETIC LEVITATION (see Figure E.3).

Figure E.3 Samuel Earnshaw (1805–1888).

Earnshaw's theorem

[computational, electromagnetism, general] A statement made by SAMUEL EARNSHAW (1805–1888) in 1839 on passive MAGNETIC LEVITATION. Based on his observations and calculations, he postulated that no stationary fixed configuration of magnets can be maintained in a stable equilibrium by any combination of static MAGNETIC FIELDS with resulting magnetic forces or respective combinations of GRAVITATIONAL FORCES. On a second interpretation in 1842, he also stated that an assembly of POINT CHARGES cannot solely be maintained in a stationary stable equilibrium resulting from the electrostatic interaction between the fixed and static charges alone. This lemma is a consequence of the GAUSS's LAW, stating that the ELECTRIC FORCE ($F_{electric}$) resulting from an electric field (and equivalently for a MAGNETIC FIELD) will always be DIVERGENCELESS, meaning that there will be no local maxima or minima. The closest to a minimum/maximum result of the LAPLACE EQUATION for the electric potential will be SADDLE POINTS. $\nabla \cdot \overline{F_{electric}} = \nabla \cdot (-\nabla U) = -\nabla^2 U = 0$, where ∇U is the DIVERGENCE of the potential, and subsequently $\nabla^2 f = \nabla \cdot \nabla f = \Delta f = \sum_{i=1}^{n} \partial^2 f_i / \partial x_i^2$ is the LAPLACE OPERATOR (second-order differential operator in the n-dimensional [axis: x_i] EUCLIDEAN SPACE, generalized function f), and U is the electrical potential. This principle will apply to molecular configuration in the same manner as on a MACROSCOPIC scale.

Earth

[general] The Earth is the third PLANET migrating outward in the sequence of CELESTIAL BODIES orbiting around the SUN in what is known as our SOLAR SYSTEM. The Earth's surface has 29% landmass and the remaining 71% is water. *Note*: This ratio is changing with respect to the climatological conditions. During the last ice-age (which ended approximately 12,000 years ago) this ratio was probably 31:69 since at that time, the sea level was approximately 85 m lower compared to current day. The Earth has a structure that is formed around an internal core, which has two components: the inner core, which is solid, and the outer

core, which is LIQUID. The core is enclosed by the mantle which in turn is covered by the crust, on which we live. The crust is only approximately 100 km thick versus the full radius of 6380 km on average (oblong ellipsoidal, compressed at the POLES). The MAGNETIC POLES of the Earth are the result of the rotation of the iron-rich outer core. The exact magnetic pole orientation is not fixed and the NORTH and South Poles have switch orientation (geomagnetic reversal) at least a documented 51 times over the past 12 million years, as recorded by the rock-sediment magnetic orientation. The GEOGRAPHIC NORTH and South Poles are fixed by geographic definition. The Earth's magnetic core may flip between 100,000 and 1 million years, and the last time was an estimated 780,000 years ago; the process itself may take 1,000–10,000 years. Magma, which is expelled in volcanic eruptions, will turn to lava when it flows over our soil. The magma originated from a depth below the surface ranging from 200 to 10 km, but still not precisely known. The Earth has an ATMOSPHERE that is composed in a layered format. The five principal layers moving outward toward space are the TROPOSPHERE (0–12 km), STRATOSPHERE (12–50 km), MESOSPHERE (50–80 km), THERMOSPHERE (80–700 km), and the EXOSPHERE (>700 km). Note that the International Space Station is in orbit at approximately 350 km. The Earth has seven continents: Africa, Antarctica, Asia, Europe, North America, Oceania, and South America; note that the North Pole theoretically does not consist of any landmass and is only frozen water hence does not form a continent or not part of any continent. The Earth's population in 2014 is approximately 7.2 billion people. The 10 countries with the largest population, in decreasing rank are China, India, the United States, Brazil, Pakistan, Nigeria, Bangladesh, Russia, and Japan (see Figure E.4).

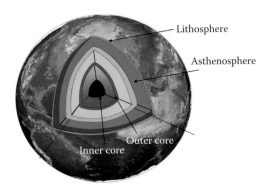

Figure E.4 Images representing various characteristics and properties of Earth.

Earth, angular momentum

[general] Looking down on the NORTH POLE from outer space will show that the EARTH is rotating counter-clockwise. The revolutions are very stable at an ANGULAR VELOCITY: $\omega = 7.27 \times 10^{-5}$ rad/s. For a solid sphere, the angular momentum is defined by the MOMENT OF INERTIA: $I_{rev} = \sum m_i r_i^2$, where m_i is the mass of each point in the volume of the sphere and r_i the radius of each respective mass segment in relation to the axis of rotation passing through the origin of the REFERENCE FRAME. For a SOLID SPHERE (*Note*: This is an approximation since the Earth is technically an ellipsoid [*see* EARTH, ASPHERICITY]) the moment of inertia is $I_{rev\,solid\,sphere} = 2/5\,mR^2$, with m the mass of the sphere and R the radius at the equator. This yields for the angular momentum: $L_{ang} = I_{rev}\,\omega = \left(9.71 \times 10^{37}\ \text{kgm}^2\right) \times \left(7.27 \times 10^{-5}\ \text{rad}/\text{s}\right) = 7.06 \times 10^{33}\ \text{kgm}^2/\text{s}$. However, the Earth also has PRECESSION and swings in a manner similar to a TOP. The precession provided the variation of the axis with respect to the SUN and hence the SEASONS.

Earth, as a conductor

[general] Generally a homogeneous solid CONDUCTOR will by definition be devoid of an electric field on the interior while all the charges are concentrated on the surface. As a result, inside a solid conductor the electric potential will be the same for the entire volume (i.e., equipotential). However, since the EARTH is not homogeneous, there will be potential differences across the surface of the Earth, although generally the connected solid surface is at the same potential and is assumed to be at zero potential, at least in reference to the power plants generating the ELECTRICITY to which we reference the Earth as "GROUND." Inside the PLANET the LIQUID core itself generates electrical potential differences and both electric and magnetic fields are created. The MAGNETIC FIELD lines extend around the planet and are distorted by solar influences to yield the EARTH'S MAGNETIC FIELD.

Earth, asphericity

[astrophysics, general, geophysics] Even though SIR ISAAC NEWTON (1642–1727) showed that at any point outside the shell of the EARTH'S CRUST, all the mass can be considered in the center of the EARTH, the fact that the Earth is not a perfect sphere has its implications. Because of the rotational MOTION, the Earth is compressed at the POLES with the equator having an approximately 43 km larger diameter than the pole to pole DISTANCE. Apart from the valleys and mountains, the NORTH Pole jets out over a significant surface area producing an Earth shape reminiscent of a short pear (see Figure E.5).

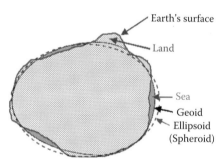

Reference spheroid and the geoid

Figure E.5 Earth's asphericity.

Earth, charge stored

[general] The only free moving charge that can reliably be stored is the electron, with an ELECTRIC CHARGE of -1.6×10^{-19} C. The EARTH in general stores a charge of approximately $-400,000$ C. Additionally, the Earth (on a clear day) also continuously leaks electrons to the ATMOSPHERE at a steady pace of 1500 C/s, while during a LIGHTNING storm these electrons are returned in quantities on average of 20 C per lightning bolt.

Earth, composition of atmosphere

[general] The ATMOSPHERE is composed of the following substances: 78% nitrogen, 21% oxygen, ~4.0% water VAPOR, 0.9% argon, 0.03% carbon dioxide, and TRACE gases, including neon, helium, krypton, and xenon (see Figure E.6).

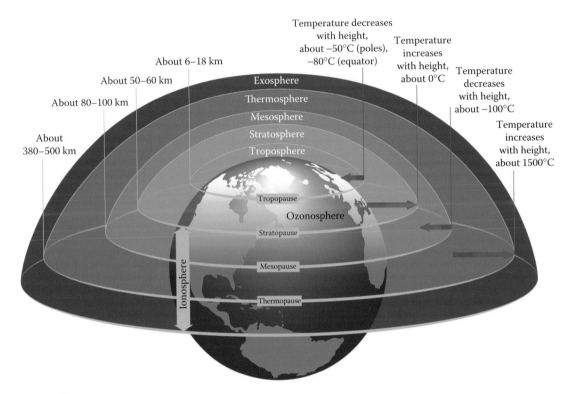

Figure E.6 Artist impression of the composition of the Earth's atmosphere.

Earth, density

[general] LORD HENRY CAVENDISH (1731–1810) concluded that the average density of EARTH based on spherical volume and the mass based on solar and lunar attraction should be 5.5×0^3 kg/m^3, more than five times denser than water at 1.0×10^3 kg/m^3.

Earth, gravitational interaction with objects

[geophysics, mechanics] As defined in NEWTON'S SECOND LAW, the attraction between the EARTH and external objects results in a force (F) that is proportional with the mass of the object (m) and the Earth's GRAVITATIONAL ACCELERATION ($a = g = 9.8$ m/s^2) expressed as: $\vec{F} = m\vec{a}$.

Earth, gyroscopic stability

[general, geophysics, mechanics] The rotation of the EARTH with associated angular momentum creates a relatively stable orbit around the SUN, with a PRECESSION effect, equivalent to a spinning top. However, the MOON with its gravitational interaction creates an external force that influence the ENERGY CONSERVATION principle and disturbance of the CONSERVATION OF ANGULAR MOMENTUM. The Earth's rotational periodicity is increased by 2.5×10^{-8} s/day, which will result in an increase of the current 24 h day by approximately 3 h over 10^9 years.

Earth, heat flow

[computational, geophysics, thermodynamics] During the day, the EARTH is generally heated by the SUN and during the night, the Earth will locally experience radiative cooling. The thermal process in the top layer of the EARTH'S CRUST satisfies the heat equation for diffusive heating and cooling. Because of the large radius, the Earth's surface may be considered flat, which reduced the radiative heat DIFFUSION to a one-dimensional process: $(\partial\theta/\partial t)-(C/c\rho)(\partial^2\theta/\partial z^2)=0$, where θ is the temperature, t is time, C is a material constant, C is the SPECIFIC HEAT of the soil (or water for that MATTER), ρ is the local density (averaged in this case), and z is the depth. The heat flow (Q) per unit time out of a surface with area A is defined as $(dQ/dt)=\int_{z_1}^{z_2}c\rho(\partial\theta/\partial t)dz\,A$. One major boundary condition that holds true for the globe is that one side of the medium is kept at a constant temperature (inner Earth). This is the DIRICHLET BOUNDARY CONDITION, in contrast to when heat is syphoned off, where the NEUMANN BOUNDARY CONDITION applies. The SOLUTION of the DIFFUSION EQUATION uses the Sturm–Liouville eigenvalues yielding a solution of the form: $\theta(x.t)=\sum_{n=1}^{\infty}Ce^{-n^2\pi^2kt/L}\sin(n\pi z/L)$, where n is the series counter for the harmonics, L is the characteristic length, and $k=C/c\rho$. Note $\pi z/L=\omega t$ with ω the ANGULAR VELOCITY and t time. The AMPLITUDE of the nth harmonic of the fluctuations decreases exponentially with increasing depth, and hence the influence of the subsequent harmonics will be less or even negligible. Because of the dampening only the first harmonic is transferred to depths greater than half the penetration of the first harmonic. Based on the first-order approximation an indication of the depth at which temperature changes are still noticeable can be made. For DIURNAL (day-night) temperature fluctuations, the depth at which there is less than 1 °K variation on average (depending on location, e.g., city vs. dessert), the temperature will be at thermal equilibrium at depths greater than 1 m, whereas the annual fluctuations are not statistically significant at depths greater than 10 m, taking the local climatological conditions into account for corrections. In general, the temperature fluctuations at depths less than the equilibrium depth are out of PHASE by one half period, whereas closer to the surface the HYSTERESIS will be lesser with decreasing depth. Because of the DAMPING effect with factor $EXP(-\sqrt{\omega/2k}\,x)$ with value $\sqrt{\omega/2k}\cong2\pi/900$ cm^{-1}, for seasonal variations the period is $T=365*24*3600=31{,}536{,}000$ s for ordinary soil $k=2*10^{-7}$ m²/s yields a depth of roughly 4.50 m at which a change in phase of π will occur. At this depth, the HEAT TRANSFER will experience a delay of $T/2=6$ months. At this point, the damping factor is approximately $e^{-\pi}\cong1/16$. On a daily fluctuation, this results in damping and PHASE ROTATION at 1.23 m for the values above (see Figure E.7).

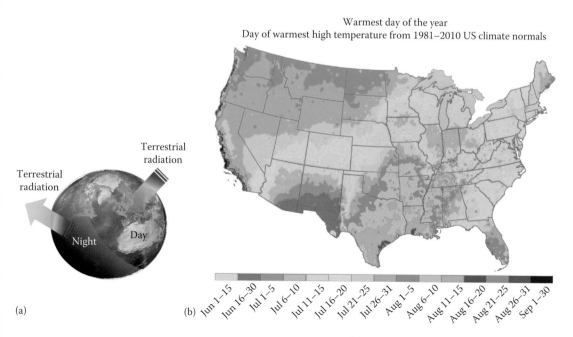

Warmest day of the year
Day of warmest high temperature from 1981–2010 US climate normals

Terrestrial radiation

Terrestrial radiation

Night   Day

Jun 1–15 Jun 16–30 Jul 1–5 Jul 6–10 Jul 11–15 Jul 16–20 Jul 21–25 Jul 26–31 Aug 1–5 Aug 6–10 Aug 11–15 Aug 16–20 Aug 21–25 Aug 26–31 Sep 1–30

(a)   (b)

Figure E.7 (a) Global temperature profile and (b) heat flow. (From National Oceanic and Atmospheric Administration, United States Department of Commerce; Silver Spring, MD, USA.)

Earth, magnetism

[general, geophysics] The way the EARTH holds a MAGNETIC FIELD was described by WILLIAM GILBERT (1540–1603) as produced by a DIPOLE in 1600, which has validity. The magnetic field may have changed direction at least 150 times over the last 170 million years as found from geological evidence (see Figure E.8).

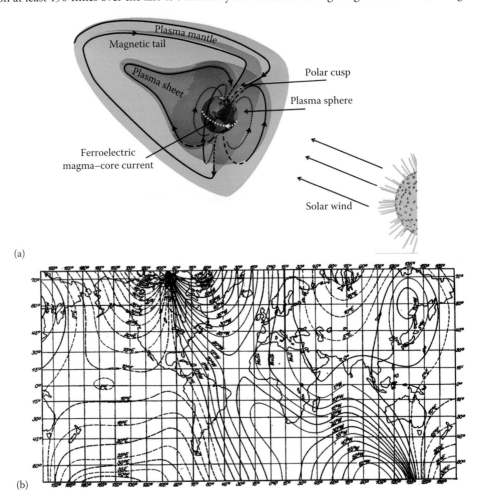

(a)

(b)

Figure E.8 (a,b) The Earth's magnetic field lines.

Earth, rotation

[general] The EARTH performs a full rotation around the SUN in approximately 365 days and 5 h, more exact: $T = 31,554,000$ s, which converts into ANGULAR VELOCITY of $\omega = 2\pi/31,554,000 = 1.9912484 \times 10^{-7}$ rad/s, whereas for the daily rotation around it own axis the angular velocity is $\omega = 2\pi/24 \times 3600 = 7.2722 \times 10^{-5}$ rad/s.

Earth satellite

[general] Satellites are used for various data transmittance purposes: television and RADIO communications and observation. Satellites are moving around the EARTH in various distinctly different motions. Certain satellites are placed in a fixed location with respect to the Earth's geometry, so that we do not have to adjust the (dish-) ANTENNA aimed at the SATELLITE used for television reception. The fixed location with respect to Earth refers to the fact that the satellite is in the GEOSYNCHRONOUS ORBIT or GEOSTATIONARY ORBIT. Spy satellites and weather satellites, on the other hand, are moving around in variable or predetermined patterns, scanning the surface in a grit-pattern. Other forms of orbital MOTION are equatorial orbit, graveyard orbit, low Earth orbit, medium Earth orbit, MOLNIYA ORBIT, polar orbit, subsynchronous orbit, and supersynchronous orbit (see Figure E.9).

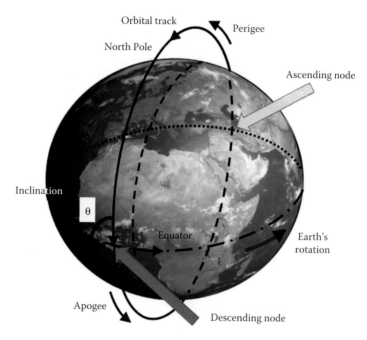

Figure E.9 Pattern of the orbit and position for satellites in orbit around the Earth.

Earthquake

[computational, fluid dynamics, geophysics, thermodynamics] It primarily means the sifting of the TECTONIC PLATES because of the continuous MOTION of the Earth's shell, however slow. The WAVE action associated with the motion of the shell is subject to DISPERSION due to the inhomogeneous composition of the mantle, specifically the LIQUID filled lithosphere (e.g., marked by volcanic eruptions). The inharmonic motion of the tectonic plates causes clashes, providing a SHOCK WAVE that propagates along the crust. The long-term slow movement of the tectonic plates is also visible in the ever-increasing transatlantic separation between Africa and South America, which at one time seem to have been joined together based on their respective coastal contour shapes. The existence of faults was first suggested by the Japanese scientist and geologist Bunjiro Koto (1856–1935) in 1891, after studying the Japanese earthquake of 1871. Bunjiro Koto may be recognized as the first seismologist, specifically with his appointment in 1893. The Koto model was refined by the study of the 1906 San Francisco earthquake by geologist Henry Reid from the United States. Harry

Fielding Reid (1859–1944) provided a shear strain model in the shell based on the continual tectonic movement in 1911 (see Figure E.10).

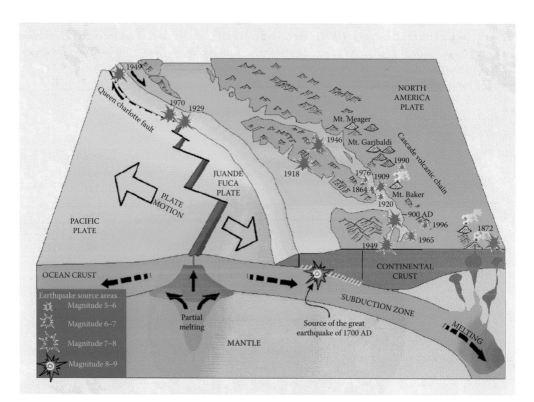

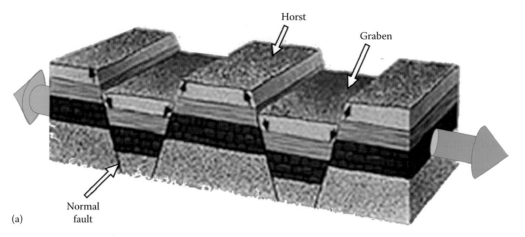

Figure E.10 Earthquake: (a) Vancouver, Canada. (Courtesy of The Canadian Geoscience Education Network (CGEN) is the education arm of the Canadian Federation of Earth Science; Ottawa, ON, Canada.) (Continued)

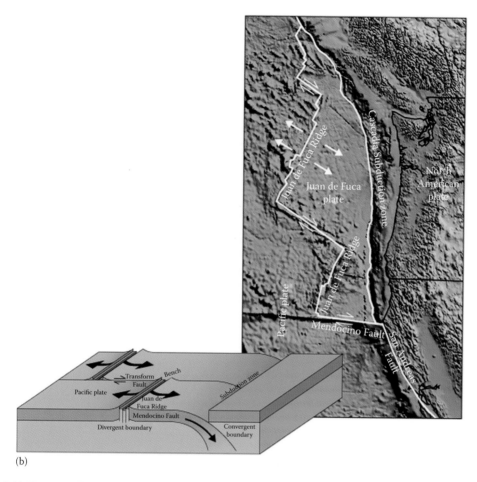

(b)

Figure E.10 (Continued) Earthquake: (b) earthquake crack after tremor in northern California. (Courtesy of Incorporated Research Institutions for Seismology, iris.edu.)

Earthquake compression wave

[general] The compression waves associated with an EARTHQUAKE are longitudinal waves. Earthquake compression waves are also called "primary waves" or "P-waves." The primary waves are preceding the secondary or S-waves. Earthquake compression waves are more pronounced resulting from underground nuclear explosions or weapons testing and allow for a mechanism to distinguish "man-made" from natural earthquakes. The bulk-modulus partially define the propagation is defined as $B_{\text{elastic}} = -[\Delta P/(\Delta V/V_0)]$, where ΔP represents the pressure gradient, ΔV the EXPANSION/contraction of an undisturbed volume V_0. The bulk-modulus is most likely a function of depth, due to the changes in density and deformability with DISTANCE to the surface and soil composition as a function of depth. The disparity in the bulk modulus will result in a DISPERSION in wave-propagation because the propagation speed is directly proportional to the bulk modulus and DENSITY (ρ) as $v = \sqrt{B_{\text{elastic}}/\rho}$; note that generally the speed of propagation of the compression still increases with increasing density due to the compounding factors affecting the bulk modulus as well. This phenomenon will be different for an earthquake-induced TSUNAMI, where $v = \sqrt{Y_{\text{elastic}}/\rho}$, where $Y_{\text{elastic}} = \text{stress/strain} = \sigma_{\text{stress}}/\varepsilon_{\text{strain}}$ represents the Young's modulus of the LIQUID, with $\sigma_{\text{stress}} = F/A$,

the force tangential to a unit surface area and $\varepsilon_{strain} = \Delta L/L_0$ with L_0 the undisturbed length and ΔL the deformation in length under the influence of an applied force (see Figure E.11).

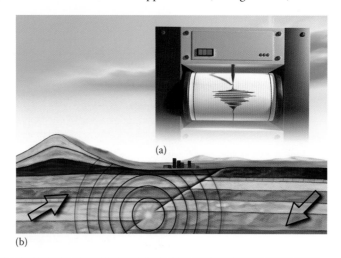

(a)

(b)

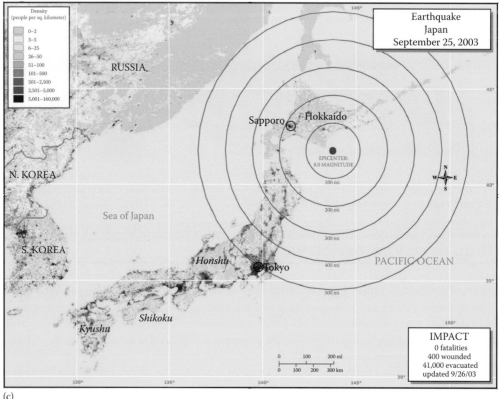

(c)

Figure E.11 (a) Graphical illustration of earthquake compression wave with seismic recording (Courtesy of USAID Office for Foreign Disaster Assistance.), (b) displacement diagram at a fault line, and (c) epicenter illustration for the radiating compression wave that represents the 2003 earthquake in Japan with magnitude 7.3 on the Richter scale, named after Charles Richter (1900–1985).

Earth's crust

[general, geophysics, nuclear] The EARTH can be thought of as being divided in four separate layers, or concentric shells. The outer part is the crust, upon what we live, with a thickness of 1–100 km. Below the crust is the mantle. The mantle can be subdivided in upper and lower mantle. The crust and the upper part of the mantle are also called the "lithosphere." Below the mantle is the inner core. The semi-rigid part of the mantle that supports the magma FLOW is called the "ASTHENOSPHERE." Inside the outer core is the inner core. The asthenosphere provides the mechanism for the movement of TECTONIC PLATES, that is, earthquakes. Most of our knowledge is from experimental observations of the crust. All other information from the inner portions is based on mathematical derivations as well as speculation. Based on various observations, it has been derived that the inner core must have temperatures and associated pressures that are so great that the constituents such as metals are squeezed together preventing movement as a liquid. The components in the inner core presumably vibrate in place as if it was a solid (see Figure E.12).

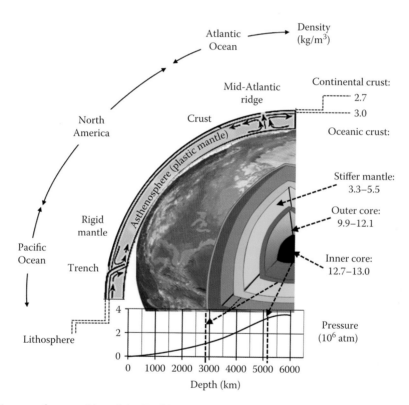

Figure E.12 Diagram of composition of the Earth's crust.

Eccles, John Carew (1903–1997)

[biomedical, electromagnetism] An Australian physiologist and scientist that worked on a formal description of the SYNAPTIC communication and encoding for NERVE CELLS and received the Nobel Prize in Physiology in 1963 together with SIR ANDREW FIELDING HUXLEY (1917–2012) and SIR ALAN LLOYD HODGKIN (1914–1998), where the latter two described the ACTION POTENTIAL in a nerve CELL (see Figure E.13).

Figure E.13 John Carew Eccles (1903–1997). (Courtesy of The John Curtin School of Medical Research, The Australian National University, Canberra City, Australia.)

Echogenic

[biomedical, imaging, solid-state] Material property referring to the ability to "reflect" acoustic pressure WAVE at an interface between two media. In biomedical imaging, soft tissue (e.g., fat and MUSCLE) is less echogenic than the abdomen, the HEART in the lung and bone. Teeth are too hard to support any imaging of the internal structure by ultrasonic means. Specific attention is given to the fetus in the womb, which is established primarily and more pronounced in the second trimester of fetal development (*also see* ULTRASOUND).

Echolocation

[biomedical, imaging] The use of SOUND to determine the DISTANCE between the source and a discontinuity in acoustic impedance. Several animals use sonar echolocation for orientation, for instance, bats and porpoises. Humans use sonar for echolocation for underwater navigations, especially by submarines, but also by blind people using specially designed sonar walking stick, next to the tried-and-true method of tapping with a regular METAL or wooden stick. Additional applications are in the sonographic location of fish and in ULTRASONIC IMAGING. Bats employ frequency modulation to increase the RESOLUTION and obtain superior signal-to-noise ratio. Porpoises (e.g., dolphins) produce ultrasonic waves in the range of several hundred kilohertz with the specially designed resonant organs just inside their blow-hole, which is subsequently emanated from the acoustic "magnaphone" in the bulge on their forehead. Bats use vocal cords in their throat. Both bats and porpoises emit the sound in short burst in the order of millisecond duration, at variable rate of emission. The emission can be composed of a well-defined FREQUENCY SPECTRUM, specifically pertaining to the frequency modulation aspect. The frequency spectrum will change under REFLECTION and hence reveal details to the BAT about the mechanical properties of the object: biological or inanimate, based on, for instance, the Young's modulus and surface characteristics. Certain bats also have the ability to send out constant frequency sound burst for which they are able to distinguish the DOPPLER shift and hence derive their relative speed with respect to prey and obstructions. Bats are known to utilize a low AMPLITUDE general scanning mode, which increases in frequency spectrum and amplitude once a target (prey) has initially been identified (see Figure E.14).

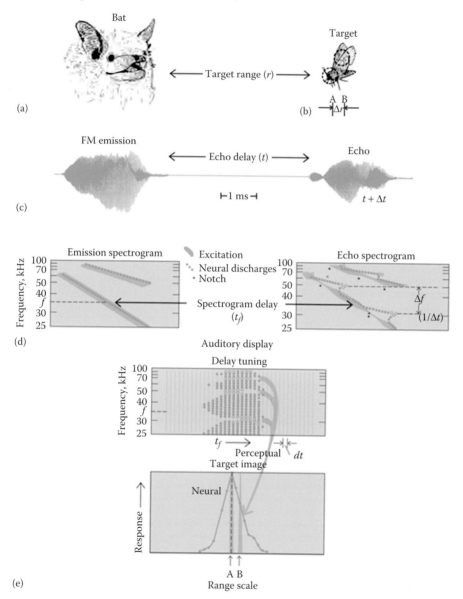

Figure E.14 (a) The bat emits a sonar sound that travels outward to impinge on an insect at some target range (r). Reflections return to the bat from various insect body parts—for example, wing and head separated by a small difference in range (Δr). These sources of reflections are called "glints," and the insect is depicted here as a 2-glint target. (b) Sound-pressure waveforms of the bat's transmitted sonar sound and the reflections from the insect's glints—wing and head. The transmitted sound is heard directly by the bat to initiate processing by establishing a zero-time origin for reception of subsequent echoes (long vertical dashed line across pictures). Echoes return to the bat's ears after a delay (t) related to target range (r) at rate of 5.8 ms/m. Reflections from the 1st and 2nd glints add together at a small time separation (Δt) to interfere with each other when they form a combined echo arriving at the bat's ears. For each transmitted sound, the acoustic stimulus received by the bat is not just the echo but a "pair" of sounds—the outgoing broadcast received directly at the moment of emission followed by the returning combined echo composed of, however, many reflections returned by the target's glints. (c) Spectrograms—or time-frequency plots—of the transmitted sound (left) and combined echo (right) showing the frequency modulated (FM) sweeps. The broadcast signal contains two sweeps—a first harmonic from 45 to 22 kHz and a second harmonic from 80 to 45 kHz—that appear as sloping ridges, one over the other. *(Continued)*

Figure E.14 (Continued) The echo also contains these two harmonics, but the sloping ridges for the combined echo spectrogram are not as smooth as those in the broadcast spectrogram because interference between the overlapping reflections from the 1st and 2nd glints creates alternating peaks (*p*) and notches (*n*) that appear as ripples or undulations. The notches are locations where energy in the echo has been canceled by interference; their locations are the principal defining characteristics of the ripples because their frequencies (f1, f2, and f3) are related to the delay separation (Δt) of the reflections from the glints (see Figure E.2a). The bat's inner ear transforms the spectrograms of broadcasts and echoes into neural spectrograms composed of spikes that register the FM sweeps in terms of the tuned frequencies of different neurons and times, or latencies, of the spikes (see Figure E.9). (d) Diagram illustrating neural spectrograms of a broadcast and echo. The spectrogram is composed of on-responses in neurons tuned to different frequencies (vertical axis). For downward FM sweeps (as in c), excitation in the bat's inner ear moves rapidly from high frequency (basal end of cochlear spiral) to low frequency (apical end of cochlear spiral). As this excitation rushes past receptors, it triggers spikes (red circles) in auditory nerve fibers from any given location (see Figure E.3a and b). (Conventional spectrograms from Figure E.1c are shown here as shades of gray.) The on-response (first spike) is sharply coincident with the onset of this rapidly moving excitation, and it registers the time of occurrence of the neuron's tuned frequency in each sweep. At any given frequency, the total duration of excitation is very brief—only a few hundred microseconds in duration—because the sound sweeps rapidly away to lower frequencies. Consequently, there is only time enough for the on-responses to occur reliably. The whole volley of on-response spikes (red circles) spread across neurons tuned to different frequencies traces the shape of the FM sweep for the broadcast and then for the echo. In the auditory nerve, if excitation is strong, cells often produce one or more additional spikes that follow the on-response, but they are less sharply synchronized to the onset of excitation, and cells exhibit spontaneous activity, too, that has no relation to stimuli. These secondary spikes are stripped out of the volley as the spikes pass through the cochlear nucleus, so that only the on-responses shown here are used to determine echo delay. Note that the frequency tuning of the neurons segregates the traces for the first and second harmonic sweeps in each sound. Because the echo contains closely spaced reflections of the broadcast that were returned from different parts of the target (Figure E.1a), interference occurs, and some frequencies are canceled out to create notches in the spectrogram. Auditory neurons tuned to frequencies of notches fail to produce on-responses, so there are gaps or "holes" (unfilled circles around f1, f2, and f3) in the neural spectrogram for the echo. In auditory nerve fibers, generation of spikes takes place on the neurons at a site very close to the organ of Corti (Figure E.3b), and the spikes have a latency of about half a millisecond. The proximity of the spike-generation site to the receptors, and the proximity of the bat's cochlea to its brainstem (see Figure E.3a and d) may make spike initiation particularly stable in time and minimize the distance spikes have to travel to reach the first synapse, where synchrony of spikes across closely spaced auditory-nerve fibers is used to sharpen temporal registration of the spectrogram before it is distributed upward throughout the auditory pathway for further processing. (e) Hypothetical traces depicting two possible versions of the image of echo delay perceived by the bat to represent target range. The horizontal axis (*r*) is a psychological scale showing the perceived value of delay. Although this now refers to events inside the bat, such values nevertheless can be measured in behavioral experiments. The two possible versions of the image differ according to how information in the spectrograms is used to estimate range from delay. Does the bat perceive an "image" corresponding to the delay of the echo spectrogram representing the overall range (*r*) to the target, shown here as a blue distribution for the combined echo as a whole? In this case, the underlying representation is a single numerical delay estimate with a potentially broad time-width corresponding to the integration time (t_i) of the spectrogram (see c). Both glints are subsumed into one delay or range perception. Or, does the bat separately perceive the glints themselves (first glint dark blue; second glint red), even though they are separated by a small interval of distance (Δr), so that their reflections are separated by an equivalently small amount of time (Δt) compared to the integration time (t_i)? If the bat perceives the first kind of image, then target range (*r*) most likely is estimated from the auditory equivalent of rightward displacement of the spectrogram of the transmitted sound until it superimposes on the spectrogram of the echo. Although the range difference between the glints (Δr) is not itself perceived as such, the presence of interference notches in the echo provides the bat with indirect information about the target's structure. However, if the bat perceives the second kind of image, in which the two glints are recognized as being at different ranges, then the means by which the bat determines the delays of the reflections must be different for the two glints because their manifestation in the combined echo spectrogram (c) is different. First, the overall delay of the combined echo most likely is estimated from the auditory equivalent of rightward displacement of the spectrogram of the transmitted sound to determine the overall range (*r*). Second, the delay separation of the reflections from the two glints most likely is estimated from the frequencies of the interference notches in the combined spectrogram to determine the range separation (Δr). Note that this latter procedure explicitly involves transforming numerical values of frequency (for notches) into an equivalent numerical value of time (for delay separation), that is, inverse Fourier transform.

Eckert, Ernst Rudolf Georg (1904–2004)

[fluid dynamics] An aeronautical engineer and physicist from the Czech Republic (erstwhile Czechoslovakia). Ernst Eckert worked on film-coating, specifically with respect to durability and efficiency related issue in aeronautical engines. Dr. Eckert may be considered one of the fathers of modern practical HEAT TRANSFER and mass transfer. His contributions are personified by the ECKERT NUMBER (see Figure E.15).

Figure E.15 Ernst Rudolf Georg Eckert (1904–2004). (Courtesy of George Grantham Bain Collection, New York City; United States Library of Congress, Washington, DC.)

Eckert number ($Ec = v^2 / [c_p(T_s - T_\infty)]$)

[fluid dynamics] A dimensionless number applicable to compressible FLOW and the phenomena of HEAT TRANSFER as well as momentum transfer. It has been named in honor of the principal contributor ERNST RUDOLF GEORG ECKERT (1904–2004). The relative ratio of the kinetic energy, marked by the VELOCITY (v) squared, with respect to the enthalpy (H) difference at the WALL. The heat transfer in a system is equal to the enthalpy change in the system: $\Delta H = \Delta Q$. The difference in enthalpy is between the main body of flow temperature ($T\infty$) and the surface temperature (T_s), with c_p being the SPECIFIC HEAT.

Eddington, Arthur Stanley, Sir (1882–1944)

[astronomy/astrophysics, general] An astronomer, physicist, and mathematician from Great Britain. Sir Arthur Eddington is most known for his contributions to the brightness of celestial objects (i.e., stars) with respect to the anticipated core and surface temperature (see Figure E.16).

Figure E.16 Sir Arthur Stanley Eddington (1882–1944). (Courtesy of United States Library of Congress, Washington, DC.)

Eddington limit

[astrophysics, general] The maximum luminosity (radiance: L_{Edd}) a celestial body (i.e., specifically geared to the emissions of the QUASAR star, discovered in 1963) can attain based on an energy balance resulting from the RADIATION FORCE acting outward and the gravitational force acting toward the core: $L_{Edd} = 3.0 \times 10^4 (m_{quasar}/M_\odot)$, where $M_\odot$ is the mass of the STAR, $L_\odot$ the radiance of the Earth's SUN, and m_{quasar} the mass of the quasar. It is named after the astronomer SIR ARTHUR STANLEY EDDINGTON (1882–1944), also referred to as the "Eddington luminosity."

Eddy current

[electromagnetism, fluid dynamics, general] Circular current in either electrical CONDUCTOR or liquids. On electric scale, the eddy refers to a system of CIRCULATION current on a MICROSCOPIC scale inside a conductor. Eddy-current loops will generate MAGNETIC FIELD as described by the BIOT–SAVART LAW, and when these magnetic moments line up the object itself forms a PERMANENT MAGNET (*also* FOUCAULT CURRENT). Eddying water currents refer to circular currents in large water bodies. The eddy is associated with VORTEX or maelstrom (whirlpool) formation. Maelstroms are named, just as hurricane and typhoons.

Edison, Thomas Alva (1847–1931)

[general] A scientist and inventor from the United States. Thomas Edison is well known for his introduction of the incandescent light bulb in 1880. In 1877, Edison invented the phonograph, a mechanical SOUND recording and replay device, the predecessor to the record player in the twentieth century, playing vinyl records. Additional efforts were on the perfection of the telephone introduced by his contemporary ALEXANDER GRAHAM BELL (1847–1922) (see Figure E.17).

Figure E.17 Thomas Alva Edison (1847–1931); Thomas Edison and his original dynamo, Edison Works, Orange, NJ, c. 1906. (Courtesy of the United States Library of Congress, Washington, DC.)

Edison effect

[atomic, nuclear] The fact that charged particles (later identified as electron) are emitted from a heated filament will migrate in VACUUM to a (grounded or positively charged) METAL plate and hence create a free-space current. This phenomenon introduced the concept of the ELECTRON TUBE. The thermal emission of electrons is governed by the work function of the material. The thermionic emitter current density ($J = I/S$, where S is the emitter surface area) was described in 1923 by SAUL DUSHMAN (1883–1954) as $J = I/S = A_0 T^2 \mathrm{EXP}[-b_0/T]$, where $A_0 = 60.2 \times 10^{-4}$ Amps$/\mathrm{m}^2 {}^\circ C^2$, while $b_0 = e(E_w/k)$ (where E_w is the WORKFUNCTION of the medium, $k = 1.3806503 \times 10^{-23}$ m^2 kg$/\mathrm{s}^{-2}/\mathrm{K}^{-1}$ the Boltzmann constant, and $e = 1.60217646 \times 10^{-19}$ C an electron charge) is a material constant, as a MATTER of fact, it is an indication of the energy requirement to get the electrons through the surface, or in thermodynamic terminology, this can be interpreted as the LATENT HEAT of EVAPORATION of the electrons for the material in question. The work of Dushman was based on the THIRD LAW OF THERMODYNAMICS.

Ehrenfest, Paul (1880–1933)

[atomic, general, nuclear, quantum, thermodynamics] A scientist and mathematician from Austria. Ehrenfest's contributions were in the field of theory of adiabatic invariants, PHASE transitions, and general THERMODYNAMICS (see Figure E.18).

Figure E.18 Paul Ehrenfest (1880–1933), with his graduate students at the University of Leiden, the Netherlands; 1926: Gerhard Heinrich Dieke, Samuel Abraham Goudsmit, Jan Tinbergen, Ralph Kronig, and Enrico Fermi.

Ehrenfest relation

[thermodynamics] The "birth–death-rate" process for SECOND-ORDER PHASE TRANSITION with respect to the number of particles (measured as partial pressure P) involved between phases I and II, occupying a volume V, with respective entropies S_I and S_{II} as a function of temperature (T), expressed as $dP/dT = \{(\partial S_{II}/\partial T)_p - (\partial S_I/\partial T)_p\}/\{(\partial S_{II}/\partial P)_T - (\partial S_I/\partial P)_T\} = \Delta c_p/\Delta \alpha_T TV$, and $dP/dT = \{(\partial V_{II}/\partial T)_p - (\partial V_I/\partial T)_p\}/\{(\partial V_{II}/\partial P)_T - (\partial V_I/\partial P)_T\} = \Delta \alpha_T/\Delta \kappa_T$, with α_T being the THERMAL EXPANSION coefficient, c_p the SPECIFIC HEAT at constant pressure, and κ_T the adiabatic bulk modulus. It is also known as the "Ehrenfest model," based on the work by PAUL EHRENFEST (1880–1933).

Eigenfunction

[atomic, computational, nuclear, quantum] A particular SOLUTION to a complex differential formulation that represent a characteristic function or value (i.e., EIGENVALUE) that can be used independently. The concept of eigenfunction originates from the use of "function space" or "Hilbert space." The eigenfunction generally represents a particular function that is directly linked to the LINEAR OPERATOR. Eigenfunctions are widely used in solving the SCHRÖDINGER EQUATION. The eigenfunction is a solution to the operator (e.g., differential operator F_{op}) for every point where the operator is defined, which gives it the eigenfunction property. In the Schrödinger equation, for example, $F_{op}\Psi = f\Psi$, the function Ψ represents the eigenfunction and f is the eigenvalue belonging to the eigenfunction and the operator applied to the function (*also see* EIGENVALUE).

Eigenvalue

[atomic, computational, nuclear, quantum] Linear projects of a vector space, associated with a function on itself, will provide a linear transformation. The projections of the points that form the line that passes through the origin provide the unique and characteristic values of universal applicability referred to as "eigenvalues." In a linear transformation of linear vector space, there is a SCALAR that is a direct SOLUTION to the function (solution to the EIGENFUNCTION). All eigenvalues correspond to the complex solution of datapoints in vector space outlined as eigenvectors. The eigenvalue times the unit matrix subtracted from the eigenfunction forms a matrix that has no INVERSION. The operator yields the "expected" value for the eigenvalue, which is a sharp value, using the SCHRÖDINGER EQUATION in the following format: $F = \int \Psi^* F_0 \Psi d\tau = \int \Psi^* f\Psi d\tau = f\int \Psi^* \Psi d\tau = f$. This ensures that the eigenvalue has a sharp value; the variance $((\Delta F)^2)$ will illustrate that the following is satisfied: $(\Delta F)^2 = \int \Psi^* (F_{op} - F)^2 \Psi d\tau = \overline{F^2} - F^2 = \int \Psi^* f\Psi d\tau - f^2 = 0$. For the Schrödinger equation in particular, this provides a series of solutions (n), specifically with respect to the energy (E) written as $H\Psi_n = E_n\Psi_n$, (H is the HAMILTONIAN OPERATOR) with respective eigenfunctions Ψ_n and eigenvalues E_n. In case two eigenfunctions have the same energy, in which case the solution is found to be energy degenerate: $E_n = E_m$. In certain situations, the eigenfunction of the Hamiltonian may simultaneously be solutions to other operators, for instance, the operator for one or more components of the angular momentum (*also see* EIGENFUNCTION).

Eigenvector

[atomic, nuclear] The linear transformations from the eigenfunctions to vector space that are outlined by the eigenvalues are transformed as eigenvectors.

Einstein, Albert (1879–1955)

[atomic, biomedical, general, nuclear] Albert Einstein used the QUANTUM THEORY model presented by MAXWELL KARL ERNST LUDWIG PLANCK (1857–1947) in 1900 to formulate the particle-wave duality of the PHOTON representing ELECTROMAGNETIC RADIATION as one of the fundamental building blocks of the theory of light, as presented in one of three publications released in 1905. The discrete energy content of the Planck model is represented by the PARTICLE aspect, in contrast to the continuous energy spectrum of the theories of ELECTRICITY and MAGNETISM by JAMES CLARK MAXWELL (1831–1879). Another article released in 1905 by Einstein introduces his concept of the THEORY OF RELATIVITY. A third presentation made by Albert Einstein in 1905 is on statistical MECHANICS, building on the works of LUDWIG EDUARD BOLTZMANN (1844–1906) and JOSIAH WILLARD GIBBS (1839–1903), primarily focusing on the movement of molecules in a FLUID media (see Figure E.19).

E Henriot E Herzen E Schrödinger W Pauli W Heisenberg
A Piccard P Ehrenfest Th de Donder J Verschaffelt R Fowler
P Debye M Knudsen L Brillouin
M Knudsen N Bohr
W Bragg P Dirac M Born
H Kramers Ch Guye
M Curie A Einstein C Wilson
M Planck H Lorentz P Langevin
I Langmuir D Richardson
A Compton L de Broglie

Figure E.19 (a) Albert Einstein (1879–1955) seated to the right of Niels Bohr (1885–1962; on Niels Bohr's left hand side) during a conversation at the home of Paul Ehrenfest (1880–1933) in Leiden, the Netherlands in 1925. (b) The Solvay attendants in 1927 including Albert Einstein. The Solvay Conferences were originally organized by the industrialist Ernest Solvay (1838–1922) from Belgium and were generally held in Brussels, Belgium.

Einstein equations

[astrophysics, computational, relativistic] Analysis of the curvature of space and time based on the work by ALBERT EINSTEIN (1879–1955) performed between 1907 and 1915 and the interpretation and contributions from the mathematician from Russia, working in Germany. Hermann Minkowski (1864–1909) found to be a generalization of rotational invariance from space to "space-time." Albert Einstein was a student of Hermann Minkowski. Hermann Minkowski defined four-dimensional space-time in theoretical format known as "Minkowski space-time." The "curvature," expressed by the Einstein tensor (G_{Einstein}) of space-time describing the relationship between the energy-momentum tensor (T_{em}; QUANTIFICATION of the "matter content") and GRAVITY expressed by the gravitational constant: G_{grav} under special relativistic conditions, with c the speed of light yields: $G_{\text{Einstein}} = (8\pi G_{\text{grav}}/c^4)T_{\text{em}}$, under general relativity this is written as $G_{\mu\nu} + \Lambda g_{\mu\nu} = (8\pi G_{\text{grav}}/c^4)T_{\mu\nu}$. Because of the multidimensional aspect, it is also known as "Einstein's field equations." The constant Λ is known as the "cosmological constant," representing the value of the energy density of the VACUUM of space. After the discoveries by EDWIN POWELL HUBBLE (1889–1953) in 1929, the cosmological constant is generally assumed to be zero: $\Lambda = 0$. More generally, the space-time mathematics of general relativity is described as $G_{\text{Einstein}} = G_{\mu\nu} = R_{\mu\nu} - (1/2)R_{\text{curv}}g_{\mu\nu} + \Lambda g_{\mu\nu} = (8\pi G_{\text{grav}}/c^4)T_{\mu\nu}$, where $R_{\mu\nu}$ is the Ricci "curvature" TENSOR (mathematician from Italy Gregorio Ricci-Curbastro [1853–1925]), under the curvature SCALAR: $R_{\text{curv}} = g_{\mu\nu}R_{\mu\nu}$, where $g_{\mu\nu}$ represents the metric tensor. The Einstein tensor is symmetric in space-time: $G_{\text{Einstein}} = G_{\mu\nu} = G_{\nu\mu}$ and is DIVERGENCELESS: $\nabla_\mu G_{\mu\nu} = 0$. These gravitational equations can be used to solve the planetary orbital trajectories under a weak gravitational attraction, resembling NEWTONIAN MECHANICS (SIR ISAAC NEWTON [1642–1727]). The Ricci tensor can be related to the Riemann curvature tensor under a more generalized mathematical approach: $R_{\mu\nu} = R_{\mu\alpha\nu}^{\alpha}$, based on the work of. The geometric description between the point in space-time is generally described by the Riemann curvature tensor, after Georg Friedrich Bernhard Riemann (1826–1866), a mathematician from Germany. The notation for the space-time dimensional representation ($\mu\nu$) is derived from Christoffel symbols, based on the work of Elwin Bruno Christoffel (1829–1900), a mathematician from Prussia (Germany). The Christoffel symbols represent three-"dimensional" (not spatial dimension, but mathematical cube array of an n-dimensional manifold: $n \times n \times n$) of real numbers, which describes, in mathematical space coordinates, the effects of parallel transportation over curved surfaces (see Figure E.20).

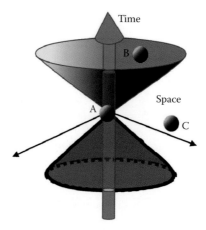

Figure E.20 Space–time convolution depicted as cones of time in a three-dimensional space.

Einstein–Lorentz transformation

[atomic] Lorentz transformation applied to special and general relativistic phenomena (*see* **LORENTZ TRANSFORMATION**).

Einthoven, Willem (1860–1927)

[biomedical] A physiologist from the Netherlands who in 1903 developed the first device to record standardized three-point electrocardiograms from the DEPOLARIZATION vector progression for the HEART MUSCLE (ECK or EKG). Willem Einthoven received the Nobel Prize in MEDICINE for his work on the study of the mechanism of action of the heart and the data acquisition with subsequent SIGNAL processing aspects in 1924 (see Figure E.21).

(a) (b)

Figure E.21 (a) Willem Einthoven (1860–1927) in 1906. (b) Einthoven with his arms and legs in the basins with electrically conducting solution acting as electrodes to record the electrocardiogram (ECG).

Ekman, Vagn Walfrid (1874–1954)

[fluid dynamics, geophysics] A geophysicist and oceanographer from Sweden. Ekman recognized the influence of Coriolis forces on the migration pattern for icebergs and corrected the movement trajectory which was initially based on the wind direction only. The MOTION is captured by the introduction of the Eckman layer (see Figure E.22).

Figure E.22 Vagn Walfrid Ekman (1874–1954).

Ekman layer

[computational, fluid dynamics] The FLUID layer underneath a floating object that balances the CORIOLIS FORCE influence on the MOTION, resulting from wind FLOW as well as pressure gradient ($\partial P/\partial x_i$) forces in the water generated during flow next to turbulent DRAG. Generally, the theoretical analysis was based on the fact that the observed movement of an iceberg is at a 20°–40° ANGLE in clockwise direction, with respect to the direction of the prevailing wind. The concept was introduced by VAGN WALFRID EKMAN (1874–1954) in 1902. The respective force balances are with respect to a flow with velocity $\vec{v} = (v_x, v_y)$ for a flow with diffuse eddy VISCOSITY $K_m = \xi'^2 |\partial v_z / \partial z|$, with ξ' the MIXING length and $\partial v_z/\partial z$ the vertical pressure gradient which is averaged, and where the Coriolis parameter is f_{Cor}, this yields for a fluid flow with density ρ_0 the following force densities: $-f_{Cor}v_y = (1/\rho_0)(\partial P/\partial x) + K_m(\partial^2 v_x/\partial z^2)$; $f_{Cor}v_x = (1/\rho_0)(\partial P/\partial y) + K_m(\partial^2 v_y/\partial z^2)$; and $(1/\rho_0)(\partial P/\partial z) = 0$, where x and y are parallel to the surface. Note that the diffuse eddy viscosity has been assumed as a constant, which is an approximation since the fluid density is not constant due to the inherent temperature gradient. The boundary conditions are as follows: $a(\partial v_x/\partial z)_{z=0} = \sigma_s^x = F/A_x$ is the shear stress in the x-direction at the surface resulting from the wind and the ice field (with force F and affected area A) is $a(\partial v_y/\partial z)_{z=0} = \sigma_s{}^y$, where a is a material constant. The solutions to these conditions are the following:

$$v_x = v_{x0} + \sqrt{2}\,(fd)^{-1}\{\sigma_s{}^x \cos((z/d)-(\pi/4)) - \sigma_s{}^y \sin((z/d)-(\pi/4))\};$$

$$v_y = v_{y0} + \sqrt{2}\,(fd)^{-1}\{\sigma_s{}^y \cos((z/d)-(\pi/4)) + \sigma_s{}^x \sin((z/d)-(\pi/4))\};$$

$$v_z = (f\rho_0)^{-1}\{[(\partial\sigma_s{}^x/\partial x)+(\partial\sigma_s{}^y/\partial y)]e^{z/d}\sin(z/d)+[(\partial\sigma_s{}^y/\partial x)-(\partial\sigma_s{}^x/\partial y)][1-e^{z/d}\cos(z/d)]\},$$

where $d = \sqrt{(2(K_m/f_{Cor}))}$ is a dimensional factor, and $v_{x;0}$ is background velocity (at a DISTANCE far enough from the iceberg to ignore local influences). The curl velocity in the z-direction (v_z) is referred to as the "Ekman spiral." Taking the integral over the displaced volume the movement is shown to be clockwise from the flow. However, this holds true only for the Northern hemisphere (Ekman's point of reference), a sign conversion is required for the southern hemisphere. The total momentum with respect to the object at DRIFT per unit surface is $p = -(iK_m/2\varpi\rho_0)$, where ϖ follows from $2\varpi v_x = \eta^*(\partial^2 v_y/\partial z^2)$ and η^* represents the coefficient of viscocity per mass density and can be derived from the one-dimensional equation of MOTION for incompressible flow (ignoring the pressure gradient; derived by CLAUDE-LOUIS NAVIER [1785–1836] and SIMÉON DENIS POISSON [1781–1840]): $\partial v_x/\partial t = \eta^*(\partial^2 v_x/\partial z^2)$; note that $\partial v_x/\partial z$ is the vorticity for the flow.

Elastic collision

[general, nuclear] Collision where no kinetic energy is lost. Both conservation of momentum and conservation of *kinetic* energy apply. In general, MACROSCOPIC collisions are not ideal elastic due to deformations energy, residual FRICTION (rolling or translational), as well as AIR resistance. Objects sliding under the influence of MAGNETIC LEVITATION will still experience some form of deformation that is gradually depleting the system of energy. The same applies to objects floating on an air cushion. Collisions of SUBATOMIC PARTICLES under the influence of ELECTROSTATIC FORCE (Coulomb law) will be most representative of elastic collisions (see Figure E.23).

(a)

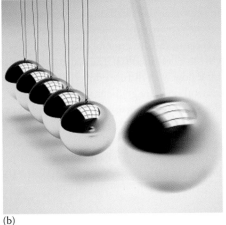

(b)

Figure E.23 (a) Example of idealized elastic collisions in the game of billiards, or pool (Courtesy of Evan Pagano.) and (b) knocking ball on a pendulum string suspension.

Elastic energy (PE$_e$)

[general, mechanics] Energy stored in a compressible medium, which can be a spring or a GAS, which is released when the spring/gas expands. For a spring, the potential energy is proportional to the material constant of the spring, the spring constant (k_s), and the displacement (s) squared as $PE_e = (1/2)k_s s^2$, whereas for a compressed gas this becomes, assuming ADIABATIC COMPRESSION and EXPANSION for an IDEAL GAS with SPECIFIC HEAT c_p (for instance, a compressor), where the temperature changes during compression from T_1 to T_2, as the pressure changes from P_1 to P_2, with a corresponding change in volume V. The compression the potential energy equals the amount of work performed, based on the change in ENTHALPY (H) and the TEMPERATURE (T) multiplied by the change in entropy (S), for a reversible process $PE_e = T\Delta S + \Delta H$; this provides $PE_e = c_p T_1 \left[(P_2/P_1)^{(\gamma-1)/\gamma} - 1 \right] = (1/(\gamma-1))P_1 V_1 \left[(P_2/P_1)^{(\gamma-1)/\gamma} - 1 \right]$ for an ISENTROPIC PROCESS, where $\gamma = (c_p/c_v)$ is the ratio of the specific heats. For an ideal isothermal compression, the potential energy stored by compression equals the work from the integral: $PE_e = \int_{V_1}^{V_2} P dV = K_T \left\{ \left[V_2^{1-\gamma} - V_1^{1-\gamma} \right]/(1-\gamma) \right\} - P_0 (V_2 - V_1)$, where for an ADIABATIC PROCESS $PV^\gamma = K_T$, with $K_T = -(1/V)(\partial V/\partial P)_T$ representing the coefficient of isothermal compressibility. Note that during combustion the change in enthalpy during the chemical reaction needs to be included to obtain the total "elastic" energy, and needs to be integrated over time to yield the change in mass for the rate of change in mass for the chemical reaction (see Figure E.24).

Figure E.24 Elastic energy in a bungee jump.

Elastic force (spring) constant (*k*)

[general] *See* SPRING CONSTANT.

Elastic limit

[general] The stretch allowed before deformation or failure sets in (see Figure E.25).

Figure E.25 The second image from the left may result in elastic return to the original configuration upon release of the applied tension.

Elastic modulus (E_*)

[general, mechanics] *See* YOUNG'S MODULUS (Y_n).

Elastic restoring force

[general] *See* SPRING.

Elastic scattering

[nuclear] *See* COMPTON SCATTERING (*also see* CROSS SECTION *and* NUCLEAR REACTION).

Elastic shear strain

[general] *See* SHEAR STRAIN.

Elastic stiffness constant

[acoustics, mechanics] *See* ELASTIC MODULUS (*also see* ELASTIC CONSTANT).

Elastic wave

[acoustics, imaging, mechanics] Longitudinal compression WAVE in solids or fluids as well as the transverse wave on a string or rope. Elastic waves are the fundamental phenomenon for transmission of SOUND waves. Temporal and spatial fluctuations in local stress (as a function of location $\vec{r}$) and strain form a system of second rank tensors, called "dyadic." Stress waves can be internal waves in isotropic or inhomogeneous media, as well as irrotational waves (longitudinal) and solenoidal (transverse), next to SURFACE WAVES. ULTRASONIC IMAGING relies on bulk waves in isotropic media (primarily in first-order approximation). These waves are described by the WAVE EQUATION for a medium of unperturbed density ρ_0, implementing HOOKE'S LAW: $(\lambda_\ell + \mu_\ell)\nabla\nabla\cdot\vec{r} + \mu\nabla\cdot\nabla\vec{r} = \rho_0(\partial^2\vec{r}/\partial t^2)$, with the Lamé coefficients $\mu_\ell = (\sigma_s/\mu_s)$, which is equal to the shear modulus, μ_s the shear strain and σ_s the shear stress, and $\lambda_\ell = B_e - (2/3)\mu_\ell$ the Lamé strain coefficient, with $B_e = P/\mu_s$ being the bulk modulus, μ_s the shear strain, and P the pressure. For irrotational waves, the wave equation becomes $\nabla\cdot\nabla\vec{r} = (1/c_\ell^2)(\partial^2\vec{r}/\partial t^2)$, where $c_\ell^2 = (\lambda_\ell + 2\mu_\ell)/\rho_0$ represents the displacement wave velocity. For solenoidal waves, the equation changes with respect to the expression for the speed of sound as $c_s^2 = \mu_\ell/\rho_0 = Y_n/2(1+\sigma)\rho_0$, where Y_n is Young's modulus, and σ = the shear stress;

under solenoidal FLOW $\nabla \cdot \vec{r}$ vanishes and hence the wave-equation becomes: $\nabla \cdot \nabla \vec{r} = \left(1/c_s^2\right)\left(\partial^2 \vec{r}/\partial t^2\right)$. For surface waves, *see* **WATER WAVE** *and* **WAVE** (see Figure E.26).

Figure E.26 Longitudinal elastic compression wave.

Elasticity

[atomic, general, nuclear] The phenomenon of elasticity was studied with great detail by GALILEO GALILEI (1564–1642), but the interest predates this era by centuries. The majority of the knowledge of elasticity is derived from the experimental and theoretical description of LATTICE vibrations. Specifically, the propagation of long WAVELENGTH ($\lambda \gg a$, where a is the characteristic dimension of the atomic lattice structure, *also* LATTICE SPACING) waves in materials has a well-defined pattern that complies with HOOKE'S LAW for the MACROSCOPIC stress and strain. In classical lattice vibrations, the ions are represented as cores with shells of electrons. The electron shell is attached to the nondeformable core by springs. Note that for the inelastic case, the model will include a dampener (e.g., DASHPOT). The springs primarily represent the COULOMB FORCES (ionic and polarized) reducing the model to a MASS-ON-A-SPRING with associated HARMONIC OSCILLATION configurations. Because most materials are anisotropic, HOOKE'S LAW forms a vector description for the stress (σ) to strain (ϵ) relationship:

$$\sigma_m = \mathbf{C}_{mn} \begin{bmatrix} \epsilon_1 & \cdots & \\ \vdots & \ddots & \vdots \\ & \cdots & \epsilon_m \end{bmatrix} = \begin{bmatrix} C_{11}\epsilon_1 & \cdots & C_{1m}\epsilon_m \\ \vdots & \ddots & \vdots \\ C_{n1}\epsilon_1 & \cdots & C_{nm}\epsilon_m \end{bmatrix}$$

or expressed as a TENSOR: $\sigma_{ik} = c_{ikjl}\epsilon_{jl}$, with σ_{ik} the dimensional components of the full-deformational stress, ϵ_{jl} the dimensional components of the full-deformational strain and c_{ikjl} the ELASTIC STIFFNESS CONSTANT for each degree of freedom (generally six DEGREES OF FREEDOM total, $m = n = 6$), and the subscript indicates the degrees of freedom, also the notation indicating the position in the MATRIX ELEMENTS. In CARTESIAN COORDINATES the displacement in the x, y, and z direction is u, v, and w, respectively. The stain tensor components are directly related to the displacement as $\epsilon_1 = \partial u/\partial x$, $\epsilon_1 = \partial v/\partial y$, $\epsilon_3 = \partial w/\partial z$, $\epsilon_4 = \left(\left(\partial v/\partial z\right) + \left(\partial w/\partial y\right)\right)$, $\epsilon_4 = \left(\left(\partial u/\partial z\right) + \left(\partial w/\partial x\right)\right)$, and $\epsilon_6 = \left(\left(\partial u/\partial y\right) + \left(\partial v/\partial x\right)\right)$. Written out for the first and the last vector ELEMENT multiplication this shows as follows: $\sigma_1^{\text{P}} = C_{11}\epsilon_1 + C_{12}\epsilon_2 + C_{13}\epsilon_3 + C_{14}\epsilon_4 + C_{15}\epsilon_5 + C_{16}\epsilon_6$, where the superscript P refers to pressure or stress,

which accounts for the first three equations. Equivalent $\sigma_6{}^S = C_{61}\epsilon_1 + C_{62}\epsilon_2 + C_{63}\epsilon_3 + C_{64}\epsilon_4 + C_{65}\epsilon_5 + C_{66}\epsilon_6$, where the superscript S refers to shear stress, which accounts for the lower three equations. In the case of a crystal with a cubic lattice, the matrix is symmetric $C_{11} = C_{22} = C_{33}$, $C_{44} = C_{55} = C_{66}$, $C_{12} = C_{23} = C_{31}$, and $C_{14} = C_{15} = C_{16} = C_{34} = C_{35} = C_{36} = C_{24} = C_{25} = C_{26} = C_{45} = C_{46} = C_{56} = 0$. In the matrix notation, the shear modulus of a cubic crystal is the coefficient of the element on the diagonal below center $S_{\text{cub}} = C_{44}$, *also* MODULUS OF RIGIDITY. In the case of a centralized force between the ions of the cubic lattice the relationship between the elastic stiffness coefficients is defined by the Cauchy relation as $C_{12} = C_{44} = S_{\text{cub}}$. WAVE propagation in elastic solids can be described with the help of these matrix terms for instance in the *x*-direction with displacement *u* as follows:
$$\rho\left(\partial^2 u/\partial t^2\right) = \left(\partial\sigma_1/\partial x\right) + \left(\partial\sigma_6/\partial y\right) + \left(\partial\sigma_5/\partial z\right) = C_{11}\left(\partial\epsilon_1/\partial x\right) + C_{12}\left(\left(\partial\epsilon_2/\partial x\right) + \left(\partial\epsilon_3/\partial x\right)\right) +$$
$C_{44}\left(\left(\partial\epsilon_6/\partial y\right) + \left(\partial\epsilon_5/\partial z\right)\right)$ in Cartesian coordinates with ρ the local density. The last part of the equation is specifically for a cubic crystal.

Elasticity number ($E\ell = \theta_r\eta/\rho R^2$)

[mechanics] Pertaining to VISCOELASTIC FLOW, the dimensionless number that relates the ratio of the elastic force to the inertial force. The viscous force is defined by the coefficient of VISCOSITY: η, with an additional term representing the RELAXATION TIME of the event: θ_r, with ρ the density of the FLUID in MOTION and R the radius of the PIPE in which the flow occurs or half the width of the canal. The viscous force provides a mechanism of dampening the inertial forces. Elasticity number provides a range of boundary conditions which yield conditions that result in EXPANSION or contraction associated with the fluid VISCOELASTICITY. This number is applicable to both NEWTONIAN and non-NEWTONIAN media.

Elasticity of gases

[hydrodynamics] Gases have an inherent elastic property due to compression mechanism (*see* ELASTIC ENERGY *and* IDEAL GAS).

Elasticity of wall of pipe

[fluid dynamics] The EXPANSION of a flexible tube with diameter D and WALL thickness t given as: $B_{\text{eq}} = tE/D$, where $t \ll D$, and the elastic modulus of the wall material E (*also see* WATER HAMMER *and* COMPLIANCE) (see Figure E.27).

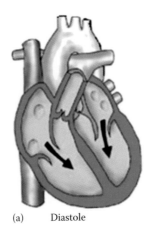
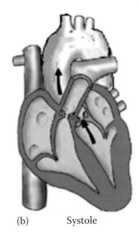

(a)   Diastole   (b)   Systole

Figure E.27 (a) Elastic expansion of the aorta during systolic filling with large quantity of blood while (b) the flow is restricted by peripheral resistance in the arterioles and capillaries.

Elastin

[biomedical, general] Connective biological tissue that has significant stretch properties, such as those found in the SKIN and in the arterial blood-vessels as well as in the veins, where the VEIN has additional smooth MUSCLE tissue. The ELASTIN provides a COMPLIANCE that is equivalent to the Windkessel effect (see Figure E.28).

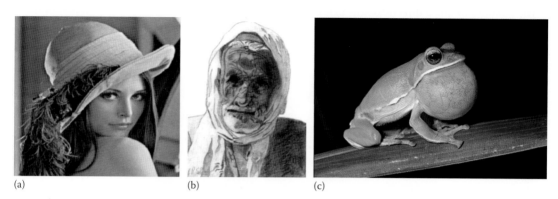

(a)   (b)   (c)

Figure E.28 Elastin of the skin reduces with age: (a) healthy complexion of young elastic skin, (b) old skin with fixed wrinkles and loss of elasticity, and (c) elastic skin of a Taiwanese tree frog.

Elastomer

[chemical, fluid dynamics, mechanics] Elastic POLYMER with high yield-strain value (i.e., resilient) and generally in solid form has a low elastic modulus (Young's modulus). Elastomers have mechanical properties that are similar to rubber, and are formed by either vulcanization (thermal setting: thermoset) providing thermal cross-linking (forming irreversible chemical bonds) or by means of thermal blending, where the material is moldable—semi-LIQUID state—above a certain temperature (the "glass temperature") while it sets to a solid state when the material cools down (thermoplastic elastomer; reversible; which makes them ideal for mold forming). Thermoplastic elastomers are frequently used in injection molding (see Figure E.29).

Figure E.29 Typical elastomer: surgical gloves.

Electric battery

[general] *See* BATTERY.

Electric car

[general] An automobile that has a primary ENGINE that consists of an electric MOTOR. Frequently, the electric motor is mounted directly on the axis of the car, making the axis a direct and integral part of the armature of the induction process. The ELECTRIC CAR requires the use of a significant BATTERY bank to provide an acceptable travel DISTANCE with having to stop for adding energy (see Figure E.30).

Figure E.30 The electric car concept: automobiles charging at a public electrical distribution center.

Electric charge

[atomic, electronics, general, nuclear] It is a net charge, primarily because of an excess or deficit in electrons; on a nuclear scale, because of positrons and protons and their respective antiparticles. The earliest documented description of electric charge dates back to 600 BC and the work of the scientist from Greece THALES OF MILETUS (c. 624–546 BC; known as Thales [Θαλῆς]), describing rubbing AMBER and cat fur to generate a form of attraction and repulsion with other objects (*see* ELECTRON CHARGE).

Electric current (*I*)

[atomic, electronics, general] Transport of electrons that are released from the atomic and molecular VALENCE BAND to become free electrons under an applied external electric field resulting from a potential difference across the CONDUCTOR. The current was initially assumed to be the result of POSITIVE CHARGE migration, however, later to be refuted as NEGATIVE CHARGE carriers; hence the electrons move in the opposite direction of the direction of the current, introducing the concept of "hole" migration providing the current in the appropriate direction. Electric current was recognized by Count ALESSANDRO GIUSEPPE ANTONIO ANASTASIO VOLTA (1745–1827) in 1800, fueled by the research conducted by his colleague LUIGI GALVANI (1737–1789) in 1780; however, the electron itself was discovered in 1897 by JOSEPH JOHN THOMSON (1856–1940). The holes represent the missing electrons in the atomic structure. A current is the end result of the movement of all electrons in a conductor, which means that the first electron out is the closest to the terminal; the electron does not have to travel from pole to pole on the potential source. The conduction process may be compared to a tube filled with marbles, where injecting one marble on one side will eject a marble on the opposite side almost instantaneously, thus no requirements for QUANTUM mechanical approach in case the electron had to approach the speed of light, hence merely ELECTRON DRIFT VELOCITY.

Electric dipole

[biomedical, chemical, condensed matter, electronics, solid-state] Charge distribution that is weighted with excess POSITIVE CHARGE on one side and excess NEGATIVE CHARGE on the opposite side. Dipoles can be molecular (e.g., H_2O), ions (missing electrons or captured electrons) and can also be on a MACROSCOPIC scale where the charge distribution in an object (isolator or semiconductor) has permanently been shifted to form a permanent DIPOLE or the influence of an external electric field temporarily moves the electrons in a CONDUCTOR to one side, hence creating a deficit (holes) on the opposite side. In a semiconductor or INSULATOR electrons may be added to the medium bay means of charge transfer from a another object. Animals may also create a whole-body dipole which they use for sensing purposes. The platypus and the Elephant fish (*Gnathonemus*) are only two examples of many animals that form a static electric field due to an induced dipole. These animals also have the ability to sense spatial changes to the electric field distribution, hence allowing for navigation as well as for seeking prey. The dipole attraction is very strong and forms one of several chemical binding mechanisms (see Figure E.31).

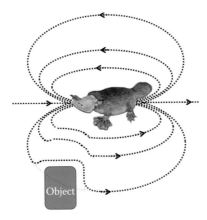

Figure E.31 Electric dipole: specifically used for navigation by certain animals.

Electric dipole moment

[atomic, condensed matter, nuclear] The moment expressed by a set of equal but opposite charges ($+q$ and $-q$) at a fixed separation DISTANCE (d) in direction $\vec{\iota}$, pointing from the positive to the NEGATIVE CHARGE in the respective frame of reference, expressed as $\vec{p} = q\vec{\iota}$.

Electric dipole transition

[atomic] Electronic transitions in the electronic energy levels or the nuclide structure that results in a nanoscopic shift in charge balance that is governed by the various selection rules for transitions.

Electric field (*E*)

[atomic, electronics, general, nuclear] The virtual lines connecting migrating out from positive, or alternatively converging on a NEGATIVE CHARGE distribution (Q), with a MAGNITUDE defined by the force per unit charge experienced at any point at a DISTANCE (r) from a charge distribution. The electric field can exist as the result of a single charge with no defined termination; keep in mind that the MAGNETIC FIELD by definition is a closed loop. The electric field is defined by Coulomb's law: $E = (1/4\pi\varepsilon_r\varepsilon_0)(Q/r^2)$, designed by CHARLES-AUGUSTIN DE COULOMB (1736–1806) in 1785, where ε_r represents the RELATIVE PERMITTIVITY and $\varepsilon_0 = 8.85419 \times 10^{-12}$ C^2/Nm^2 is the permittivity of free space. The work of Coulomb was followed by the work of MICHAEL FARADAY (1791–1867) on ELECTRIC FORCE.

Electric field in conductor

[atomic, electromagnetism, nuclear] It is a function of radius of curvature of the surface contour. Free ELECTRONS are drawn into the electric field in the CONDUCTOR, with a field distribution that depends on the radius of curvature (r_c), smaller radius higher field due to the higher SURFACE CHARGE-DENSITY ($\rho_e(r)$) as illustrated by $\vec{E}(\vec{r}) = (1/4\pi\epsilon_0) \sum_j (\rho_e(\vec{r}_j)(\vec{r} - \vec{r}_j) dV_j / |\vec{r} - \vec{r}_j|^3)$, where V_j is the localized electrical potential in location $\vec{r}_j$ and ϵ_0 is the DIELECTRIC permittivity. Note that in a conductor the free electrons repel and they all want to be on the surface of the conductor making the description a surface charge density.

Electric field intensity

[atomic, electronics, general, nuclear] MAGNITUDE of the intensity of an electric field at a particular point, equal to the force that would be exerted upon a unit POSITIVE CHARGE placed in the field at that point. The direction of the electric field is the direction of this force.

Electric force

[electromagnetism, general] The fact that opposite (unlike) ELECTRICAL CHARGES (positive vs. NEGATIVE CHARGE) will attract or similar charges repel each other with a force proportional to the inverse of the square of the separation DISTANCE between the net charges. This phenomenon was described in 1760 by DANIEL BERNOULLI (1700–1782) and BENJAMIN FRANKLIN (1706/1705?–1790) in circa 1750. The formal equation was postulated by CHARLES-AUGUSTIN DE COULOMB (1736–1806) in 1785: $F_e = (1/4\pi\varepsilon_r\varepsilon_0)(Q_1 Q_2/r^2)$, where Q_1 and Q_2 are the net charges between which the electric field and resulting electric force is formed, $\varepsilon_0 = 8.8541878 \times 10^{-12}$ C^2/Nm^2 the DIELECTRIC permittivity of VACUUM, ε_r the RELATIVE PERMITTIVITY (sometimes also called the "dielectric constant," which is $\varepsilon_r = 1$ for vacuum) and r the separation between the two net charges. The net charge in one location is the sum of all the charges gathered in that one volume in space considered as a unique location.

Electric force, Coulomb's law

[general] *See* COULOMB'S LAW.

Electric potential (V)

[computational, electromagnetism, general] The amount of work needed to transport a positive ELECTRIC CHARGE of 1 C from infinity to the point of electric interest. When a cumulative charge is located at this point then work is performed, otherwise no work is performed and no electric potential exists in this point. In case an artificial potential difference is maintained between two points, a GALVANIC MECHANISM needs to maintain a current from the higher electric potential to the lower electric potential, note current is the FLOW of charge. An electric potential can, for instance, be generated by means of the PHOTOELECTRIC EFFECT, using ELECTROMAGNETIC RADIATION to release electrons from their atomic/molecular orbit and hence providing the mechanism of action (the work) to create a potential difference. Other means involve making surface contact between two or more metals (bimetal). The metal-plate electric potential series is a function of both the nature of the metals and the localized temperature. The electric potential exists under the condition of no current (*also see* BATTERY *and* THERMOCOUPLE). Animals such as the electric eel and electric manta-ray create a high electrical potential by chemical means, reaching several hundred volt. A similar electrochemical process is used in NERVES and MUSCLE cells (*see* NERNST POTENTIAL) for SIGNAL propagation by electrical potentials and muscle contraction.

Electric quadrupole moment (Q_{elect})

[atomic] The NUCLEUS generally does not possess a static DIPOLE moment, but does possess a dynamically fluctuating QUADRUPOLE MOMENT. The measure of eccentricity of the ellipsoidal surface of the nucleus is defined as $Q_{elect} = (1/e) \int (3z^2 - r^2) \rho d\tau$, where $e = 1.60217657 \times 10^{-19}$ C is the ELECTRIC CHARGE, z is the location on the major axis of the ellipsoid, r is the DISTANCE from the geometric center, ρ is the charge density of the respective nucleus, and τ is the spatial variable. As an example, the electric quadrupole moment for the $_1^2$H hydrogen nucleus is $Q_{elect} = +0.00282 \times 10^{-24}$ cm^2, and for $_{92}^{235}$U uranium: $Q_{elect} = \pm 4.1 \times 10^{-24}$ cm^2.

Electricity

E

[atomic, general, nuclear] Property of ELECTRIC CHARGE concentrations, either stationary or in MOTION, primarily pertaining to the mobility of free electrons, which can be released from VALENCE electrons bands in conductors and SEMICONDUCTORS. The phenomenon of electricity was studied with great detail by GALILEO GALILEI (1564–1642), but the interest predates this era by centuries. The earliest recorded description with respect to static electricity dates back to THALES OF MILETUS (c. 624–546 BC) a scientist from Greece, known as Thales [Θαλῆς]). Thales described rubbing AMBER and cat fur to generate a form of attraction and repulsion with other objects. Modern concepts of electricity include the generation of electric power by generators and the current FLOW between a potential difference as defined through the COULOMB POTENTIAL (see Figure E.32).

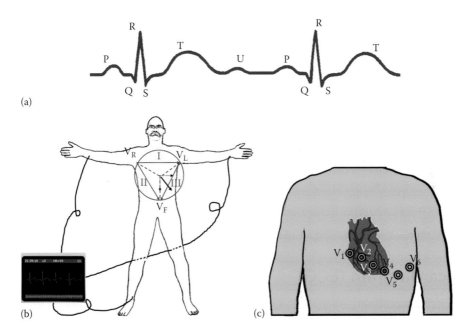

Figure E.32 (a) ElectroCardioGram (ECG; sometimes also referred to as EKG [electrokardiogram]) with the P-wave from the depolarization of the atrium, followed by the ventricular depolarization QRS-complex. (b) Generally, the ECG is collected from electrodes placed on the skin; either, the wrists and ankles (Einthoven electrode placement), or (c) more detailed information obtained with electrodes placed on the chest (Goldman electrode placement). (*Continued*)

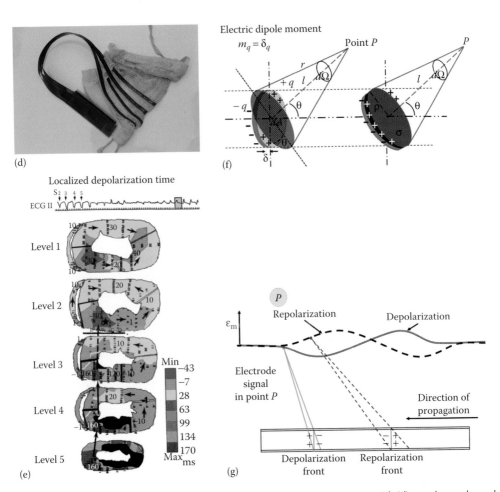

Figure E.32 (Continued) Further detail can be achieved during open-chest surgery with (d) an, electrode sock placed around the heart or by needles inserted into the heart. The needles and the sock circumvent the volume conduction aspect and the associated time and location averaging of the cellular depolarization superposition when measuring a great distance based on the current produced as a result of the time-dependent electrical potential difference of the action potential of the heart muscle cells. (e) The location and time specific electrical depolarization is presented for the electrode configuration outlined in (c), (f) the registration of the single cell electrical depolarization is affected by volume conduction where the electrode is placed in point P, (g) the collected signal is the result of an electrical dipole of the depolarizing cell outlined in with the "front" depolarization and "back-side" repolarization providing a biphasic and triphasic potential.

Electrocardiogram (ECG)

[biomedical, electromagnetism] When MUSCLE cells depolarize during contraction, specifically for the cardiac muscle (i.e., HEART), the cell's intrinsic DIPOLE LAYER reverses polarity, creating a change in the electrical potential at a DISTANCE from the electric bilayer. It is also found as ElectroKardioGram (EKG) on historical basis, as designed by WILLEM EINTHOVEN (1860–1927), the scientist and physiologist from the Netherlands. This process is periodic and repeats every heartbeat, which is approximately once every second, but in reality ranging from approximately 1 BEATS per minute to over 200 beats per minute, depending on the size of the hemodynamic system of the animal. Larger animals will by definition have a slower heart rate; babies have a heart rate over 100 beats per minute, whereas the average healthy adult has at approximately 75 beats per minute at rest. Electrodes placed on the surface of the chest of the animal in location P measure the propagation of the DEPOLARIZATION dipole layer at an arbitrary distance away from the ACTION POTENTIAL. By the time the electrical potential as a function of time reaches the respective electrodes, the conduction volume between the source and the detection will have dispersed and attenuated the vectorial profile of the electrical potential as a function of the conductive, semiconductive, and insulating (e.g., AIR in lungs) components in the path of the depolarization WAVE that traverses the inhomogeneous tissues with the mathematical description defined by the TELEGRAPH EQUATION. Electrodes can be placed in various numbers and in a variety of placement configurations, ranging from 3 (wrists and one ankle, or patches on the chest) to 12 in a well-designed pattern on chest and back that allows for reproducible and interpersonal analysis and comparison. The larger the number of electrodes, the higher the accuracy in recreating the vector displacement of the electrical potential. The volume conduction of the electrical impulses will influence the delays in action potential as a function of location, as will it affect the way the depolarization WAVEFRONT is view: depolarizing in positive or negative direction based on the dipole moment of the axon, or cluster of cells and the respective configuration in view of the distant ELECTRODE. In the computational approach, several assumptions need to be made, the first being that the medium surrounding the location as a function of time for the action potential is infinitely spread out in all directions. The second assumption implicates that the CELL that produces the action potential has ideal cylinder symmetry (both for a single muscle cell as well as for an axon of a nerve), and that the MEMBRANE thickness is negligible in comparison to the cell dimensions. The membrane is a dipole layer with surface charge density σ_{charge}, and dipole moment per unit surface area: $m_{dipole} = (q\delta/A) = \sigma_{charge}\delta$, now δ the "plate separation." The cell at rest has a potential of -90 mV. The potential in the external point P is influenced by two oppositely charged dipole layers, with equal contributions. The SOLID ANGLE of influence and the dipole moments per surface area are identical. Hence, the front and back sides of the cell cancel each other out, and no potential is registered in P for a cell in steady-state potential and cannot be detected. During the depolarization process, the chemical ION transfer is not considered to be influencing the development of the action potential geometrically, no gradients in Na^+ and subsequent K^+ ions. The resulting depolarization propagation now has two discontinuities, the depolarization front and the repolarization front. Both move transporting along the length of the cell cylinder. In the propagation process, the depolarization will be limited to a short cell segment only, with the distal portion of the nerve cell still in EQUILIBRIUM STATE and the trailing end going through the process of repolarization and subsequent equilibrium. The depolarized section of the cell is defined by the dipole moment per unit area $m_{dipole,2}$, and the equilibrium section of the cell has dipole moment per unit area $m_{dipole,1}$. Only within the narrow solid angle that encloses the transition ring on the tube geometry that is converting from equilibrium to depolarized state will provide an electrical dipole and associated potential that can be detected: $dV_p = K(\overrightarrow{m_{dipole}} \cdot \overrightarrow{dA}/r^3) = K(m_{dipole}\cos\theta dA/r^2) = Km_{dipole}d\Omega$, where Ω is the solid angle, K is a constant, and r is the distance between the dipole and point P, subsequently integrated over the full solid angle for total effect. The electrical potential in point P as a result of the depolarization is formed by the combined effect of all electrical activation within a solid angle of view of the detector: both "front" and "back" side of the cell under depolarization, as well as the sides. The total electrical influence on the electrode in position P is the combined effect of the proximal (Ω_p) and distal (Ω_d) contributions, formulated as $V_p = K(m_{dipole,1} + m_{dipole,2})\Omega_p - K(m_{dipole,1} + m_{dipole,2})\Omega_d$. The cross-sectional projection can be represented as two cross-sectional charge disks in the cylindrical cell within the same solid angle, simplifying the mathematical description of a propagating depolarization wave front. The distal segment, captured within Ω_d, will generate a negative potential because a net NEGATIVE CHARGE is on the side of the electrode. During the transport of the depolarization wave front, the distance separating the depolarization front and the repolarization front can be determined from the duration (t) of the action potential combined with the speed of propagation (v_{signal}) of the SIGNAL along the CELL MEMBRANE ($x = vt$). For instance, if the depolarization process takes place over a time frame of 1 ms ($t = 1$ ms), and the propagation velocity is $v_{signal} = 2$ m/s, the distance becomes $h = 2$ mm. Because of the propagation of the depolarization front along the WALL of the cell, the angle with respect to the electrode in point P of the dipoles will continuously change in time. In the discussion, there is no difference whether the depolarization front moves with respect to the in situ electrode, or if the electrode moves and the depolarization remains stationary, although in practice the latter is not practical. The depolarization wave front and the associated dipole migration can be described by a MOTION algorithm. The repolarization follows an identical process. Using the angle with respect to the normal to the cell wall as a reference, the solid angle capturing the event will initially increase and subsequently decrease and will be nonexistent when the depolarization is directly underneath the normal pointing at P. Based on this mathematical feature as well as the distance aspect, the electrical dipole potential initially increases in approach to the electrode, and subsequently decreases to zero. After passing the normal position with respect to position *P*, the electrical potential renders a negative value. This type of fluctuation in electrical potential is referred to a "biphasic action potential." The superposition of both the depolarization and repolarization wave front generates a triphasic action potential (see Figure E.33).

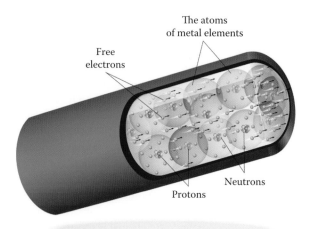

Figure E.33 Schematic of electric current.

Electrochemical equivalent

[atomic, general, nuclear, quantum, thermodynamics] The amount of MATTER involved in ELECTROLYSIS under the FIRST LAW OF FARADAY that transports one (1) coulomb of ELECTRIC CHARGE, or the ionic quantity (e.g., acids, bases, and SALT solutions) that is transported between the CATHODE and ANODE in a LIQUID under exposure to 1 AMPERE for 1 s. For instance, the silver in the electrochemical equivalent is 1.118 mg. Similar conditions apply for molten salts.

Electrode

[biomedical, electronics, general] Conducting pole attached to a wire that leads to a recording device. The electrode is designed to collect electrical ACTIVITY (conduct current to measure an electrical potential) for diagnostic purposes. The electrodes may be used to pinpoint electronic behavior of an electronic circuit, or perform nondestructive testing on CONDUCTOR and semiconductor-based materials with applied current frequency range from 0 Hz to 2 MHz. Additional use of electrodes is in biomedical applications with respect to any cellular DEPOLARIZATION, ranging from MUSCLE to brain activity to the rhythm analysis of the HEART as well as nerve pulse propagation and sensory validation from the organ to the spinal cord to the brain, respectively. Biological defects may be identified this way as well as electronic design flaws and component failure. Generally, electrode are designed for external use as a pin or pad; however, additional applications may require a needle electrode for invasive studies. The external electrodes for biological use can be attached to a sticky tape or a miniature suction cup. Generally, a reference electrode is required to provide a baseline to which the diagnostic MEASUREMENT is checked for validity. Whole EARTH GROUND will, by definition, provide a reference potential of 0 V, whereas on the HUMAN BODY, the reference electrode frequently needs to be placed in an inconspicuous place, where in theory, all time-dependent signals will cancel out, hence only a relative measurement is performed in contrast to absolute measurements in ELECTRONICS.

Electrodynamics

[general] The field of electrical ENGINEERING pertaining to the influence of electric currents on each other and the energetic content of electric phenomena. This electrodynamics is in contrast to ELECTROSTATICS and is a subcategory in PHYSICS or electrical engineering.

Electroencephalogram (EEG)

[biomedical] Electrical potential distribution recording resulting from brain ACTIVITY. Because of the complex three-dimensional structure of the brain and the excitation at multiple locations in a simultaneous or dissimilar fashion, the retracing effort will depend on retracing the volume discharges as a function of location in time recorded over the surface of the skull (*see* VOLUME CONDUCTION and the description of electrical mapping during an ELECTROCARDIOGRAM) (see Figure E.34).

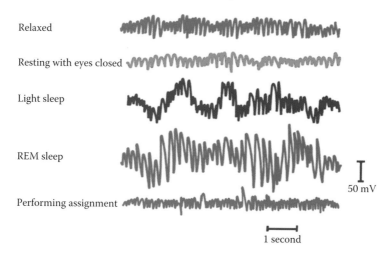

Relaxed

Resting with eyes closed

Light sleep

REM sleep

Performing assignment

50 mV

1 second

Figure E.34 The collection of brain wave patterns, grouped by frequency pattern.

Electroluminescence

[general] The phenomenon of emission of visible light from an object during or shortly after exposure to electric discharge and associated electric field (high MAGNITUDE), irradiation by CATHODE RAYS, alpha, or beta RADIATION. A special case is the scintillation of materials such as zinc sulfite and diamond under alpha radiation exposure. Electroluminescence is the mechanism resulting from recombination of electrons and depleted atomic or molecular structures in a material, frequently of semiconductor nature. The accelerated electrons that are captured or excited in the process release their energy as photons, which is the standard mechanism of action (see Figure E.35).

Figure E.35 Electroluminescence on the front panel of an electrode beam tube (cathode ray tube: CRT).

Electrolysis

[atomic, electromagnetism, general, quantum, solid-state] The chemical separation of ELEMENTS by means of passing an electrical current through the medium, primarily in SOLUTION. The electric current FLOW between two conductors called the CATHODE and the ANODE in a solution. The class of chemical substances that allow this process to take place is called an "ELECTROLYTE." The electrolyte conducts the electric current by means of ionic migration (ION-flow, or a form of displacement current). The FIRST LAW OF FARADAY states that the quantity of released material (i.e., ionic gas or solution) is directly equivalent to the applied current and time duration. This law is supported by the electron valance of the molecules and atoms involved in the process, which also correlates to the transported mass. The transported MATTER during electrolysis is called the ELECTROCHEMICAL EQUIVALENT.

Electrolyte

[biomedical, general, quantum] Mixture of constituents that is separated under passing an electric current through the substance. The substance can be SALT solutions and watery acids or bases as well as molten salts. The electrolytes are the split constituent molecules or atoms containing positive or NEGATIVE CHARGE: ions.

Electrolytic capacitor

[atomic, electromagnetism, general] CAPACITOR that has a capacitive value that is directly proportional to the surface area of the opposing plates and inversely proportional to the DISTANCE between the plates. The metallic plates are suspended in ELECTROLYTE solution. Under initial current, the plates form a molecular oxide layer that acts as INSULATOR, preventing further current passage. The separation distance is usually reduced to the combined thickness of the oxide layers. Electrolytic capacitors are rendered inactive under alternating current due to the breakdown of the oxide layer and hence only operate under direct current.

Electrolytic dissociation

[atomic, electromagnetism, general] The splitting of a mixture in its respective constituents when exposed to a medium, for instance, in SOLUTION. SVANTE AUGUST ARRHENIUS (1859–1927) developed a theoretical description of the chemical process.

Electromagnet

[electronics, general] Device designed to produce a MAGNETIC FIELD distribution of a specific configuration based on the passage of a current through a wire that has a particular geometry and layout. Electromagnets are generally designed from coils with multiple windings where the number of loops provides a mechanism to control the MAGNITUDE of the magnetic field. The first electromagnet was built in 1825 by WILLIAM STURGEON (1783–1850) from Great Britain, and this was later improved by JOSEPH HENRY (1797–1878) and released in 1831 (see Figure E.36).

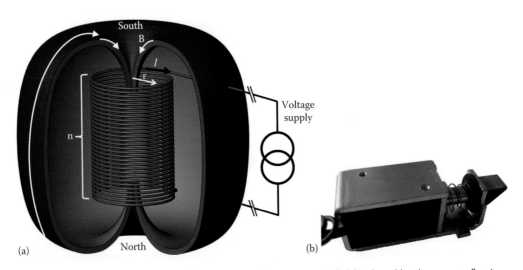

(a) (b)

Figure E.36 (a) The design of an electromagnet, generating a magnetic field induced by the current flowing through the coil. The blue surface indicates an isomagnetic field contour. (b) Coil magnet-operated lock.

Electromagnetic charge

[general] The fact that ELECTRIC CHARGE has two "distinct electricities" was discovered by the botanist CHARLES FRANÇOIS DE CISTERNAY DU FAY (1698–1739) from France in 1734. The earliest documented description of electric charge dates back to 600 BC when THALES OF MILETUS (c. 624–546 BC, known as Thales [Θαλῆς]), a scientist and philosopher from Greece, described rubbing AMBER and cat fur to generate a form of attraction and repulsion with other objects. However, the physical charge of the electron was not identified until in 1897 by JOSEPH JOHN THOMSON (1856–1940) (*see* ELECTRIC CHARGE *and* ELECTRIC CURRENT).

Electromagnetic energy

[atomic, electronics, optics, thermodynamics] The energy contained in the combined electric and magnetic fields associated with the broad spectrum of ELECTROMAGNETIC RADIATION. The two components of electromagnetic RADIATION are never separated and cannot be considered as independent; however, the electric and the MAGNETIC FIELD contributions can be analyzed separately, whereas the combined effect is represented by the Poynting vector. The electric field has an energy density (u_e) that is directly proportional to the ELECTRIC FIELD (E) strength squared, in equivalence to the "intensity" for other phenomena such as acoustic WAVE ENERGY: $u_e = \left(1/2\right)\varepsilon_0 \vec{E}^2$, where $\varepsilon_0 = 8.85419 \times 10^{-12}$ C^2/Nm^2 is the permittivity of free space. Similarly, the magnetic field energy density (u_b) is provided by the square of the magnetic field AMPLITUDE (B) as $u_b = \left(1/2\right)\left(\vec{B}^2/\mu_0\right)$, where $\mu_0 = 4\pi \times 10^{-7}$ H/m is the permeability of free space (*see* **POYNTING VECTOR** *and* **ELECTROMAGNETIC RADIATION**) (see Figure E.37).

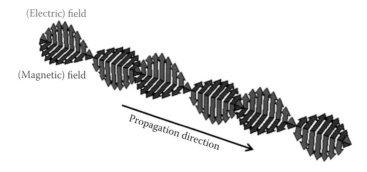

Figure E.37 Graphical representation of the perpendicular positioning of the electric field with respect to the magnetic field in electromagnetic radiation.

Electromagnetic field

[thermodynamics] Interrelated COHESION of electric and magnetic fields, based on mutual dependence. The changing electric field indirectly generates a synchronized changing MAGNETIC FIELD and vice versa. Only an *accelerated* charge will provide the conditions for a *changing* electric field and hence a changing magnetic field. The theoretical interrelation between the two fields is described by MAXWELL EQUATIONS (*see* ELECTROMAGNETIC RADIATION).

Electromagnetic radiation

[atomic, general, nuclear, optics] RADIATION composed of an electric field based on Coulomb law and GAUSS's LAW for electric fields, which changes as a function of time due to an accelerating charged PARTICLE, which in turn induces a changing MAGNETIC FIELD, in principle based on FARADAY LAW and AMPÉRE's LAW, where the magnetic field is required to satisfy Gauss' law for magnetic fields. The total electric and magnetic field theory for an accelerated charged particle is captured by slight modifications to the aforementioned laws, next known as "MAXWELL EQUATIONS." On the repercussion side, the charged particle itself will again be subject to electric and magnetic fields during its motions, hence there are constraints to the source outlined by the Lorentz force law.

The currently known electromagnetic radiation spectrum spans a broad wavelength (FREQUENCY SPECTRUM) range from at least as small as 10^{-18} m ~ 10^{25} Hz as gamma radiation to 10^8 m ~ 1 Hz, known as "long-wave radio." The visible spectrum is approximately from 3.3×10^{-7} m, blue; to 6.6×10^{-7} m, RED. The speed of propagation of electromagnetic radiation was initially determined within 30% of the currently accepted value in 1675 by OLAUS ROEMER (1644–1710). Electromagnetic radiation can be organized in a predictable fashion using a range of theoretical tools developed over the centuries. The laws of Snell with regard to REFLECTION and REFRACTION were introduced around the turn of the sixteenth century by WILLEBRORD SNELL (1580–1626). The HUYGENS' PRINCIPLE introduced in the seventeenth century by CHRISTIAN HUYGENS (1629–1695) predates the acceptance of the WAVE theory but is phenomenologically correct. The Huygens' principle introduces the hypothesis that every point in the path of a propagating wave forms a new source, while connecting all the synchronized sources forms a wave front. The blue sky (the background is black because the SUN is only in one location) results from RAYLEIGH SCATTERING as described by JOHN WILLIAM STRUT, THE THIRD BARON RAYLEIGH (1842–1919). Additional more complex RADIATIVE TRANSFER theory describing both galactic propagation and biomedical imaging-related OPTICS was introduced by SUBRAHMANYAN CHANDRASEKHAR (1910–1995). Another interesting theoretical contribution to the PERCEPTION of light was introduced by GUSTAV MIE (1869–1957); although his efforts were initially geared to the quality of printed paper, his work provided significant insight on the interaction of light with large particles, most prominently clouds in the sky. The overcoupling Maxwell theories were introduced by JAMES CLERK MAXWELL (1831–1879) in 1861 (*see* **ELECTROMAGNETIC WAVES**). Electromagnetic radiation does not require a medium for propagation, in contrast to ACOUSTIC WAVES. Electromagnetic radiation is emitted from every object with a temperature above ABSOLUTE ZERO due to the thermal agitation on an atomic and molecular level (charge VIBRATION, translation, rotation, and dipole-scissor action). Additional sources of emission are energetic DECAY to a lower energy state by EXCITED STATES of radioactive isotopes, which continues until the final GROUND STATE is eventually reached. CENTIMETER scale waves are generated in a MICROWAVE OVEN, which brings water molecules in RESONANCE, which results in an increase of kinetic energy that is a temperature rise. Long wavelength radiation can be generated by an electronic circuit that sends an oscillating current into an ANTENNA, which is subsequently broadcast as television and RADIO waves. Electromagnetic radiation transposes energy and hence one can attribute momentum. The momentum of electromagnetic radiation may be interpreted partially as SOLAR WIND. The momentum of electromagnetic radiation is the energy divided by the speed of light $\left(c = \sqrt{\mu_0 \varepsilon_0}^{-1} = (\mu_0 \varepsilon_0)^{-(1/2)} = 2.99792458 \times 10^8 \text{ m/s}\right)$, or the POYNTING VECTOR $(\vec{S})$ divided by the speed of light squared yields the momentum per unit volume: $\vec{p} = \vec{S}/c^2$. Keep in mind that the momentum of a single PHOTON is $p_{\text{phot}} = \hbar k = h/\lambda$, for wavelength λ and PLANCK'S CONSTANT: $h = 6.626070040(81) \times 10^{-34}$ m^{2}kg/s(Js). Equivalently, the radiation pressure is the energy squared multiplied by the permittivity of free space, normalized by division by 2, or $P_{\text{rad}} = a\left(\left|\vec{S}\right|/c\right)$, where $a = 1$ for total absorption and $a = 2$ for total reflection (see Figure E.38).

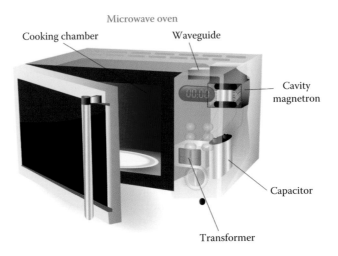

Figure E.38 Microwave "oven" (magnetron), emitting electromagnetic radiation in the GHz range.

Electromagnetic spectrum

[general, nuclear] A synopsis of the EMISSION SPECTRUM of ELECTROMAGNETIC RADIATION is outlined in a graph (see Figure E.39).

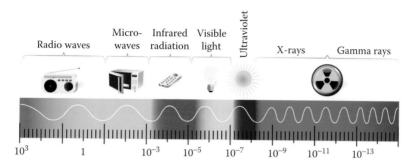

Figure E.39 Representation of the electromagnetic spectrum.

Electromagnetic theory

[atomic, general, imaging, nuclear, optics] *See* MAXWELL THEORY.

Electromagnetic (EM) waves

[computational, electromagnetism, general] Transverse ENERGY WAVES constructed of interrelated periodically changing ELECTRIC FIELD and MAGNETIC FIELD that migrate in synchronized fashion from location to location while remaining in a fixed PHASE relationship. The field lines are always perpendicular to each other in a transverse manner. The interrelation between the electric and magnetic phenomena is defined by the MAXWELL EQUATIONS. No real CURRENT is required to initiate the electric field; a DISPLACEMENT CURRENT resulting from a changing electric field can in turn produce a magnetic field, which inherently changes in a synchronized fashion. EM waves propagate with the SPEED OF LIGHT in the respective MEDIUM. EM waves range from ultra-short GAMMA RAYS to long-wave RADIO WAVES. Visible LIGHT is one specific segment of the total EM SPECTRUM. EM waves are identified by either FREQUENCY (ν) or WAVELENGTH (λ) as all waves (*see* ELECTROMAGNETIC RADIATION) (see Figure E.40).

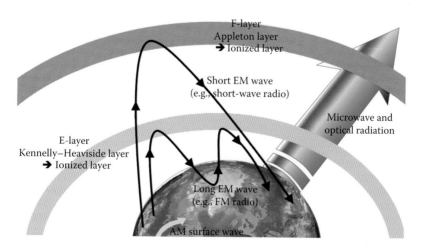

Figure E.40 Color as a function of the electromagnetic wave reflection or absorption.

Electromotive force (ε_{emf})

[general] Based on the Faraday's law of induction a changing magnetic FLUX can generate a voltage known as the "electromotive force." The MAGNETIC FLUX (Φ_M) is a function of both the magnetic field MAGNITUDE (B) and the area (A) that captures the fieldlines: $\varepsilon_{emf} = -N(\partial\Phi_M/\partial t) = -N((\partial\vec{B}\cdot\vec{A})/\partial t)$, where the wire loop has N windings. Additional contributions to the electromotive force are chemical potential and thermal effects, which yield for the total electromotive force: $\varepsilon_{emf} = \oint_C(E+\vec{v}\times\vec{B})d\ell + (1/q)(G/[C])(1/F) + \varepsilon_{emf,Therm}$. The chemical potential can be found by using the NERNST EQUATION based on the reaction processes, or from the HALF-CELL potential using the Gibbs free energy (G), with the current density J, expressed as $G = q\varepsilon_{emf,Chem}[C]F$, where $[C]$ is the chemical concentration with charge q in a volume V and F the FARADAY CONSTANT. The thermal electromotive force is more difficult to find and has as foundation the heat of the reaction Q_{chem}, while the temperature is determined by the chemical reaction or influences the chemical reaction as a catalyst, defined as $Q_{chem} = -q[C]F(\varepsilon_{emf,Therm} - T(\partial\varepsilon_{emf,Therm}/\partial T))$. *Also see* **SEEBECK EFFECT** (named after the German scientist THOMAS JOHANN SEEBECK (1770–1831), who introduced this in 1821) or thermoelectric effect, and *see* **PELTIER EFFECT**. Other factors may also contribute to the total electrical potential.

Electromyogram

[biomedical, electromagnetism, imaging] MEASUREMENT and display of the electrical DEPOLARIZATION as the initiation process for muscular ACTIVITY, that is, contraction. Depolarization sequence of the MUSCLE measured by electrodes placed on the SKIN or by means of ELECTRODE needles inserted in the muscle. The electromyogram will rely on volume conduction and spatial RESOLUTION is dependent on the number of electrodes and their respective placement (see Figure E.41).

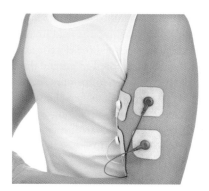

Figure E.41 Recording of skeletal muscle depolarization with electrodes place on the skin, specifically the biceps.

Electron (e−)

[atomic, energy, general, nuclear, solid-state] Atomic elementary PARTICLE, FERMION in the category of LEPTON. A negatively charged particle which is a constituent of every ATOM. Electrons are subject to the PAULI EXCLUSION PRINCIPLE due to their half-integer SPIN. The fermion nature forms the foundation of the atomic structure of atoms consisting of multiple electrons, which covers all ELEMENTS apart from hydrogen: ^{1}H (*also see* E).

Electron capture

[atomic, nuclear] Specific nuclear DECAY mode in which an orbital ELECTRON (e^-) is captured by the NUCLEUS as part of one of the options in a beta decay process with simultaneous emission of either a NEUTRINO or an antineutrino. Both the NEUTRINO (v') and the ANTINEUTRINO ($\bar{v}'$) were postulated by WOLFGANG PAULI (1900–1958) in 1931 to account for the energy balance in the decay process and the angular momentum conservation. The electron capture is favored by PROTON (p) rich nuclides, next to β^+ emission, while NEUTRON rich nuclides will predominantly have β^- emission. The process is schematically represented as: $p + \bar{e} \rightarrow n + v'$.

Electron decay

[nuclear] *See* BETA DECAY.

Electron drift velocity

[electronics, general] Migration VELOCITY (v_e) of a single electron in a CONDUCTOR under the influence of the local temperature (T), the applied ELECTRIC FIELD (E) resulting from the POTENTIAL DIFFERENCE (V): $E = Vd$, applied over a DISTANCE d. Additional factors are the material composition, such as the electrical density (defined by the MOLAR MASS M and physical density ρ) and charge distribution within the conductor, defined by the equivalent RESISTANCE: resistivity ρ_R, temperature coefficient of resistivity α_T, the respective length over which the interaction takes place ℓ, and the number of free electrons that can potentially be released from the prevailing ATOM (the average electron release) f_e, expressed as $v_e = MV/(dN_a \ell e f_e \rho_R (1 + \alpha_T T))$, with Avogadro's number: $N_A = 6.022137 \times 10^{-23}$ J/K and electron charge: $e = 1.60217657 \times 10^{-19}$ C (*also see* ELECTRIC CURRENT).

Electron energy state

[electromagnetism, energy, general, thermodynamics] The electron-binding states under atomic and molecular configurations under equilibrium conditions for a FLUID or solid are characteristic for the ISOLATED SYSTEM they belong to. The GROUND STATE energy configuration is generally sharp for a fluid, and relatively broadened on the outer shells for a solid. The respective energy states have associated WAVE functions that follows from the SCHRÖDINGER WAVE EQUATION. For solids, the energy configuration for crystalline structures is the most stable and most predictable, in contrast to the more amorphous energy band structure in glassy substances. The energy structure in crystals is considered an ordered solid where the Hamiltonian for the SCHRÖDINGER EQUATION ($H\Psi = E\Psi$) has periodic solutions $H(\vec{r} + \vec{R_i}) = H(\vec{r})$, with fundamental solutions $\Psi(\vec{k}, \vec{r}) = \Psi(\vec{r}) e^{i\vec{k} \cdot \vec{R_i}}$, with k the wave vector for the virtual wavelength "λ" of the SOLUTION, considered a QUANTUM NUMBER for the electron in this case, partially governed by the Fermi–Dirac rules. For metals, the highest energy band (VALENCE BAND) provides a CONTINUUM with solutions $E_n(\vec{k_F}) = E_F$, where $\vec{k_F}$ is the Fermi energy wave factor and E_F the Fermi energy associated with the Fermi band of the electron distribution. The Gibbs energy equivalent (G) for the electron conglomerate filling the BRILLOUIN ZONE within the unit volume of a CELL outlining a single unit (i.e., ATOM or molecule: V_Ω) can be shown to satisfy $G(E) = (V_\Omega/4\pi^2)(m^*/\hbar^2)^{3/2} E^{1/2}$, with m^* being the effective mass for the grouping, $\hbar = h/2\pi$:: where $\hbar = h/2\pi$, with $h = 6.626070040(81) \times 10^{-34}$ m^2kg/s (Js) the Planck's constant, and all energy states in the Brillouin zone are equal: $E = E_n(\vec{k})$. The description of the energy configuration for a disordered system is complex and can be solved analytically only based on the Bloch theorem conditions. In case the periodicity is not presented, the system becomes ill-defined because the wavevector no longer has a sharp value. Chaotic systems such as alloys and heated liquids are not confined due to the free exchange of electrons between states and between systems.

Electron lens

[atomic, nuclear] Magnetic LENS used in ELECTRON MICROSCOPE to bend the path of electrons to achieve a focal irradiation (see Figure E.42).

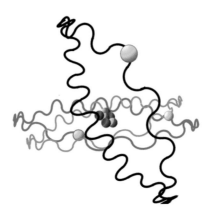

Figure E.42 Schrödinger wave electron energy orbital fit to energy band.

Electron microscope

[atomic, biomedical, imaging, nuclear] Imaging system that uses ballistic electrons (mass: $m_e = 9.10939 \times 10^{-31}$ kg; Electron charge: $e = -1.60217657 \times 10^{-19} C$) in a VACUUM to engage in a SCATTERING interaction on a molecular level. Electrons are ELEMENTARY PARTICLES with NEGATIVE CHARGE discovered by JOSEPH JOHN THOMSON (1856–1940) in 1897 (for which he received the Nobel Prize in PHYSICS in 1906). These particles are composed from the CATHODE RAYS. Two specific elementary types of electron microscopy are available: transmission electron microscopy (TEM) and scanning reflection electron microscopy (SEM), with various versions. The electron microscope used the DE BROGLIE WAVELENGTH ($\lambda = h/p$, where PLANCK'S CONSTANT: $h = 6.626070040(81) \times 10^{-34}$ m²kg/s (Js), and $p = m_e v$ is the electron momentum, where m_e is the electron mass moving with velocity v, recognized by PRINCE LOUIS VICTOR PIERRE RAYMOND DUC DE BROGLIE [1892–1987] in 1924) of a fast moving electron to obtain high-resolution imaging based on the RAYLEIGH CRITERION: $D_{res} = D/1.22\lambda$, where D is the characteristic APERTURE of the imaging device. In 1927, Hans Busch (1884–1973) provided the theoretical evidence that electron beams can be focused by means of an inhomogeneous MAGNETIC FIELD. In his work, he described how the focal length of such a magnetic ELECTRON LENS could be continuously adjusted through adjustment of the coil current. In 1931, Ernst Ruska (1906–1988) and Max Knoll (1897–1969) constructed a magnetic lens, confirming the theoretical premise and continued to construct the first TEM instrument. In 1938, MANFRED VON ARDENNE (1907–1997) constructed a scanning TEM by adding additional coils that provide a steering mechanism to a TEM. In 1942, Vladimir Kosmo Zworykin (1889–1982), James Hillier (1915–2007), and R.L. Zinder (?–?) developed an electron microscope based on SCATTERING, primarily BACKSCATTER, eliminating the need for extremely thin samples. Backscatter describes the SEM. The scattering events result from the WAVE phenomenon according to the Bragg law describing the interaction with "planes" of atoms, which provides the relation between interplanar DISTANCE d and diffraction ANGLE θ as $n\lambda = 2d \sin \theta$. The wavelength of the electron beam is a direct result from the accelerating POTENTIAL DIFFERENCE (V) in the electron microscope, expressed as $\lambda = h/\sqrt{(2m_e eV)}$, where $h = 6.62606957 \times 10^{-34}$ m²kg/s the Planck's constant. At the acceleration voltages used in TEM, relativistic effects have to be taken into account, providing: $\lambda = h/\sqrt{\{2m_e eV[1+(Ev/(2m_e/C^2))]\}}$, with $c = 2.99792458 \times 10^8$ m/s the speed of light. Just as a light MICROSCOPE, the electron microscope consists of a source collimators and lenses, and additionally requires an detector system for the collection of the electron and subsequent processing for IMAGE reconstitution. The mechanism of operation is purely electronic/magnetic, in contrast to the GLASS/POLYMER lenses in an optical microscope. The control mechanism for the electrons is based on Lorentz force: $\vec{F} = -e\left(\vec{E} + \vec{v} \times \vec{B}\right)$, where $\vec{B}$ represents the applied magentic field and $\vec{E}$ is the electric field vector. Based on the inherent process of focusing the resolution (D_{res}) is described based on the electronic configuration of the objective "LENS" that can be defined by a device specific SPHERICAL ABERRATION constant (C_s) and a general NORMALIZATION constant A as: $D_{res} = AC_s^{1/4}\lambda^{3/4}$. The magnification attainable with SEM is in the order of $100,000 \times$ and respectively for the TEM in excess of $1,000,000 \times$ (compared to an effective magnification in the order of $1000 \times$ with an optical microscope). Whereas an SEM primarily provides topographical information with RESOLUTION in the order of 10 nm, the TEM can show details about cellular composition, in biological imaging, with resolution better than 0.24 nm. The sample preparation requirements for TEM are much more strict than for SEM, down to 50 nm, whereas SEM can be used to observe intact animal specimen (see Table E.1).

Table E.1 Electron wavelengths at some acceleration voltages used in TEM

V_{acc}[kV]	B[PM]	A[PM]
100	3.86	3.70
200	2.73	2.51
300	2.23	1.97
400	1.93	1.64
1000	1.22	0.87

V_{acc}: accelerating voltage; A: non relativistic wavelength; B: relativistic wavelength.

The interaction process of the electron with the medium under investigation relies on several mechanisms of action (dominated by Coulomb forces), ranging from COMPTON SCATTERING to elastic and inelastically (consequence: increase in wavelength) scattered as well as AUGER ELECTRONS and gamma RADIATION emission, next to unscattered transmitted electrons. The gamma and X-RAY emission resulting from atomic and nuclear excitation can provide spectral information referring to the internal energy configuration. Factors influencing the electron penetration depth and the interrogated volume are the angle of incidence, the current MAGNITUDE and the accelerating voltage of the beam, next to the material properties such as the average ATOMIC NUMBER (Z) of the sample as well as the localized specific atomic number distribution. The excitation volume is punch-ball-shaped region with penetration ranging from 1 to 5 μm. Because of the potential for interaction of the electrons with any molecule or ATOM, electron microscopy generally needs to be performed under low to high VACUUM conditions. The imaging pressure ranges from 10^{-4} Pa to 10^{-9} Pa, while the electron gun may operate under 10^{-11} Pa, but can be as low as 10^{-7} Pa for modern microscopes. One aspect of electron microscopy imaging is sample preparation, making the "non-conductive" sample susceptible to electron interaction. The sample preparation generally involves dehydration and coating with aluminum, gold, or silver (see Figure E.43).

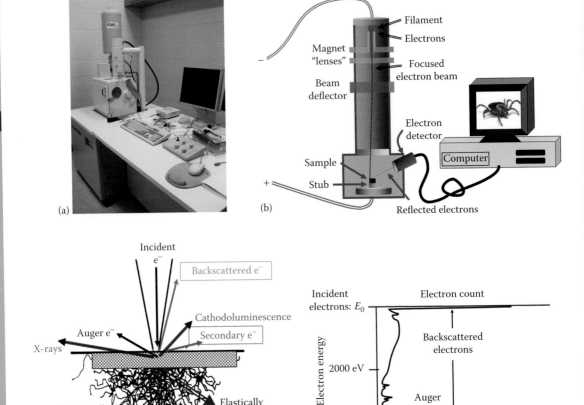

Figure E.43 (a) Diagram of the operation of a scanning electron microscope, (b) electron scattering distribution, (c) electron energy configuration, and (d) influence of surface contour on electron re-emission distribution.

Electron shells

[general] *See* **BOHR ATOMIC MODEL.**

Electron spin

[electromagnetism, energy, general, thermodynamics] Every electron has an associated SPIN, magnetic moment as well as mass and charge. The magnetic moment is defined with respect to the spin $\vec{s} = \hbar s$ for an electron with spin QUANTUM NUMBER s (a half-integer correlated to the angular momentum; note that particles with half spin are fermions and particles with whole integer spin are bosons) and where $\hbar = h/2\pi$, which is Planck's constant, $h = 6.626 \times 10^{-34}$ J/Hz divided by 2π as $\overrightarrow{\mu_{mag}} = -g_e \beta_m \vec{s}$, where $g_e = 2.0023$ the Landé factor, $\beta_m = \hbar |e|/2m_e$ ($m_e = 9.10939 \times 10^{-31}$ kg the electron mass) the Bohr magneton (see Figure E.44).

Figure E.44 The two possible electron spin configurations with $s = +(1/2)$ and $s = -(1/2)$.

Electron spin resonance

[electromagnetism, energy, general, thermodynamics] An electron can be placed in an external magnetic field with a steady-state value $(\vec{B_0})$ and a perturbation $\vec{B}$ (oscillation). The steady-state magnetic field provides the "Zeeman energy" of the electron as: $W_Z = -\overrightarrow{\mu_{mag}} \cdot \vec{B_0}$, with the magnetic moment $\overrightarrow{\mu_{mag}} = -g_e \beta_m \vec{s}$, where $g_e = 2.0023$ the Landé factor or g-factor, $\beta_m = \hbar |e|/2m_e$ ($m_e = 9.10939 \times 10^{-31}$ kg the electron mass) the Bohr magneton, and $\vec{s} = \hbar s$ the spin (in this case, specifically for an electron with spin quantum number s and where $\hbar = h/2\pi$, which is Planck's constant, $h = 6.626 \times 10^{-34}$ J/Hz divided by 2π). This situation has two stationary states: α, associated with the quantum number for the spin projection $m_s = +1/2$ and β, associated with the quantum number for the spin projection $m_s = -(1/2)$. The steady-state magnetic field induces precession with angular velocity: $\omega_0 = \gamma_e \vec{B_0}$, where $\gamma_e = g_e |e|/2m_e$ is the gyromagnetic ratio. Under the influence of a homogenous magnetic field, the electrons will precess in phase. When a perturbation with fixed frequency ($\nu = (\omega/2\pi) = (c/\lambda)$), with wavelength λ and the speed of light: $c = \sqrt{\mu_0 \varepsilon_0}^{-1} = (\mu_0 \varepsilon_0)^{-(1/2)} = 2.99792458 \times 10^8$ m/s) is applied with frequency ν_0 that matches the resonance condition: $h\nu_0 = g_e \beta_m B_0$ the transitions between the α and the β state will flip in resonance.

Electron tube

[electromagnetism, energy, general, thermodynamics] It is the predecessor of the transistor. The vacuum tube device was introduced by Thomas Alva Edison (1847–1931) in 1883 and was refined by Sir John Ambrose Fleming (1849–1945) in 1904. The vacuum tube has an anode (+) and a cathode (−) as well as a set of conductive plates that can influence the migration of cathode rays. The original cathodes were filaments that emitted electrons on heating by means of passing a current through the filament. Several electron tubes (also known as "vacuum tubes") were made to perform a variety of electronic functions, such as diode, triode, transistor, photo-multiplier, and a high-frequency triode (transistor equivalent) called the "pentode." The triode has gain (amplification) that results for the effective internal electronic configuration with an equivalent circuit representing induction, resistance, and capacitance. The photomultiplier tube is also referred to as "photocell" or "photoemissive tube," which uses the electrode work function in response to external electromagnetic irradiation to induce a current that provides the means to quantify the radiance incident on the tube.

Electron volt (eV)

[general] It is the amount of energy gained by an electron charge (either positive or negative) in passing through a potential difference of 1 V.

Electron wake

[atomic, nuclear] Variations in electron density for the free-electron distribution in a CONDUCTOR, semiconductor as well as INSULATOR after the passage of an ION with charge (Ze) traveling with high VELOCITY (v) through the SOLID-STATE medium (condensed MATTER). The WAKE will be formed with a spatial resonant frequency (ω) with spatial period: $d = 2\pi v/\omega_0$, also known as the "WAVELENGTH" (λ). The local AMPLITUDE of the wake OSCILLATION has a maximum just within half a wavelength from the trajectory amounting to: $V_{max} = Ze(\omega_0/4\varepsilon_0 v)$, rapidly dampened by INELASTIC COLLISIONS of charges. The amplitude of the wake reduces to e^{-1} within 10λ behind the traveling ion and laterally $r_{e^{-1}} = v/\omega_0$. For conducting media the resonant frequency (ω_0) is the PLASMA FREQUENCY of the METAL: $\omega_p = \sqrt{(\rho_{e,0}e^2/\varepsilon_0 m'_e)}$, with $\rho_{e,0}$ the electron density, $e = 1.60217657\times10^{-19}\,C$ the electron charge, m'_e the effective electron mass, and $\varepsilon_0 = 8.85419\times10^{-12}\,C^2/Nm^2$ the permittivity of free space. At speeds greater than the Bohr speed (i.e., $v_M = 2.2\times10^6$ m/s, which is the velocity of an electron in the hydrogen GROUND STATE in the Bohr model) the atoms in the path are stripped of their VALENCE electrons, thus corresponding to an equivalent Mach speed. Under these conditions, the created ions are freely migrating. The ion wake formed under these conditions is subject to INTERFERENCE of traveling ion clusters. The Coulomb forces between the heavy ions creates conditions that support collisions and an "explosion" will be effected (known as a "Coulomb explosion"). In analog-to-fluid dynamic principles the ion will also generate a "bow" WAVE of electrons in front of the ion location, spreading radially from the ion location on the trajectory.

Electronegativity

[atom, chemical, general, solid-state] The ability of a molecular ATOM to attract electrons and bind to them. This phenomenon was introduced by LINUS CARL PAULING (1901–1994) in 1932. The electronegativity is a relative indication of the electron attraction within the molecule. The electronegativity is considered between two atoms and is a function of the intermolecular bonds (strong bonding within the respective atoms) and intramolecular bonds (covalent binding between atoms 1 and 2). The bonding energy is measured by dissociation energy (E_d; measured in eV) for the respective attractions. The difference in electronegativity between $atom_1$ (χ_1) and $atom_2$ (χ_2) is defined as the weighted average for all chemical bonds as: $\chi_1 - \chi_2 = C\sqrt{(E_{d:12} - ([E_{d:11} + E_{d:22}]/2))}$, where C is a unit NORMALIZATION constant. Fluorine is the most electronegative ELEMENT (3.98 on the Pauling electronegativity scale). The definition was later refined by Robert Sanderson Mulliken (1896–1986). In 1934, Robert Mulliken suggested the use of IONIZATION ENERGY (E_i) and electron affinity (E_{ea}) for a single element, hence defining an absolute scale: $\chi = (E_{ea} + E_i)/2$. The final revision was made in 1989 by Leland Cullen Allen (1926–2012) using the VALENCE energy of the free electrons in the respective element using the one-electron s- and p-energy for the isolated atom (ϵ_s and ϵ_p) and the respective number of electrons in either state (n_s and n_p): $\chi = c'((n_s\epsilon_s + n_p\epsilon_p)/(n_s + n_p))$, where c' is a normalization scaling factor (energies expressed in J vs. eV).

Electronics

[electromagnetism, energy, general, thermodynamics] Configuration of devices based on components that can manipulate the MAGNITUDE and frequency (spectrum) of electron currents. Electronics provides the basis for the production of test and MEASUREMENT devices as well as television and RADIO. Electronics uses components ranging from resistors, capacitors, and inductors to transistors and INTEGRATED CIRCUITS. Electronics may involve the specific design of a device, the operating conditions and output parameters versus input parameter requirements as well as designing a complex diagram connecting multiple components on a circuit to provide a specific outcome based on the available input parameter and conditions (see Figure E.45).

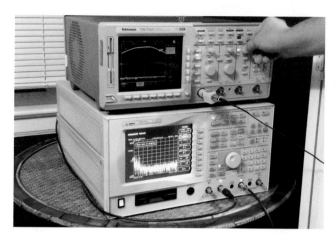

Figure E.45 Example of electronics: oscilloscope (top) and electromagnetic field spectrum analyzer (bottom).

Electro-optical effect

[electromagnetism, energy, general, thermodynamics] *See* **Kerr effect**.

Electrophysiology

[biomedical, chemical, electronics] The biological functionality associated with cellular function. Specific functions are the excitation and contraction process for the HEART muscle, nerve excitation and propagation, and SKELETAL MUSCLE coordination and contraction, next to brain ACTIVITY and a broad range of chemical and electronic processes in the human and animal body that can be measured with scientific tools, quantified, and theoretically modeled. The electrical, chemical, and physiological functions associated with these functions range from a single cellular MEMBRANE function to intercellular and intracellular communications and conduction and anatomical composition. In most cases, the data recording will need to be performed noninvasively to preserve the integrity of the phenomenon and to avoid interfering with the biological and physical responses. The measured data can be used to TRACE back the anatomical origin of certain signals and the initiations source process. The diagnostic mechanism and the theoretical analysis can provide valuable information about the source of a deviation in the anticipated biological system; taking into account that the biological system is an integral system, all factors are communicating with each other and subsequently influence each other. One striking example is the fright and flight response where a sensory input (e.g., flash or light, LOAD NOISE) is converted in a chemical release (e.g., adrenaline) followed by a range of activities that assist in a removal from suspected harm. Those activities may include an increase in heart rate, increased VENTILATION rate, skeletal muscle contraction, and enhanced sensory attention (supposed redirection of the emphasis of brain activity). Examples are the recording of the ECG, EMG, EEG, as well as specific functional SIGNAL acquisition such as VISION, HEARING, touch, smell, and taste; for animals with special sensory devices, the range extends to MEASUREMENT of the muscle contraction of a potential pray (e.g., SHARK) (*also see* VOLUME CONDUCTION) (see Figure E.46).

Figure E.46 Electrophysiologic monitoring of heart rate (ECG), respiration rate, blood oxygenation, brain function (EEG), and additional vital signs at the bed side of a patient.

Electroscope

[atomic, general, nuclear] Device used to visualize the ELECTRIC CHARGE content on a device either by POLARIZATION or by charge transfer. Generally, the device consist of a VACUUM chamber with a movable leaf that is in static and dynamic equilibrium and can move freely around a central axis. Once electric charge is applied, the leaf and the fixed counterpart plate are charged with the same charge and will create a repulse force between the like charged panels, forcing the leaf on the axis to turn away from its counterpart, also found as "electrometer" (see Figure E.47).

Figure E.47 Electroscope used to provide indication of static charge on a device by touching it with the electrode from the electroscope.

Electrostatic field

[electronics, general] Electric field line density terminating on or merging from an ELECTRIC CHARGE distribution. The electrostatic field is stationary and steady-state and obeys Coulomb's law $E_e = Q/4\pi\varepsilon r^2$, with Q being the total charge assembled in a point of reference, where outside the charge distribution, the charge may be considered in the center of the system, measuring the DISTANCE as the radius from the center r and $\varepsilon = \varepsilon_r\varepsilon_0$, with ε_r being the material constant RELATIVE PERMITTIVITY and $\varepsilon_0 = 8.85419 \times 10^{-12}\,C^2/Nm^2$ the permittivity of free space. The field inside the charge CARRIER will be zero for a CONDUCTOR. For an INSULATOR, the field changes with the volumetric charge distribution and follows from GAUSS's LAW, calculating the electric FLUX through the surface enclosed up to the point of interest. The charge is now expressed as the VOLUME CHARGE DENSITY (ρ_e; for a sphere with radius R: $\rho_e = Q/((4/3)\pi R^3)$, assuming a homogeneous charge distribution), with respect to the point of interest at location r ($r \le R$) from the center of the charge distribution. The electric field in the insulator is now: $E_e = Qr/4\pi\varepsilon R^3$, keeping in mind that the charge outside the GAUSSIAN SURFACE does not provide in inward electric field. Note that at the boundary this reverts to the original electric field postulated by CHARLES-AUGUSTIN DE COULOMB (1736–1806) (see Figure E.48).

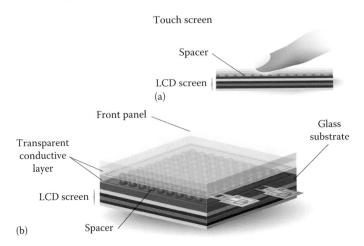

Figure E.48 Illustrations of two kinds of static electric field configurations: (a) touch screen that relies on capacitive charge for on/off commands and (b) the formation of lightning by means of the build-up of a static electric field between the clouds and the Earth's surface, alternatively between respective clouds.

Electrostatic force

[electromagnetism, general] A force between two static and steady-state charges q_1 and q_2 respectively at a separation DISTANCE r: expressed as: $F_e = q_1 q_2 / 4\pi\varepsilon r^2 = q_2 E_e$, where $\varepsilon = \varepsilon_r \varepsilon_0$, with ε_r the material constant RELATIVE PERMITTIVITY and: $\varepsilon_0 = 8.85419 \times 10^{-12} \, C^2/Nm^2$ the permittivity of free space, E_e the electric field (*see* ELECTRIC FORCE).

Electrostatic generator

[general] *See* VAN DER GRAAFF GENERATOR *and* TESLA COIL.

Electrostatics

[electromagnetism, energy, general, thermodynamics] A field of science and ENGINEERING dealing with the effects of steady-state charge distributions, involving positive and NEGATIVE CHARGE configurations. All phenomena can be described by coulomb fields and coulomb forces. Static ELECTRICITY forms a special case in electrostatics. The branch of electricity science that applies to objects and electrical phenomena at rest, in contrast to ELECTRODYNAMICS. Static electricity is one of the most well-known examples of electricity at rest.

Electrostriction

[thermodynamics] A decrease in volume associated with an increase in electric field strength in a specific regional volume within a thermodynamic system volume operating at constant temperature. The Gibbs energy (G) for the confined volume (V_c) within the larger VOLUME (V) is expressed as: $G = U - TS + PV - V_c\epsilon^*\mathcal{E}^2$, where U is the internal energy, S is the entropy of the system, P is the pressure, $\mathcal{E}$ is the uniform ELECTROSTATIC FIELD strength in the volume V_c and $\epsilon^* = \epsilon^*(T,P)$ the DIELECTRIC constant in the volume. For a system with n constituents and in this geometry and with the specified configuration, the following two conditions apply illustrating the electrostriction phenomenon: $(\partial S/\partial\mathcal{E})_{T,P,V_c,n} = 2\mathcal{E}V_c\left(d\epsilon^*/dT\right)_P$ and $(\partial V/\partial\mathcal{E})_{T,P,V_c,n} = -2\mathcal{E}V_c\left(d\epsilon^*/dT\right)_T$.

Electrotonic length (λ_ℓ)

E

[biomedical, electromagnetism] In a myelinated nerve CELL (axon), this is the length covered by a Schwann cell in electronic definition. Over this length, no ION exchange is possible due to the insulation and hence has no local action-potential: $\lambda_\ell = \sqrt{r_1^2\pi R_m/2\pi r_1 R_\ell} = \sqrt{r_1 R_m/2R_\ell}$, with r_1 being the radius of the axon, R_m the MEMBRANE resistance, and R_ℓ the linear conductive RESISTANCE over the length of the Schwann cell. Because in a MYELINATED NERVE cell, the SIGNAL "jumps" from NODE OF RANVIER to another node of Ranvier over the segment covered by the Schwann cell the DEPOLARIZATION potential needs to be larger for long electrotonic length than for a short electrotonic length (e.g., unmyelinated axon) to sustain the depolarization to be carried over the long segment through conduction instead of depolarization. Note that resistive conduction is a passive process and is lossy, whereas depolarization is an active process and is lossless. The determination of the depolarization potential and the electrotonic length is an important factor when considering multiple sclerosis, which destroys the myelin.

Electroweak epoch

[astronomy/astrophysics, general, quantum] A time frame in the period from 10^{-34} to 10^{-11} s after "Genesis" in the Big Bang creation of the UNIVERSE playbook where the universe "consisted" in a small volume with only leptons, bosons photons, quarks, vector bosons, and gluons, and the interaction was dominated by the ELECTROWEAK FORCE under an approximate temperature of 3×10^{15} K. As the universe cooled, the electroweak force diminished, the vector bosons acquired mass, and the FOUR FORCES became the ruling interaction: GRAVITY, WEAK FORCE, electromagnetic force, and strong force. Note that the vector bosons W^-, W^+, and Z^0 were experimentally verified in 1983.

Electroweak field

[general] Convoluted merger of the "weak" force and ELECTROMAGNETIC FIELD (*also see* HIGGS FIELD).

Electroweak force

[general, high-energy] Force structure based on the nonrenormalizable WEAK INTERACTION theory of ENRICO FERMI (1901–1954), postulated in 1932. The neutron DECAY defined under this needed additional details, which was added by HIDEKI YUKAWA (1907–1981), a theoretical physicist from Japan in 1934, introducing a "messenger particle." The physicist Julian Seymour Schwinger (1918–1994) from the United States introduced the W-vector bosons as the representative messenger PARTICLE in 1956 under the developing GAUGE THEORY. The electroweak force defines the interaction between quarks, by definition, transforming the "flavor" without changing the "COLOR." The WEAK FORCE depends on the handedness of the particles involved in the interaction process as well as the alignment, antialignment of the linear momentum and SPIN angular momentum for the fermions, and quarks and leptons involved.

Electroweak theory

[general] QUANTUM gauge theoretical incorporating four-gauge QUANTA; the vector bosons W^-, W^+, Z^0 as well as the forth massless force CARRIER the PHOTON. This theory was designed by SHELDON LEE GLASHOW (1932–), a physicist from the United States, STEVEN WEINBERG (1933–), a scientist from the United States, and Abdus Salam (1926–1996), a physicist from Pakistan (initially working independently but merging thoughts; he shared the Nobel Prize in PHYSICS on this), based on the work of Gerhard van't Hooft (1947–), a graduate student from the Netherlands, with respect to the Yang-Mills gauge fields performed in 1969. The unified force acts between leptons and between hadrons over an effective range in the order of 10^{-17} m. The electroweak theory combines the "WEAK FORCE" and the "electromagnetic force." The vector bosons have three CHARM conditions (–1, 0, and 1) also referred to as v, μ, and τ (see GAUGE THEORY). The gauge is the phenomenon part of the interaction and is defined by $\mathcal{L}_{gauge} = -(1/4)F^i_{\mu\nu}F^{\mu\nu i} - (1/4)B_{\mu\nu}B^{\mu\nu}$, where $i = 1, 2, 3$, with the field strength tensors defined by the respective "MAGNITUDE" ("W^{ll}_{kk}") "PHASE" ("B^{ll}_{ii}") as $F^i_{\mu\nu} = \partial_\mu W^i_\nu - \partial_\nu W^i_\mu + g^* \epsilon_{ijk} W^j_\mu W^k_\nu$ (j, $k = 1, 2, 3 \cdots$), with W^{xx}_{yy} representative of the "weak force," g^* the gauge-coupling coefficient for condition $SU(2)(U(1))$ linking the "phases" W^0 of the W vector bosons to the "phases" for B in gauge classification $U(1)$ as referenced by the weak neutral current: $\mathcal{L}_{SU(2)\times U(1)} = \mathcal{L}_{gauge} + \mathcal{L}_\phi + \mathcal{L}_f + \mathcal{L}_{Yukawa}$, ϵ_{ijk} a symbol that provide total asymmetry in the gauge configuration, and $B_{\mu\nu} = \partial_\mu B_\nu - \partial_\nu B_\mu$.

Element

[atomic, chemical, energy, quantum] A pure substance consisting of ATOMS of the same ATOMIC NUMBER, which cannot be subdivided by ordinary chemical means. The designation "element" was introduced in 1661 by ROBERT BOYLE (1627–1691). Evidence leading to the definition of an element was based on the work by JOHN DALTON (1766–1844), who postulated the first CHEMICAL REACTION statement in 1808, including that every pure CHEMICAL is made up of same elementary components, forming the element foundation. The ELEMENTS themselves were also considered to be composed of a unique grouping of identical components, being atoms. Additional work in this field was made by AMEDEO AVOGADRO (1776–1856) in 1811, who concluded that not all elements consist of single atoms, some are biatomic (i.e., MOLECULES), which provided supporting evidence with regard to the chemical mass statements made by Dalton (see Figure E.49).

Periodic table of the elements

Atomic # → ⎤ 29 +2,1 ← Ions commonly formed
Atomic symbol → | Cu
English element name → ⎦ copper
63.55 ← Atomic mass (rounded)

Legend: **1** Gases | Liquids | **3** Metalloids

Period	1 I A																	18 VIII A
1	1s	1 ±1 H hydrogen 1.008																2 He helium 4.003

		2 II A										13 III A	14 IV A	15 V A	16 VI A	17 VII A	
2	2s	3 +1 Li lithium 6.941	4 +2 Be beryllium 9.012									5 +3 B boron 10.81	6 −4 C carbon 12.01	7 −3 N nitrogen 14.01	8 −2 O oxygen 16.00	9 −1 F fluorine 19.00	10 Ne neon 20.18

3	3s	11 +1 Na sodium 22.99	12 +2 Mg magnesium 24.31	3 III B	4 IV B	5 V B	6 VI B	7 VII B	8 VIII B	9 VIII B	10 VIII B	11 I B	12 II B	13 +3 Al aluminum 26.98	14 −4 Si silicon 28.09	15 −3 P phosphorus 30.97	16 −2 S sulfur 32.07	17 −1 Cl chlorine 35.45	18 Ar argon 39.95

| 4 | 4s | 19 +1 K potassium 39.10 | 20 +2 Ca calcium 40.08 | 3d | 21 +3 Sc scandium 44.96 | 22 +4,3,2 Ti titanium 47.87 | 23 +5,2,3,4 V vanadium 50.94 | 24 +3,2,6 Cr chromium 52.00 | 25 +2,3,4,6,7 Mn manganese 54.94 | 26 +3,2 Fe iron 55.85 | 27 +2,3 Co cobalt 58.93 | 28 +2,3 Ni nickel 58.69 | 29 +2,1 Cu copper 63.55 | 30 +2 Zn zinc 65.41 | 31 +3 Ga gallium 69.72 | 32 +4,2 Ge germanium 72.64 | 33 −3 As arsenic 74.92 | 34 −2 Se selenium 78.96 | 35 −1 Br bromine 79.90 | 36 Kr krypton 83.80 |

| 5 | 5s | 37 +1 Rb rubidium 85.47 | 38 +2 Sr strontium 87.62 | 4d | 39 +3 Y yttrium 88.91 | 40 +4 Zr zirconium 91.22 | 41 +5,3 Nb niobium 92.91 | 42 +6,3,5 Mo molybdenum 95.94 | 43 +7,4,6 Tc technetium 98 | 44 +4,3,6,8 Ru ruthenium 101.1 | 45 +3,4,6 Rh rhodium 102.9 | 46 +2,4 Pd palladium 106.4 | 47 +1 Ag silver 107.9 | 48 +2 Cd cadmium 112.4 | 49 +3 In indium 114.8 | 50 +4,2 Sn tin 118.7 | 51 +3,5 Sb antimony 121.8 | 52 −2 Te tellurium 127.6 | 53 −1 I iodine 126.9 | 54 Xe xenon 131.3 |

| 6 | 6s | 55 +1 Cs cesium 132.9 | 56 +2 Ba barium 137.3 | † 5d | 71 +3 Lu lutetium 175.0 | 72 +4 Hf hafnium 178.5 | 73 +5 Ta tantalum 180.9 | 74 +6,4 W tungsten 183.8 | 75 +7,4,6 Re rhenium 186.2 | 76 +4,6,8 Os osmium 190.2 | 77 +4,3,6 Ir iridium 192.2 | 78 +4,2 Pt platinum 195.1 | 79 +3,1 Au gold 197.0 | 80 +2,1 Hg mercury 200.6 | 81 +1,3 Tl thallium 204.4 | 82 +4,2 Pb lead 207.2 | 83 +3,5 Bi bismuth 209.0 | 84 +4,2 Po polonium 209 | 85 At astatine 210 | 86 Rn radon 222 |

| 7 | 7s | 87 +1 Fr francium 223 | 88 +2 Ra radium 226 | ‡ 6d | 103 +3 Lr lawrencium 262 | 104 Rf rutherfordium 261 | 105 Db dubnium 262 | 106 Sg seaborgium 266 | 107 Bh bohrium 264 | 108 Hs hassium 277 | 109 Mt meitnerium 268 | 110 Ds darmstadtium 281 | 111 Rg roentgentium 272 | 112 Cn copernicum 285 | 113 Uut ununtrium 284 | 114 Fl flerovium 289 | 115 Uup ununpentium 288 | 116 Lv livermorium 292 | 117 Uus ununseptium 293 | 118 Uuo ununoctium 294 |

lanthanides (rare earth metals) † 4f	57 +3 La lanthanum 138.9	58 +3,4 Ce cerium 140.1	59 +3,4 Pr praseodymium 140.9	60 +3 Nd neodymium 144.2	61 +3 Pm promethium 145	62 +3,2 Sm samarium 150.4	63 +3,2 Eu europium 152.0	64 +3 Gd gadolinium 157.3	65 +3,4 Tb terbium 158.9	66 +3 Dy dysprosium 162.5	67 +3 Ho holmium 164.9	68 +3 Er erbium 167.3	69 +3,2 Tm thulium 168.9	70 +3,2 Yb ytterbium 173.0

actinides ‡ 5f	89 +3 Ac actinium 227	90 +4 Th thorium 232.0	91 +5,4 Pa protactinium 231.0	92 +6,3,4,5 U uranium 238.0	93 +5,3,4,6 Np neptunium 237	94 +4,3,5,6 Pu plutonium 239	95 +3,4,5,6 Am americium 243	96 +3 Cm curium 247	97 +3,4 Bk berkelium 247	98 +3 Cf californium 251	99 +3 Es einsteinium 252	100 +3 Fm fermium 257	101 +3,2 Md mendelevium 258	102 +2,3 No nobelium 259

3 Metals

Figure E.49 Example of one outline for the periodic table of elements. (Copyright 2007–2012 Jeff Bigler.)

Elemental species

[chemical, thermodynamics] The complete set of independent ingredients of a chemical reaction. The respective independent constituents become involved in a reaction that conforms to the conservation of atomic nuclei and new constituents can, and will be formed during the reaction process (i.e., product), while the reaction cannot consist of only the primary constituents. An elemental species by definition is formed from a single group of atomic nuclei and the molecular structures formed during the reaction are the most stable kind. All process are considered at normal temperature ($T = 25\,^{\circ}\mathrm{C}$) and pressure ($P_0 = 1\,\mathrm{atm} = 1.01325 \times 10^5\,\mathrm{Pa}$).

Elementary particles

[atomic, electromagnetism, energy, general, high-energy, nuclear, solid-state, thermodynamics] Building blocks of atomic PHYSICS, with a subdivision to particles smaller than the electron, PROTON, and NEUTRON. The elementary particles are divided in leptons and hadrons. The leptons have two specific pairs: the ELECTRON (e^-)–electron-NEUTRINO (ν_e) pair and the MUON (μ^-)–muon-neutrino (ν_μ) pair, respectively. The following categorical discrimination in elementary particles can be made: first, second, and third generation elementary particles, which are subdivided in quarks, leptons, and HADRONS. The detailed subdivision is accordingly arranged per group. Hadrons are further subdivided in mesons and BARYONS. The mesons consist of PION and KAON particles, whereas the baryons are made up of the following: lambda (Λ^0), neutron (n), omega (Ω^-), proton (p), sigma (Σ^+, Σ^0, and Σ^-), and xi units (Ξ^0 and Ξ^-). Leptons have the following constituents: electron (e^-), muon (μ), NEUTRINO (e, ν_e; μ, ν_μ; τ, ν_τ), and tau (τ) entities. In turn, all particles also have a so-called anti-particle (*see* HADRON *and* LEPON).

Elements

[general, nuclear] A chemical ELEMENT is an elementary material with specific unique physical and chemical properties that are equal to the homogeneous substance consisting of a single element. The element is labeled by a single atomic constituent. All known elements are grouped in the PERIODIC TABLE OF ELEMENTS. At standard conditions, elements can be LIQUID, solid, or GAS. There are several classifications of elements, ranging from metals, metalloids, and nonmetals, to a more detailed groping in families of: alkali EARTH, alkaline Earth, transition METAL, rare Earth, other metal, metalloid, and nonmetal halogen to NOBLE GAS.

Ellis, Charles Drummond (1895–1980)

[general, nuclear] A scientist from Great Britain who designed a micro-calorimeter in 1927 in order to establish the energy exchange with respect to the NEUTRINO and beta DECAY for radium E decaying to polonium, and showed the potential for fundamental PARTICLE validation; however, he failed in the initial attempt. Nevertheless, the initial failure lead to the discovery of the neutrino based on the apparent energy loss. In 1933, WOLFGANG PAULI (1900–1958) suggested the new neutrino as the source of the missing energy, which as later identified by LISE MEITNER (1878–1968) during her repeated experiment in 1957. Charles Ellis' collaborator was WILLIAM ALFRED (PETER) WOOSTER (1903–1984) (see Figure E.50).

Figure E.50 Charles Drummond Ellis (1895–1980). (Courtesy http://nuclphys.sinp.msu.ru/.)

Elster, Julius Johann Phillipp Ludwig (1854–1920)

[geophysics, nuclear] A physicist from Germany. The work of Julius Elster involved the investigation of atmospheric ELECTRICITY phenomena and ionizing RADIATION. Julius Elster invented the photocell in collaboration with HANS FRIEDRICH GEITEL (1855–1923) in 1893 (see Figure E.51).

Figure E.51 Julius Johann Phillipp Ludwig Elster (1854–1920) and Hans Friedrich Geitel (1855–1923). (Courtesy Museum im Schloss Wolfenbüttel.)

Elster and Geitel experiment

[atomic, nuclear, quantum] Experimental verification of the theoretical description of electrification of RAIN drops by JULIUS JOHANN PHILLIPP LUDWIG ELSTER (1854–1920) and HANS FRIEDRICH GEITEL (1855–1923) in 1900. This work supported the novel general concepts of QUANTUM THEORY. Additional conclusions about RADIOACTIVITY could be derived from this work in the early stages of the discovery of radioactivity.

Embolism

[biomedical, fluid dynamics] Obstruction in the FLOW, particularly in BLOOD vessels. In MEDICINE, an EMBOLISM can be a blood clot, fat, plaque, or a BUBBLE that is formed rather abruptly. One form is the release of an AIR bubble (either during clinical procedures or as a result of the quick release of gasses from the vessel WALL under CAISSON DISEASE) that becomes trapped in a narrow diameter blood vessel, a part of thrombus, or plaque that becomes dislodged and occludes a distal vessel (see Figure E.52).

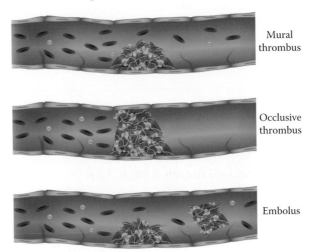

Mural thrombus

Occlusive thrombus

Embolus

Figure E.52 An embolism in a blood vessel.

Emission coefficient

[general] The emitted radiance from an object within a narrow wavelength bandwidth indicating the ability of a body to emit RADIATION as a fractional reference. A perfect emitter will have a emission coefficient $\varepsilon_\lambda = 1$.

Emission spectrum

[general, optics] The optical emission over a broad range of wavelengths from an excited GAS. The emission lines are representative of specific atomic transitions and can hence be used to identify the constituents as well as the energy configuration, such as the influence of neighboring atoms and molecules (see Figure E.53).

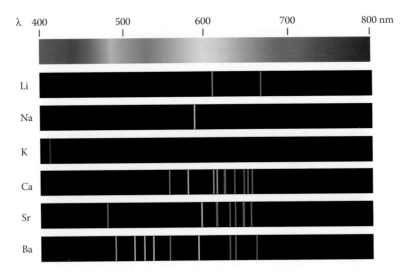

Figure E.53 Emission spectra in the visible range for several elements: lithium (Li), sodium (Na), potassium (K), calcium (Ca), strontium (Sr), and barium (Ba).

Emission theory

[general] Conceptual interpretation of light as a stream of infinitely small particles. Under this premise, the EYE emits light and captures that same beam of particles that is reflected from the objects in sight, documented as proposed by the Greek philosopher, mathematician, and scientist Plato (427–347 BC). The emission theory provided a plausible explanation for REFLECTION. It is also known as the "theory of extramission," where the eye emits a "probing beam" without external light. The EXTRAMISSION could however not account for the lack of VISION at night. The first attempt to discredit the emission theory or extramission was by ARISTOTLE (384–322 BC).

Emissivity

[atomic, energy, nuclear, optics] Emission ability relative to a BLACK BODY at the same TEMPERATURE (T) as a function of WAVELENGTH (λ) expressed as a fraction: $0 \leq \varepsilon_\lambda (T) \leq 1$. The emissivity relates to the Kirchhoff law introduced in 1859 (*also see* BLACK BODY RADIATION).

Empirical valence bond (EVB)

[biomedical, chemical, general, nuclear, thermodynamics] In biological systems, the conversion processes between various forms of energy often take place on an atomic level. One form in particular converts PHOTON energy into CHEMICAL ENERGY, for cellular consumption, specifically involving BACTERIORHODOPSIN (*also see* PN-JUNCTION, TRANSISTOR, MECHANOSENSITIVE ACTION, *and* MEMBRANE).

Endergonic process

[thermodynamics] A process where the energy of the initial state is lower than the energy of the final state, also known as "endoergic process."

Endocytosis

[biomedical, chemical] The active process of cellular consumption of both solid and LIQUID material, defined as PHAGOCYTOSIS (eating) and PINOCYTOSIS (drinking) next to TRANSCYTOSIS for large molecule transport. The CELL MEMBRANE forms an enclosure (vesicle) that engulfs the material of interest for consumption and transports the filled "BUBBLE" through the cell MEMBRANE. The vacuole will open (by means of chemical breakdown) once inside the cell to release the content or will be transported through the cell for release as a whole enclosure outside the cell intended for consumption by other cells. The other processes available for ingestion are attachment of molecules in specifically designed transmembrane pores that offer selective attachment to fatty substances (lipid, providing lipophilic transport mechanism) and WATER (hydrophilic molecules, loosely attaching to the water molecules for subsequent release under changing polarity resulting from chemical conditions, which are different in the intercellular environment from the extracellular milieu). The LIGAND gated channels providing molecular transport form a selective mechanism of chemical exchange. The transport of the lipophilic and hydrophilic molecules with LOAD attached, is primarily facilitated through chemical and electrical gradients across the cell membrane (e.g., facilitated DIFFUSION). Endocytosis also plays an important role in neural transmission in the space of a SYNAPSE (synaptic cleft), linking one or more nerve cell from axon to dendrite(s). The cellular membrane as well as intracellular organelles are composed of lipids and proteins arranged in a dynamic lipid bilayer. The term "dynamic" means that it can actively adjust to external conditions (e.g., stress and strain) as well as actively form a segment of bilayer that has a specific function (*also see* CELL MEMBRANE). Endocytosis is complemented by EXOCYTOSIS for the removal from the cell (see Figure E.54).

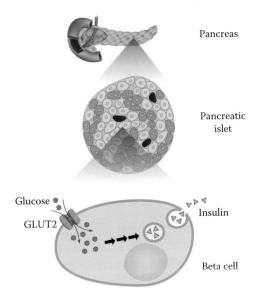

Figure E.54 The process of endocytosis highlighted by means of cartoon interpretations of the chemical and mechanical interactions at the cellular membrane.

Endoergic reaction

[atomic, chemical, condensed matter, nuclear] A reaction that absorbs energy. A process with a net positive change in Gibbs energy, specifically referring to nuclear reactions that absorb energy, also found as endergonic reaction.

Endoscope

[biomedical, general, optics] An optical device constructed with fiber-optics or LIGHT GUIDE to be introduced into the HUMAN BODY for examination of internal structures of organs (see Figure E.55).

Figure E.55 Illustrations of an endoscopic examination.

Endothermic process

[atomic, nuclear, thermodynamics] Chemical or thermodynamic process that requires the addition of heat in order for the process to initiate and proceed. One specific example of an endothermic process is vaporization (see Figure E.56).

Figure E.56 Example of endothermic process. The electrically powered nicotine distribution in the electronic cigarette with turbulent smoke (vapor).

End-point energy

[atomic, nuclear] Energy content of a system at the termination of nuclear DECAY. The endpoint energy is defined as the difference in energy between the initial and the final states.

Endurance

[biomedical, general, mechanics, solid-state] In biomedical applications, endurance refers to the ability to sustain a certain level of ACTIVITY, for example, running. In material science, endurance refers to the resilience of a material to repeated exposure to stress and strain. In both cases, the end result is FATIGUE, either MUSCLE fatigue or material failure. In material failure, there will be crack formation, which can be captured as a the growth of the length of the crack ℓ_c, which is a function of the applied stress. The crack propagation under the influence of repetitive stress is defined as $d\ell_c/d\omega_N = C_{mat}K_F^n$, with ω_N being the total number of stress cycles, C_{mat} a material constant (found in a tabulated form for the final configuration of the material as used in analysis for the strength of materials and the planning for final requirements of the end-product composition), $K_F \cong S_\sigma\sqrt{(\pi\ell_c)}$ the intensity factor of the cyclic stress, S_σ the peek MAGNITUDE of stress (the AMPLITUDE of the harmonic stress function), and n a proportionality factor that ranges from 2 to 6, generally 4. This expression of crack propagation is known as the "Paris equation." The number of cycles required to reach failure is expressed by the Coffin–Manson law as a function of the plastic strain AMPLITUDE: $\epsilon_p = \epsilon_s - \epsilon_e$, where ϵ_e is the elastic strain, where failure limit due to repetitive MOTION is expressed as $\omega_N^q = C_p/\epsilon_p$, with $C_p \cong \epsilon_{sf}/2$ a material constant, with ϵ_{sf} the PLASTIC STRAIN at failure under static LOAD, and q a procedural constant, generally $q = 1/2$, under the condition $\epsilon_p \gg \epsilon_s$ (see Figure E.57).

Figure E.57 Example of physical endurance during mountain climbing.

Energetics of fission

[nuclear] In NUCLEAR FISSION, the NUCLEUS will be broken down into smaller size, and lighter, nuclei. The difference in mass between the initial composition of the PARENT NUCLEUS "P'" and the end-product DECAY siblings "S'" can be expressed in a relativistic energy balance providing the energy release during FISSION using $E = mc^2$, with m being the mass of the respective component and c the speed of light with respect to the reaction process: ${}_b^a P' + \text{instigator} \rightarrow {}_d^c S_1' + {}_f^e S_2' + \text{radiation} + \text{elementary partile} + \text{energy}$ (see Figure E.58).

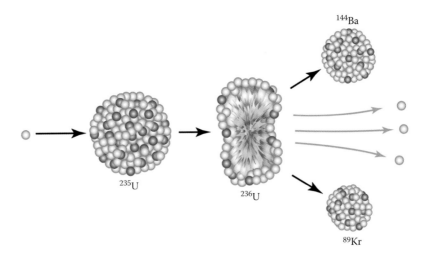

Figure E.58 Fission process with particle product.

Energy

[biomedical, nuclear, thermodynamics] The term was introduced by Thomas Young (1773–1829) in 1807 and was applied to thermodynamic principles postulated by Lord Kelvin (1824–1907) in 1852. Energy is defined as the ability of an object to perform work. Energy can either be supplied or taken away from the system in order for the processes, as well as changes in processes, to occur, and is the capacity for doing work. Potential energy is the energy inherent in a body because of its position with reference to other bodies.

Kinetic energy is the energy possessed by a mass because of its MOTION. The energy transferred into the system or released by the system is called "work." Energy is required to sustain all forms of ACTIVITY, including life in all its forms. One of the most well-known and most powerful natural sources of energy for the EARTH is the SUN. The amount of energy that reaches the Earth's surface is 1366 W/m², which represents ELECTROMAGNETIC RADIATION of all wavelengths irradiated by the Sun. This energy is used to sustain activity (both chemical and molecular) as well as life in every aspect. One specific example of the contribution of energy to life and chemistry is in PHOTOSYNTHESIS in plants, where visible light is essential in the conversion of carbon dioxide to oxygen; other examples are in activation of the photosensitive cells in the EYE, next to the regulation of the circadian cycle by the hypothalamus. Energy is also used for powering biometric communications such as electrical SIGNAL conduction and neurotransmitter release in nerve cells, and on a MICROSCOPIC scale, the importing/exporting and processing ions and molecules in cells. Energy is stored in biological cells in a similar fashion as in a BATTERY (potential energy) in the form of bonds between chemicals, in biology the phosphate groups in the ADENOSINE TRIPHOSPHATE (ATP) molecule. When the ATP bonds are broken and phosphor is released (forming ADENOSINE DIPHOSPHATE [ADP]), energy is made available for other endergonic (energy gaining) reactions in the CELL and HOMEOSTASIS (see Figure E.59).

Figure E.59 Collage of various energy resources.

Energy, internal

[thermodynamics] Introduced by Rudolf Clausius and William Rankine in the late 1800s, replacing "internal work" and "inner work" as thermodynamic indicators of molecular ACTIVITY in THERMODYNAMICS.

Energy bands in solids

[atomic] In comparison to a GAS where the individual atoms have their own respective energetic configuration; in a solid, the atoms will be closely separated providing an opportunity to share low-energy external electrons. The sharing of electron across atoms in a solid results in an energy split. The energy split results in energy band configuration for the solid instead of energy levels as found in the GAS PHASE. The WAVE functions for the energy band (known as the transition from the VALENCE BAND to the CONDUCTION BAND in conductors) will permeate across neighboring atoms, essentially filling the entire sample. For insulators, the energy gap to the energy band above the filled single-atom atomic band is too great to allow for transitions to the sharing state of electrons, leaving the conduction band unoccupied. In semiconductor, the valence band and conduction band are widely separated, but the external influences of thermal or electronic energy can bring electrons in the shared state conduction band.

Energy conservation

[general] *See* CONSERVATION OF ENERGY.

Engine

[mechanics, thermodynamics] Device designed to perform work, generally by means of artificial source of energy, above and beyond human MUSCLE power. Several types of engines have been designed and developed of the centuries with a variety of energy sources used to provide the mechanism of operation. Examples of engines are Steam engine, combustion engine (internal and external), JET engine, electric engine (electric MOTOR), wind-powered engine, spring-action (e.g., clocks made between the early fifteenth and twentieth century, mainly replaced by ELECTRONICS in the twenty-first century), hydropower, SOLAR POWER, NUCLEAR FISSION and nuclear FUSION, and refrigeration to name but a few. Engines can be described by various processes that are specific to the mechanism and operation. Examples of thermodynamic models are the CARNOT CYCLE, the Rankin cycle, the OTTO CYCLE, the DIESEL CYCLE, the JOULE–BRAYTON CYCLE, the STIRLING CYCLE, and the LINDE–HAMPSON LIQUEFACTION CYCLE (see Figure E.60).

Figure E.60 Example of the engine displayed at North Carolina State University. A cross-sectional view of a mishmash of different mechanisms of action in the design of a combustion engine, presented here in a V-twin format.

Engineering

[general] A classification of science, mathematics, and technological applications related to the practical implementation of scientific concepts in order to design, construct, and validate devices and processes. Engineering also involves the integration of the economic impact (marketing) and practical application (ergonomics, social, and ease-of-use constraints) into the system development, quality control, and COMPLIANCE with standards for health and safety. The intended design applications are to forecast the behavior and operational performance under specific operating conditions with respect to its intended function. Software engineering is a generalized virtual application that integrates math, science, and technology in an algorithm that can simulate the process or function. Other applications of engineering are in the data acquisition and data processing aspects of technology, economics, biology, and THERMODYNAMICS and any aspect of integration and overlap between these functional segments. The first-known use of the term and concept of engineering has been documented in 1720. It is a cross-functional multidiscipline integrating FLUID DYNAMICS, MECHANICS, thermodynamics, material sciences, OPTICS, chemistry, biocompatibility, biology, design, draftsmanship, ACOUSTICS, ELECTRONICS, software code, geology, METEOROLOGY, astronomy, HYDRAULICS,

pneumatics, and nuclear, but not limited to the number of fields and including a range of other disciplines, such as textiles and agriculture as well as architecture (see Figure E.61).

Figure E.61 Examples of engineering.

Enthalpy (*H*)

[fluid dynamics, thermodynamics] The sum of all energy contributions in chemical reactions both required for the breaking of bonds and needed for forming new bonds. When a reaction is endothermic, the change in entropy is positive, whereas for an EXOTHERMIC REACTION, $\Delta H < 0$. In case the reaction is endothermic, the change in enthalpy is positive ($\Delta H > 0$). Enthalpy is also defined as the cumulative contributions of INTERNAL ENERGY (U) and work performed on the external environment (W), which is primarily confined to the volumetric change under constant pressure, written in incremental format as $H = U + W = U + PV$, with P and V being the pressure and volume of the system, respectively. The heat-transfer in a system is equal to the enthalpy change in the system: $\Delta H = \Delta Q$. This make the heat of vaporization the difference between the enthalpy of the LIQUID phase and the VAPOR phase. In the case of constant pressure, the change in enthalpy is equal to the HEAT (Q) added to or removed from the system (see Figure E.62).

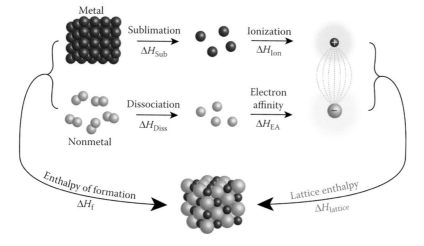

Figure E.62 Schematic representation of the lattice enthalpy of formation.

Entropy (S)

[computational, fluid dynamics, thermodynamics] Measure of order or disorder in a system. It was introduced by RUDOLF JULIUS EMANUEL CLAUSIUS (1822–1888) in 1865. The entropy of a system is the partial derivative change in free energy (F) with respect to changing temperature (T): $S = \left(\partial F / \partial T \right)_{N,V}$ under constant atomic content (N) and constant volume (V). Entropy is a measure of molecular randomness or disorder, defined by the total number of (energetic-) states in a system expressed by the Boltzmann equation using the thermodynamic PROBABILITY p, written as $S = k \ln p$, where k is the Boltzmann constant. Systems that are well ordered have a low entropy, primarily due to the low probability of the system being so well ordered. When HEAT (Q) is added under constant temperature (T), the system becomes more disordered due to kinetic interaction between molecules resulting in breaking of bonds (covalent or chemical) and will transition in a new state with higher entropy: $\Delta S = \Delta Q / T$. A clear example is melting ice where the amount by which entropy has increased is a direct function of the heat provided by external sources. Another way of expressing the entropy is with respect to internal temperature: $S(A) = \int_O^A dQ/T = \int_O^T (C_P(T)dT)/T$, yielding the difference in entropy between two energetic states of a system, where O is an arbitrarily chosen initial state, T the ABSOLUTE TEMPERATURE, dQ the HEAT TRANSFER into/out of the system and A the final state of the medium. For solids, the second part holds true with, $C_P(T)$ the heat capacity of the system at constant pressure. In a different form as an absolute value the entropy is also described as a function of the number of dynamic states of a system at a given thermodynamic state, also called the "probability of state," π, defined as the Boltzmann relation, $S = k^{10}\log(\pi)$, with k the Boltzmann constant. This can be written for an IDEAL GAS as $S = C_v {}^{10}\log T + R {}^{10}\log V + a$, with V and a being the volume of the ideal gas and the ENTROPY CONSTANT OF THE GAS, respectively. The entropy change in an ideal gas that is exposed to a abrupt EXPANSION is $S_2 - S_1 = -(W_{rev}/T) = N \ln(V_2/V_1)$, with W_{rev} being the reversible work while changing volume (V), while changing from state$_1$ to state$_2$, and N being the number of gas atoms/molecules. A special case can be seen for ^{4}He at the melting temperature of 1.4 K; the entropy of the LIQUID is almost identical to that of the VAPOR. Even more remarkable is the solid–liquid curve for another QUANTUM LIQUID: ^{3}He, which has a negative slope (virtually horizontal for ^{4}He and positive for all general materials) that can be explained by the fact that liquid ^{3}He acts as a FERMI GAS, which as its atomic momentum states, ordered in the FERMI SPHERE, whereas the solid is not as well defined energetically. The entropy constant of a monoatomic gas is, for instance, $a = R {}^{10}\log\left(\left(2\pi M R^{3/2}\omega e^{5/2}\right)/\left(h^2 + A^4\right)\right)$ (see Figure E.63).

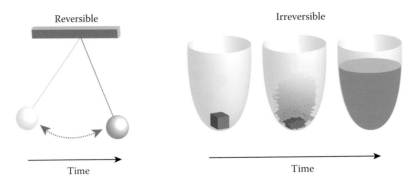

Figure E.63 Schematic representation of reaction entropy.

Eötvös number (Eo)

[fluid dynamics, mechanics] dimensionless number representing the quotient of the dependency between the gravitational force over the SURFACE TENSION. In the ratio, the following definitions apply: ρ the density of the droplet, ρ_f the density of the surrounding FLUID medium, L a characteristic length (size of droplet, confinement space, etc.), and σ_{surf} the surface tension: $Eo = (\rho - \rho_f)L^2/\sigma_{surf}$. The Eötvös number describes the momentum transfer in fluid atomization, as well as the classification of the calculations of the MOTION of BUBBLES and droplets. The Eötvös number has significant similarities to, and identical applications as the Bond number.

Epitaxial growth

[electronics, general, solid-state] In semiconductor construction (e.g., CHIP: silicon crystal designed to perform select electronic functions), layers are evaporated and deposited that will form a crystallographic structure that is consistent and a reproduction of the substrate layer (i.e., wafer or silicon structure). Note that crystallographic defects are similarly reproduced within each layer. Deposition can be facilitated by laser.

Epitaxial layer

[electronics, general, solid-state] In semiconductor manufacturing, the layer formed during deposition under EPITAXIAL GROWTH (see Figure E.64).

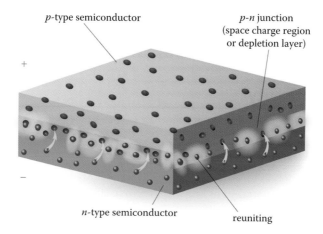

p-type semiconductor

p-n junction
(space charge region
or depletion layer)

n-type semiconductor

reuniting

Figure E.64 The epitaxial layer in a p-n junction in semiconductor construction.

Epithermal neutron

[atomic, nuclear, thermodynamics] Energy configuration for NEUTRON that is released from a FISSION process and the neutron has slowed down but has not degraded to the THERMAL ENERGY level yet.

Eptesicus pulse

[acoustics, biomedical, theoretical] Pulse echo emitted by the family of *Eptesicus* bats (Big Brown BAT) during cruising flight (search mode), with four to five pulses being emitted per SECOND. During hunting mode, the number of acoustic pulses emitted by the *Eptesicus* bat increase to hundreds per second with as little as 5 ms interval between pulses, compared to 200 ms intervals during surveillance mode. This pattern is very different from the MYOTIS PULSE. The *Eptesicus* bat is found in Europe, the mid-East/northern Arab nations, southern Russia, and China (see Figure E.65).

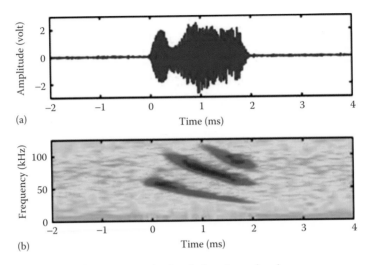

(a)

(b)

Figure E.65 (a,b) Spectral shape of the *Eptesicus* bat's echolocation pulse shape.

Equation of continuity

[general] *See* CONTINUITY *and* CONSERVATION LAWS.

Equation of Kármán–Nikuradse

[fluid dynamics] FLOW description by THEODORE VON KÁRMÁN (1881–1963) combined with the work of Johann Nikuradse (1894–1979), a student of LUDWIG PRANDTL (1875–1953). Correction to the FRICTION in the viscous sublayer or WAKE region: $1/\sqrt{(C_f/2)} = 2.46 \ln \left\{ C_f / (2Re_D \sqrt{(C_f/2)}) \right\} + 0.29$, where $C_f = C/Re_D^{1/m}$ the coefficient of friction, Re_D the REYNOLDS NUMBER ($Re_D > 4000$), C a constant that is linked to the flow conditions, m a constant linked to the flow conditions, which represents the specific friction factor associated with the values tabulated for specific PIPE parameters, also known as "Prandtl's universal law of friction" for smooth pipes. The equation incorporates specific aspects of TURBULENCE, and the respective influence on the viscous friction with respect to the WALL (see Figure E.66).

Figure E.66 The wake behind a ship created by the turbulence from the boat (object) moving through liquid, with relative velocity.

Equation of radiative transfer

[astronomy/astrophysics, biomedical, energy] An equation system describing the transport and interaction with media of ELECTROMAGNETIC RADIATION introduced by KARL SCHWARZSCHILD (1873–1916) for the theoretical description of electromagnetic wave propagation through galactic dust: $\vec{s} \cdot \nabla L(\vec{r}, \vec{s}) = -(\mu_a + \mu_s) L(\vec{r}, \vec{s}) + \mu_s \int_{4\pi} p(\vec{s}, \vec{s'}) L(\vec{r}, \vec{s}) d\omega'$, where $L(\vec{r}, \vec{s}) = u(r_b) \Psi$ defines the spatial radiance of the electromagnetic RADIATION propagating through space in three dimensional format, with Ψ the radiation power received, transferred or emitted from a finite (small) sphere, $u(r_b)$ defines the spatial distribution of the source, ω is the SOLID ANGLE in space, $p(\vec{s}, \vec{s'})$ is the SCATTERING PHASE FUNCTION, μ_s and μ_a are the local SCATTERING and absorption coefficient of the medium, respectively, $\vec{r}$ is the direction and $\vec{s}$ is a segment of space, with $\vec{s} \cdot \vec{s'} = \cos \theta$ the scattering angular distribution (*see* $L(\vec{r}, \vec{s}, t)$ and RADIANCE) (*also see* DIFFUSION, FLUENCE RATE, PHOTO-ACOUSTIC MICROSCOPY, RADIANCE, RADIANT ENERGY FLUENCE RATE, *and* TRANSPORT EQUATION) (see Figure E.67).

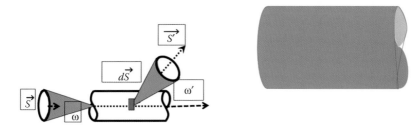

Figure E.67 Outline of geometric configuration for the redistribution process in the equation of radiative transfer.

Equation of state

[thermodynamics] The expression of the interrelation between PRESSURE (P), TEMPERATURE (T), and VOLUME (V) of a system indicating the interdependence as $f(T, P, V) = 0$, signifying that the remaining condition can be derived when two are known. The IDEAL GAS LAW is the most well-known equation of state for a single-phase system. Additionally, PHASE transitions can be incorporated as well. Specific applications of the equation of state are the Van der Waals equation, the Dieterici equation, the Beattie–Bridgeman equation and Benedict–Webb–Rubin equation. The Van der Waals equations reads as $\left(P + (a/V^2)\right)(V - b) = RT$, where $R = 8.3144621(75)$ J/Kmol is the universal gas constant and a and b are constants that represent the specific properties of the substance. The expression for the compressibility of a substance also falls under the equation of state. The Dieterici equation is defined as $P = (RT/(V - b))e^{-(a/RTV)}$, which is derived from the Van der Waals equation and provides a correction for deviations with respect to the critical point. The Beattie–Bridgeman equation uses a slightly different approach as an extended iteration, expressed as $P = RT/V\left(1 - (c'/VT^3)\right)\left[1 + (B_0/V)\left(1 - (b'/V)\right)\right] - (A_0/V^2)\left(1 - (b'/V)\right)$, where A_0, B_0, and a', b', and c' are constants that represent the specific properties of the substance. The most intricate is the Benedict–Webb–Rubin equation: $P = (RT/V) + (1/V^2)\left(B_0'RT - A_0' - (C_0'/T^2)\right) + (1/V^3)\left(b''RT - A_0'\right) + (a''\alpha^*/V^6) + (1/V^3)(c''[1 + (\gamma^*/V^2)]/T^2)e^{-(\gamma^*/V^2)}$, with A_0', B_0', C_0', and a'', b'', c'', α^*, and γ^* constants that are tabulated, which represent the specific properties of the substance. Under certain conditions is has value to introduce the acentric factor: $\omega_S = -^{10}\log\left((P_{sat}(0.7T_c))/P_c\right)$, where P_c is the critical pressure, P_{sat} is the SATURATION (vapor-)pressure at a specific temperature, T_c is the temperature at the critical point, or critical

E

temperature. The equation of state is directly linked to the Gibbs energy, with the fall out to the PRINCIPLE OF CORRESPONDING STATES (see Figure E.68).

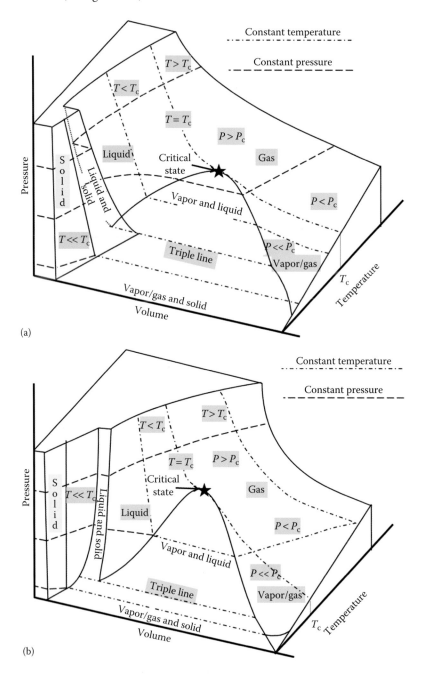

Figure E.68 Graphical representation of the conditions for the equation of state (a) H2O. (b) Generalized medium that does not expand during the phase change from liquid to solid. (c) Phase diagram for generalized medium outlined by specific entropy and specific internal heat vs. specific volume. *(Continued)*

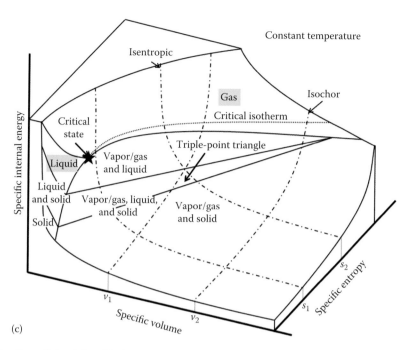

Figure E.68 (Continued) (a–c) Graphical representation of the conditions for the equation of state.

Equation of state of ideal gas

[thermodynamics] Combining the laws of BOYLE, GUY–LUSSAC, and AVOGADRO applied to each of the constituents i of the gas-mixture: $P_i V_i = (m/M_i)RT$, where P_i is the ABSOLUTE PRESSURE in the respective GAS, V_i the volume occupied by the gas m is the mass of the gas, M_i is MOLECULAR WEIGHT of the particular ELEMENT or molecule, T the ABSOLUTE TEMPERATURE in KELVIN, and $R = 8.3145$ J/mol K is the universal gas constant.

Equations of Blasius

[fluid dynamics] With respect to the boundary FLOW for a semi-infinite flat plate, the two-dimensional Navier–Stokes equations are used with respect to incompressible flow under the assumptions of negligible buoyance and a constant free stream velocity. This reduces the problem statement to a two-dimensional situation (x, y). Under these boundary conditions, $v_x = Ua(\eta_K)$, where $\eta_K = y/(\delta(x))$, $\delta(x) = (\eta_K x/U)$, v_x the flow velocity in the x-direction, U the limit of the velocity parallel to the plate at infinite DISTANCE $(v_x \underset{y \to \infty}{\to} U)$, $\eta_K = \eta_D/\rho$ the KINEMATIC VISCOSITY, $\eta_D = (F/S_A)(b/U)$ the dynamic VISCOSITY (also found as μ), F the FRICTION force, S_A the surface area, b the thickness of the viscous layer (BOUNDARY LAYER), ρ the density, and $a(\eta) = df_{\text{flow}}/d\eta_K$ the acceleration, where $f_{\text{flow}} = \phi(x, y) + i\psi(x, y)$, $\phi(x, y)$ represents the VELOCITY POTENTIAL, and $\psi(x, y)$ represents the stream potential, or STREAM FUNCTION, describing streamlines that are by definition tangent to the velocity vector of the flow (*Note*: $v_x = \partial\psi/\partial y$; $v_y = \partial\psi/\partial x$), providing a flow function for a potentially irrotational and solenoidal vector field. The Blasius equation, named after PAUL RICHARD HEINRICH BLASIUS (1883–1970), scientist from Germany, now yields $(1/2)f_{\text{flow}}(d^2 f_{\text{flow}}/d\eta_K^2) + (d^3 f_{\text{flow}}/d\eta_K^3)$. The Blasius equation can generally not be solved analytically and requires a NUMERICAL approach, as outlined in the field of computational FLUID DYNAMICS.

Equations of Itaya

[fluid dynamics] The generalized Burger's equation for a compressible FLOW of a viscous FLUID. The equations are as follows in a one-dimensional system: $(\partial v(x,t))/\partial t = (\eta_{\text{visc}}/\rho(x,t))(\partial^2/\partial x^2)v(x,t) - v(x,t)(\partial v(x,t)/\partial x) - (K/\rho(x,t))(\partial\rho(x,t)/\partial x)$, respectively $(\partial\rho(x,t))/\partial t + (\partial/\partial t)\rho(x,t)v(x,t) = 0$, where t represent time, K is a constant: $K\rho = P$, ρ the density, P the pressure, $v(x, t)$ the velocity, and η_{visc} the VISCOSITY. Introduced by Nobutoshi Itaya (1933–) in 1970.

Equations of momentum

[fluid dynamics, mechanics] In general, the conservation of moment applies, next to the force integral over time to provide the momentum. However, FLOW momentum situations with particular applications in flow approximation specific to porous media can be resolved using FLUID DYNAMICS specific equations of momentum (*see* CAUCHY MOMENTUM EQUATION).

Equations of motion

[biomedical, quantum, thermodynamics] Set of differential equations (plural) that outline the conditions and operation of a system as a function of time specific to the particular state of the system. In classical MECHANICS, the equations of motion are described by NEWTON'S SECOND LAW. A system consisting of N particles can be analyzed for the individual particles as well as the system of particles on average. The equations arrange from displacement to energy. In displacement, the velocity is the first-order derivative with respect to location (r) as a function of time (t): $\vec{v} = d\vec{r}/dt$, generally defined as a vector with MAGNITUDE and direction with respect to the frame of reference. The acceleration is the derivative of velocity and the second order to time for location: $\vec{a} = d\vec{v}/dt = d^2\vec{r}/dt^2$, which may be repeated for individual components (ti) up to the total number N. Tying the force aspect in yields $\vec{F} = \sum_{i=1}^{N} m_i \left(d^2\vec{r}_i/dt^2 \right)$. Under QUANTUM MECHANICS, the equation of MOTION is outlined by the SCHRÖDINGER EQUATION, as introduced by Erwin Schrödinger (1887–1961). Alternatively, under the Heisenberg approach, the state vectors are constant as a function of time, while the observables do change. The Heisenberg representation regards the changes in operators (H, HAMILTONIAN OPERATOR) or "observables" (A^*), as introduced by WERNER KARL HEISENBERG (1901–1976) in 1925. The Heisenberg PICTURE or Heisenberg representation with respect to the equations of motion is $(d/dt) A^*(t) = (i/\hbar)[H, A^*(t)] + (\partial A^*(t)/\partial t)$, where $\left[H, A^*(t) \right]$ is the commuter of the two operators: H and $A^*(t)$. For DIFFUSION as a function of concentration $([C_i])$ of the respective constituents the equation of motion is $(1/\vec{r}^2)(d/d\vec{r})(\vec{r}^2 (d[C_i]/d\vec{r})) = 0$. For FLOW, the equation of motion can be generalized as $(\partial \psi(\vec{r})/\partial t) + \nabla \cdot (\psi(\vec{r})\vec{U}) + \nabla \cdot \overline{\chi(\vec{r})} = \xi$, where $\vec{U}$ is the velocity field, ξ a state constant (characterizing external influences), which is zero under specific boundary conditions, $\overline{\chi(\vec{r})}$ represents a diffusive transport term, and $\psi(\vec{r})$ represents the stream potential, or STREAM FUNCTION, describing streamlines that are by definition tangent to the velocity vector of the flow (*Note*: $v_x = \partial \psi/\partial y$ and $v_y = \partial \psi/\partial x$).

Equations of Nikuradse

[fluid dynamics] See EQUATION OF KÁRMÁN–NIKURADSE.

Equilibrium state

[thermodynamics] When a system is in equilibrium, the processes are generally in steady state with no net force acting on the system, confined under statics. There are many equilibrium conditions to be considered. Mechanical equilibrium applied to a system that is rotationally stable as well as translationally stable, meaning no changes in velocity or direction; this also applies to the geometry of an object and the way it may be placed on a three-dimensional surface. For instance, a ball placed on top of an incline will not be in equilibrium because there are no forces preventing the ball from rolling down the hill. A see-saw will be in equilibrium when none of the people seated use their muscles to change their position or push against the floor. Thermal equilibrium pertains to the fact to there is no heat exchange with the environment. A chemical equilibrium will result in a constant concentration for all the constituent, but resource and product. In a nuclear equilibrium the ATOM will not exchange particles (mass) or energy. A person sitting in a car that is not moving and does not have the ENGINE running while eating and drinking is not in equilibrium; however, the car and content (without the RADIO playing, no lights on and without the air-conditioning or heat running) are most likely in a situation that can be confined to an EQUILIBRIUM STATE. For a mechanical system, the first condition of equilibrium is that the sum of all the forces is zero $\left(\sum \vec{F}_i = 0\right)$, and the second condition of equilibrium means that the sum of the torques in zero $\left(\sum \vec{\tau}_i = 0\right)$. The fact that there cannot be heat exchange is an additional requirement for equilibrium with the outside of the system $\oint \dot{Q} dA \, dt = 0$. Under the constraints of equilibrium, a LIQUID may not evaporate if the liquid is considered the confined medium; however, when the liquid is in a tank with room for AIR, the tank may be in equilibrium as a whole.

Equilibrium theory of tides

[fluid dynamics] The EARTH as a whole with the MOON and the SUN as its components are considered the system when the equilibrium conditions for the tides are evaluated. The tides on the Earth are affect by gravitational attraction from Earth as well as both the Sun and the Moon, captured by the net gravitational potential Φ_g, which is a three-dimensional matrix. Other forces result from centripetal forces as well as wind, next to perturbations of melting ice (which can be a slow process that in most cases can be discarded as a second- or third-order term) and precipitation (e.g., RAIN and snow) next to EVAPORATION may be a small influence that can be captured in a generalized term (Ω). Equilibrium "sea level" is defined by $\Phi_g + \left(1/2\right)\omega^2\vec{r}^2 + \Omega = 0$, where ω is the Earth's rotation and $\vec{r}$ the DISTANCE to the sea surface with respect to the center of the Earth. Technically, each parameter has both spatial and temporal perturbations (see Figure E.69).

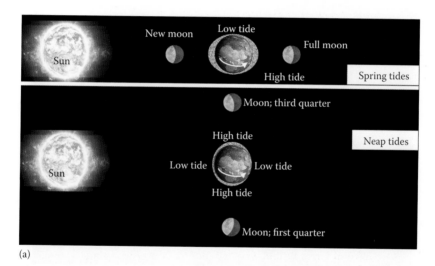

(a)

(b)   (c)

Figure E.69 (a) Outline of the influence of astrophysical objects (e.g., Sun and Moon, but potentially other celestial objects as well) on the tide level. (b) The tides may also affect commercial activity, a significant number of fish boats are grounded at low tide in this harbor. (c) Visual of how the tides affect the water migration process and create turbulent flow pattern. The turbulent flow is exemplified by the foam captured in the current, resulting from the capillary waves.

Equinox

[geophysics] A cross-sectional plane traversing the longitudinal outline of the equator where the average path of the Sun touches on its azimuth increases or decreases to the Tropic of Cancer in the northern hemisphere or the Tropic of Capricorn in the southern hemisphere. Because the Earth is slanted at an angle with the rotational plane, the North Pole will be closer to the Sun on one half of its revolution around the Sun. The Sun will have its highest position (closest to the North Pole) when "above" the Tropic of Cancer. During the Earth's elliptical path around the Sun, the cross-sectional plane outlined by the Earth will make the Sun pass North over the equator during the vernal equinox. The vernal equinox defines the beginning of spring on the Gregorian calendar. The vernal equinox is around March 20 (give or take up to 2 days earlier or later, depending on calendar adjustments). When the Earth's track places the Sun above the Tropic of Cancer, the calendar beginning of summer for the Northern hemisphere is initiated. Alternatively, the average position of the Sun's geographical location crosses the equator in southern direction around September 22 (give or take up to 2 days either way, depending on calendar adjustments) indicates the beginning of autumn. Passing the equator this time named the autumnal equinox. After which the Sun proceeds onward for its average position before it reached the Tropic of Capricorn as the calendar onset for winter. *Note*: Seasonal changes geared to the Northern hemisphere (see Figure E.70).

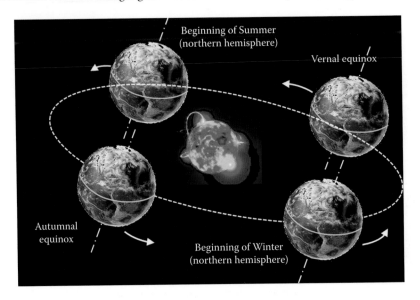

Figure E.70 Position of the Earth with respect to the Sun defining the respective equinox orientation.

Equipartition theorem

[atomic, computational, general, nuclear] Generalization of the temperature of a classical system to its energy content. All forms of energy are shared equally over the various degrees of freedom of the system. For a medium at temperature above absolute zero, the distribution of vibrational to rotational to translational energy (contributing to the temperature) should be 1/3 to 1/3 to 1/3. The equipartition theorem predicts that for a monoatomic gas the energy will be the average kinetic energy: $KE = (3/2)k_bT$, where $k_b = 1.3806488 \times 10^{-23}$ m^{2}kg/s^{2}K the Boltzmann coefficient, T is the temperature (in Kelvin). For an electron gas interaction with phonons, the equipartition theorem provides the following $n(\epsilon_F)\hbar v_F q'^2 \phi_{q'}^2 = k_bT$, with ϵ_F the Fermi energy, $\hbar = h/2\pi$:: where $\hbar = h/2\pi$, with $h = 6.62606957 \times 10^{-34}$ m^{2}kg/s the Planck's constant, v_F the Fermi velocity, q' the, $\phi_{q'}$; all within the Brillouin zone. This may, for instance, apply to lattice structures thermal lattice vibrations.

Equipotential surfaces

[atomic, general, nuclear] The electrical POTENTIAL (V) resulting from an ELECTRIC FIELD (E) as a function of DISTANCE from one location with a fixed potential (r) is defined as $V = Er = (1/4\pi\varepsilon_r\varepsilon_0)(Q/r)$, connecting all locations with the same potential will yield the equipotential surface for the charge distribution Q.

Equipotential volume

[atomic, general, nuclear] The volume of a CONDUCTOR enclosed by the equipotential surface obtained by Gauss's law, with no internal electric field.

Equivalence principle

[general, relativistic] Equality of the inertial mass (m_i is the mass of a body that makes it resist MOTION, only when a force [F] is applied, yielding an acceleration [a]: $F = m_i a$) and gravitational mass (m_g); the mass that provides a body with weight due to interaction with other bodies [e.g., EARTH]), which is a central concept in the GENERAL THEORY OF RELATIVITY. The general theory of relativity proposes that GRAVITY is the cause of INERTIA.

Erg

[general] Unit of work done by a force of 1 dyne acting through a DISTANCE of 1 cm. The unit of energy that can exert a force of 1 dyne through a distance of 1 cm. CGS units: dyne-cm, or gm-cm^2/s^2.

Ergodic hypothesis

[computational] In statistical PHYSICS, the ergodic theory describes the dynamic system with an invariant measure that exhibits the same behavior for the ensemble averaged as the same system averaged over time for all system states involved (*also see* **MARKOV CHAIN**).

Ether

[general, geophysics, optics] *See* AETHER.

Euclid (c. 325–270 BC)

[general] A mathematician from Greece. Euclid's work inspired the introduction of Euclidian space. Euclid is often referred to as the "father of geometry" (see Figure E.71).

Figure E.71 Impression of what Euclid (c. 325–270 BC) may have looked like.

Euclidean geometry, axiom of

[computational] Postulation of equality documented approximately in 300 BC by EUCLID (c. 325–270 BC), stating that "parameters that are equal to the same phenomenon are inherently equal to one another."

Euclidean space

[computational, general] A three-dimensional space describing rotational and translational vectors in Euclidean geometry (the geometry that has been the basis of "*modern geometry*", omitting the "Euclidean" denomination part due to the fact that there have been no other comprehensive definitions of geometric space since EUCLID [c. 325–270 BC]). Euclidean space is based on the following six definitions and concepts: point, line, ANGLE (skew lines), parallel, orthogonal, and congruence. A point is undefined apart from the location of another point in reference to the first point (generally a point is set as the reference and is called the "origin"), two points connect to form a line with a length and a DISTANCE, two lines that are not coplanar and intersect are at an angle (skewed) which lies in a plane that conforms to both lines, lines that are coplanar and never intersect are parallel (only one parallel line can pass through a single point that is not on the first line), lines that are passing through one single point and have the same angle on either side are orthogonal to each other, figures that are made from three or more lines that form a closed shape are congruent when there are two distances between two extreme points on each diagram that are identical. The following postulates define the Euclidean space: (1) a unique line must be drawn between two points; (2) any line between two points can be extended to infinity; (3) a circle is defined by a central point and a radius; (4) all orthogonal angles are equal; and (5) a third line intersecting with two lines will form a connecting set of line (i.e., triangle) if the angle the third line makes with both lines is smaller on the same side that on the opposite side. The following axioms are associated with Euclidean geometry: (1) "parameters that are equal to the same phenomenon are inherently equal to one another"; (2) the sum of equals are equal to the sum of identical equal numbers of line definitions and equal angles; (3) the whole of a collection of parameters is greater than the individual parts. In support of Euclidean space, the practical use of the value of π may have been known for centuries prior, but ARCHIMEDES (287–212 BC) made his formal derivation and definition around 250 BC; one of the most appropriate approximations is 22/7 (see Figure E.72).

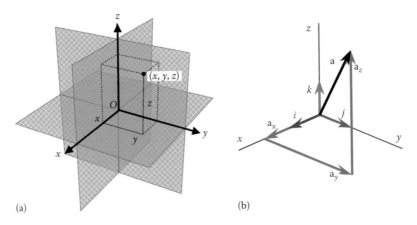

(a)   (b)

Figure E.72 Outline of (a) Euclidean space and (b) Euclidean vector.

Euler, Leonhard (1707–1783)

[acoustics, computational, fluid dynamics, general, mechanics, thermodynamics] A scientist and mathematician from Switzerland. Leonhard Euler's work introduced the exponential dependency named after his last initial: "$e = 2.718281828\ldots$," The number e plays an important role in numerous biological and natural phenomena specifically as exponent and NATURAL LOGARITHM (see Figure E.73).

Figure E.73 Leonhard Euler (1707–1783). (Engraving by B Holl and painting by Jakob Emanuel Handmann.)

Euler angle

[astrophysics, energy, fluid dynamics] Set of three ANGLES in EUCLIDEAN SPACE that identify the orientation of a rigid body, introduced by LEONHARD EULER (1707–1783). The angles define the rotational matrices, rotating around the three respective axes: x, y, and z. The rotation matrices can be interpreted as follows. For a rotation of θ around the x-axis:

$$R_x\left(\theta\right) = \begin{bmatrix} 1 & 0 & 0 \\ 0 & \cos\theta & -\sin\theta \\ 0 & \sin\theta & \cos\theta \end{bmatrix};$$

respectively a rotation ψ around the y-axis:

$$R_y\left(\psi\right) = \begin{bmatrix} \cos\psi & 0 & \sin\psi \\ 0 & 1 & 0 \\ -\sin\psi & 0 & \cos\psi \end{bmatrix};$$

and around the z-axis with EULER ANGLE ϕ:

$$R_z\left(\phi\right) = \begin{bmatrix} \cos\phi & -\sin\phi & 0 \\ \sin\phi & \cos\phi & 0 \\ 0 & 0 & 1 \end{bmatrix}.$$

Euler number (E$_n$)

[computational] In number theory, this represents a sequence of integers defined by a Taylor series EXPANSION as: $\left(\cos ht\right)^{-1} = \left(2/\left(e^t + e^{-t}\right)\right) = \sum_{n=0}^{\infty}\left(E_n/n!\right) \times t^n$. In this description, the hyperbolic secant gives the secant numbers or zig numbers representing the number of odd alternating permutations.

Euler number (Eu = ($\Delta P/\rho v^2$))

[fluid dynamics, mechanics] A representation of the ratio between pressure and inertial force. The pressure gradient $\Delta P = P_{upstream} - P_{stream}$ is the pressure drop over, for instance, a VALVE or other constriction, and the KINETIC ENERGY density (kinetic energy per unit volume) is expressed by the DENSITY (ρ) times the characteristic FLOW VELOCITY (v) squared. Even though this resembles the CAVITATION number, the functionality is entirely different. The Euler number can be used to rank the frictional losses in the momentum of the flow, where Eu = 1 represents frictionless flow.

Euler relation

[thermodynamics] The energy content of a system with r constituents is defined as $U = TS - PV + \mu_1 n_1$, where U is the internal energy, T the temperature in kelvin, S the entropy, μ_i the chemical potential of constituent i and n_i the respective quantities; named in honor of LEONHARD EULER (1707–1783) (*also see* GIBBS ENERGY).

Euler's equation

[acoustics, computational] The primary definition of the Euler principle is in the complex notation of points in a two-dimensional space as $e^{ix} = \cos x + i \sin x$. Additional application with respect to the equation of MOTION with respect to a body moving around its axes (e.g., TURBULENCE and governing INVISCID FLOW) is $\vec{N} = (\partial \vec{L}/\partial t) + \vec{\omega} \times \vec{L}$, where $\vec{L} = \vec{\omega}\left(\sum_i m_i |r_i|^2\right) - \sum_i m_i r_i (r_i \cdot \vec{\omega})$ is the angular momentum, r_i is the DISTANCE and direction of the respective masses m_i to a point of reference, and $\vec{\omega}$ is the angular velocity. The Euler equations use the moment of INERTIA ($I = \sum_i m_i |r_i|^2$, with respective angular momentum: $\vec{L}_k = I_k \omega_k$) to yield the Euler equations: $\vec{L}_k = I_k(\partial \vec{\omega}_k/\partial t) + (I_m - I_l)\omega_l \omega_m$ {k, l, m (1, 2, 3)}. It is named after LEONHARD EULER (1707–1783). The Euler equations are of particular importance in the analysis of inviscid flow by adding the equations for CONSERVATION OF MASS, conservation of momentum, and CONSERVATION OF ENERGY (E) without VISCOSITY and without HEAT TRANSFER, in corresponding to the Navier–Stokes equations. This corresponds to the following set of equations for FLOW with density ρ, pressure P, and velocity function $\vec{u} = (u, v, w)$ as: $(\partial \rho/\partial t) + \nabla \cdot (\rho \vec{u}) = 0$; $(\partial(\rho \vec{u})/\partial t) + \nabla \cdot (\vec{u} \otimes \rho \vec{u}) + \nabla P = 0$; and $(\partial E'/\partial t) + \nabla \cdot (\vec{u}(E' + P)) = 0$, where $\otimes$ is the convolution or TENSOR product for the energy density $E' = \rho \hat{e} + (1/2)\rho(u^2 + v^2 + w^2)$ where $\hat{e}$ is the energy density.

Euler–Cromer method

[computational] A mathematical expression used to dissect the processes of a harmonic OSCILLATOR as a refinement to the original ideas of LEONHARD EULER (1707–1783) implemented by Tom Cromer (?–?) in the 1980s. The original Euler method introduces a set of incremental steps to solve the differential equations analytically based on the following process. The ANGULAR VELOCITY (ω) is defined as the first-order derivative with respect to TIME (t) for the changing ANGLE (θ): $\omega = d\theta/dt$, which in turn is converted in a set of series: $\omega_{n+1} = \omega_n - \theta_n \Delta t$, respectively: $\theta_{n+1} = \theta_n + \omega_n \Delta t$, $t_{n+1} = t_n + \Delta t$, with energy $E_{n+1} = (1/2)\left(\omega_{n+1}^2 + \theta_{n+1}^2\right) = E_n\left(1 + \Delta t^2\right)$, which is apparently ever increasing. The Cromer correction is as follows: $\theta_{n+1} = \theta_n + \omega_{n+1}\Delta t$, which yields for the energy: $E_{n+1} = E_n + (1/2)\left(\omega_n^2 - \theta_n^2\right)\Delta t^2 + O\left(\Delta t^3\right)$, providing a converging series instead (*also see* RUNGE–KUTTA METHOD).

Eutectic composition

[thermodynamics] System of two or more constituents that can all coexist with both the LIQUID and the solid phases at the same temperature (i.e., the eutectic temperature).

Ev

[electromagnetism, energy, general, thermodynamics] Electron-volt, the energy (also expressed in joule, unit J) of an electron under the influence of an electrical potential of 1 V: $1\,eV = 1.6 \times 10^{-19}$ J.

Evanescent wave

[acoustics, biomedical, general, imaging, mechanics, optics, quantum] WAVE phenomenon that is remaining within a close proximity to the interface between two media. Generally, the evanescent wave is formed when a wave enters into a BOUNDARY LAYER at grazing ANGLE and propagates reflecting of the boundary layer and the bulk medium with bulk REFLECTION coefficient at angles in excess of the CRITICAL ANGLE (i.e., TOTAL INTERNAL REFLECTION between the external medium 1 and the internal medium 2: $R_0 = n_1/n_2$). The bulk reflection for a (half-infinite) optical medium is defined by Kubleka–Munk RADIATIVE TRANSFER theory introduced by the Austrian physicist Paul Kubelka (1900–1953?) and the German (Prussia) scientist Franz Munk (1880–1951) in 1931. The bulk medium reflection is defined as $R_\infty = 1 + (K/S) - \sqrt{[((K/S)+1)^2 - 1]}$, where λ represents the wavelength of ELECTROMAGNETIC RADIATION, K the Kubleka–Munk absorption CROSS SECTION, and S the Kubelka–Munk scattering cross section. The Kubelka–Munk theory was developed for the quality assessment of painted surfaces. The Kubelka–Munk parameters relate to the universal optical absorption (μ_a) and scattering coefficient (μ_s) as follows: $K = 2\mu_a$, and the SCATTERING cross-sectional factor has a more intricate connection, involving the reduced scattering coefficient, which involves the scattering anisotropy factor (mean cosine of scattering angle: $g = \cos\theta$) as $\mu'_s = (1-g)\mu_s$, which can be measured directly, providing $(4/3)S + (1/6)K = (1-g)\mu_s$. For light, the media interface and the bulk tissue under grazing angle of incidence will form the equivalent of two opposing mirrors, trapping the light to travel along the surface with exponential DECAY in radiance with respect to depth (i.e., DISTANCE from the interface), not sinusoidal. In ACOUSTICS, the evanescent wave can be formed in several ways; for instance, in ATMOSPHERIC LAYERS, a thin layer of AIR can force itself between two layers of air that are stacked and have distinctly different densities (e.g., INVERSION layer). The air forced between the two stacked layers will form a flutter (similar to blowing air between your lips: "whistle"), creating a spatial frequency pattern of condensed air forming strips of clouds. Evanescent wave also apply to certain conditions in SCHRÖDINGER EQUATION. The evanescent profile can be used for selective ablation (removal of thin layers of media by performing a process similar to LITHOGRAPHY, ranging to molecular ablation) as well as diagnostics due to the specific characteristics associated with the evanescent propagation, and the inherent media parameters represented (*also see* LAMB WAVE) (see Figure E.74).

Figure E.74 Example of evanescent wave in the atmosphere. The pressure troughs create condensation, manifesting as rows of clouds.

Evans, Robley Dunglison (twentieth century)

[nuclear] A physicist from the United States who provided a detailed description of the construction of the atomic NUCLEUS.

Evaporation

[general] Conversion from LIQUID to gaseous state, called VAPOR, by administering heat. The heat requirement (Q) is set by the heat of vaporization: L_v and the total affect mass m as $Q = mL_v$.

Evaporation rate

[thermodynamics] The rate of conversion of LIQUID into VAPOR, measured in seconds ($t = 1s$). This mechanism will provide a mechanism of COOLING by using the HEAT OF VAPORIZATION (L_v) for the liquid to be extracted from the substance covered by liquid with MASS m, yielding a POWER (P) output of the body based on the HEAT EXCHANGE (Q) per SECOND: $P = Q/t = mL_v/1$. The rate of VAPORIZATION is given by $m_{liquid}/t = Q/L_v$. In human cooling, sweat serves as the mechanism of action, only when evaporated, not when dispelled as liquid as drops or wiped.

Ewald sphere

[solid-state] The spherical configuration with respect to electromagnetic or PARTICLE wave (*Note*: DE BROGLIE WAVELENGTH) interrogation of a LATTICE structure. The vector length resulting from crystalline INTERFERENCE with respect to the reciprocal lattice point outlines a spherical shape described by PAUL PETER EWALD (1888–1985), a physicist from Germany. The graphical representation by means of the Ewald sphere applies to electron diffraction, NEUTRON diffraction as well as X-RAY crystallography and is a variation of Bragg's law for lattice diffraction defined by drawing a sphere with radius reciprocal WAVELENGTH (λ): $R = 1/\lambda$, originating from the "real crystal," where the circumference intersects with the reciprocal lattice for the transmitted beam. The DISTANCE between the planes of the crystal is defined by the cord (d_{hkl}) for the location of the reciprocal lattice point on the sphere with respect to the direction of the interrogating beam as $1/d_{hkl}$ (see Figures E.75 and E.76).

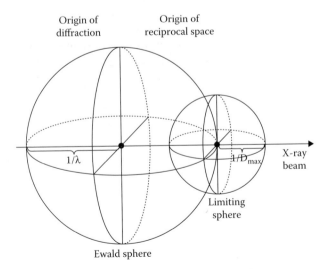

Figure E.75 Geometric outline of the Ewald sphere configuration. The Ewald sphere allows the visualization of a diffraction process; diffraction can only occur in the geometry where a reciprocal lattice point intersects the Ewald sphere.

Figure E.76 Paul Peter Ewald (1888–1985). (Courtesy of International Union of Crystallography, Chester; *Acta Cryst.*, A42, 1–5, 1986.)

Exchange interaction

[computational, electronics, quantum] QUANTUM mechanical symmetry relation linked to the intrinsic SPIN of electrons in their interaction with atoms, specifically discontinuities in the energy distribution resulting from LATTICE impurities. The exchange interaction J_{exch} is an indication of the symmetry between all atomic particles. At a specific temperature, the scattering CROSS SECTION for electrons suddenly has a discontinuity, with an increase defined by the KONDO TEMPERATURE: $T_K = \epsilon_F J_{exch} e^{-(1/n_s J_{exch})}$, with ϵ_F being the Fermi energy (~kinetic energy of free electrons) and n_s the number of quantum states (also known as the "density of states") (*see* KONDO EFFECT *and* THERMOCOUPLE EFFECT).

Excited states

[general, nuclear] *See* ENERGY LEVELS.

Exclusion principle

[atomic, general, nuclear] Exclusion of conditions with respect to state. For instance, fermions within the same system cannot occupy the same state with the same QUANTUM numbers. The states are conditionally defined by the four principle quantum numbers n (principal quantum numbers: $1, 2, 3, \ldots$), ℓ (ORBITAL ANGULAR MOMENTUM QUANTUM NUMBERS: $0, 1, 2, 3, \ldots, (n-1)$), m_ℓ (orbital magnetic quantum numbers: $0; \pm 1; \pm 2; \ldots; \pm \ell$), and m_s (spin angular momentum QUANTUM NUMBER: $\pm(1/2)$). As the NUCLEUS is constructed with energy levels tightly stacked from the bottom up to the highest kinetic energy level, the FERMI LEVEL (Fermi energy), there are few opportunities for nuclei to change states in this configuration. Each NUCLEON may be represented by a WAVE function that has the DE BROGLIE WAVELENGTH ensuring a discrete filling with subsequent waveforms and SPIN configurations, excluding two nucleons to occupy the same conditions at any time because the energy content under those conditions would approach infinity. Only two neutrons and two protons can occupy the same energy level, as long as their respective spins are opposite (up vs. down); this applies to any and every energy level. Hence, the nuclear structure is very stable indeed. Specifically, nuclides (NEUTRON and protons) group in quantities referred to as the "magic numbers" are extremely stable (resembling the atomic NOBLE GAS electron structure) with $Z = 2; 8; 20; 28; 50; 82; 126$. The exclusion principle indicates that the NUCLEAR FORCE has SATURATION levels. (*also see* PAULI EXCLUSION PRINCIPLE.)

Exergonic

[chemical, thermodynamics] Spontaneous chemical reaction where the Gibbs free energy is negative ($G < 0$).

Exocytosis

[biomedical] The active process of cellular expulsion of both solid and LIQUID material. The CELL MEMBRANE forms an enclosure (vesicle) that engulfs the material of interest to be discarded and transports the filled "BUBBLE" out through the cell membrane.

Exoergic reaction

[chemical, thermodynamics] Chemical reaction that liberates energy.

Exosphere

[general, geophysics] The external layer of the Earth's ATMOSPHERE, beyond the THERMOSPHERE, ranging from approximately 640 km out to 10,000 km. The major components are hydrogen and helium, escaping from the thermosphere, at very low densities (see Figure E.77).

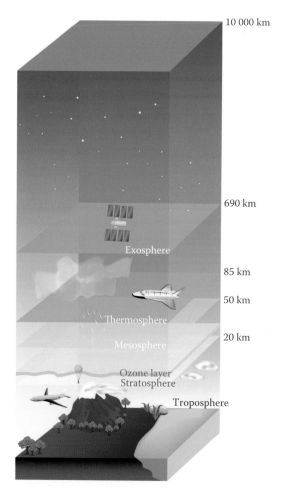

Figure E.77 The relative shell configuration for the Exosphere in the Earth's atmospheric arrangement.

Exothermic reaction

[atomic, nuclear, thermodynamics] Process, chemical, or otherwise that produces heat. A prime example is OXIDATION and related general combustion. For an exothermic process of a system with r components, the enthalpy reaction $\Delta h(T)$ is negative: $\Delta h(T) = T\Delta S + V\Delta P + \mu_1 \Delta n_1 + \mu_2 \Delta n_2 + \cdots + \mu_r \Delta n_r$, where T is the temperature, V is the volume, P is the pressure, S is the entropy, μ_i is the chemical potential of constituent i, and n_i is the respective quantities (see Figure E.78).

Figure E.78 Fire as an exothermic reaction, oxidation.

Expanding universe

[astronomy/astrophysics, nuclear] *See* **BIG BANG THEORY**.

Expansion

[general, fluid dynamics, mechanics] Most object and liquids expand under rising temperature, with the exclusion of, for instance, water, which has its highest density at 4 °C. Certain metals have different expansion profile with respect to temperature, encouraging the use of alloys that have an expansion coefficient that is not as large as the individual metals alone. This type of alloy is used in the construction of laser systems to ensure thermal stability. The expansion can be considered linear, two dimensional, or three dimensional (volumetric) depending on the specific application or interest, although the expansion in reality is always three dimensional. The linear expansion coefficient is the change in length (ℓ) with respect to the initial length as a function of temperature: $\alpha_{ex} = (\Delta\ell/\ell)\Delta T$; similarly, the volumetric expansion is $\gamma_{ex} = (\Delta V/V)\Delta T = 3\alpha_{ex}$. The following definitions apply: T is the temperature and V is the volume.

A mercury or ALCOHOL thermometer relies on expansion of the LIQUID in the GLASS tube for indication of the temperature, assuming that the radial expansion is negligible with respect to the longitudinal expansion of the glass tube in the process (see Figure E.79).

Figure E.79 Example of the impact of expansion with respect to a long single beam of railway iron track. The gap in between sequential segments prevents deformation during expansion.

Expansion wave

[fluid dynamics] *See* ACOUSTICS.

Expiratory reserve volume (ERV)

[biomedical, fluid dynamics] The quantity of air that can be ejected by maximum effort exhaling in succession to exhalation during gentle TIDAL BREATHING, Expiratory reserve volume and other tidal breathing parameters can be measured with a spirometer.

Exposition du système du monde (Exposition of the System of the World)

[general] A publication by the French mathematician PIERRE-SIMON DE LAPLACE (1749–1827) in 1798 describing the theoretical concept of a celestial object with a size 250 times that of the SUN with similar density as the EARTH would provide such a large gravitational attraction resulting in an escape velocity exceeding the speed of light in which case light would not be able to escape. Laplace thus provides the first hypothesis of the BLACK HOLE, without its observation or verification. The theoretical verification was provided in great detail under relativistic terms by KARL SCHWARZSCHILD (1873–1916) in 1916 and was experimentally verified in 1961 (almost 200 years later) by the binary system in the GALAXY Cygnus X-1.

Extramission

[biomedical, optics] Vision concept postulated by EUCLID (c. 325–270 BC), PTOLEMY (323–283), and PLATO (427–347 BC) in various versions, where the EYE is presumed to be capable of emitting rays that reflect from object and return to the eye for rendering the interpretation of the outside world. This concept could not explain the need for ambient light (no VISION in darkness). ARISTOTLE (384–322 BC) had already postulated that the eye captures light reflecting from light sources such as the SUN or a candle. However, the extramission theory was not abandoned until the scientific explanation by LEONARDO DA VINCI (1452–1519) of the workings of the eye in close resemblance to our current understanding. The experimental and theoretical OPTICS work by ALHAZEN (965–1040) formed the foundations for Da Vinci's revelations.

Eye

[biomedical, chemical, optics] In mammals, this refers to an anatomical feature constructed of an oblong shape constructed with a LENS (CRYSTALLINE LENS) at one side and a RETINA on the opposite side, separated by the VITREOUS HUMOR and an iris directly in front of the lens for the mechanical control of light intensity directed toward the retina. The lens is separated from the outside world by the AQUEOUS HUMOR covered by the cornea. Insects have a different construction of the mechanism of VISION called a "compound eye." The various groups of animals defined by amphibians, birds, fish, insects, mammals, and reptiles; they all have specific design features that offer optimal adaptation to the conditions these groups live and seek for nourishment. The various components of the eye have the following respective optical characteristics, specifically for the INDEX OF REFRACTION (on average over the entire VISIBLE SPECTRUM of ELECTROMAG-NETIC RADIATION): $n_{cornea} = 1.351$; $n_{aqueous\ humor} = 1.337$; for the lens the following index of REFRACTION applies to the optical axis $n_{crystalline\ lens,0} = 1.437$; and the main body, the vitreous humor in general: $n_{vitreous\ humor} = 1.337$, while there is an optical "canal" with a different index of refraction with respect to the vitreous humor (lower) that confines the light, channeling it from the lens to the FOVEA (on the optical axis). Keep in mind that WATER has an index of refraction of $n_{water} = 1.333$ and for AIR $n_{air} = 1.000277$ at standard temperature (20°C) and pressure (1 atmosphere). The lens is shaped to correct for CHROMATIC ABERRATIONS in addition to a design of graded index of refraction (GRIN lens) as a function of radius that makes the lens more effective in focusing next to the accommodation process. The accommodation process applied force on the lens by means of the ciliary MUSCLE that stretches the lens and makes it thinner to focus on far off objects, whereas the lens is thick and rounded in the relaxed state for nearby IMAGE acquisition. The index of refraction of the lens of the eye as a whole in approximation has an index of refraction that is parabolic in nature as a function to the DISTANCE with the optical axis (r) as $n(r) = n_0\left(a + br - cr^2\right)$, where $n_{crystalline\ lens,0} = 1.437$ and a, b, and c are positive constants. The eye as a whole is subject to age-related changes, one being that the central part of the lens starts hardening after the AGE of approximately 25. At this point, the lens will configure a portion of the gel-like structure into a NUCLEUS. With advances in age, this nucleus in the lens progressively becomes harder. The final outcome of the hardening process is cataract, with an inherent deterioration of transparency. Cataract will require replacing the lens with an artificial lens. Age-related changes to the composition and structure of the eye frequently result in corrections by means of spectacles (glasses) or refractive surgery (e.g., LASIK) (see Figure E.80).

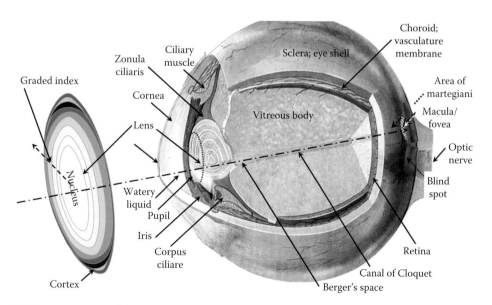

Figure E.80 Representation of the composition and components of the eye.

Eyeglasses

[biomedical, general] Optical prosthetic devices to correct for flaws and imperfections in VISION (see Figure E.81).

Figure E.81 Eyeglasses for corrective vision.

Eyring absolute rate equation

[chemical, condensed matter, energy] Equation describing the changes in energy configuration for a system based on the fact that the energy for each reaction is subject to QUANTUM THEORY because of the inherently discrete nature of the chemical reactions. The Gibbs energy change (ΔG) associated with the activation of the chemical reaction is defined as $\Delta G = RT\{\ln(k_b/h) - \ln(k_r/T)\}$, $R = 8.3144621(75)$ J/Kmol is the universal GAS constant, using the Boltzmann constant $k_b = 1.3806488(13)\times 10^{-23}$ J/K, and PLANCK'S CONSTANT $h = 6.626070040(81)\times 10^{-34}$ m2kg/s Js yields $k_b/h = 2.08358\times 10^{10}$ K^{-1} s^{-1}, and k_r is the rate constant for the respective chemical reactions (function of the choice of concentration units, which directly correlates to the choice of thermodynamic GROUND STATE) and temperature (T) is expressed in Kelvin. The change in disorder associated with the increase in entropy ($\Delta S = S_{final} - S_{innitial} > 0$) is inherently linked with the entropy of a system as defined by LUDWIG BOLTZMANN (1844–1906), as the natural log of all the possible configurations for a system ω_c is multiplied by the Boltzmann constant k_b: $S = k_b \ln \omega_c$, defining the maximum attainable entropy for a system. Under equilibrium, $\delta S = 0$. The change in internal energy (ΔU) as a result of external influences is concurrently defined by the change in enthalpy of the system: ($\Delta H = U + \Delta(PV)$), where the change in PRESSURE (P) multiplied by the VOLUME (V) represents the work performed by the system. Applying the IDEAL GAS LAW to the constituents of the dilute system, this can be rewritten as $\Delta H = U + \Delta nRT$, where Δn represents the change in specific types of molecules because of the chemical reaction and R is the universal gas constant, temperature expressed in Kelvin. Under constant pressure, the work is volume labor only: $W = P\Delta V$. The enthalpy change for an EXOTHERMIC REACTION of a system is negative, that is, the chemical reaction produces heat, vice versa for an endothermic reaction. In this format, the Eyring absolute rate equation defines the rate constant as $k_r = (k_b T/h)\exp(-(\Delta G^2/RT))$. Using the standard enthalpy of activation $\Delta_+^{\ddagger} H^0$ and the standard entropy of activation ($\Delta_+^{\ddagger} S^0$), solving for the molecular quantity change Δn yields $k_r = (k_b T/h)\exp(\Delta_+^{\ddagger} S^0/R)\exp(-(\Delta_+^{\ddagger} H^0/RT))$, which provides the "linear" Eyring equation: $\ln(k_r/T) = -(\Delta_+^{\ddagger} H^0/RT)(1/T) + \ln(k_b/h) + (\Delta_+^{\ddagger} S^0/R)$. During this process, the HEAT TRANSFER of the system is equal to the change of enthalpy in the system: $\Delta H = \Delta Q$. The standard enthalpy of activation $\Delta_+^{\ddagger} H^0$ represents the enthalpy change derived from the thermodynamic form of the rate equation obtained from conventional transition state theory. In simplified form, the rate equation can be written as $k_r = A_1\exp(-(E_1/RT))$, relying on the rate of change to take place at $k_b T/h$, with empirically derived coefficients $A_1 = (k_b T/h)\exp(\Delta_+^{\ddagger} S^0/R)$ and $E_1 = \Delta\Delta_+^{\ddagger} H^0$. These constants are per molecular context and will need to be determined experimentally. Combining the respective definitions of entropy and enthalpy provides the change in Gibbs free energy of a system (G) as $G = \Delta U - T\Delta S - S\Delta T + \Delta nRT + nR\Delta T$. In a biological system, thermal process can provide denaturation of proteins (boil an egg, bake a steak, etc.). The denaturation process provides irreversible changes to the protein structure of a biological system, that is, coagulation. The coagulation process provides a change in internal energy associated with all the molecules that make up the system. The SECOND LAW OF THERMODYNAMICS states that for an irreversible process in an ISOLATED SYSTEM, the entropy always increases: $\Delta S_{den} = k_b \ln\left(\omega_{c,denatured} / \omega_{c,native}\right)$. The denaturation process is given energetically by $\Delta G_{den} = \Delta H_{den} - T\Delta S_{den}$, which belongs the respective chemical reaction: $a[A] + b[B] \leftrightarrows c[C] + d[D]$, with $[C_i]$ representing the physical concentration of the reaction constituents, and a, b, c, and d are reaction coefficients. This provides $\Delta G_{den} = \Delta G_{native} + RT \ln\left(\left([C]^c [D]^d\right) \big/ \left([A]^a [B]^b\right)\right)$. The activation energy for the denaturation process can be external from a fire, or laser in biology (clinical treatment) or chemical. The physical expression of the denaturation will be a biological (tissue) volume, which has been destroyed. The volume of denaturation can be derived by means of the DAMAGE INTEGRAL. The denaturation process is identified under MICROSCOPIC examination by histological markers of nuclear (i.e., CELL NUCLEUS) destruction (*also see* DAMAGE INTEGRAL) (see Figure E.82).

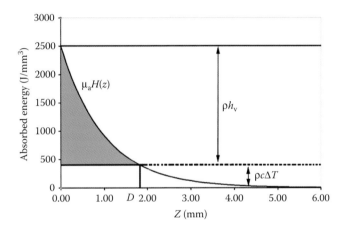

Figure E.82 Graphical outline of the Eyring absolute rate equation in a chemical process.

$F = m * a$

[general] FORCE (F) is the product of the net acceleration ($a = \partial v / \partial t$, rate of change in velocity [v] with respect to time [t]) of an object with its mass m (see Figure F.1).

Figure F.1 Illustration of force, overcoming gravity by means of muscle action.

Fabry, Maurice Paul Auguste Charles (1867–1945)

[electronics, optics] Physicist from France, most known for the introduction of the INTERFERENCE principle in SPECTROSCOPY, primarily in collaboration with his colleague ALFRED PÉROT (also "Perot") (1863–1925) (see Figure F.2).

Figure F.2 Maurice Paul Auguste Charles Fabry (1867–1945). (Courtesy of George Grantham Bain Collection, New York City; United States Library of Congress, Washington, DC.)

Fabry–Pérot interferometer

[atomic, optics] High-resolution INTERFEROMETER, designed for high-resolution spectral detection. The spectral RESOLUTION of the Fabry–Pérot interferometer is a direct function of the reflectiveness (R_{refl}) of the surfaces of the plate as well as the WAVELENGTH (λ) as: $\lambda/\Delta\lambda = m\pi\sqrt{R_{refl}}/(1 - R_{refl})$, where m is the order of the INTERFERENCE. Interferometer named after the French inventors CHARLES FABRY (1867–1945) and ALFRED PÉROT (1863–1925), introduced in 1899. The Fabry–Pérot interferometer uses an optically transparent plate (or multiple plate configuration; i.e., aligned reflective surfaces), referred to as the Fabry–Pérot étalon, where "étalon" is the French word for "gauge of measuring," "benchmark," or "standard." This interferometer is also used inside laser cavities for spectral filtration (see Figure F.3).

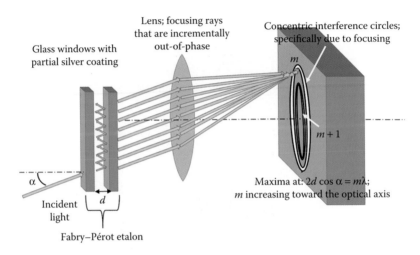

Figure F.3 Fabry–Pérot interferometer.

Fahraeus–Lindqvist effect (Fåhræus–Lindqvist effect)

[biomedical, fluid dynamics] The material effect where the VISCOSITY of a FLUID is a function of the dimensions of the FLOW system. This is applied in particular to colloid solutions, for example, BLOOD and slurry. Most well-known consequence is AGGREGATION of blood in small vessels, specifically the clumping of red blood CELL, both in ROULEAUX FORMATION under specific chemical conditions (hormonal and diseases related) as well as in CAPILLARY stacking. Since the RED blood cells have a diameter that is greater than the diameter of the capillaries, the red blood cells will stack as dishes on the counter of the dishwasher in a restaurant, primarily due to equilibrium force. RED blood cells will be transported with the central axis lining up perfectly with the axis of the vessel, scraping the WALL in circumference. Because of the FÅHRÆUS–LINDQVIST EFFECT the blood viscosity will drop sharply when the vessel diameter will decrease below a specific diameter ($D = 2R$), approximately 1 mm. In these arterioles, the flow RESISTANCE is close to the highest in the CIRCULATION; hence, the Fåhræus–Lindqvist effect provides a significant advantage for the blood PERFUSION. The apparent viscosity under the Fåhræus–Lindqvist effect can be expressed as follows: $\eta_{app,FL} = \eta_p\left[1 - (1 - (\delta/R))^4 (1 - (\eta_p/\eta_{bl}))\right]^{-1}$, with δ the cell-free PLASMA layer at the wall of the vessel, η_p the KINEMATIC VISCOSITY of the blood plasma, and η_{bl} the central viscosity of blood (non-Newtonian). The phenomenon is named after the scientists (pathologist and hematologist, respectively) ROBIN SANNO FÅHRÆUS (1888–1968) and JOHAN TORSTEN LINDQVIST (1906–2007), both from Sweden (see Figure F.4).

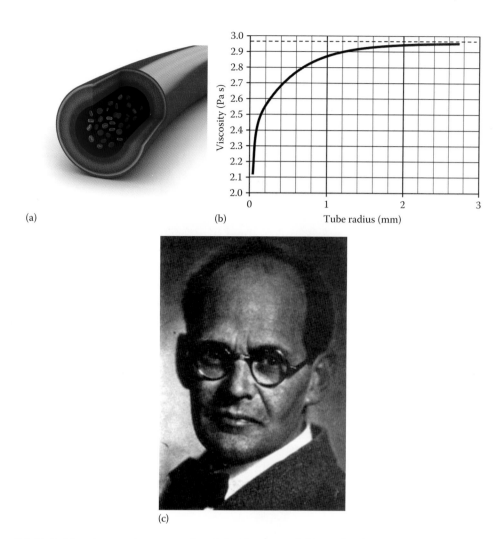

(a)

(b)

(c)

Figure F.4 (a) Cell-free layer at the outer radius of the blood vessel, (b) graphical representation of the Fåhræus–Lindqvist effect, and (c) Robin Fåhræus (1888–1968), photograph of Hans Christophersen. (Courtesy of Stiftelsen Upplands Museum, Uppsala, Sweden: Upplandsmuseet/Uppsalabild.)

Fahrenheit, Gabriel Daniel (1686–1736)

[general, thermodynamics] Physicist from the Prussian Empire (now his birthplace is in Poland), known for his introduction of the FAHRENHEIT TEMPERATURE SCALE. Gabriel Fahrenheit followed the inspiration left by his predecessor in the world of MEASUREMENT, CARLOS RENALDINI [RINALDINI] (1615–1698) who performed his temperature measurements almost 50 years prior. Another contemporary metrologist was ANDERS CELSIUS (1701–1744), whose temperature scale is used prevalently in the world apart from the United States. In the United Kingdom, the Fahrenheit scale has officially been discontinued since 1962, but common use persisted until the mid-1980s, with colloquial use of Fahrenheit still reported in current day newspapers and thermometers often still provide a dual scale. Gabriel Fahrenheit was also a glassblower, this skill provided him with the tools to increase the accuracy of the temperature measuring devices (see Figure F.5).

Figure F.5 Gabriel Daniel Fahrenheit (1686–1736). (Courtesy of Ogarnij Ogarną.)

Fahrenheit temperature scale

[general, thermodynamics] Temperature scale devised in 1724 by GABRIEL DANIEL FAHRENHEIT (1686–1736) using the FREEZING point of a mixture of SALT and water to provide the zero point calibration and the body temperature of his wife as 96°. The scale was later adjusted to the HUMAN BODY temperature at 98°. The conversion between the CELSIUS SCALE (°C) and the Fahrenheit scale (°F) is accepted globally, the latter being primarily used in the United States, as $x\ °C = \left(y\ °F - 32 \right)(5/9)$ (see Figure F.6).

Figure F.6 Fahrenheit scale.

Fanno flow

[fluid dynamics] FLOW system that is adiabatic and does not perform work, however, is subjected to WALL FRICTION. The flow takes place under steady-state conditions with constant cross-sectional area. Since no work is performed, the reference height (h_0) remains constant, as well as the MASS FLUX (ρv, for FLUID flow with density ρ under flow velocity v), hence, $h_0 = h + ((\rho v)^2 / 2\rho) =$ Const. A "Fanno line" in the entropy versus height diagram connects locations with the same h_0 and ρv ("momentum density"). Fanno flow can be considered under two conditions: subsonic flow and SUPERSONIC FLOW. Supersonic flow will result in a SHOCK WAVE.

Farad (F)

[general] "Système International" (SI) unit for the MAGNITUDE of capacitance. The unit is derived from MICHAEL FARADAY (1791–1867), in honor of his contributions to electrical theory.

Faraday, Michael (1791–1867)

[atomic, electronics, general] Physicist and chemist from England. In 1831 Faraday published the time varying MAGNETIC FIELD effects, where a moving magnetic bar (MAGNET) induces current, known as the FARADAY LAW. In addition, Faraday introduced several other concepts in ELECTRICITY and MAGNETISM such as the unit of charge: 1 Faraday $= 96,485.3415$ Coulomb. In the early 1820s, both Michael Faraday in England and JOSEPH HENRY (1797–1878) in the United States described the intercorrelation between electric fields and magnetic fields, which were used by JAMES CLERK MAXWELL (1831–1879) to develop the electromagnetic WAVE theory and combining the four equations of Henry, Faraday, and JOHANN CARL FRIEDRICH GAUSS (1777–1855). The MAXWELL EQUATIONS outlined the fact that all ELECTROMAGNETIC RADIATION travels with the speed of light. In 1824, Faraday almost accidentally invented the rubber balloon as a result of GAS LAW experiments (see Figure F.7).

Figure F.7 Michael Faraday (1791–1867). (Courtesy of J. Cochran.)

Faraday (F)

[nuclear] Unit of collective charge associated with one MOLE of electrons (each with charge e), named after MICHAEL FARADAY (1791–1867), $F = N_a e$, where the Avogadro number $N_a = 6.02214 \times 10^{23}$ molecules/mole.

Faraday, first law of

[atomic, general, nuclear, quantum, thermodynamics] During ELECTROLYSIS in a LIQUID ionic solution, the quantity of material transported is directly proportional to the applied current and the time exposure. The transported amount is also proportional to the amount of GAS released from the SOLUTION under applied current over the working time. Named after MICHAEL FARADAY (1791–1867), a chemist and physicist from Great Britain.

Faraday, second law of

[atomic, general, nuclear, quantum, thermodynamics] The ratio of the ELECTROCHEMICAL EQUIVALENT of two constituents is directly proportional to the ratio of the atomic weights of the components divided by their respective valences. For instance the valance of silver $(A = 108)$ is unity and the valance of zinc $(A = 65)$ is two, yielding for the electrochemical equivalent for zinc the derived SOLUTION from what is known for Silver by means of the mass ratio silver to zinc participating in the ELECTROLYSIS reaction as $(108/1):(65/2)$. The QUANTUM explanation states that a material of VALENCE 2 can use halve the quantity to transport the equivalent amount of charge with respect to a material with valence 1. Named after the English chemist and physicist MICHAEL FARADAY (1791–1867), from England.

Faraday cage

[general] Conductive shell that has a closely knit wire DISTANCE, or solid CONDUCTOR plating, respectively, which will effectively provide the ideal conditions for the absence of an electric field inside the enclosure. Because of GAUSS's LAW there can be no electric field line inside the confines of a conductive enclosure (including inside a solid). This is solely due to the fact that all charges will be spread out over the outermost surface, making the integration of the enclosed field lines over any surface that is smaller than the outer surface of the conductor that has no encapsulated charges. Additionally, in case of LIGHTNING strike, the surface of the conductor quickly spreads the charges over the surface leaving no net gradient of charges on the surface (see Figure F.8).

Figure F.8 Faraday cage, person wearing a conductive wire mesh that dissipates electrical discharge arcs.

Faraday constant

[atomic] Constant linking the electrical work (W_e; in contrast to mechanical work, $W_m = P\Delta V + V\Delta P$, with P the internal pressure and V the volume of the system) to an applied electrical potential (ΔV_e is the maximum difference in the electrical potential resulting from the transferred charges) as, $W_e = -nF_a\Delta V_e$, with n is the number of charges transferred in the process, and $F_a = 96,485$ C/mol the Faraday constant.

Faraday dark space

[atomic, nuclear] In the process of the formation of the CATHODE ray in the Crookes rarefied GAS tube (after the English physicist SIR WILLIAM CROOKES [1832–1919]), an arc is formed at the location of the cathode (NEGATIVE CHARGE applied), while immediately in front of the cathode there is no LUMINESCENCE (absence of arc), referred to as the Faraday dark space, also named the Crookes dark space. Supposedly, the phenomenon was described first by MICHAEL FARADAY (1791–1867), and confirmed or documented by Sir William Crookes in 1870. The absence of discharge is due to the repulsion of charges by the cathode, removing the PLASMA charges away from the cathode charge concentration. Crookes also refined the observation by describing the total disappearance of the dark space when the GLASS tube was totally evacuated, after migrating first toward the ANODE. Once the glass tube is totally under VACUUM conditions, the glass itself on the anode side will begin to glow.

Faraday law

[atomic, electronics, general] Law of electromotive induction describing the electromotive induced voltage (ε_{emf}) in a circuit by a changing MAGNETIC FIELD through an open surface (S) defined as $\varepsilon_{emf} = -(d/dt)(N\phi_m) = -N(d/dt)(\phi_m) = -N(d/dt)(\int_S \vec{B} \cdot \vec{n}\, da)$, indicating the rate of change (dt) of the MAGNETIC FLUX (ϕ_m) from a magnetic field ($\vec{B}$) incident perpendicular ($\vec{n}$) to the surface S of the loop bounded by an electrical circuit, where N denotes the number of loops of the circuit. Named after the physicist MICHAEL FARADAY (1791–1867), from England (see Figure F.9).

Figure F.9 The device used by Michael Faraday to conceive the Faraday's laws of induction.

Faraday's induction law

[general] *See* **FARADAY LAW.**

Far-field diffraction

[general] *See* **DIFFRACTION, FAR-FIELD.**

Farsightedness

[biomedical, general, optics] Conditions known for humans that are not able to accommodate the LENS (the lens will not "bulge" out to its full relaxed state or small radius of curvature) of the EYE in order to provide a clear focus of an object at close proximity. Under these conditions the IMAGE of an object at close proximity will be formed behind the RETINA of the eye. The "unwillingness" of the lens to form a fully relaxed small radius of curvature is due to the loss of COMPLIANCE of the lens, primarily as aspect of loss of collagen with increasing AGE. Additionally, shortening of the eye ball itself can be a cause for this defect as well. This condition can be corrected by placing a POSITIVE LENS in front of the lens of the eye, or by artificially modifying the focal length of the eye by refractive surgery (e.g., LASIK). It is also known as hyperopia (in contrast to "presbyopia," near-sightedness) (see Figure F.10).

Normal vision and hyperopia

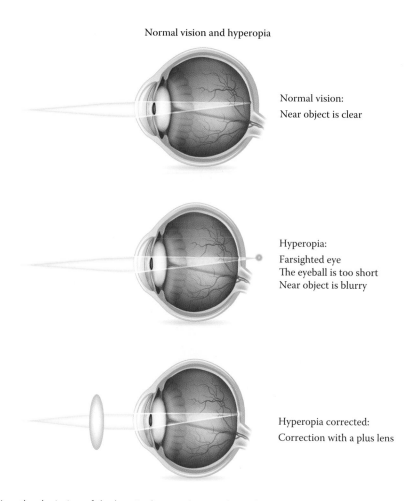

Normal vision:
Near object is clear

Hyperopia:
Farsighted eye
The eyeball is too short
Near object is blurry

Hyperopia corrected:
Correction with a plus lens

Figure F.10 How the deviation of the lens in the eye that produces hyperopia can be corrected.

Fast Fourier transform (FFT)

[computational] SEPARATION OF VARIABLES with respect to a frequency broadband SIGNAL in the respective base and harmonic waves, based on an algorithm and methodology introduced by the French mathematician BARON JEAN BAPTISTE JOSEPH FOURIER (1768–1830). FFT is an accelerated signal processing algorithm based on the discrete Fourier transform techniques, with the best acceptable resolution and amount of detail with respect to the processing time, based on several assumptions of regularity, reducing the number of computations required for signal processing of N data points from $2N^2$ to $2N \log_2 N$ computations. The FFT concept itself dates back to JOHANN CARL FRIEDRICH GAUSS (1777–1855) in 1805, and discussed in greater detail with elaborated principles by the mathematicians from the United States, SIR JAMES WILLIAM

Cooley (1926–) and John Wilder Tukey (1915–2000) in 1965, and implemented in refined format in 1999 by the Australian mathematician and physicist Ronald Newbold Bracewell (1921–2007) based on the work he started in 1983. The Fourier transform principles use the conversion of time to frequency and vise versa as the basis. Both the Nyquist or Shannon theorem prescribe that a periodic signal must be minimally sampled at twice the fastest frequency phenomenon of the signal or IMAGE (spatial frequency). One specific example is the deconvolution of the heart DEPOLARIZATION signal: the ELECTROCARDIOGRAM (ECG), on its base signal in the range of 0.75 Hz (approximate average value at rest) to 3 Hz (running), with up to the first eight harmonics to be able to fully reconstruct the depolarization pattern. The MAGNITUDE of the harmonics can provide a quick analysis of potential clinical problems that require further attention, which will also require careful and meticulous analysis of the transient ECG pattern, specifically with isolation of ectopic BEATS (i.e., single beats that have a very different period from the "standard" HEART rate). It is also known as FFT (*also see* FOURIER TRANSFORM) (see Figure F.11).

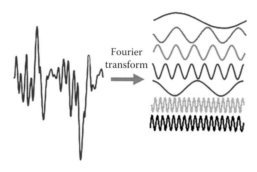

Figure F.11 Outline of the separate frequencies compiling the convoluted signal on the left under Fourier transform.

Fatigue

[general, mechanics] Failure mode of object subjected to cyclic stress (see Figure F.12).

Figure F.12 Failure mode of object that has been bent back and forth several times.

Fault line

[general, geophysics] Identified geologic feature that extends over a DISTANCE and is associated with the relative movement of landmasses with respect to each other, specifically under an EARTHQUAKE. Fault lines are defined based on their respective azimuthal orientation as well as the dip of the fault, outlining the surface demarcation of a landmass fault. Several different types of faults can be identified. There is a strike-slip fault, which performs a lateral MOTION of the hanging WALL with respect to the footwall and have virtually no vertical displacement. Dip-slip faults slide over each other at a shear ANGLE with respect to the horizontal plane, acting as a ramp, making the hanging wall move at an incline angle with respect to the footwall. A specific configuration of the dip-slip fault is the THRUST fault, with a ramp angle of less than 45°, known as the most vigorous earthquake forming mechanisms (large Richter scale classification earthquakes). Another format is provided by the oblique-slip fault mechanism, also positioning the hang wall at an angle of incline with respect to the footwall and moving in both lateral and supine motion. Other fault formats are listric, ring, and synthetic/antithetic faults. Listric faults are identified by a curved fault separation between the hanging and the footwall. Ring faults are not specifically associated with tectonic plate, but more so with volcanic craters and meteoric impact sites. Fault lines can on a larger scale be correlated to the TECTONIC PLATES of the EARTH'S CRUST. The attributes of tectonic interactions are described by the following eight characteristics: (1) LATITUDE, (2) longitude and (3) depth (providing the epicenter); (4) origin time (providing the focus combined with the epicenter information; (5) MAGNITUDE OR ENERGY; (6) strike or azimuth; (7) dip of the relative movement; and (8) plunge (vector motion). Examples of specific fault lines are located all over the world. The San Andreas Fault, identified in California in 1895, extends over approximately 1300 km forming the fissure boundary between the North American tectonic plate and the Pacific plate; identified as strike-slip fault. The daily ACTIVITY of the San Andreas Fault is taken as part of everyday life in San Francisco, not discrediting the large number of more powerful earthquakes, including the one in 1906 with an estimated magnitude of 7.8 on the Richter scale. The Richter scale was introduced in 1935, developed under collaborative effort between physicist CHARLES FRANCIS RICHTER (1900–1985) from the United States and seismologist BENO GUTENBERG (1889–1960), originally from Germany, both working in the California Institute of Technology. The Richter scale uses the LOGARITHM of the AMPLITUDE of the SEISMIC recordings using a calibrated SEISMOMETER (mechanical vibrations) corrected for distance to the source, using three-point averaging between several stations. The Garlock Fault in California, extends over 250 km, is a left-lateral strike-slip fault located on the northeast–southwest of the NORTH margins of the Mojave Desert of Southern California. The Garlock Fault is under continuous motion, moving at 2–11 mm/year with additional 7 mm slip, but is not a majorly active earthquake fault. Another large fault line is near Charleston, South Carolina. Other fault lines in the United States are in Hawaii and Alaska. The South American Pichilemu and Liquiñe-Ofqui Fault fault lines are posing a significant threat to structural integrity and human lives. In China, the Longmenshan Fault (a thrust fault) extends over a length of 250 km and resulted in a 7.9 magnitude earthquake in 2008. Additionally, the Longmenshan Fault raises Tibet up by 4 mm/year. In Turkey, there is North Anatolian Fault (right-lateral strike-slip fault), which passes south of Istanbul and runs along the northern sea shore of the country. The North Anatolian Fault has endured many greater than 6.8 magnitude earthquakes (12 in the past 80 years). In Australia, the Moyston and Selwyn Faults have shown the capability of producing earthquakes with magnitudes between 7.0 and 7.5 on the Richter scale. Close to 100 different faults can be distinguished over the world. Apart from the Richter scale, the prior classification of earthquakes was made on the Mercalli scale (1931), based on a mixture of emotional impact and mechanical destructive impact. The energy in a tectonic shift (earthquake) ranges from approximately 10^8–10^{15} J. Activity in the region of fault lines can be recognized by animals prior to any mechanical activity is registered by the seismometric instrumentation. Specific characteristics associated with fault activity can be smells, tastes (water flavor changes due to MIXING of underground wells), and sounds. Additionally, during the movement, apart from the mechanical tremors there have been observations of light activity. Light activity observed has been glow in the sky, aurora's, next to "MICROSCOPIC" cavitation arcs at the soil level as well as frictional IONIZATION. The atmospheric generation of light can be the results of density changes in ATMOSPHERIC LAYERS next to the rapid changes in local MAGNETIC FIELD density. The study of faults is SEISMOLOGY (see Figure F.13).

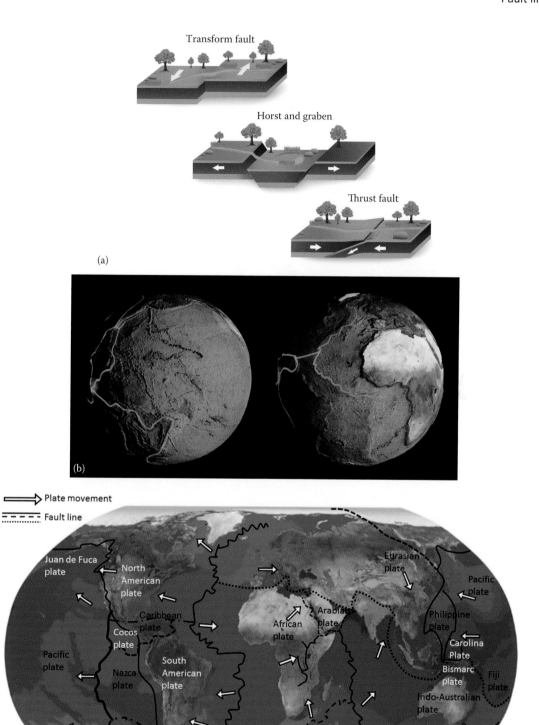

Figure F.13 (a) Types of fault lines and (b) the outline of the fault lines for the tectonic plates of the world. (c) Fault lines on a geographic map. (Courtesy of Maps of World.)

FDA

[biomedical] United States (US) Food and Drug Administration. Regulatory institution on governmental basis designed to oversee and control the safety and efficacy of medical devices and devices with medical and consumer aspects of medical intention. It is also called as USFDA.

Fermat, Pierre de (1601–1665)

[computational, general, optics] Lawyer and mathematician from France. Pierre de Fermat is best known for his treatise on OPTICS, the FERMAT PRINCIPLE is based on the geometric analysis of the rays of optics. Pierre de Fermat also worked closely with another contemporary French mathematician BLAISE PASCAL (1623–1662). The work of both Pascal and Fermat formed the mathematical foundation for the theory of PROBABILITY (see Figure F.14).

Figure F.14 Pierre de Fermat (1601–1665).

Fermat principle

[optics] PHYSICS principle closely related to the action principle in classical mechanics. Analogously related to other physics principles, it describes all the appropriate PHYSICS LAWS pertaining to the GEOMETRICAL OPTICS principles to take place under the constraints of minimal time, derived by the French mathematician PIERRE DE FERMAT (1601–1665). In OPTICS it defines the path traversed by a beam of light between two points to proceed in the least amount of time. Applied to the WAVE theory of light, this also entails that small deviations in the path of light will not result in first-order deviations from the transit time. The extremes in optical path length are either minimal or maximal, as a function of the applied and persistent. One particular extreme providing a maximal path length will be realized under gravitational lensing. This ties into Snell's REFRACTION law from WILLEBRORD SNEL (SNELLIUS) VAN ROYEN (1580–1626) as well as the diffraction statement by AUGUSTIN-JEAN FRESNEL (1788–1827), more than century later. The Fermat principle is clearly visible under conditions of a heated asphalt or desert sand surface in the DISTANCE, producing a "mirage" also known as "Fata Morgana." Under specific thermal conditions, a density gradient provides an optical path that may carry an IMAGE for tens of kilometers, beyond the edge of the horizon, or reflect an object in distorted view.

Fermi, Enrico (1901–1954)

[atomic, nuclear, solid-state] Italian physicist. He made major theoretical contributions to the description of the atomic model. His work was based on his use of neurons in the creation of isotopes. The theoretical modeling included the introduction of SPIN orbit restrictions: PAULI EXCLUSION PRINCIPLE, and ENERGY quantization. He also played a key role in the development of the ATOMIC BOMB (see Figure F.15).

Figure F.15 (a) Picture of Enrico Fermi (1901–1954). (Courtesy of National Archives.) (b) Enrico Fermi attending the 1933 Solvay conference in Brussels, Belgium.

Fermi (unit of length)

[general] Equivalent to a femtometer ($1\,fm = 10^{-15}\,m$) in honor of the Italian scientist ENRICO FERMI (1901–1954) as a characteristic length on nuclear basis, with respect to his introduction of the "NEUTRINO" as a definition for a nuclear PARTICLE in 1930, meaning "little neutral one." Later the neutrino came to be named electron antineutrino, due to the fact that it was emitted from the NUCLEUS under CONSERVATION OF ENERGY along with an electron, based on the experimental observations of WOLFGANG PAULI (1900–1958).

Fermi decay

[nuclear] The process of nuclide beta DECAY is the most common in RADIOACTIVE DECAY, primarily due to the fact that a large portion of nuclides does not reside in the "valley of stability." The beta decay process can be identified by the ORBITAL ANGULAR MOMENTUM that is disposed of as electron and NEUTRINO (i.e., electron

antineutrino) are emitted, next to the intrinsic change in PARITY as well as the respective alignment of the electron neutrino spins. When the ELECTRON SPIN is aligned with that of the electron antineutrino, the process is described by the GAMOW–TELLER DECAY, whereas when they are antiparallel the process is described by Fermi decay (as described in preliminary fashion by ENRICO FERMI [1901–1954] in 1934). The Fermi decay is the most common decay process, where the ALLOWED TRANSITION is confined to a transfer of zero (0) orbital angular momentum ($S_\beta = 0$. S_β spin of beta decay). Under these conditions, the decay constant is roughly proportional to the decay ENERGY to the fifth power. The QUANTUM mechanical equivalent decay process defined by Enrico Fermi uses similar concepts used in the electromagnetic decay process under classical MECHANICS, adapted to an "electron neutrino field." The decay constant is defined based on a perturbation potential (ΔV) and the number of nuclide states (N_n) per unit energy (E), expressed as $dN_n/dE = (p^2(dp/dE)/2\pi^2\hbar)\ell^3$ (momentum $p = (\hbar\pi/\ell)n$; POTENTIAL WELL with dimension ℓ with energy $E = (\pi^2\hbar^2/2m_e\ell^2)(n_x^2 + n_y^2 + n_z^2)$, n an integer constant; $\hbar = h/2\pi$, with $h = 6.62606957\times10^{-34}$ m^{2}kg/s the Planck's constant), which yields for the DECAY CONSTANT (τ_F) as a function of the eigenfunctions to the SCHRÖDINGER EQUATION (ψ_i; complex conjugate ψ_i^*, ψ_i the excited NUCLEUS; ψ_f the final nucleus, after decay): $\tau_F = (2\pi/\hbar)\left|\int_{\text{Nuclides Volume}} \psi_f^*(\text{system}) \Delta V \psi_i(\text{system})dxdydz\right|^2 (dN_n/dE)$. The measured half-life for nuclide decay under the beta-decay process ranges from 10^{-3} s to 10^{16} years. The "valley of stability" $Q_a = ((M_D + M_\alpha)/M_D)KE_\alpha \cong (A_P/(A_P - 4))KE_\alpha \equiv -S_{a,D}$ referring to the DAUGHTER NUCLEUS, p pertaining to the "parent," KE_α the kinetic energy for the ALPHA PARTICLE, and S_a representing the separation energy for an alpha particle, with respect to the PARENT NUCLEUS.

Fermi distribution

[energy, nuclear, quantum] High temperature (T) correction for the Schrödinger EIGENVALUE potential well values $\left(\text{well with dimension } \ell \text{ with ENERGY } E = (\pi^2\hbar^2/2m_e\ell^2)(n_x^2 + n_y^2 + n_z^2)\right): k_i\ell = n_i\pi, i = x, y, z$; the WAVE number $k = 2\pi/\lambda$ for wavelength λ; n an integer constant; $\hbar = h/2\pi$, with $h = 6.62606957\times10^{-34}$ m^{2}kg/s the Planck's constant, and m_e the electron mass for the electrons in an "electron–gas" configuration. The correction is provided as: $F(E, T) = 1/(\delta + e^{\alpha_F}e((E - E_0)/k_bT))$, with the Boltzmann coefficient $k_b = 1.3806488 \times 10^{-23}$ m^{2}kg/s^{2}K; classical (BOLTZMANN) DISTRIBUTION $\delta = 0$, $e^{\alpha_F} = N = \int_0^\infty n(E)dE$ the total number of particles; for bosons $\delta = -1$, $\alpha_F = NB$ derived from the NORMALIZATION $\int P(E)dE = 1$; for fermions $\delta = +1$, $\alpha_F = NB$. It is also known as Fermi PROBABILITY distribution (see Figure F.16).

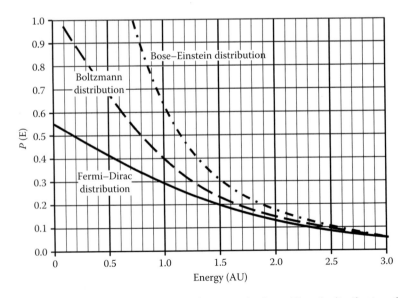

Figure F.16 Graphical representation of the difference between the Bose–Einstein distribution, the Boltzmann distribution, and the Fermi–Dirac distribution.

Fermi energy (E_F)

[atomic, chemical, general, quantum, solid-state] ENERGY gap between the unoccupied states of electron orbital configuration in the BOHR ATOMIC MODEL from the occupied orbital energy bands in atomic and molecular structure. The Schrödinger equations provide the tools to calculate the atomic and molecular energy levels that are available to bring electrons in a state that will allow them to move freely in a volume of medium made up of a single ELEMENT or a mixture. Schrödinger's conditions will quantize these energy levels for electrons. In the Bohr model, each energy level can be occupied by two fermions, where both will have opposite SPIN (spin up vs. spin down). At ABSOLUTE ZERO, all ATOMIC ENERGY levels are filled reaching a system energy for the element that is referred to as the CHEMICAL POTENTIAL as well as the Fermi energy. For an electron in a box with equal side (cube), the energy levels can be defined as $E_n = (\pi^2\hbar^2 n^2/2m_e\ell^2) = E^*$, where $n^2 = n_x^2 + n_y^2 + n_z^2$, $\hbar = h/2\pi$ with $h = 6.6260755\times10^{-34}$ Js the Planck's constant, n are the filled energy levels, m_e the electron mass, and ℓ the length of the sides (also referred to as the depth of the POTENTIAL WELL). In terms of Fermi energy levels this can be rewritten at the Fermi QUANTUM numbers n_F, $E_F = (\pi^2\hbar^2 n_F^2/2m_e\ell^2) = 3^{2/3}(\pi^{4/3}\hbar^2/2m_e)(N/V)^{2/3}$, with the volume of charges $V = \ell^3$ and the respective number of electron states operating at energy lower than E^*, $N = (\ell^3/3\pi^2\hbar^3)(2m_eE^*)^{3/2}$, defining $n_{max} = (\ell/\pi\hbar)(2m_eE^*)^{1/2}$. The Fermi energy indicates the occupancy cut-off for the Fermi–Dirac system, under conditions of $T = 0$ (*also see* CHEMICAL POTENTIAL) (see Figure F.17).

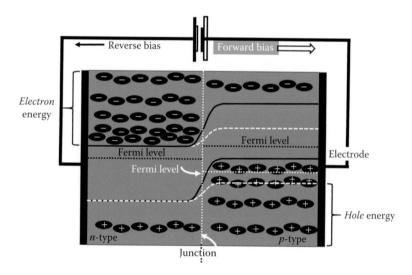

Figure F.17 The shift in Fermi energy for a semiconductor material *pn*-junction under forward bias (white) with respect to reverse bias (black). The *pn*-junction indicates semiconductor material doped with electrons for the *n* material, and depleted of electrons in the *p* material.

Fermi function

[nuclear] The penetrability with respect to a POTENTIAL BARRIER, resulting from nuclear and Coulomb forces, $V = 2e^2Z_D/r$, Z_D the PROTON number (atomic number) for all the charges within the barrier boundaries of the daughter, $e = 1.60217657\times10^{-19}$ C the electron charge, and r the radius from the center of the NUCLEUS. The PROBABILITY for penetration is described as $P_F \approx e^{-\gamma_F}$, where $\gamma_F = (2/\hbar)\int_R^b [2M_0((ze^2Z_D/r) - Q_\alpha)]^{1/2} dr$ and $M_0 = M_\alpha M_D/(M_\alpha + M_D)$ represents the REDUCED MASS due to recoil and b the "barrier thickness." Where the DECAY time correlates as $\tau_F \approx (v_{in}/R)P_F$, with v_{in} the velocity of the ALPHA PARTICLE inside the PARENT NUCLEUS.

Fermi gas

[atomic, computational, energy, fluid dynamics, thermodynamics] Collective of fermions. Fermions are named after ENRICO FERMI (1901–1954), and are particles that are conforming to the Fermi–Dirac statistical distribution. The FERMI GAS is the simplistic approach to free electron description in a CONDUCTOR. In a Fermi gas, the lowest possible ENERGY corresponds to zero kinetic energy, whereas for a standard METAL with free electrons the lowest energy state resides at the "bottom" of the CONDUCTION BAND.

Fermi hole

[energy, nuclear, quantum] QUANTUM mechanical description of the electron and specifically the lack of electron locations in a many-atom system. The ENERGY configuration is bound by Pauli's EXCLUSION PRINCIPLE. In particular the occurrence of paired electrons with parallel SPIN will interact with other electrons by means of repulsion, making the "seated" electrons part with their respective energy configuration. Additionally, these electrons with parallel spin will repel each other, forming what is known as Fermi holes; hence reducing the repulsive force between them. Nevertheless, all electrons are equal and indistinguishable, rendering the concept convoluted. In contrast, electrons traveling in pair possessing opposing spin will group together, forming what is known as an "electron heap." For Helium ($1s2p$), this can be described as a WAVEFUNCTION (Schrödinger SOLUTION) with spatial antisymmetry (Fermi hole), splitting the wavefunction in a spatial part (ψ_0) and a spin part (ψ_s), $\psi_{\uparrow\uparrow} = \psi_0\psi_s = \left[1s(1)2p(2) - 1s(2)2p(1)\right] \times \uparrow(1)\uparrow(2)$, compared to the antisymmetric situation (Fermi heap) with spin antisymmetry, $\psi_{\uparrow\downarrow} = \left[1s(1)2p(2) + 1s(2)2p(1)\right] \times \left[\uparrow(1)\downarrow(2) - \uparrow(2)\downarrow(1)\right]$.

Fermi level

[atomic, general] Both for the electron SHELL MODEL of the ATOM as well as the PROTON and NEUTRON configuration in the NUCLEUS, the highest occupied level, with respect to the associated kinetic ENERGY in the waveform fitting that orbital design. Electrons can occupy the same state for at most two electrons, provided their SPIN is in the opposite direction. For the nucleons the shells are defined by the DE BROGLIE WAVELENGTH ("circumference" to be a match to a multiple of wavelengths), with a grouping of closely separated energy levels forming one shell. Under QUANTUM mechanical considerations, these shells are constructed of a system of closely separated subshells. The major shell are spaced farther apart as was described by MARIA GOEPPERT-MAYER (GÖPPERT) (1906–1963) and JOHANNES HANS DANIEL JENSEN (1907–1973) in 1947. The standing WAVE fitting the orbit conforms to a QUANTUM NUMBER. Neutrons and protons do not influence each other's energy configuration and can share the same states, while at most two protons and two neutrons can occupy the same state. When the respective major shells are filled the atomic nucleus is more stable then when not, such as the filled configuration for Tin with 50 protons and 50 neutrons, can support 10 stable isotopes.

Fermi quantum number (n_F)

[atomic, energy, nuclear] Dimensional expression for the "radius" of a sphere $(\overrightarrow{n_F})$ in K-SPACE ($k = 2\pi/\lambda$ for wavelength λ of the Schrödinger WAVEFUNCTION) that contains all ENERGY states, with MAGNITUDE: $n_F = (3N/\pi)^{1/3}$, using the respective number of electron states operating at energy lower than E^*, $N = (\ell^3/3\pi^2\hbar^3)(2m_e E^*)^{3/2}$, where an electron in a box with equal side (cube, with ℓ the length of the sides; the depth of the POTENTIAL WELL) the energy levels can be defined as having energy $E_n = (\pi^2\hbar^2 n^2/2m_e\ell^2) = E^*$, where $n^2 = n_x^2 + n_y^2 + n_z^2$, $\hbar = h/2\pi$ with $h = 6.6260755 \times 10^{-34}$ Js the Planck's constant, n are the filled energy levels, and m_e the electron mass.

Fermi shape

[atomic, energy, nuclear] Shape of the semiempirical nuclear POTENTIAL WELL density distribution: $f(r) = 1/(1 + e^{(r-R_0)/a_f})$, where r is the radius, $R_0 = r_0 A^{1/3}$ is the half-way radius (A the atomic number and r_0 the root-mean-square nuclear Fermi radius [of a uniformly charged sphere], order of femtometer), and a_f the diffuseness of the charge distribution.

Fermi sphere

[atomic, computational, energy, fluid dynamics, thermodynamics] In QUANTUM THEORY all particles, including atoms and molecules, can be considered as fermions or bosons. The Fermi sphere describes the ENERGY outline in K-SPACE for fermions (*see* **FERMI SURFACE**).

Fermi surface

[atomic, computational, electronics, quantum, solid-state] Concept used with respect to "density of states" relating to electron charges in conductive solids and liquids, defined in the K-SPACE (wave number space). Every conceivable state of the electrons is defined in reference to the three vector components of the WAVE NUMBER ($k = 2\pi/\lambda$, where $\lambda = c/v$ the wavelength for the WAVEFUNCTION, c the speed of light, and v the frequency). The electrons will fill all the energies to the Fermi ENERGY when at ABSOLUTE ZERO temperature, using the Fermi–Dirac distribution for half-integer SPIN particles, with momentum $p = k\hbar$, $\hbar = h/2\pi$, with $h = 6.62606957 \times 10^{-34}$ m^{2}kg/s the Planck's constant. The total number of electrons (N) will occupy a volume in k-space with radius: $k_F^3 = 3\pi^2 N$, defining the spherical surface outline. Thus the FERMI SURFACE is a sphere with radius k_F. The polyhedron that outlines the Fermi surface is referred to as the Brillouin zone. The Brillouin zone defines the Bragg REFLECTION planes, the periodic crystalline LATTICE planes structure providing a diffraction structure for the electron waves. Additional reference to the BRILLOUIN ZONE may also be required for full definition (see Figure F.18).

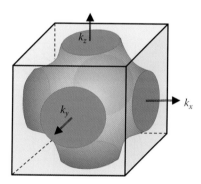

Figure F.18 Representation of the Fermi surface for a simple cubic atomic lattice.

Fermi wavevector (k_F)

[solid-state, thermodynamics] In a one-dimensional electron GAS there are free electrons that can roam freely through a material (CONDUCTOR). The Fermi wavevector determines the Fermi ENERGY (ϵ_F) of the electron gas as $\epsilon_F = (\hbar^2/2m)(N_0\pi/2L_e)^2 = (\hbar^2/2m_e)k_F^2$, where L_e is the length of the electron chain at the FERMI SURFACE, $\hbar = h/2\pi$, with $h = 6.62606957 \times 10^{-34}$ m^{2}kg/s the Planck's constant, N_0 the total number of electrons, m the PARTICLE mass (i.e., electron; m_e), and the Fermi wavevector: $k_F = (N_0\pi/2L_e) = N_e\pi$, where N_e is the total number of electrons per unit length per SPIN direction. The density of spin states for one spin direction is $n(\epsilon) = (L_e/\pi\hbar)\sqrt{(m_e/2\epsilon_F)} = L_e/\pi\hbar v_e$, where the electron speed follows from $v_e = \hbar k_b/m_e$, $k_b = 1.3806488 \times 10^{-23}$ m^{2}kg/s^{2}K is the Boltzmann coefficient.

Fermi workfunction

[atomic, computational, energy, fluid dynamics, thermodynamics] The MINIMUM ENERGY requirement to satisfy the removal of an electron from a solid when residing at the Fermi energy level. $W_F = E_v - E_F$, with E_v the energy of the electron at rest once removed from the solid (VALENCE energy) and Fermi energy within the solid $E_F = \pi^2 \hbar^2 n_F^2 / 2m_e \ell^2$, $\hbar = h/2\pi$ with $h = 6.6260755 \times 10^{-34}$ Js the Planck's constant, n is the filled energy levels, m_e the electron mass, and ℓ the length of the sides of the WAVE number space (also referred to as the depth of the POTENTIAL WELL), and the Fermi QUANTUM numbers n_F (*also see* WORKFUNCTION).

Fermi–Dirac statistics

[atomic, nuclear] Statistical description of the ENERGY configuration for a nuclear system that conforms to the PAULI EXCLUSION PRINCIPLE. The statistical description was introduced by ENRICO FERMI (1901–1954), with corrections by P. Dirac, incorporating the fact that the total WAVE function (i.e., SOLUTION to the SCHRÖDINGER EQUATION) may also be antisymmetric. For a system of n_i occupation levels with associated PARTICLE energies ϵ_i, with g_i respective nondegenerate QUANTUM states (only one particle allowed per state, respectively two when considering the SPIN condition; consequently $n_i \leq g_i$) the PROBABILITY of finding the respective energy levels occupied is given as $W_P = \prod_i (g_i!/(g_i - n_i)! n_i!)$, note that $\sum n_i = N$ the total number of particles, and $\sum n_i \epsilon_i = E$ the total energy. This also yields the FERMI DISTRIBUTION defined as $n_i = g_i/(A^{-1} \epsilon^{\epsilon_i/k_b T} + 1)$, as a function of (absolute) temperature T, with the Boltzmann coefficient $k_b = 1.3806488 \times 10^{-23}$ m²kg/s²K, and A a constant derived from the energy constraints. Compared to the BOLTZMANN DISTRIBUTION, this is written as $W_P^B = \prod_i g_i^{n_i} / n_i!$ (see Figure F.19).

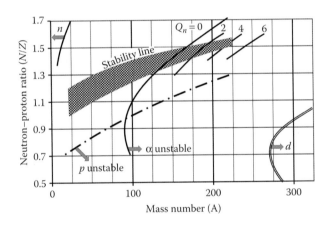

Figure F.19 Graphical representation of Fermi decay.

Fermi–Pasta–Ulam problem

[computational] Apparent paradox in QUANTUM MECHANICS and chaos theory with regard to the definition of LATTICE oscillations and the respective periodic behavior, based on the contributions from ENRICO FERMI (1901–1954; Italy–the United States), JOHN R. PASTA (1918–1984; the United States), and STANISŁAW MARCIN ULAM (1909–1984; Austria/Hungary/Poland–the United States; "inventor" of the MONTE CARLO METHOD), introduced in 1953. The computational analysis of an OSCILLATION introduced nonlinear terms with quadratic and cubic components with respect to a "vibrating string."

Fermion

[atomic, general, quantum, solid-state] Class of particles that obeys the FERMI–DIRAC STATISTICS, including the electron, PROTON, NEUTRON, and NEUTRINO (BARYONS). This confines the SPIN to half-integer values only, in contrast to bosons with whole integer spin (e.g., ALPHA PARTICLES and PHOTON).

Ferrite

[general] Conductive materials, primarily labeled as the oxides of magnesium, zinc, manganese, or nickel. Note that the ELEMENT iron is ferrum ($^{56}_{26}$Fe). Ferrites are used to form permanent magnets based on their inherent EDDY CURRENT composition. Aligning the eddy currents by means of an externally applied electric field will transpose the medium in a PERMANENT MAGNET. Ferrites are also popular in TRANSFORMER cores and SOLENOID-based mechanisms as the MAGNETIC FIELD confining medium (see Figure F.20).

Figure F.20 Ferrite example, clamp on signal cable to reduce noise on the line (e.g., found on computer monitor cables).

Ferroelectric

[electronics, material] Material property describing the electric properties of a medium that behave in similar fashion to ferromagnetic for MAGNETISM. Ferroelectric indicates a material with high DIELECTRIC constant that can sustain permanent electrification and additionally displays electrical HYSTERESIS. Ferroelectric media are found to have an ELECTRIC DIPOLE MOMENT for the entire medium in EQUILIBRIUM STATE.

Ferromagnet

[electronics, material] IRON and iron-like substances that can be magnetized. Ferromagnetic materials are iron, cobalt, nickel, gadolinium (rare EARTH) as well as certain alloys of manganese and chromium, whereas the latter two are not ferromagnetic by themselves. Substances with properties of FERROMAGNETISM are called ferromagnetic (general) (see Figure F.21).

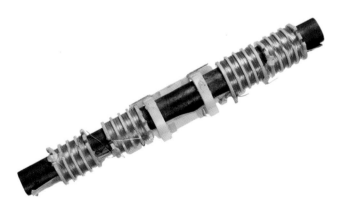

Figure F.21 Old-fashioned AM radio antenna with ferromagnet rod inside.

Ferromagnetism

[atomic, general, nuclear] Any substance with free electrons has the capability to generate local, MICROSCOPIC magnetic fields based on the MOTION of the free electrons based on the Lorentz force expressed on a moving charge and the MAGNITUDE can be found from AMPÈRE'S LAW (for a linear moving charge) and BIOT–SAVART LAW (for a circular current loop). Primarily, the orbital motion of electrons around the NUCLEUS of an ATOM in the BOHR ATOMIC MODEL has an inherent MAGNETIC FIELD. Combining the atomic fields resulting from alignment due to free electrons can produce a resultant external magnetic field. Electrons in second place possess a SPIN with associate angular momentum and linked magnetic moment. The magnetic moment of the individual electrons is oriented directly opposing the angular momentum vector due to the NEGATIVE CHARGE.

Fessenden, Reginald Aubrey (1866–1932)

[acoustics, imaging] Scientist and electrical engineer from Canada, known for the first patent on ECHOLOCATION mechanism of action in 1912, as applied to ULTRASONIC IMAGING and underwater sonar tracking (see Figure F.22).

Figure F.22 Reginald Aubrey Fessenden (1866–1932).

Feynman, Richard Phillips (1918–1988)

[atomic, energy, general, mechanics, molecular, quantum] Theoretical physicist from the United States. Dr. Feynman is best known for his popular teachings of PHYSICS and his contributions to the development of the ATOMIC BOMB. Richard Feynman uncovered several inconsistencies in the known theories on QUANTUM ELECTRODYNAMICS and his revisions were rewarded with the Nobel Prize of Physics in 1965 (see Figure F.23).

Figure F.23 Richard Feynman (1918–1988). (Courtesy of Tamiko Thiel.)

Feynman diagram

[general, quantum, theoretical] Graphical representation of the virtual interactions between ELEMENTARY PARTICLES pertaining to QUANTUM ELECTRODYNAMICS. The illustrations are designed to elude to the basic principles of the presumed interactions, based on theoretical predictions. The "AMPLITUDE" of a PARTICLE being in a certain location can be illustrated as a vector in three-dimensional space, alternatively the associated PROBABILITY of an interaction at that location is represented by the amplitude squared. The processes for these types of constituents are convoluted and cannot be isolated as individual process, hence the "virtual process" reference. The processes of elementary particles have not been empirically established due to the limitations in current experimental technology as well as the inherent uncertainty associated with these processes based on the Heisenberg principles. Some of the better verified interactions are with respect to photon–electron interactions, photon–photon interactions, and electron–electron processes. For a given reaction, the object and interaction are all represented by amplitudes; say the migration of a PHOTON from point $\vec{x_1}$ to point $\vec{x_2}$ is represented by amplitude $B(\vec{x_1}, \vec{x_2})$, while the undisturbed track for an electron from point $\vec{x_1}$ to point $\vec{x_2}$ is represented by amplitude $C(\vec{x_1}, \vec{x_2})$, and associated interaction amplitude of the photon with the electron $A_{(1)} = eD$, e the electron charge. Alternatively, additional events may take place, such as the resonant interaction of the photon with the electron (for instance, in location $B(1,0)$) resulting in a "virtual" photon, which may subsequently be reabsorbed ($C(1,2)$), rendering the electron in $B(0,2)$, represented by amplitude $A_{(2)}{}^a = e^3 \int DB(1,0) DB(0,2) DC(1,2)$. Each occasion is referred to as a "vertex," effectively the DEGREES OF FREEDOM of the system. The complexity may progress with additional interactions, adding additional nodes in the Feynman diagram, and resulting superposition in the amplitude integration. The total amplitude is provided by the summation of all the interaction, $A = A_{(1)} + A_{(2)} + A_{(3)}{}^a + A_{(3)}{}^b + \cdots$, with associated PROBABILITY $P = |A|^2$. The potential interactions are strong interaction, WEAK INTERACTION, electromagnetic and gravitational for regular (electron, NEUTRON, PROTON), and elementary particles (quarks, etc.) as well as RADIATION. The complexity of the Feynman diagram increases with the number of events and components (see Figure F.24).

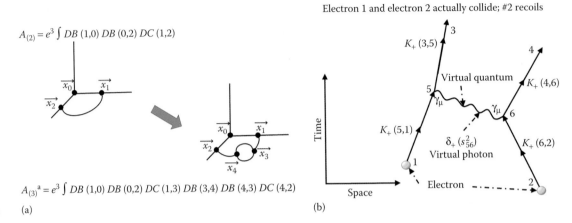

Figure F.24 (a) Feynman diagram for photon–electron interaction with reabsorption in phase 2 on the right and (b) Feynman diagram for electron collision and virtual photon emission based on the energy exchange.

FFT

[computational] *See* **FAST FOURIER TRANSFORM.**

f-function

[nuclear, relativistic] Gauge force space with respect to strong and weak nuclear forces. Based on the realization that NEUTRON and protons are not FUNDAMENTAL PARTICLES, but consist of quarks with strong force holding together to appear a single unit, the force gauge invariance principles can be illustrated by means of the introduction of an auxiliary force space: the *f*-space. The *f*-space is outlined as a sphere (circle). The *f*-function can take values between 0 and 2π (circle) without affecting the transform applied to the WAVE function: ψ (SCHRÖDINGER WAVE EQUATION SOLUTIONS); with transform $\psi'(\vec{r},t) = e^{if(\vec{r},t)}\psi(\vec{r},t)$, in both space ($\vec{r}$, respectively $[x, y, z]$) and time (t). When considering the MAXWELL EQUATIONS for a four-vector field potential ($\varphi_x, \varphi_y, \varphi_z, \varphi_t$,) instead of the electric (E) and magnetic (B) field, the following transformation can be applied: $\varphi'_t = \varphi_t + (\partial f / \partial t)$, respectively: $\varphi'_{x_i} = \varphi_{x_i} + (\partial f / \partial x_i)$, yielding the transformed vector field: ($\varphi'_x, \varphi'_y, \varphi'_z, \varphi'_t$,). The exponential value of the EIGENFUNCTION remains unchanged by addition of multiples (fractional values) of 2π. In order to define the gauge forces in theory a total of eight strong force *f*-functions are required in addition to four WEAK FORCE *f*-functions would be required, however due to the symmetry only one *f*-function coupling constant is needed instead of 12 next to coupling constant for the weak force and one additional for the strong force; three total. This becomes more evident when considering the Yang–Mills theory of the NUCLEAR FORCE.

Fiber-optic

[general, optics] A light conduit concept that is based on TOTAL INTERNAL REFLECTION principles. The general concept of the fiber-optic is a transparent, round core surrounded by another transparent medium with a lower index of REFRACTION (n_j), the "cladding" (ensuring the conditions for perfect grazing incidence reflection). The outside shell is providing a protective layer and is referred to as the jacket. Fiber-optic diameter ranges from a few micrometer to several hundred micrometer. The larger the fiber-optic core diameter, the more rigid the fiber structure, with increasing bending radius. A 100-μm fiber-optic will have a bending radius in the order of 2 cm, whereas a 5-μm fiber-optic can be wrapped in a radius of a few millimeter. The transmitting medium can be made of various materials, while high-end fiber-optics use silica GLASS or fused silica for both the core and the cladding, while the jacket is made of plastic or acrylic materials and in some cases a varnish. The transmission of light with the reflective aspects at respective interfaces with different index of refraction can be described by solving the MAXWELL EQUATIONS for both the ELECTRIC FIELD (E) and the MAGNETIC FIELD (B). At the interface between two layers of different index of refraction, the electric field is confined by the following conditions: (1) the electric field needs to be perpendicular to the interface, or the electric field is zero on a conducting WALL (specifically for a hollow waveguide) and (2) in the direction of propagation the magnetic field is restricted to being parallel to the interface. Considering the waveguide in first-order approximation for small segments of the conduit with CARTESIAN COORDINATES, z is the direction of propagation, y is the radial direction toward the core, and x is the direction of the circumference parallel to the surface the general solutions are a transverse electric field and a transverse magnetic field, respectively. The transverse electric field SOLUTION with WAVE numbers $k_i = 2\pi/\lambda_i$, where $\lambda = c/\nu = c/(\omega/2\pi)$ (speed of light $c = \sqrt{\mu_0 \varepsilon_0}^{-1} = (\mu_0 \varepsilon_0)^{-(1/2)} = 2.99792458 \times 10^8$ m/s in VACUUM, for the medium the VELOCITY (v_ℓ) is restricted by the index of refraction as $n = c/v_\ell$; ν the frequency, ω the ANGULAR FREQUENCY, and λ the wavelength) represents the wavelength (illustrating the confinement as well as DISPERSION in directional propagation, with disparity in group velocity and PHASE VELOCITY), defined by an electric field $E_z = -(k_y/k_z)E_0 \sin(k_y y)\sin(\omega t - k_z z)$, $E_y = E_0 \cos(k_y y)\cos(t - k_z z)$, and $E_x = 0$, and magnetic fields $B_z = B_y = 0$ and $E_y = (\omega/k_z c^2)E_0 \cos(k_y y)\cos(t - k_z z)$, respectively. The transverse magnetic field solution is written as $E_x = E_0 \sin(k_y y)\cos(\omega t - k_z z)$; $E_z = E_y = 0$, respectively $B_z = -(k_y/\omega)E_0 \cos(k_y y)\sin(\omega t - k_z z)$, $B_y = -(k_z/\omega)E_0 \sin(k_y y)\cos(\omega t - k_z z)$, and $B_x = 0$. Keeping in mind that the electric and magnetic fields are freely rotating as they travel along the length of the optical fiber, the Cartesian approach, including the angular component will add additional complexity. The solutions for the Maxwell equations can provide a multitude of waveforms for a large core fiber, making the solution process rather complex. In case special constraints are applied only a few or a single specific wavelength can be transmitted, with a well-defined solution for the Maxwell equations. The special case relates to the "single-mode" fiber, which is specifically designed to fit a particular laser WAVELENGTH (λ); using shorter wavelengths will provide "few" mode transmission. Generally, a fiber-optic is defined by the core diameter and the NUMERICAL APERTURE of the core design. The NUMERICAL aperture describes the maximum opening ANGLE (θ_{max}) for the collection of light, $NA = n_i \sin\theta_{max} \cong n_i \sqrt{(n_{core}^2 - n_{cladding}^2)}$, per core index with external index of refraction n_i. Because of the confined space in the diameter of the transparent rod the light will be interfacing at an angle of incidence greater than the critical angle, hence providing perfect REFLECTION. Generally, large core fibers are multimode, describing the loss of any phase-related information of a coherent beam of light due to multiple paths with various lengths that provide "mode-scrambling." A special format of fiber-optic is the *single-mode fiber*, which is designed to provide PHASE velocity-confined transmission, hence the light emitted on the distal end will be in full coherent synchronicity with the light entering on the proximal side of the fiber-optic. Based on the solutions to the Maxwell equations, the transmitted waveforms are restricted to the LP_{01} (linearly polarized mode of order 1; approximate), or respectively the exact HE_{11} mode ("helical" path, with respect to a skew ray; H indicating the magnetic field and E the electric field mode, hence primary wave only) only. The quality of a single-mode fiber-optic is described by the V-number: $V_n = ka_r NA$, with $k = 2\pi/\lambda$ the wave number and a_r the fiber core radius. For instance, $V_n < 2.405$ defines an appropriate value for a single-mode fiber, defining the "cut-off" wavelength λ_c. Generally, the core diameter is also restricted to only a few. The total number of guided modes is derived as $M_{guided} \approx (V_n^2/2)$ (which can amount to several thousands for a multimode fiber). Single-mode fibers are particularly useful for coherent imaging, for example, OPTICAL COHERENCE TOMOGRAPHY (OCT). Coupling light into a multi-mode fiber-optic is relatively easy, whereas a single-mode fiber-optic has specific requirements that need to be satisfied for efficient and low loss coupling; focusing the waist of the laser beam to at least within 110% of the core diameter, while simultaneously considering the numerical aperture (angle of incidence) requirements for the fiber-optic. One specific application is the use of a graded index of refraction for the core of the optical rod. The graded index of refraction (GRIN) provides a means of focusing the light, hence increasing the preservation of the phase and AMPLITUDE features of the light, avoiding dispersive distortion. The graded index can be step-wise or continuous/smooth. The optical path in a GRIN fiber is defined through SNELL'S LAW; WILLEBRORD SNEL VAN ROYEN (1591–1626). The increases in requirements for fiber-optic communications and data transmission have resulted in the introduction of high-quality graded-index POLYMER optical fibers (PLASTICS). An alternative approach is the use of a hollow fiber, for instance, a glass or polymer tube coated with reflective material, specifically geared to the transportation of INFRARED and ULTRAVIOLET light where most glass, plastic, and semiconductor media (silicon, germanium, phosphorus) have limited transmissivity and may hence result in substantial wavelength-dependent attenuation per unit length. Fiber quality is generally indicated by the attenuation per unit length in logarithmic form (dB/m). Fiber-optics operating in the ultraviolet require a high water content ("high OH," e.g., used for Excimer laser operating in the 308 nm range), whereas infrared transmission requires low water content ("low -OH"). Splicing refers to the process of joining two fiber-optic fibers together. A mismatch in core and/or cladding dimensions will result in coupling losses. The arrangement of a bundle of fibers for IMAGE transfer needs to have a one-to-one correlation between the input and output locations of the fibers, and is referred to as a coherent bundle. When the fibers are randomly arranged the bundle is said to have an incoherent configuration, generally applies to ENERGY transfer (e.g., fiber-optic catheter for atherectomy: plaque removal from BLOOD vessels). Applications of fibers are in imaging systems (endoscopes and safety/quality inspection for commercial/industrial use), clinical therapeutic devices (catheters: atherectomy, laser treatment of cardiac arrhythmias [irregular HEART rate]; lithotripsy [breaking-up KIDNEY stones]; interstitial cancer treatment, and many more), as well as industrial applications for cutting and a significant amount of fiber-optic cabling is used for optical communications (digital as well as coherent). For some applications the use of LIQUID light guides can be preferred due to the large core and hence substantial potential for light transmission. The liquid LIGHT GUIDE uses a liquid core rather than a solid plastic or glass core, capped on either end with a solid optical rod. Liquid light guides are, for instance, used for illuminating the sample under a MICROSCOPE next to the illumination of a surgical site as a headlamp for a surgeon and in automotive repairs for inspection of engines and other hard to reach places (see Figure F.25).

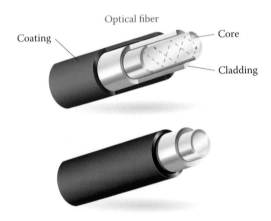

Optical fiber

Coating

Core

Cladding

Figure F.25 Illustrations of fiber-optic light guide designs and applications.

Fibonacci sequence

[atomic, biomedical, chemical, computational, general, quantum] Number sequence. The number sequence is found in many biological events and conditions, such as the idealized population EXPANSION for mice or rabbits, next to the geometric structure of a snail shell, the branches of a tree, and many more phenomena as well as computational applications. Certain chemical compositions are formed based on the Fibonacci sequence, specifically uranium and chromium oxides next to other chemical chains; additional examples are found in LATTICE structures for crystals and some may say that the electron arrangement of an ATOM in internal transitions follows a Fibonacci sequence. The number sequence is defined starting out with zero and one and the subsequent numbers are found by adding the last number to the number before that: $0,1,1,2,3,5,8,...$; formally introduced in 1202. Even though the history of this number series can be traced back to Pingala (scientist from Asian/Arabic geographic location; India). Pingala lived approximately in the second century BC, associated with the formal introduction of a binary mathematical structure. There are additional rooted traces to the history of the recognition of the importance of this number sequence in Indian/Arabic mathematics well before the western calendar ("before Christ"). The NUMERICAL sequence is named after the mathematician from Italy attributed with official documented applications: LEONARDO PISANO BIGOLLO (c. 1170–1250), whose alias was "Fibonacci," also known as Leonardo Pisano, or Leonardo Fibonacci.

Fick, Adolf Eugen (1829–1901)

[biomedical] Physiologist from Germany, known for his contributions in osmotic GAS exchange and the calculation of CARDIAC OUTPUT. Other contributions to modern MEDICINE by Fick were in ANESTHESIOLOGY (see Figure F.26).

Figure F.26 Adolf Fick (1829–1901). (Courtesy of Anton Klamroth in 1897.)

Fick, Adolf Gaston Eugen (1852–1937)

[biomedical, optics] Physiologist and physician from Germany who has on record developed the first contact LENS (see Figure F.27).

Figure F.27 Adolf Gaston Eugen Fick (1852–1937).

Fick's equation

[biomedical, chemical, computational] Description of the migration of small molecules across a BARRIER that can be chemical, mechanical, or electrical (POTENTIAL BARRIER), primary concepts were developed by ADOLF EUGEN FICK (1829–1901). It is also known as Fick's first law (*see* DIFFUSION EQUATION).

Fine structure

[atomic, energy, optics, solid-state] An electron in orbit around the atomic nucleus will perform PRECESSION around an applied external MAGNETIC FIELD in discrete configurations. The number of discrete magnetic moments is defined by the ORBITAL MAGNETIC QUANTUM NUMBER m_ℓ. The orbital QUANTUM NUMBER ranges in integer values from negative ORBITAL ANGULAR MOMENTUM QUANTUM NUMBER (ℓ) to positive value (where ℓ has values between 0 and $n-1$, n being the PRINCIPAL QUANTUM NUMBER. The split in configuration with respect to the absence of an external field provides a multitude of spectral lines within close values of the natural state. This split in spectral lines is referred to as the fine structure. The fine structure ENERGY configuration (E_{fs}) can be derived to be as follows, specifically for the HYDROGEN ATOM: $E_{fs} = -(hcZ^2R/n^2)\{(\alpha_{fine}Z^2/n)[1/(j_d+1/2)-(3/4n)]\}$. With PLANCK'S CONSTANT $h = 6.626070040(81)\times10^{-34}$ m^{2}kg/s (Js), FINE STRUCTURE CONSTANT ($\alpha_{fine} = (e^2/4\pi\hbar c) = 1/137$), j_d the "DOUBLET" split for the angular momentum: $j_d = \ell + (1/2)$; $j_d = \ell - (1/2)$ (*Note:* for $\ell = 0$ only $j_d = \ell + (1/2)$ is allowed.), R = 10973731.6 m^{-1} the RYDBERG CONSTANT, Z the ELECTRIC CHARGE for the NUCLEUS, and $c = 2.99792458\times10^8$ m/s the speed of light. When the nuclear magnetic moment becomes involved, the spectral profile splits into HYPERFINE STRUCTURE (*also see* ZEEMAN EFFECT) (see Figure F.28).

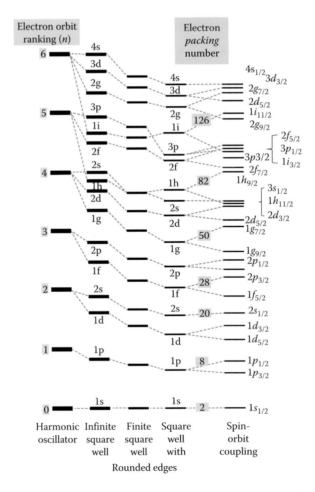

Figure F.28 Finely split structure of the electron orbit under autonomic conditions. The diagram illustrates the impact of the choice of energy model for the nuclear potential on the electron configuration, while obeying all exclusion laws.

Fine structure constant ($\alpha_{fine} = (e^2/4\pi\hbar c) = 1/137$)

[atomic, general, quantum] Dimensionless quantity used in description of the QUANTUM theoretical description of the atomic events, where $\hbar = h/2\pi$ is PLANCK's CONSTANT $h = 6.626070040(81) \times 10^{-34}$ m^{2}kg/s (Js) divided by 2π, c the speed of light, and e the electron charge. In quantum ELECTRODYNAMICS, α_{fine} represents the electromagnetic coupling strength. The FINE STRUCTURE constant relates to Coulomb's law ($F_e = q_1 q_2/4\pi\varepsilon_0 r^2$, where q_1 and q_2 are the respective charge in value of Coulomb, $\varepsilon_0 = 8.8541878 \times 10^{-12}$ C^2/Nm^2 the DIELECTRIC permittivity in VACUUM, and r the separation between the centers of the two charge concentrations, respectively).

Finesse ($\mathcal{F}$)

[fluid dynamics, optics] Qualification factor for an optical cavity, alternatively in AERODYNAMICS it defines the ANGLE of the AIRFOIL with respect to the FLUID stream (in aviation; the glide ratio) and resulting LIFT potential. Examples of *optical* cavities this applies to are Fabry–Pérot étalon pertaining to INTERFEROMETER construction. Finesse defines the ratio between the "FREE SPECTRAL RANGE" (FSR; $FSR = c/2L$, c the speed of light, L characteristic length for the phenomenon) and the "full-width half-maximum" (FWHM) of the beam, respectively, resonator. The FSR is the wavelength spacing for the maxima or minima created by two wavelengths in REFLECTION or transmission. Finesse is described by the parameter: $= FSR/FWHM$; high "Q" value represents "high finesse," and low "Q" value indicates "low finesse." The cavity is relatively short compared to a laser cavity; however, the principal goal remains filtering with respect to wavelength based on cavity configuration. In case the path length between the two reflective surfaces (often referred to as optical flats), with reflectivity R, forming the cavity has a match to a whole number of wavelengths the transmitted light will be in PHASE and amplified, whereas if out of phase destructive INTERFERENCE will provide the conditions for a minimum in transmission. Assuming the light with wavelength λ enters the étalon with refractive index n and thickness ℓ at an angle θ, the resulting phase shift after each successive reflection will be $\delta = (2\pi/\lambda)2n\ell\cos\theta$. On transmission T_{etalon}, the yields for the étalon are $T_{etalon} = (1-R)^2/(1+R^2 - 2R\cos\delta) = 1/(1 + F_{finesse}\sin^2(\delta/2))$, with $F_{finesse} = 4R/(1-R)^2$ the coefficient of finesse. The Fabry–Pérot étalon will have several closely spaced transmission lines across the optical spectrum, separated by the free-spectral range FSR: $\Delta\lambda$, a device constant. The finesse can now be written as $\mathcal{F}_{optic} = \Delta\lambda/\delta\lambda = \pi/\left(2\arcsin\left(1/\sqrt{F_{finesse}}\right)\right)$. The flats enclosing the cavity are often shaped in the form of a wedge to avoid multiple internal reflections with the boundary. In case the reflectivity is high (> 0.6), the interference pattern will produce concentric circles that are highly visible and sharply demarcated, described by a high finesse, whereas low contrast is defined by low finesse. In *aerodynamics*, the finesse indicates the cotangent of the downward angle of the WING of a bird or plane, expressed as the ratio of forward speed ($v_{forward}$) and downward draft VELOCITY (v_{down}) for an unpowered airplane or soaring bird-of-prey (e.g., vulture or eagle): $\mathcal{F}_{flight} = L/D = \Delta s/\Delta h = v_{forward}/v_{down}$, where L/D is the lift over DRAG ratio, Δs

displacement, and Δh the incline of the wing. If the AIR rises faster than the downward draft or rate of sink the object will remain airborne and will in fact climb (Table F.1).

Table F.1 **Examples of glide ratio—finesse in aerodynamics**

OBJECT IN FLOW (FLIGHT)	CONDITIONS	L/D RATIO; GLIDE RATIO
Apollo CM	Re-entry	0.368
Gimli glider	For example, Boeing 767–200 running low on fuel	~12
Hang glider	Airborne under own lift	15
Northern flying squirrel	Floating through the air	1.98
Paraglider	High-performance model	11
Parachute that is under external power	Rectangular Elliptical parachute	3.6 5.6
Sail-plane	Airborne under own lift	45–70 (function of wingspan)
Space shuttle	Landing approach to earth base	4.5
Space shuttle	Hypersonic in outer space (e.g., orbit around the Moon)	1.0
Wingsuit	Drifting through atmosphere after jump	2.5

Finsen, Niels Ryberg (1860–1904)

[biomedical, optics] Physicist and physician from Iceland, Denmark. Niels Finsen received the Nobel Prize in Medicine in 1903 for his contributions to the evolution in solar therapeutic applications. Nonetheless, the benefits of sunlight to the body for general health, for therapeutic applications to SKIN diseases (e.g., acne, psoriasis as well as the treatment of rubella), and general benefits to mood were known for millennia by the Egyptians, and healers in India and China. The photodynamic therapy applications for rubella (mumps) and smallpox was described by the physician HENRI DE MONDEVILLE (1260–1320). More in-depth work on photodynamics interaction was described by OSCAR RAAB (1886–1986) and his work under the guidance of Hermann von Tappeiner (1847–1927) based on the PHOTOTOXICITY by Marcacci in 1888 (see Figure F.29).

Figure F.29 Niels Ryberg Finsen (1860–1904).

First law of Faraday

[atomic, general, nuclear, quantum, thermodynamics] *See* **FARADAY, FIRST LAW OF**.

First law of thermodynamics

[biomedical, energy, general, nuclear, solid-state, thermodynamics] That the change in the INTERNAL ENERGY (U) of a system is the net result of the HEAT (Q) added to, or removed from, the system and the work (W) done by the system on the surroundings as given by $\Delta U = \Delta Q - \Delta W$. The work done can be mechanical (W_M), and/ or electrical (W_{elec}) (*also see* **THERMODYNAMICS, FIRST LAW**). The LAWS OF THERMODYNAMICS were introduced by JOSIAH WILLARD GIBBS (1839–1903), RUDOLF JULIUS EMANUEL CLAUSIUS (1822–1888), HERMANN LUDWIG FERDINAND VON HELMHOLTZ (1821–1894), NICOLAS LÉONARD SADI CARNOT (1796–1832) and PETER GUTHRIE TAIT (1831–1901).

First postulate of relativity

[general] All laws of PHYSICS are identical with respect to the uniformly advancing observer. This PRINCIPLE OF RELATIVITY was introduced by ALBERT EINSTEIN (1879–1955).

Fission

[nuclear] Primarily referring to NUCLEAR FISSION, the splitting of the atomic nucleus into smaller ISOTOPE(s) accompanied by RADIATION consisting of particles (alpha, beta, NEUTRON, etc.) and/or ELECTROMAGNETIC RADIATION. The first experimentally observed fission was in 1934, based on the work by ENRICO FERMI (1901–1954; in his day referred to as "The Pope" based on his "new faith" with respect to QUANTUM MECHANICS), where uranium bombarded by thermal neutron generated beta rays. The German chemist IDA NODDACK (1896–1978) postulated in 1934 that the uranium nucleus can be deconstructed into several fragments that constitute isotopes of known ELEMENTS with additional emissions. This concept was initially rejected by the entire PHYSICS community, but was confirmed through a follow-up experiment by the German/Austrian group of scientists consisting of LISE MEITNER (1878–1968), OTTO HAHN (1879–1968), and FRIEDRICH WILHELM STRASSMANN (1902–1980; Straβmann) in 1937. The term "nuclear fission" was introduced in 1938 by LISE MEITNER (1878–1968) and her nephew OTTO ROBERT FRISCH (1904–1979) in conversation with NIELS BOHR (1885–1962). The initial fission principles were described based on the liquid-drop atomic model of Niels Bohr and JOHN WHEELER (1911–2008), formally shared in 1939. Under the LIQUID-DROP MODEL, the introduction of a kinetic PARTICLE provides a perturbation, which is interpreted as a "jelly-like" VIBRATION (also referred to as "liquid-drop") resulting in a separation of the blob into two or more jelly-plum-puddings (also remember the "raisin-plum-pudding" atomic model by Lord Kelvin; WILLIAM THOMSON, 1ST BARON KELVIN [1824–1907], and JOSEPH JOHN THOMSON [1856–1940]). In this model the perturbation provides the means for the repulsive forces between the similar charges within the NUCLEUS to take the upper hand. Additional work by Robert Frisch provided the theoretical concept for the development of the ATOMIC BOMB in 1940, along with a German scientist, RUDOLF ERNST PEIERLS (1907–1995), while both in the United Kingdom. In 1942, the first fission REACTOR power plant was tested by the group of Enrico Fermi in a new facility at the University of Chicago, Illinois. The ENERGY released by the process of nuclear fission is the net surplus with respect to the relativistic masses of all the constituents. In 1963, a more energy-based description of the fission concept was introduced by the Polish/Swedish physicist Wladyslaw J. Swiatecki (1926–2009), including the "POTENTIAL BARRIER" model of the nuclear shell, based on his investigations of fission half-lives dating back to the early 1950s. The fission process will be required to exceed a potential BARRIER before the attractive forces are overcome by the repulsive forces (*also see* **NUCLEAR FISSION**) (see Figure F.30).

F

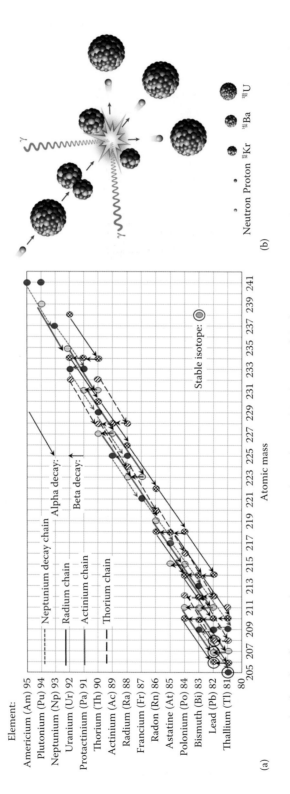

Figure F.30 Example of nuclear fission: (a) plutonium (americium) decay chain and (b) diagram of the fission process in nuclear reactors, generating electrical power as a result of steam created from the release of heat during the fission process.

FitzGerald, George Francis (1851–1901)

[atomic, general, nuclear] Physicist from Ireland. FitzGerald is most known for his collaboration with HENDRIK LORENTZ (1853–1928) for their work on the theory of special relativity and LENGTH CONTRACTION: the LORENTZ–FITZGERALD CONTRACTION (see Figure F.31).

Figure F.31 George Francis FitzGerald (1851–1901).

FitzGerald–Lorentz hypothesis

[atomic] *See* **LORENTZ–FITZGERALD CONTRACTION**.

Fizeau, Armand Hippolyte Louis (1819–1896)

[general] French physicist. Fizeau made remarkable advances in the determination of the speed of light in 1849. He used a sprocket with more than 100 teeth that periodically interrupted the path of a light beam, rotating at several hundred revolutions per SECOND. Fizeau measured the time the light took to travel to and from a fixed MIRROR through the same gap between the sprocket teeth at a DISTANCE of over 8.5 km. This crude method resulted in a value for the speed of light as 3.133×10^8 m/s in AIR, which is within 5% of the accepted standard of 2.99792458×10^8 m/s in VACUUM (see Figure F.32).

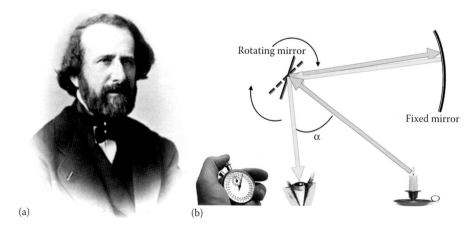

(a)   (b)

Figure F.32 (a) Armand Hippolyte Louis Fizeau (1819–1896) and (b) Fizeau's experimental design.

Flavors of quarks

[energy, general, quantum, relativistic] Designation of the "attitude" of a QUARK as proposed by MURRAY GELL-MANN (1929–) and GEORGE ZWEIG (1937–). There are six flavors: up, down, strange, charmed, beauty (or bottom), and top (see Figure F.33).

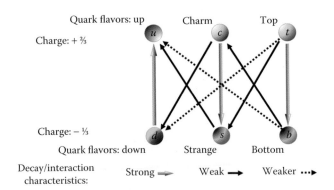

Figure F.33 The flavors for quarks.

Fleming, John Ambrose, Sir (1849–1945)

[electromagnetism, solid-state] English physicist and electrical engineer. Fleming laid the foundation for the VACUUM tube in 1907. Fleming's work inspired the development of the RADIO transmitter by GUGLIELMO MARCONI (1874–1937). Sometimes referred to as the "father of modern ELECTRONICS" (see Figure F.34).

Figure F.34 John Ambrose Fleming (1849–1945) in 1890. (Courtesy of JDR, *Electrical World*, vol. XVL, no. 10, McGraw-Hill, New York, 1890.)

Fleming's right-hand rule

[electronics, general, mechanics] Right-hand rule as introduced by the physicist SIR JOHN AMBROSE FLEMING (1849–1945) from the United Kingdom (*see* RIGHT-HAND RULE) (see Figure F.35).

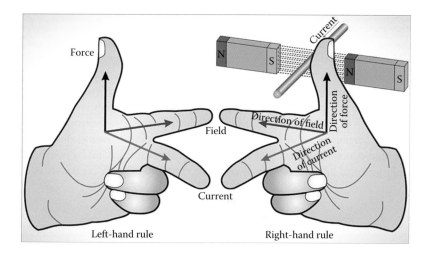

Figure F.35 Fleming's right- and left-hand rule.

Fletcher, Harvey (1884–1981)

[general] Physicist from the United States and student of ROBERT MILLIKAN (1868–1953), who systematically analyzed thousands of charged droplets to provide the electron charge in the Millikan OIL-DROP EXPERIMENT. Additional work by Harvey Fletcher was in audio and ACOUSTICS, specifically stereophony (sometimes referred to as the "father of stereophonic sound") (see Figure F.36).

Figure F.36 Harvey Fletcher (1884–1981) on Y mountain, Utah, in 1912.

Flow

[acoustics, fluid dynamics, general] Flow is characterized as fluids in MOTION with specific features which were first defined by LEONHARD EULER (1707–1783). Each point within the FLUID can be defined and is associated with a range of physical parameters as a function of time and location. Some physical parameters are VELOCITY (v), DIRECTION ($\vec{s}$), MASS (m), DENSITY (ρ), and PRESSURE (P). Flow can be under specific conditions; examples of flow conditions are steady and nonsteady, compressible and incompressible, rotational and irrotational, as well as viscous and nonviscous. Each flow condition is subject to specific boundary conditions. The path along which fluid flow is known as a streamline. Under steady flow, at every point in the fluid the velocity is independent of time, while the opposite is true for nonsteady, or turbulent, flow. When a fluid in motion has a density that is independent of position and time it is incompressible, which holds true for most liquids, whereas gasses can have a variable density in both time and place under flow and are considered compressible. Rotational flow is characterized by changing directional flow, such as seen in whirlpools and tornadoes, in contradiction to IRROTATIONAL FLOW with parallel, or diverging respectively converging streamlines, but no rotational flow patterns. In case frictional forces are influencing the streamlines, the fluid exhibits VISCOSITY (i.e., viscous flow) whereas an ideal fluid is nonviscous or frictionless. Fluid flow is generally governed by the EQUATION OF CONTINUITY next to BERNOULLI'S EQUATION (see Figure F.37).

Figure F.37 General flow representation.

Flow, pulsatile

[biomedical, computational, fluid dynamics] Pulsatile FLOW can present in many different formats, the BLOOD flow is one of the more complex expressions. The HEART beat and resulting flow generates a complex flow rate pattern, as illustrated in Figure F.38. FOURIER ANALYSIS of this SIGNAL will show that the WAVE pattern can be approximated by the first eight sinusoidal harmonics, with the base wave the heart rate. The frequency-dependent flow requires the introduction of a frequency component in the flow parameter. Using the axial flow $u(r)$ as the base for steady-state flow, the frequency-dependent flow can be developed in a harmonic EXPANSION with the use of a Poisson series and can be described as $u(r) = u'(r)e^{i\omega_n t}$, where ω_n the angular frequency harmonics for the pulsating flow driven at frequency v. Substitution of the time-dependent harmonically oscillating flow in the POISEUILLE FLOW equation provides the FLUID dynamic equivalent of the TELEGRAPH EQUATION for wave transmission as the real value of the complex SOLUTION described by Bessel function of the first order $J_0(x)$: $u(r,t) = \mathrm{Re}\left\{-(ia_n/\omega_n)e^{i\omega_n t}\left[1-(J_0(\sqrt{(-i\alpha)}(r/R))/J_0(\sqrt{(-i\alpha)}))\right]\right\}$, where the pulsatile flow can be developed in a FOURIER SERIES with coefficients a_n for the harmonic contributions, $\alpha^2 = Re S r$, Re the REYNOLDS NUMBER, and $S_r = \omega D/v_{avg}$ the Strouhal number, where v_{avg} the average flow velocity and $D = 2R$ the vessel diameter. The Strouhal number is the ratio of the TURBULENCE-induced fluidic forces over the internal forces rendering local accelerations within the velocity profile. The Strouhal number is critical in estimating the timescale for formation of VORTEX flow. The Strouhal number provides a determining factor in the contribution for the respective frequency parameters. Thus the flow velocity profile is critically linked to the vessel diameter, PUMP frequency, and flow rate. For small α (long timescale for turbulence), the flow velocity becomes parabolic under steady-state Poiseuille flow, $u(r,t) = a_n\left\{[1-(r^2/R^2)]\cos(\omega_n t)\right\}$ (see Figure F.38).

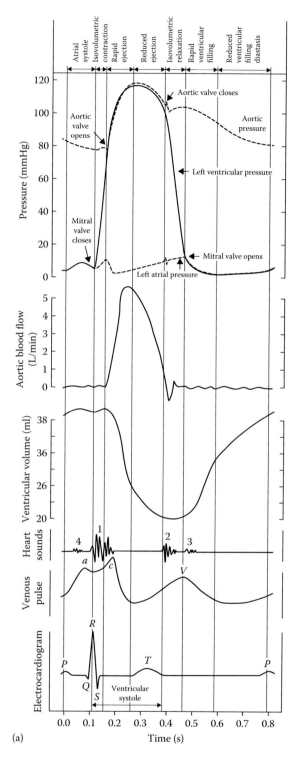

Figure F.38 Pulsatile flow representation: (a) human heart flow pulsation and pressure form.   (*Continued*)

F

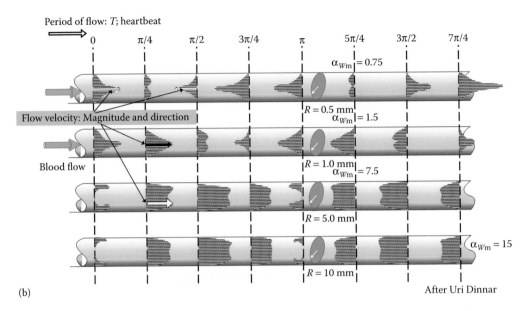

(b)

After Uri Dinnar

Figure F.38 (Continued) Pulsatile flow representation: (b) pulsatile vascular flow velocity profile (relative magnitude) for non-Newtonian liquid, based on the work of Uri Dinnar (1939–2013). $\alpha^2 = \omega_n R^2/v$, for a pump frequency v, in a tube diameter R with the nth harmonic flow velocity $\omega_n = 2\pi v_n$. This is generally valid in blood vessels ranging from $R = 100\ \mu m$ to $R = 15$ mm. (From Dinnar, U., *Cardiovascular Fluid Dynamics*, CRC Press, Boca Raton, FL, 1981.)

Fludeoxyglucose (FDG)

[atomic, biomedical, chemical, imaging, nuclear, solid-state] A RADIOPHARMACEUTICAL used in positron emission TOMOGRAPHY fluoride (^{18}F) is the active component in FDG, RADIONUCLIDE half-life, and inherent ENERGY ^{18}F : 110 min, $E_{18_F} = 0.64$ MeV, which decays as $^{18}_9$F $\rightarrow$ $^{0}_{+1}e + ^{18}_8$O, where oxygen is released, $^{18}_8$O. The fluoride is incorporated in biological process and decays involving the annihilation process of the positron ($^{0}_{+1}e$) or BETA PARTICLE, producing subsequent gamma RADIATION that can be captured to locate the chemical events and hence map the physiological ACTIVITY as a function of volume location in the biological medium. Specifically, FDG acts as a GLUCOSE equivalent hence illustrating the metabolic rate, which can specifically indicate cancer as well as brain activity.

Fluence ($\vec{\Psi}$ (r, θ, φ))

[astrophysics/astronomy, biomedical, electromagnetism, energy, optics, thermodynamics] Exposure of PHOTON or MECHANICAL ENERGY to a surface, vector expression of radiant exposure. The light energy per unit area is expressed by the average POYNTING VECTOR ($\vec{S} = (1/\mu_0)(\vec{E} \times \vec{B}) = \vec{E} \times \vec{H} = (\sqrt{\varepsilon_0\mu_0}/4\pi)(\vec{E} \times \vec{B})$, $c = \sqrt{\varepsilon_0\mu_0}$ the speed of light): $\vec{\Psi}(r,\theta,\phi) = \vec{S} = ((\vec{E} \times \vec{B})/\mu)$ (μ the DIELECTRIC permeability of the medium, ε the permittivity of the medium). In general, the energy is the integral of the photon or ionizing particle FLUX over the time of exposure in a specific location. Alternatively, the particle flux can be counted by detectors. The fluence concept is based on the work by SIR FRANZ ARTHUR FRIEDRICH SCHUSTER (1851–1934) from Germany in 1905 and KARL SCHWARZSCHILD (1873–1916) from Germany in 1906. The exposure time can be a limited duration PARTICLE stream, ionizing ELECTROMAGNETIC RADIATION or laser pulse (e.g., alpha, electron next to gamma, ULTRAVIOLET, visible, and INFRARED, respectively). The units for fluence are [J/m²], alternatively for particle stream [m⁻²]. The fluence in a location determines the thermal effects that will be achieved over a period of exposure to electromagnetic radiation, in particular, exposure to laser light (see Figure F.39).

(a)

(b)

Figure F.39 (a) The fluence on the podium and the audience during a rock concert. (b) The fluence from the distant Carina Nebula, scattered by space debris, captured by the Hubble telescope. (Courtesy of ESO/T. Preibisch.)

Fluence rate (Φ (r, θ, ϕ, t))

[biomedical, chemical, electromagnetism, energy, optics, thermodynamics] When presented as radiant ENERGY fluence rate, it represents the intensity equivalent for ELECTROMAGNETIC RADIATION. Intensity of electromagnetic radiation is ill-defined since intensity by definition is the AMPLITUDE of a phenomenon squared and electromagnetic radiation has two paired phenomena: (1) alternating electric field strength and direction and (2) alternating MAGNETIC FIELD strength and direction. The fluence rate $\Phi(\vec{r})$ is the radiance $L(\vec{r}, \vec{s})$ (also described as irradiance) in location $\vec{r}$ with direction $\vec{s}$ integrated over SOLID ANGLE (4π): $\Phi(\vec{r}) = \int_{4\pi} L(\vec{r}, \vec{s}) d\omega$. The fluence rate identifies the RADIATION power received, transferred, or emitted from the surface of a volume, a (small) sphere of medium with finite but small dimensions at the location $\vec{r}$. The concept of fluence rate was introduced by the biologist from the United States, Claud Stan Rupert (1919–) in 1974 for the description of influence of light on the repair mechanism of DNA. The units for fluence rate are (W/m^2). Fluence rate and irradiance have the same units but have a fundamentally different meaning in light exposure. Irradiance describes the linear impact of electromagnetic radiation on a plane surface, not on a spherical surface. This term is commonly used to describe the light distribution inside a turbid medium. The difference with irradiance is in fact energy FLUX crossing the surface of a volume of medium, which is only equal to the fluence rate when the incident flux is orthogonal to the surface. The location-specific generation of heat in a medium under irradiation is the local absorption coefficient multiplied by the fluence rate as a function of location. The chemical effects of light interaction are a direct result of the fluence rate. For instance, in PHOTOSYNTHESIS, or alternatively in photodynamics therapy, the fluence rate determines the efficacy for the process, provided through dose–response curves. Photodynamics therapy is the treatment of cancer by excitation of singlet oxygen to form a reactive component that initiates CELL death, primarily through the disintegration of the local vasculature. Photosynthesis is the conversion of light energy in a chemical reaction that converts carbon dioxide into oxygen and carbon waste by-product. Also found as $E_{p,0}$. It is also referred to as SCALAR radiance (*also see* EQUATION OF RADIATIVE TRANSFER).

Fluid

[biomedical, chemical, fluid dynamics] State of a medium that is not solid, while GAS and LIQUID phase are fluids. The atoms, molecules, and ions, respectively, are allowed to freely move about with respect to each other. In a liquid the atoms or molecules do maintain a relatively fixed DISTANCE to each other, whereas in a gas the atoms and molecules have no particular correlation in distance. A gas will fill a volume to all the boundaries, while a liquid will only conform to the five walls of a container. Fluids can be ideal or nonideal (not adhering to the standard laws and continuity equations), for instance, the case of the IDEAL GAS. Furthermore, fluids can be compressible or incompressible, which affects the theoretical treatise of the problem in significant ways. Frequently, for simplicity, a fluid is considered inviscid (absence of VISCOSITY) and incompressible. In case the FLOW itself changes the ENERGY and entropy content, the process can become rather complex from a computational point, hence adiabatic flow is preferred from a first-order point of view. From a dynamic perspective fluids can be classified as Newtonian or non-Newtonian. The NON-NEWTONIAN FLUIDS can be subclassified as being independent of time and shear and those who are dependent. The various formats of shear rate of flow are dilatant fluid (shear thickening), Newtonian, pseudoplastic (shear thinning), and Bingham fluid. A special phenomenon of fluid flow is described by the FÅHRÆUS–LINDQVIST EFFECT, which applies for instance to BLOOD flow as a function of tube diameter. One parameter of particular interest for fluids is the viscosity. Various laws and DIMENSIONLESS NUMBERS are involved in defining the behavior of fluids. In particular, fluid flow creates boundary layers at the interfaces with other media, moving or stagnant walls. One phenomenon that exemplifies the BOUNDARY LAYER is the collection of a layer of dust on the blade of a fan PROPELLER, or algae growing on the blade of the propeller of a ship. Some laws of interests include Bernoulli's law, Navier–Stokes, and Poiseuille profile. For gases the Boyle–Gay-Lussac law is important for the description of equilibrium conditions, with modifications such as the Van der Waals equation. During gas exchange with liquids, the VAN'T HOFF EQUATION comes into play to describe the partial pressure of dissolved gasses in a liquid. One force of particular interest with respect to flow is the CORIOLIS FORCE. Another phenomenon that comes into question with regard to flow is TURBULENCE and vorticity; with particular attention to the Von Kármán vortices (e.g., VON KÁRMÁN VORTEX STREET) in various flow situations. Flow can be laminar or turbulent, with a special case for COUETTE FLOW. Some of the dimensionless numbers are ARCHIMEDES number, Atwood number, Bagnold number, Bejan number, Bond number, Brinkman number, Brownell–Katz number, CAPILLARY number, Colburn J factors, Darcy friction factor, DEAN NUMBER, Deborah number, Eötvös number, Ericksen number, Euler number, Fanning friction factor, Froude number, Galilei number, Görtler number, Graetz number, Grashof number, Hagen number, hydraulic gradient, Iribarren number, Karlovitz number, Keulegan–Carpenter number, Knudsen number, Kutateladze number, Laplace number, LIFT COEFFICIENT, (Love numbers), Lundquist number, MACH NUMBER, NUSSELT NUMBER, Ohnesorge number, Péclet number, pipeline parameter, POWER NUMBER, Prandtl number, REYNOLDS NUMBER, Richardson number, Roshko number, Rossby number, Rouse number, Schmidt number, SHAPE FACTOR, (Sherwood number), Sommerfeld number, Stanton number, Stokes number, Strouhal number, Taylor number, Ursell number, Vadasz number, VAN'T HOFF FACTOR, VISCOSITY (static, kinetic, various conditions), Weber number, Weissenberg number, and WOMERSLEY NUMBER. In biology, specific definitions are applied to fluids with respect to their role and anatomical location. Interstitial fluids in biological entities fill the space surrounding cells as well as organs, while extracellular fluid is part of this fluid system and only involves the immediate perimeter of the CELL. The cell itself is filled with intercellular fluid. The chemical differences between these fluids provide specific chemical, mechanical, nutritional, and transportation functions. The fluid balance for a biological system, fluid intake, and fluid excretion form the HOMEOSTASIS. Blood flows, while sweat, urine, and amniotic fluid have a discrete migratory pattern. Sweat, for instance, plays a critical role in the thermal regulation of the body for many animals; a thermodynamic mechanism. Synovial fluid provides LUBRICATION in JOINTS. Oils are a fluid, serving a role in lubrication, as is grease. Grease is a non-NEWTONIAN FLUID.

Fluid dynamics

[fluid dynamics, general] Field in ENGINEERING and PHYSICS that concerns itself with the dynamic aspects of phases, ENERGY states, transitions, motions, and parameters of fluids. Fluid dynamics also is involved in the description of flight as well as FLOW in canals, tubes, and waves coming on shore at the beach (see Figure F.40).

Figure F.40 Fluid dynamic aspect of mixing in a blender.

Fluidization

[fluid dynamics, solid-state] Transition process for a colloidal/particulate aggregate from static solid conditions to a fluidic state under dynamic MOTION. The medium can also be considered a non-NEWTONIAN FLUID, such as a concrete or cement mix, next to a straightforward PARTICLE aggregate such as coffee grinds being ejected from the espresso grinder, or gumballs flowing from a candy machine, or the infamous quicksand next to the mechanism of operation for the hourglass (see Figure F.41).

(a)

Figure F.41 Visualization of fluidization: (a) ground coffee flowing from dispensing unit.

(Continued)

(b)

(c)

Figure F.41 (Continued) Visualization of fluidization: (b) dump truck unloading sand and (c) flow of wheat grain during harvest. (Courtesy of Charles Knowles.)

Fluidization number

[fluid dynamics, solid-state] Ratio between the voids in a bed of two powders, or a powder and a FLUID, that indicates a transition during MOTION from powder mixture to separate powders, expressed as $N_{fl} = \rho_d^3 d_p^4 g^2 / \eta_{visc}^2 E_Y$, where ρ_d is the density of the solids in the fluid motion (fluidized), d_p the particulate diameter, η_{visc} the VISCOSITY of the fluid (respectively, complementary PARTICLE) surrounding the particles, E_Y the elastic modulus, and g the GRAVITATIONAL ACCELERATION, which needs to be corrected for rotational effects as $g_{app} = \sqrt{(g^2 + \omega^4 r^2)}$ (r the particulate radius; ω the CIRCULATION rotational ANGULAR VELOCITY).

Fluorescence

[atomic, general, nuclear, optics] Re-emission of light from a substance when irradiated with light at a different wavelength, specifically a longer (i.e., lower ENERGY) wavelength. The amount of photons that are emitted from a single volume of the specimen seeded with fluorochrome material can be quantified as described by the number of photons emitted per fluorochrome n_a as $n_a \cong (p_0^2 \delta / \tau_p f_p^2) [\pi(NA)^2 / hc\lambda]$. The parameters of this formula are as follows, p_0 is the average excitation light power (preferably delivered by narrow laser line width matching the excitation conditions), f_p is the repetition rate of the delivery of light, the pulse width of the light delivery is given by τ_p, λ is the excitation wavelength of the light delivered by the laser, NA is the NUMERICAL APERTURE, $h = 6.6260755 \times 10^{-34}$ Js the Planck's constant, with c the speed of light and the photon CROSS SECTION is represented by δ. When measuring the fluence emitted as a function of ANGLE and location the origin of the volume with the highest emission can in theory be derived. The fluorescent emission as a function of location can provide details about the distribution of certain molecular concentrations (see Figure F.42).

(a)

(b)

Figure F.42 Fluorescence: (a) fluorescent body art. (Courtesy of Beo Beyond, www.beobeyond.com.) (b) Fluorescent light bulb.

Fluorescence resonance energy transfer (FRET)

[biomedical, chemical, imaging] Atomic excitation process, particularly important for the localization of nucleic acids, specifically with respect to the identification and localization of genetic encoding by means of labeled proteins on a nanometer scale (e.g., within the DNA molecule). The RESONANCE ENERGY transfer takes place between two light-sensitive molecules in living cells. The energy transfer process by means of DIPOLE–DIPOLE INTERACTION avoids chemical interaction per se. FRET is used to map molecular configurations, molecular transfer phases (stages), enzyme ACTIVITY, as well as DNA hybridization (active process traced in time) and the DYNAMICS of membranes. FRET transfer is a specific form of FÖRSTER RESONANCE ENERGY TRANSFER, as described by the scientist from Germany THEODOR FÖRSTER (1910–1974) in 1946.

Flux

[atomic, electronics, fluid dynamics, general, nuclear] Migration of energetic units through a fixed surface. The flux describes the line density of electric and/or MAGNETIC FIELD lines (the line density is a direct representation of the MAGNITUDE of the field strength), particulate transport, PHOTON fluence, or generally the "FLOW" component perpendicular to a surface (generally captured by the "dot" product) such as FLUID flow. In generalized terms, the flux of a vector field is described by the integral of the field distribution over the surface of impact. In ELECTROMAGNETIC THEORY, GAUSS's LAW defines the electric field flux through a closed surface ($\Phi_e = \oint \vec{E} \cdot d\vec{A}$), and the FARADAY LAW yields the magnetic flux through a bounded surface ($\Phi_m = \int_{\text{surface}} \vec{B} \cdot d\vec{A}$).

FM transmissions

[general] Frequency modulation, primarily pertaining to ELECTROMAGNETIC RADIATION. FM RADIO is the most popular form of wireless transfer of information, and can be applied in STEREO, compared to AM and long-wave radio that is single channel and low bandwidth.

f-number

[optics] Number representing the ANGLE subtended by the entrance opening at the axis of the object. For a CAMERA LENS, this is defined by the ratio of the focal length to the pupil diameter. A camera lens with adjustable APERTURE will have the f-number indicated on a ring of the lens that can be rotated for adjustment. An f-number smaller than 1 is considered to enhance brightness. It is also called f-stop (*also see* NUMERICAL APERTURE) (see Figure F.43).

Figure F.43 *f*-stop adjustment ring on telephoto lens of a camera.

Focal length (*f*)

[electromagnetism, general, imaging] The DISTANCE from the focusing instrument where the concentration of field lines is the greatest; in OPTICS the combined ELECTROMAGNETIC FIELD density. The focal length is the distance of the FOCAL POINT with respect to the focusing device (optical lens; magnetic lens) on the optical axis, which is the cylindrical axis of symmetry. In an ELECTRON MICROSCOPE, the electron FLUX density will be greatest in the focal point provided by magnetic steering units (magnetic coil lenses; used both in TRANSMISSION ELECTRON MICROSCOPE and REFLECTION electron microscope/scanning electron microscope). An object located in the focal point can be analyzed with the highest RESOLUTION. In addition, curved reflective surfaces also provide a focusing action, with associated focal length, specifically the use of parabolic mirrors in head lights of automobiles, and dish antennas for television and galactic RADIO (MICROWAVE emissions from atomic transactions) SIGNAL collection. Multiple lens systems, such as those in cameras and projectors provide a focusing action that allows for adjusting the focal length in order to form an IMAGE on the film at fixed distance from the LENS (image distance: d_i) from objects at selective distance (the object distance: d_o). The focal length (f) is correlated to the image distance and the object distance as $1/f = (1/d_i) + (1/d_o)$, where the image distance for a virtual image has a negative value; hence the focal length can become negative, specifically for a DIVERGING LENS. For a diverging optical lens, generally, the center (at the optical axis) is thinner than toward the edges. However, an object placed in front of a converging lens at a distance smaller than the focal length will also form a virtual image (i.e., cannot be projected) (see Figure F.44).

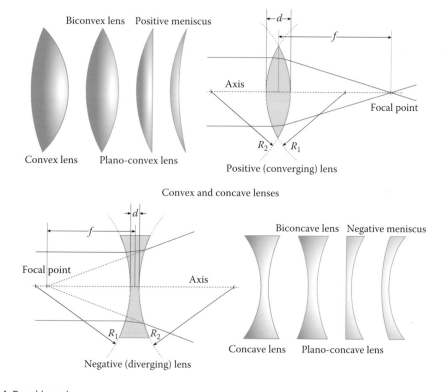

Figure F.44 Focal length.

Focal plane

[general, optics] Plane in which the IMAGE is formed from infinite DISTANCE, either on the optical axis or off the optical axis. The image formation for an object (or location on an object with finite dimensions) that is not on the optical axis yields a convergence of rays that does not coincide with the optical axis. The focal plane for a projector is generally adjusted to coincide with the location of the screen, or WALL for in-focus viewing.

Focal point

[general] The infinitely small point on the optical axis of a device where all the incident rays parallel with the optical axis converge. The focal point is located at the focal DISTANCE from the optical device (e.g., optical lens, MAGNET, MIRROR). An infinitely small LIGHT SOURCE placed in the focal point of a perfect LENS system will form a ray that is parallel to the optical axis leading to infinity. Any object of finite size placed in the focal point constitutes the FOCAL PLANE with respect to that object for the imaging device. The fact that the focal point of the instrumentation used for focusing does not coincide with the focal plane when moving away from the optical axis will result in distortions in the IMAGE (i.e., SPHERICAL ABERRATION). Similarly, the focal point ("in focus") for a perfectly configured lens (system) at a single wavelength (without spherical aberration) can have differences in index of REFRACTION for various wavelengths and hence resulting in different focal points for different wavelengths. The disparity in wavelength focusing (DISPERSION) is referred to a chromatic aberration, yielding COLOR distortions in the image formation. When an image that is formed from an object that is not on the optical axis is distorted, it is referred to as "coma." A particular form of coma is ASTIGMATISM, in which an image is contorted by nonradial symmetry in the optical focal length for the lens system (*also see* THIN LENS AND THICK LENS *and* LENS EQUATION).

Forbidden transitions

[atomic, nuclear] Both on an atomic electron level as well as with respect to energetic migrations of nuclides of an ATOM there are specific ENERGY constraints that provide ALLOWED TRANSITIONS as well as define forbidden transitions. The primary constraint in subatomic PARTICLE and elementary particle energy transitions is associated with the SPIN (and respective PRECESSION) of the rotating particle and particularly with respect to other energetic particles of the same class. An allowed transition obeys the selection rules that are primarily associated with the solutions to the SCHRÖDINGER EQUATION, the complement are the forbidden states. Forbidden transitions thus violate the selection rules. One particular exclusion rule is the fact that two particles in the same system with the same energy cannot have identical spin. Another example of an allowed transition is the change in angular momentum limited by unit change in the orbital angular momentum QUANTUM NUMBER (*also see* PAULI EXCLUSION PRINCIPLE *and* SELECTION RULES).

Force (F)

[general, fluid dynamics, mechanics, nuclear, thermodynamics] A vector of influence in the direction of the acceleration (a) of a body with mass m. Force is defined by NEWTON'S SECOND LAW, where the sum of the forces equals the mass of the body times the acceleration: $\vec{F} = \Sigma \vec{F_i} = m\vec{a}$, the push or pull which tends to initiate MOTION to a body at rest, the product of the mass of the body (m), and its acceleration ($\vec{a}$ with both a MAGNITUDE and a direction). Also, the influence applied on a body to increase or diminish the speed or change the direction of a body already in motion. Other definitions are used in specific external conditions, which still entail COMPLIANCE with mass multiplied by acceleration. In ELECTRICITY and MAGNETISM, the Lorentz force law is defined as $\vec{F} = q\vec{E} + q\vec{v} \times \vec{B}$, with q the ELECTRIC CHARGE of the body in motion in an electric field $\vec{E}$, while traveling at velocity $\vec{v}$ and when exposed to an additional MAGNETIC FIELD $\vec{B}$. Conservative force is defined by the fact that the work performed by the force is independent of the path (s) over which the work is performed. The conservative force is identified by the change in potential ENERGY (PE), defined in magnitude as $F = dPE(s)/ds$ (*also see* FOUR FORCES).

Forced oscillation

[energy, fluid dynamics, general, mechanics, thermodynamics] Periodic perturbation from equilibrium as the result of external influences. A forced oscillation of a mass m, with inherent resonant functionality obeys the harmonic OSCILLATOR equation for displacement z as a function of time t under the influence of applied force (F; fluctuating with ANGULAR FREQUENCY ω) expressed as $m(d^2z/dt^2) + b(dz/dt) + kz = F\sin\omega t$, where k is the spring constant, b the DAMPING. Note that the undamped, unforced VIBRATION will have a NATURAL FREQUENCY $\omega_0 = \sqrt{(k/m)}$. The forced oscillation will have maximum AMPLITUDE at the resonant frequency $\omega_R = \sqrt{((k/m) - (b^2/2m^2))}$. An electrical OSCILLATION can be provided by the circuit construction with CAPACITOR, INDUCTOR, and RESISTOR; for instance, in a RADIO STATION, emitting ELECTROMAGNETIC RADIATION. One particular forced OSCILLATION is the MICROWAVE energy-induced molecular vibration of water molecules in a MICROWAVE OVEN, under the ELECTROMAGNETIC FIELD resulting from a magnetron. On a swing the forced oscillation is fueled by the mechanical MOTION of swinging of the legs while leaning and changing the location of the CENTER OF GRAVITY on a periodic basis. An army platoon will need to pass a bridge out-off formation, to reduce the chances of generating a RESONANCE FREQUENCY that may actually break the structural integrity of the bridge. An example of a mechanical vibration is the combination of a spring, a damping pod, and the equivalent of a mechanical inductor, or a true electrical inductor; for example, Woofer, VOICE COIL; *see* MASS-SPRING SYSTEM (*also see* OSCILLATION *and* DRIVEN DAMPED HARMONIC OSCILLATOR) (see Figure F.45).

Figure F.45 Snapshot of forced oscillation: second string from top on guitar after being "plucked."

Forced vibration

[dynamics, fluid dynamics, general, mechanics] In contrast to self-excited vibrations there is an external force applying the oscillatory ENERGY fluctuations. The OSCILLATION frequency is directly linked to the rate of change of the driving force.

Forced vortex

[engineering, fluid dynamics] Rotational MOTION of a medium or a container filled with a LIQUID where the rotational velocity is a function of radius due to the internal VISCOSITY and rotational momentum (more mass on the outer limits than toward the center). A bottle of liquid placed on a rotating table (turntable) will produce an exemplary forced vortex. FLUID FLOW along a rotating device also induces a forced vortex. The pressure for a GAS flow in a forced vortex adheres to $P_o/P_i = \{1 + ((K_r \omega R_i)^2 / 2c_p T_i)[(R_o/R_i)^2 - 1]\}^{\gamma_p/\gamma_p - 1}$, where K_r is the core swirl (ratio of transfer of rotation of the core with radius R_i and ANGULAR VELOCITY ω to the surrounding gas; $0 \leq K_r \leq 1$), in a volume with radius R_o (also, for the pressure [P], the inner P_i and outer [P_o] values are the ones affected), T_i the internal temperature, γ_p the pressure correction factor, and c_p the SPECIFIC HEAT at constant pressure. A forced vortex will also be found in the vicinity of a PROPELLER, both in liquid or in gas (*also see* TORSION WAVE *and* TORSIONAL OSCILLATIONS) (see Figure F.46).

Figure F.46 Forced vortex resulting from stirring in chocolate syrup into milk.

Förster, Theodor (1910–1974)

[biomedical, chemical, imaging] Scientist from Germany. Theodor Förster provided significant contributions in the experimental investigations and theoretical description of the ATOMIC ENERGY transfer between nonradiative states through dipole–dipole coupling, that result in fluorescent DECAY with inherent details about the ENERGY configuration on a molecular level. The emitted RADIATION is a direct function of the molecular spacing, with the energy transfer (and hence to emission wavelength; spectral line) inversely proportional to the DISTANCE between the donor CHROMOPHORE and acceptor chromophore to the sixth power. This is also of influence in the Franck–Condon effect (see Figure F.47).

Figure F.47 Theodor Förster (1910–1974). (Courtesy of the University of Düsseldorf, Düsseldorf, Germany.)

Förster resonance energy transfer

[biomedical, chemical, imaging] Nonradiative dipole–dipole CHROMOPHORE ENERGY transfer as described by scientist THEODOR FÖRSTER (1910–1974) from Germany. The detection of the fluorescent spectrum and fluence provides details about the inter-dipole spacing and energy configuration of the constituents. When the two light sensitive chromophores are providing fluorescent signaling. The process is more generally referred to as FLUORESCENCE RESONANCE ENERGY transfer (*see* **FLUORESCENCE RESONANCE ENERGY TRANSFER [FRET]**).

Fosbury flop

[general, mechanics] The curved backward roll over a high bar for jumpers. The curved body actually places the CENTER OF GRAVITY below the bar, hence reducing the required ENERGY to reach a higher jump level (see Figure F.48).

Figure F.48 Fosbury flop.

Foucault, Jean Bernard Léon (1819–1868)

[general, optics] Physicist from France. In the year 1850, Foucault derived the speed of light in water and compared it to the established speed of light in AIR derived by ARMAND HIPPOLYTE LOUIS FIZEAU (1819–1896). Foucault did a similar MEASUREMENT as Fizeau, but with a rotating MIRROR. He measured the returning ANGLE of the reflected light over a DISTANCE of 18 m to the mirror and back, due to the rotation of the mirror, and concluded that the speed of light had to be 2.99796×10^8 m/s in air, in the year 1850, within 10^{-3} % of the accepted standard in VACUUM: 2.99792458×10^8 m/s. Foucault was also the first person to conceive and construct a GYROSCOPE (see Figure F.49).

Figure F.49 Jean Bernard Léon Foucault (1819–1868).

Foucault current

[energy, general, electromagnetism] *see* EDDY CURRENT. The Foucault name remains attached to this phenomenon due to the theoretical contributions provided by JEAN BERNARD LEON FOUCAULT (1819–1868).

Foucault's pendulum

[mechanics, geophysics] Device designed to visually illustrate the rotation of the EARTH, introduced in 1851. A pointed, heavy ($m \sim 28$ kg) object suspended from an excessively long cable (67 m) ("bob") is moving in pendulum MOTION with the sharp point indicating the path with respect to a geometric map outline directly below it, angular outline. The PENDULUM, as originally placed in Paris, France, will perform a full circular crossing over a period of 32.7 h, rotating the direction of swing in 11° increments, clockwise, per hour. These experiments were performed by JEAN BERNARD LÉON FOUCAULT (1819–1868). The direction in which the pendulum swings changes in clockwise rotation from the first linear path, and continues to make an ANGLE with the initial path on each swing, ever increasing. The forces acting on the swinging bob are following

Coriolis forces, in longitudinal and latitudinal directions, respectively, $F_{C,x} = 2m\omega_e(dy/dt)\sin\varphi$, where ω_e represents the angular velocity for the rotation of the Earth; note that the CORIOLIS FORCE is a function of the LATITUDE ANGLE φ; $F_{C,y} = -2m\omega_e(dx/dt)\sin\varphi$. The equation of motion expressed in complex notation $z = x + iy$ yields $(d^2z/dt^2) + 2i\omega_e(dz/dt)\sin\varphi + \theta^2 z$, where θ represents the rotation of swing direction. The solutions are in first-order approximation: $z = e^{-i\omega_e\sin\varphi t}\left(C_1 e^{i\theta t} + C_2 e^{-i\theta t}\right)$ (*also see* **CORIOLIS FORCE**) (see Figure F.50).

Figure F.50 Foucault's pendulum.

Four forces

[general] Four forces of nature GRAVITY, weak nuclear, (the prior two are part of general ELECTROWEAK FORCE), electromagnetic, and strong nuclear.

Fourier, Jean Baptiste Joseph, Baron (1768–1830)

[acoustics, computational, imaging, optics, theoretical] French mathematician and scientist. Joseph Fourier is best known for his computational approach to complex function analysis, specifically the FOURIER TRANSFORM (*also see* JOSEPH, JEAN BAPTISTE FOURIER) (see Figure F.51).

Figure F.51 Jean Baptiste Joseph Fourier (1768–1830).

Fourier analysis

[atomic, computational] Mathematical analysis introduced in 1822 by the French mathematician BARON JEAN BAPTISTE JOSEPH FOURIER (1768–1830) using the decomposition of functions in a series of trigonometric functions: $F(\omega) = \sum_{n=0}^{\infty} \left(A_n \cos(n\omega_0 t + \alpha_n) + B_n \sin(n\omega_0 t + \alpha_n) \right)$, where t is time when operating in the frequency domain (time and frequency are equivalent with respect to the alternative: location), α_n is the PHASE shift, C_n constant for term n, and ω_0 the FUNDAMENTAL FREQUENCY for the phenomenon. In case the phenomenon is not harmonic, the series is not necessarily an arithmetic sequence: $F(\omega) = \sum_{n=0}^{\infty} C_n \cos(\omega_n t + \alpha_n)$, where the coefficients are derived from the original function $f(t)$ (*see* FOURIER TRANSFORM).

Fourier conduction law

[thermodynamics] the rate of heat FLOW (Q/t, where Q is the heat and t the time) as a result of a temperature gradient ($\Delta T/d$, where ΔT represents the temperature gradient, specifically perpendicular to the surface: $\partial T/\partial n$, and d the length) across an area (A) expressed as $\dot{Q} = Q/t = dQ/dt = k_T A(\Delta T/d)$, where k_T is the THERMAL CONDUCTIVITY of the medium.

Fourier number (Fo = κt/cρL²)

[thermodynamics] Dimensionless number representing the ratio of the heat conduction per unit time to the rate of thermal storage. The Fourier number mainly applied to solids, with L the characteristic length of the solid, t is time, c_p the SPECIFIC HEAT CAPACITY, ρ the DENSITY of the solid, and κ the THERMAL CONDUCTIVITY. The number is used, for instance, in an alternating or transient conductive COOLING and HEATING situation, converting the thermal CONDUCTION equation in a nondimensional expression. The Fourier number provides a dimensionless time for the period over which a temperature change will occur. Considering that the unsteady DIFFUSION EQUATION is mathematically equivalent to the unsteady conduction equation, the same principles can be used to solve mass transfer situations. For mass transfer the partial differential and continuity equations will have definitions and variables that are different than for HEAT TRANSFER, such as DIFFUSION versus THERMAL CONDUCTIVITY and so on.

Fourier number, mass transfer (Fo$_m$ = $D_{AB}t/L^2$)

[computational, fluid dynamics, thermodynamics] Dimensionless number used in FLUID DYNAMICS. The ratio of the DIFFUSION rate of the constituents in SOLUTION to rate of permanent solution, with D_{AB} the diffusion rate, t time, and L the characteristic dimensional constant. Generally, the moisture content decrease with increasing MASS TRANSFER FOURIER NUMBER. The mass transfer Fourier number can be used as an indication of the drying time for a SOLUTE. This is the mass equivalent for the FOURIER NUMBER used in thermal diffusion. Generally, the mass transfer Fourier number provides a dimensionless time for the period over which a chemical diffusion will occur.

Fourier optics

[optics] Physical OPTICS branch within PHYSICS that analyzes the propagation of light through an optical system (e.g., diffractive media, turbid media) by means of linear system theory and harmonic analysis and deconvolution. The FOURIER ANALYSIS in IMAGE analysis uses a two-dimensional WAVE EQUATION. The two-dimensional (2D) wave equation forms a superposition of an array of harmonic functions with specific spatial frequencies and complex AMPLITUDES: $f(x, y) = \Sigma C_n exp[-i2(\omega_{x,n} x + \omega_{y,n} y)]$, where $\omega_{x,n}$ and $\omega_{y,n}$ are the harmonic frequencies in the x and y directions, respectively. The amplitude (C_n) of each function (Fourier component), in complex form, is derived from the FOURIER TRANSFORM. The PLANE WAVE in 2D optics is the primary form of the projection of the three-dimensional wave process in image formation, providing an analysis of an arbitrary complexity profile. The 2D projects of the plane wave are referenced as wavefronts. Each respective WAVEFRONT progresses through space defined by the wavevector $\vec{k} = (k_x, k_y, k_z)$, $k = 2\pi/\lambda$, wavelength λ. The superposition principle applies to the wave expression as $f(x, y) = (2\pi)^{-2} \iint F(k_x, k_y) \exp[-i(k_x x + k_y y)] dk_x dk_y$, $F(k_x, k_y)$ the Fourier transform of the image function $f(x, y)$. In Fourier optics, a pinhole performs a physical Fourier transform of an image on the rays passing through the pinhole, projecting a diffraction pattern on a screen (*see* FOURIER TRANSFORM).

Fourier plane

[acoustics, computational, imaging] Dissection planes in the reconstruction process for three-dimensional IMAGE forming the correlation mechanism between the dissection lines and the planes as part of the "slice theorem." This theorem states that the one-dimensional Fourier transform of the projection of an object is the same as the values of the two-dimensional Fourier transform of the object along a line drawn through the center of the two-dimensional Fourier transform. In general, the (Fourier) SLICE theorem relates the two-dimensional image to the one-dimensional projection, for instance, in computed TOMOGRAPHY (X-RAY) and ULTRASONIC IMAGING. In MRI, the process of the associated two-dimensional Fourier transform is transferred to what is known as "K-SPACE." The reconstruction of the image following backprojection ($f(\xi, \phi)$, with coordinates (ξ, ϕ, ρ) with $1/\rho$ being the FOURIER TRANSFORM for $1/r$) on each ANGLE in the respective Fourier plane is formulated as $g(r, \theta) = \int_0^\pi \int_{-\infty}^\infty \{ \int_{-\infty}^\infty f(\xi, \phi) e^{-i2\pi\rho\xi} d\xi \} |\rho| e^{i2\pi\rho r \cos(\theta - \phi)} d\rho d\phi$. The image reconstruction furthermore requires the interpolation between the one-dimensional Fourier transform derived from the source–detector scan to form a two-dimensional image in the Fourier plane. This is subsequently followed by the inverse Fourier transform in the Fourier plane to reconstruct the two-dimensional slice image. This process is an alternative to two other processes: backprojection and rho filter.

Fourier series

[atomic, computational, imaging, optics] Sum series of sinusoidal/cosine functions that are the transform of the original changing process, decomposed in processes of repetitive nature with multiples of the FUNDAMENTAL FREQUENCY of the original periodic phenomenon. The periodicity can be in the spatial domain or in the frequency domain, specifically pertaining to the analysis of objects and images, next to data streams and optical regularities (e.g., DOUBLE-SLIT EXPERIMENT). A function of a phenomenon with periodicity L as a function of DISTANCE x^* (with the fundamental repetition pattern with wavelength $\lambda = 2L$) can be expanded in a Fourier series as $f(x^*) = \sum_{n=0}^{\infty} \left(A_n \cos(2\pi n x^*/L) + B_n \sin(2\pi n x^*/L) \right) = \sum_{n=-\infty}^{\infty} C_n e^{i 2\pi n x^*/L}$ (all functions $e^{i 2\pi n x^*/L}$ are orthogonal on the interval $x^* = -(x/L)$ to $x^* = +(x/L)$; orthogonality of functions is defined through $\int_{x_1}^{x_2} \varphi_m^* \varphi_n dx = 0$, φ_m^*, the complex conjugate switching the sign of all terms with $i: -/+$), where in most cases $A_n = 0$ due to the boundary conditions (general condition: $A_n = (1/\pi) \int f(x^*) \cos(nx^*) dx^*$), and B_n is the AMPLITUDES of the harmonics; $B_0 = (1/2\pi) \int_{-(L/2)}^{L/2} f(x^*) dx^*$ and $B_n = (1/\pi) \int_{-(L/2)}^{L/2} f(x^*) \sin(nx^*) dx^*$ with $n = 1$ representing the fundamental "wave" (see **FOURIER TRANSFORM**).

Fourier theorem

[nuclear] Computational behavior for several specific conditions and applications with respect to perturbations and involving more than one function: $f(x^*)$ and $g(x^*)$, with respective Fourier functions $F(v^*)$ and $G(v^*)$. Rules of transaction force functions and their respective Fourier transforms. Based on the fundamental FOURIER TRANSFORM principles, several inherent relations can be postulated for ease of use calculations. The theorems applied are for the following processes: addition, $f(x^*) + g(x^*) \rightarrow F(v^*) + G(v^*)$; AUTOCORRELATION, $\int_{-\infty}^{\infty} f(\chi) f(\chi + x^*) d\chi \rightarrow |F(v^*)|^2$; convolution, $f(x^*) * g(x^*) \rightarrow F(v^*) G(v^*)$; derivative, $df(x^*)/dx^* \rightarrow i 2\pi v^* F(v^*)$; modulation, $f(x^*) \cos(2\pi f_0(x^*)) \rightarrow (1/2) F(v^* + f_0) + (1/2) F(v^* - f_0)$; shift, $f(x^* - x_0) \rightarrow F(v^*) e^{i 2\pi v^* x_0}$; similarity, $f(c_1 x^*) \rightarrow |c_1|^{-1} F(v^*/c_1)$; and additionally, Rayleigh's theorem, $\int_{-\infty}^{\infty} |f(x^*)|^2 dx^* = \int_{-\infty}^{\infty} |F(v^*)|^2 dv^*$.

Fourier transform

[astronomy/astrophysics, atomic, biomedical, computational, general, imaging, mechanics, optics, quantum] Probably the most commonly used SIGNAL transform applied to the analysis of signals and in IMAGE processing, introduced in 1822 by BARON JEAN BAPTISTE JOSEPH FOURIER (1768–1830). Joseph Fourier was investigating the propagation of heat and realized that the algorithm could significantly be simplified when the physical parameters were decomposed in terms of sets of trigonometry functions. The transform changes the spatial domain to the frequency domain, or respectively from the time domain to the frequency domain. The Fourier transform deconvolves a changing or periodic signal ($f(x^*)$) in an infinite number of harmonics with wavelength: λ (frequencies v: v_0 the fundamental and harmonics $2v_0$, $3v_0$, …; note WAVE number $k = 2\pi/\lambda$), based on the one-dimensional formulation: $F(v^*) = FT\{f(x^*)\} = \alpha \int_{-\infty}^{\infty} f(x^*) e^{i\beta v^* x^*} dx^*$, where α and β are NORMALIZATION constant, both may frequently be found to be one, or a multiple of pi (π). The parameters x^* and v^* should be dimensionally inverse; v^* indicative of the frequency variable. Most frequently the transform is from spatial location to spatial frequency. The Fourier function can be written as the MAGNITUDE ($|F(v^*)|$) multiplied by the complex ANGLE function with PHASE ($\angle F(v^*)$) as: $F(v^*) = |F(v^*)| e^{i \angle F(v^*)}$. When the reverse transform can also be applied the functions $F(v^*)$ and $f(x^*)$ are considered a Fourier pair ($f(x^*) = \alpha \int_{-\infty}^{\infty} F(v^*) e^{\pm i\beta v^* v^*} dv^*$). In two-dimensional format the transform is expressed in spatial frequencies (v^* and ξ^*) as a function of location (x^* and y^*) by the equation: $F(v^*, \xi^*) = \alpha' \int_{-\infty}^{\infty} \int_{-\infty}^{\infty} f(x^*, y^*) e^{-i 2\pi (v^* x^* + \xi^* y^*)} dx^* dy^*$. Fourier transforms are convenient computational tools for complex situations such as for instance found in atomic crystals; ionic crystals; magnetic doublets; mechanical systems of point masses; electrostatic point charge distributions; optical configurations, in particular the use of narrow slits for spectral isolation, specifically applied to astronomy; and the PARTICLE distribution in QUANTUM MECHANICS. An example of the time DECAY $f(t) = e^{-t}$ ($t \geq 0$) yields the Fourier transform $F(v) = 1/(1 + i 2\pi v)$. There are certain rules that apply to the Fourier function, defined by theorems. The theorems are addition, $f(x^*) + g(x^*) \rightarrow F(v^*) + G(v^*)$; AUTOCORRELATION, $\int_{-\infty}^{\infty} f(\chi) f(\chi + x^*) d\chi \rightarrow |F(v^*)|^2$; convolution, $f(x^*) * g(x^*) \rightarrow F(v^*) G(v^*)$; derivative, $df(x^*)/dx^* \rightarrow i 2\pi v^* F(v^*)$; modulation, $f(x^*) \cos(2\pi f_0(x^*)) \rightarrow (1/2) F(v^* + f_0) + (1/2) F(v^* - f_0)$; shift, $f(x^* - x_0) \rightarrow F(v^*) e^{i 2\pi v^* x_0}$; similarity, $f(c_1 x^*) \rightarrow |c_1|^{-1} F(v^*/c_1)$; and additionally, Rayleigh's theorem, $\int_{-\infty}^{\infty} |f(x^*)|^2 dx^* = \int_{-\infty}^{\infty} |F(v^*)|^2 dv^*$. The Fourier transform of the Gaussian signal $f(x) = e^{-\pi x^2}$ has a GAUSSIAN DISTRIBUTION as well ($F(v') = e^{-\pi v'^2}$), which has implications for the analysis of ULTRASONIC IMAGING, in particular considering nondiffracting waves. This concept applies to either location or time. A special application of the Fourier transform is the discrete Fourier transform. The discrete Fourier transform is applied when the function/phenomenon is a collective of events, expressed as $f(n)$, where $n=0,1,\ldots,N$, such as the slats of a (endless) picket fence. The Fourier transform for this situation is defined as the repetition rate by means of $F(k^*) = \sum_{n=0}^{N-1} f(n)e^{-i(2\pi k^* n/N)t}$, where the number of samples in the frequency domain is the same as the number of points in time domain (N), with $k^* = 0,1,2,\ldots,N-1$. Respectively with the inverse transform: $(n) = (1/N)\sum_{k=1}^{N-1} F(k^*)e^{i(2\pi k^* n/N)t}$, $n=0,1,2,\ldots,N-1$. In the SAMPLING of the signal one specific condition applies, the sampling frequency of the signal needs to be minimally twice the highest frequency observed in the signal itself. This sampling constraint is referred to as Nyquist or Shannon theorem. This ensures that no critical information is lost by analyzing with at least half the shortest "wavelength"/pattern of the phenomenon. The half-wavelength criterion ensures capturing an equilibrium condition and an extreme; either a crest or a trough. The INNER EAR, based on its design, in fact performs a physical Fourier transform by means of the location-specific sensitivity to frequency of the cochlea (*also see* FAST FOURIER TRANSFORM *and* LAPLACE TRANSFORM) (see Figure F.52).

F

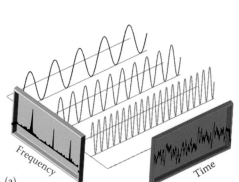

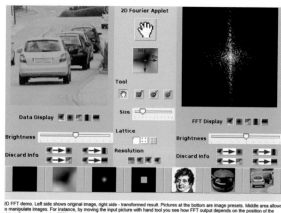

(a) (b)

Figure F.52 (a) Outline of the process of the Fourier transform and (b) screenshot of a software program developed by UltraStudio (Fällanden, CH-8117, Switzerland; http://ultrastudio.org) that is designed to perform a spatial Fourier transform of the intensity frequency profile of an image.

Fovea

[biomedical, chemical, optics] Optical center point on the RETINA of the EYE, on the optical axis, consisting of CONES only. The fovea is the central part of the macula. The fovea is meant for high-resolution COLOR viewing, primarily under bright conditions; under dim lighting the RODS will provide imaging based on radiance only (grayscale). The cone density of the fovea is 6.5–7 million cones in a spot with a diameter of 1.5 mm. The macula also has rods mixed in, whereas the fovea consists solely of cones. In comparison, the entire lining of the retina is covered over better than 2π SOLID ANGLE with 120–130 million rods, respectively with highest density at $20°$ from the optical axis (*also see* EYE). Because of the high density of cones the RESOLUTION for an object placed in the NEAR-POINT of the eye provides an angular resolution of 0.3 arc min = 0.0000872 rad apart, resulting in a spatial resolution at the center wavelength 550 nm, applying a focal length of 2.2×10^{-2} m (the length of the HUMAN EYE), to an object placed in the near-point of the HUMAN EYE (~0.25 m; range varies per person and is dependent on AGE) yielding approximately 88.65 μm on average (*see* EYE *and* RETINA) (see Figure F.53).

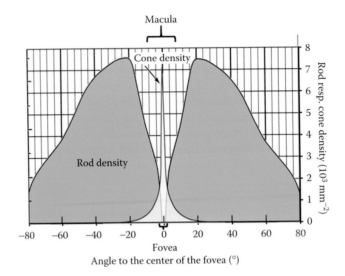

Figure F.53 Graphical representation of the location and size of the fovea in a cross-sectional view of the human eye and illustration of the distribution of the rods (radiance view; magnitude only), and cones (color: red, green, blue) optical detector cells on the retina and specifically how the fovea is "privileged" in the concentration of cones.

Franck, James (1882–1964)

[biomedical, chemical, optics, quantum] Physicist and chemist from Germany. Franck is known for his theory on the origins of fluorescent transitions and the PHOTOSYNTHESIS. Additionally, James Franck worked with GUSTAV HERTZ (1887–1975; nephew of HEINRICH RUDOLF HERTZ [1857–1894]) on electron collisions and collisions of electrons with atoms of gasses. James Franck along with Gustav Hertz received the Nobel Prize in Physics in 1926 for their work on the confirmation of the BOHR ATOMIC MODEL, demonstrating the discrete ENERGY states of excited electrons in the ATOM.

Franck–Condon principle

[optics, quantum] Description of the dissociation ENERGY transfer of molecular states, first described for a diatomic molecule in 1925 by the physicist and chemist JAMES FRANCK (1882–1964) from Germany, with contributions in 1926 by the QUANTUM mechanical physicist EDWARD UHLER CONDON (1902–1974) from the United States. The classical interpretation involves the consideration that there are no changes in the nuclear configurations associated with the electron energy transitions. The nuclear position changes relatively slow in comparison to the electron transfer. The dissociation can be accompanied by RADIATIVE TRANSFER, providing a fluorescent marker for the transition. The Franck–Condon principle provides a selection rule for spectroscopic analysis, specifically a tool in quantum chemistry. The transitions under the Franck–Condon principle relate the energy interactions between vibronic transitions during PHOTON absorption. The energy transitions take place based on the Schrödinger WAVE solutions for the molecular binding quantum well, which provides the mechanism for the photon interaction. Preferred transitions are between like-wise orientations of the waveform, but at "higher harmonic" levels. The POTENTIAL WELL for the nuclear and electron orbit energy configurations with respect to molecular bindings is shaped according to a parabolic distribution where the "matching" states are quantum harmonic OSCILLATOR waveforms at the various excitation levels ranging from level zero at GROUND STATE to infinity for a free NUCLEON. Exchanges between vibrational energy levels for the nuclear coordinates are favoring minimal change in nuclear coordinates. The fluorescent excitation process will use incident light of a certain high energy wavelength, which will lose some of its energy at interaction with the Franck–Condon state transition, hence resulting in a lower energy photon emission. The energy loss and resulting longer wavelength is known as the Stokes shift. The energy loss can be measured and provides a direct indication of the energy levels and the "dimensions" of the transition in the molecular configuration. This process can as such be used to map the molecular construction. One example of the Franck–Condon principle with respect to detectable FLUORESCENCE marking from experimental investigational standpoint is found in VISION. Rhodopsin (MOLECULAR WEIGHT ~35,200) is the macromolecule in the RODS (CHROMOPHORE) of the EYE that has a photonic transition at location C-11 (with maximum sensitivity at 500 nm) that takes place on a femtosecond scale. The rhodopsin photon interaction, with temporary energy storage, can be described based on the Franck–Condon principle. The photon interaction with associated changes in ABSORPTION SPECTRUM for an illumination investigating source as part of the chemical processes in rhodopsin can be revealed by absorption SPECTROSCOPY. For rhodopsin, the Franck–Condon energy transfer of the photon interaction is converted in a chemical transition to BACTERIORHODOPSIN within several picoseconds after the photon interaction. This photochemical interaction provides the initiation for the onset of the generation of an ACTION POTENTIAL from a single rod (see Figure F.54).

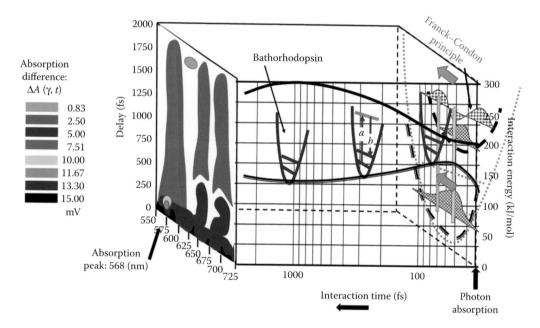

Figure F.54 Franck–Condon principle applied to rhodopsin of biological rod photosensors during interaction of photons within the first 3 fs of the photochemical process leading up to the formation of the action potential with respect to vision. Vibrational transitions in the energy level associated with torque (torsional vibrations) within the C11–C12 position for the rhodopsin molecule (molecular weight: ~35,200).

Franck–Hertz experiment

[atomic] Experimental verification of the discrete electron ENERGY levels by the physicists JAMES FRANCK (1882–1964) and GUSTAV HERTZ (1887–1975), both from Germany. The two scientists were successful in proving the BOHR ATOMIC MODEL and received the Nobel Prize in Physics in 1926 for their efforts.

Franklin, Benjamin (1706/1705?–1790)

[general] American founding father who started out as a printer and became a scientist and statesman for the forming of the United States. Benjamin Franklin acquired significant notoriety as an inventor and scientist. One of his most questionable experiments is the launch of a kite during a thunderstorm (1752) with the intention of attracting LIGHTNING, which could have killed him. These experiments lead to his introduction of lightning RODS on the tops of buildings as a practical deterrent for damage resulting from lightning strikes. During his experimentations, he concluded that there are two types of charges and he separated them as positive and negative. Based on Franklin's assumptions current flows from positive to negative, however the electrons actually migrate from negative to positive potential only. The observation of current as defined by Benjamin Franklin based on the charges applied to GLASS after rubbing it with fur, defined as positive charges by Franklin. Later the current was found to be carried by electrons (negative charges). Additionally, the work of Ben Franklin implied the LAW OF CONSERVATION OF CHARGE, but was not explicitly defined by him (see Figure F.55).

Figure F.55 Benjamin Franklin (1706/1705?–1790). (Courtesy of J. Thomson.)

Franklin

[general] As a unit of charge indicates a force of $F = 1$ dyne ($= 1 \times 10^{-5}$ N) for two charges of 1 Franklin at 1 cm spacing (Coulomb law in Gaussian system: $F = q_1 q_2 / r^2$): 1 Fr = 1 statC $\approx 3.33564 \times 10^{-10}$ C. It is also referred to as the statcoulomb (*statC*).

Fraunhofer, Joseph von (1787–1826)

[general, optics] Physicist and optician from Germany (Prussian empire). Joseph von Fraunhofer is best known for his documentation of the absorption lines in the solar spectrum, identifying molecular and atomic species. Additionally, the lenses made by Joseph von Fraunhofer were unchallenged in quality. His lenses were used to build perfectly achromatic telescopes (see Figures F.56 and F.57).

Figure F.56 Joseph von Fraunhofer (1787–1826).

Figure F.57 Artist impression of Joseph von Fraunhofer (1787–1826) with his telescope. (Courtesy of Richard Wimmer, *Essays in Astronomy*. D. Appleton & Company, New York City, 1900.)

Fraunhofer diffraction

[general, optics] Diffraction pattern created in far-field approximation primarily with respect to a single slit. Using Huygens's principle that each location in the space of the slit, with width a, constitutes a source, the sources will form INTERFERENCE. The destructive interference (dark lines) is produced as a function of the illumination WAVELENGTH (λ) in discrete intervals (integer m) with respect to the central optical axis under an ANGLE θ adhering to $\sin\theta_{dark} = m(\lambda/a)$. The incident rays of ELECTROMAGNETIC RADIATION, or mechanical waves, create an interference at a large DISTANCE where the paths of the rays are considered to be parallel. Compare with the near-field Fresnel diffraction. The secondary sources in within the area of the opening each have a PHASE angle that is slightly different from the neighboring POINT SOURCE increasing with distance from the edge, providing a vector addition theorem yielding a cord with a length representing the AMPLITUDE of the interference pattern at a specific angle with the normal to the opening (see Figure F.58).

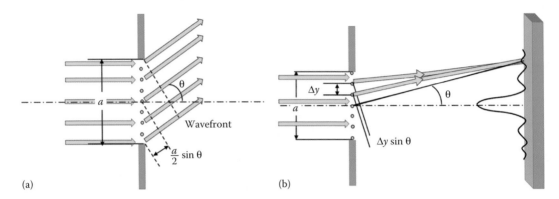

(a)   (b)

Figure F.58 Geometric representation of the Fraunhofer diffraction slit used to define the Fraunhofer diffraction principle. (a) Single slit where each point in the opening represents a secondary light source and the conditions for positive wave superposition and (b) interference pattern projected on a projection screen.

Fraunhofer lines

[astronomy/astrophysics, atomic, optics] Spectral absorption lines in the solar emission named after JOSEPH VON FRAUNHOFER (1787–1826), who documented these findings in 1814. These spectral omission lines are found ranging from the ULTRAVIOLET, starting at approximately 180 nm far into the INFRARED up to 20 μm. Each specific lines represent certain atomic and molecular transitions with respect to ENERGY transferred during absorption. The recognized shift in emission wavelength can be used as an indication of surface temperature effects, for instance, the shift for the carbon absorption at 538.03 nm (*also see* ZEEMAN EFFECT) (see Figure F.59).

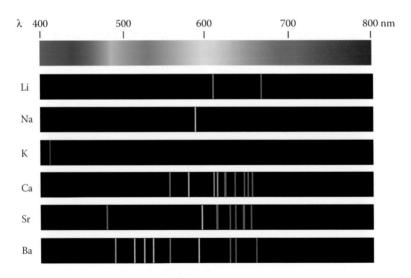

Figure F.59 Fraunhofer lines emitted by various elements.

Free charge

[electronics, general] The classical model of electrical conduction in metals based on the free migration of electrons was first proposed by Paul Drude (1863–1906) in 1900. This model was followed by OHM'S LAW, which describes resistivity related to the MOTION of atoms in place with respect to the free movement of electrons in conductors, in particular metals. The thermal atomic/MOLECULAR MOTION increases the RESISTANCE to higher incidence of collisions with the electron migration. Conduction in a material with free electrons relies only on the DRIFT VELOCITY of the electrons, however since the CONDUCTOR is filled with electrons, the relation to an emerging electron resulting from an electron entering from an electromotive force source is virtually instantaneous over any length. Free electrons can move from the VALENCE BAND to the CONDUCTION BAND of the ENERGY configuration of the material with little effort, or are occupying the conduction band due to very loose forces holding the external electron(s).

Free spectral range

[optics] In interferometry, specifically applied to a Fabry–Pérot étalon, the full-width half maximum spectral line separation ($\Delta\lambda$) between consecutive INTERFERENCE maxima is a device constant written as $\Delta\lambda = \lambda_0/(2n\ell\cos\theta + \lambda_0) \cong \lambda_0/2n\ell\cos\theta$, where the central wavelength of the maximum is λ_0, the étalon has refractive index n and thickness ℓ and light enters at an ANGLE θ.

Freezing

[general, thermodynamics] Temperature where solid and LIQUID are in equilibrium. The process of freezing requires the removal of ENERGY from a liquid, defined at the enthalpy of SOLIDIFICATION, or enthalpy of FUSION (h, generally a positive value, except for helium, which requires the addition of heat to fuse), also known as the LATENT HEAT of FUSION. For water the fusion heat is $h_{f,H_2O} = 333.55$ kJ/kg. Adding a SOLUTE to a liquid (the SOLVENT) will decrease the freezing point. This "freezing point depression" is widely used in deicing of planes and on the road surface during snow or freezing RAIN. The freezing point depression (ΔT_F) can be described in first-order approximation based on the solvent nature (expressed by the cryoscopic factor $K_F = RT_F m_M/\Delta h_f{}^m$, where $R = 8.3144621(75)$ J/Kmol the GAS constant, T_F the freezing point of the solvent, m_M the MOLAR MASS of the solvent, and $\Delta h_f{}^m$ the molar heat of fusion change for the solvent), the VAN'T HOFF FACTOR (Π_{osm}), the molar concentration of the solvent ($[S]$), as $\Delta T_F = K_F \Pi_{osm}[S]$. A more detailed analysis yields $\Delta T_F = \Delta h_f{}^m - 2RT_F \ln(a_{liq}) - \sqrt{[2\Delta c_p{}^{\ell s} T_F{}^2 R \ln(a_{liq}) + (\Delta h_f{}^m)^2]}/2((\Delta h_f{}^m/T_F) + (\Delta c_p{}^{\ell s}/2) - R\ln(a_{liq}))$, where $\Delta c_p{}^{\ell s}$ is the change in SPECIFIC HEAT (under constant pressure between the liquid and the solid PHASE and a_{liq} represents the ACTIVITY of the SOLUTION, a function of the ionic length (determined from the Pitzer model, which ties the activity to the GIBBS FREE ENERGY (G) and the CHEMICAL POTENTIAL (μ_j) of the respective constituents, as $G = \sum_j \mu_{ij} + RT \ln(b_i a_{liq})$, where b_i is the molality) (see Figure F.60).

Figure F.60 (a) Water vapor frost deposition of window. (b) Illustration shows frozen water has a density of 0.9 times that of liquid water as the iceberg floats partially submerged (90%).

Frequency (ν)

[general] The number of cycles, revolutions, or vibrations completed in a unit of time, specifically per SECOND, expressed in Hertz (Hz). Frequency is the reciprocal period of a recurring (e.g., OSCILLATION) event. The unit Hertz is derived from the physicist HEINRICH HERTZ (1857–1894) from Germany, who introduced pioneering contributions to the understanding and description of RADIO waves. Frequency is associated with a real WAVE or the graphical representation of the periodic phenomenon yields a waveform. Frequency is correlated to the WAVELENGTH (λ) and speed of propagation (v; for light $c = 2.99792458 \times 10^8$ m/s) as $\nu = v/\lambda$. Whistling, using only the lips of the mouth, can produce a single frequency acoustic wave in most cases. Frequency also applies to discrete phenomena that occur at regular intervals, such as the delivery of the mail as once per day. Many phenomena are indicated by the base frequency in order to define the recurrence, such as the average HEART rate for a person at rest at approximately 65 beats per minutes, which provides a frequency of slightly greater than 1 Hz, while the *frequency spectrum* is defined by at least the first eight harmonics for the electrical ECG SIGNAL associated with the heart rate (*also see* DOPPLER EFFECT) (see Figure F.61).

Figure F.61 Milkman delivering the milk at a frequency of once per day.

Frequency, angular (ω)

[general] The expression of a repetitive phenomenon in graphical format provides a sine WAVE, which is directly linked to a circular MOTION. Expressing the temporal AMPLITUDE evolution in FREQUENCY (ν) is hence directly tied to the angular velocity, expressed in radians per SECOND $\omega = 2\pi\nu$. The angular velocity is linked to the tangential VELOCITY (v) as a function of the radius to the axis of rotation (r) as $v = r\omega$ (see Figure F.62).

Figure F.62 A record player with vinyl record with engraved musical pattern that rotates with constant angular velocity, the tangential velocity on the outer radius will be greater than the tangential velocity closer to the center of the rotating disk.

Frequency, resonant (ν_0)

[general] Frequency at which a system will oscillate when it is not driven. Examples of OSCILLATION frequencies for specific systems are PENDULUM of length L: $\nu_0 = (1/2\pi)\sqrt{(g/L)}$, where g is the GRAVITATIONAL ACCELERATION; MASS (m) on a spring with spring constant k: $\nu_0 = 1/2\pi\sqrt{(k/m)}$; CAPACITOR ($C$)–INDUCTOR ($L$) circuit $\nu_0 = (1/2\pi)(1/\sqrt{(LC)})$; and the resonant frequencies of a string with linear mass density $\mu_\ell = m/\ell$ (mass m; length of string ℓ), under tension F_T providing base frequency ($n = 1$) and higher harmonics ($n = 1, 2, 3, \ldots$): $\nu_n = (n/2\ell)\sqrt{(F_T/\mu_\ell)}$. Note that on a string instrument the frequency of the VIBRATION can be altered by changing the length of the string under placement of the fingers, or by changing the tension as applied for tuning purposes. The frequency aspects of the string with regard to tension were described by the French scientist and priest MARIN MERSENNE (1588–1648) (the "father of ACOUSTICS") in 1627. Alternatively, the resonant frequency defines the frequency at which the AMPLITUDE will reach excessive large MAGNITUDE when driven by external force. It is also referred to as "NATURAL FREQUENCY."

Frequency hopping spread spectrum (FHSS)

[communication, computational, general, signal] In frequency hopping transmission, the electromagnetic emission source (RADIO wave, telecommunications; mobile phone) applies a specific (predetermined) "hopping" pattern, continuously changing the CARRIER frequency. The changes in transmission frequency are applied in a pseudorandom sequence, applied to both the transmitter and the receiver stations. The advantage is that the SIGNAL uses a different transmission channel with an accompanying different set of interfering signals during each hop. This process has the following two main advantages: (1) it avoids failing communication at a particular frequency, because of a fade or a particular interferer and (2) the receiving station is dedicated to the transmitter. The "dedication" avoids any other party from "tuning in" to the data transmission/conversation, thus providing protection from unauthorized access. In a similar fashion, this process prevents a mobile phone user to listen to the conversations of anyone standing in any proximity to the sender or receiver, respectively, obtain no unauthorized details from a data transmission. The concept was introduced by the movie STAR HEDY LAMARR (1914–2000) in 1941, during World War II (1940–1945) to provide guidance during the launch and trajectory of a torpedo, ensuring and maintaining and correcting course and guarantee a delivery to the target location. The early mechanism was based on musical instruments, specifically the mechanical organ, or player PIANO (pianola), electronic carillon, or other various types of orchestrion. The orchestrion (Carillion; barrel organ hurdy gurdy) uses a role of paper with holes punched in a certain pattern that engage the keys on the instrument for a specific note when the hole passes the trigger switch. The "music role" encoding concept can be traced back to approximately 1877, Cambridgeport, MA. Other mechanisms using spikes/bumps on a METAL drum or disk, such as used in a music box, which was reportedly first used in the early ninth century in the Persian Empire (now known as Iraq) (see Figure F.63).

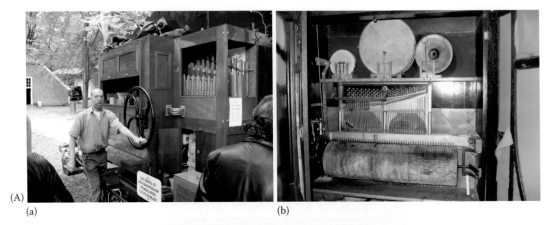

(A)
(a)   (b)

(B)

Figure F.63 (A) Musical instrument that uses the principle of "frequency-hopping" based on the fact that a roll with holes will engage a single instrument with the passing of a hole, hopping from note to note. The paper roll is advanced by rotating a wheel attached to an axle that winds the roll. This formed the basis of the concept introduced by Hedy Lamarr (1914–2000) and George Antheil (1900–1959) in 1942. (B) Musical instrument, "music-box" that uses the principle of "frequency-hopping" based on the fact that a disk with knobs engages a single chime with a distinct respective note on passing.

Frequency modulation (FM)

[electromagnetism, general] Principle of data encoding used in RADIO transmissions and other electromagnetic, and acoustic WAVE emissions as well as imaging based on superposition of a range of frequencies on top of a CARRIER WAVE (alternatively, inscribed in a baseline pattern). Frequency modulation primarily applies to analog data transfer. FM was introduced by an electrical engineer, Edwin Howard Armstrong (1890–1954), from the United States in 1933. A decoder is used to tune in to the carrier frequency and subsequently the imbedded SIGNAL is filtered out. Radio-frequency (RF) modulation is using carrier waves with a frequency range of 85–108 MHz. Television broadcasts were generally frequency modulated signals but in several countries the analog encoding has been changed to a digital format. In the transmission of radio signals, the use of frequency modulation provides a high bandwidth frequency encoding with the ability to send two channels within the same carrier wave bandwidth in order to achieve stereophonic SOUND after electronic decoding by the RADIO RECEIVER. STEREO encoding for radio transmission and listening was introduced in the late 1950s. In acoustical imaging frequency modulation of the pressure wave signal applied by the piezoelectric crystal provides the means for analysis of material properties based on DISPERSION effects. Piezoelectric crystals will act as adjustable INDUCTOR–capacitor circuits (LC tuner), with respective resonant frequency tuning range. The modulated frequency (v_m) pattern on carrier frequency v_c will adhere to the following expression: $v_{FM} = v_c - v_m \beta_m \cos(2\pi v_m t)$, where β_m is the modulation index (~efficacy). The piezoelectric tuning is used both for recording (MICROPHONE) and ULTRASONIC IMAGING. Frequency modulation can be generated by means of a voltage-controlled OSCILLATOR in the form of an integrated circuit. The signal collection is generally performed by a inductor–capacitor circuit, with variable capacitance (tuner dial turns the parallel plate CAPACITOR to change the plate area and hence changes the capacitance; other digital tuning mechanism are available through INTEGRATED CIRCUITS). A frequency DEMODULATOR involves, for instance, the use of a PHASE-LOCKED LOOP, also incorporated in integrated circuit design (see Figure F.64).

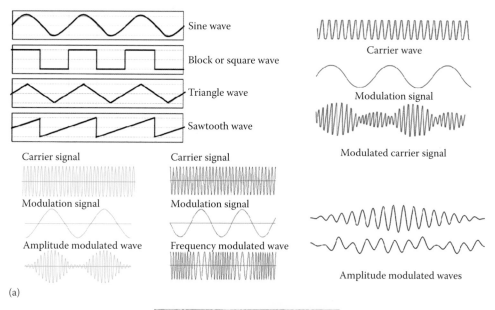

(a)

(b)

Figure F.64 (a) Graphical illustration of frequency modulation. Frequency outline: waveforms ranging from sinusoidal wave, sawtooth, and block wave. The sawtooth and block wave are composed of a broad range of (sinusoidal) frequencies (v); with associated wavelengths: $\lambda = v/v$, v the velocity of propagation for the wave; that can be decomposed by Fourier transform and (b) Edwin Armstrong (1890–1954).

Frequency spectrum

[general] The range of RADIATION from sources in various modalities, ranging from electromagnetic, to pressure (acoustic). The ELECTROMAGNETIC SPECTRUM ranges from 300 EHz in the gamma radiation to single digit Hz in the extreme low frequency band, beyond the general spectrum of RADIO waves. The acoustic frequency spectrum has a range of decimal Hertz (rumbles in the EARTH'S CRUST), to the echo scanning by bats (several hundred kilohertz) and to several hundred megahertz for ULTRASONIC IMAGING and vibrational disintegration (e.g., breaking KIDNEY stones) (*also see* ELECTROMAGNETIC SPECTRUM, EAR, *and* HEARING).

Fresnel, Augustin-Jean (1788–1827)

[general, optics] Physicist from France. Fresnel verified the WAVE phenomenon for light (ELECTROMAGNETIC RADIATION) in 1818. Fresnel's work on diffraction and INTERFERENCE provided the fundamentals for modern coherent and noncoherent imaging as well as general optical ray-tracing techniques. Fresnel performed experimental verification of the POLARIZATION concepts introduced by Christiaan Huygens (1629–1695), based on the transverse-wave phenomenon introduced by Huygens (see Figure F.65).

Figure F.65 Augustin-Jean Fresnel (1788–1827).

Fresnel diffraction (1788–1827)

[general, optics] Diffraction pattern created in near-field approximation. The incident rays of ELECTROMAGNETIC RADIATION, or mechanical waves, create an INTERFERENCE at close DISTANCE where the paths of the rays are considered to originate from a POINT SOURCE, or the diffraction is observed in a small observation point. Compare with the near-field FRAUNHOFER DIFFRACTION. The Fresnel LENS yields a significant consequence of the work of AUGUSTIN-JEAN FRESNEL (1788–1827), specifically applied to lighthouse lights for focusing large beams (see Figure F.66).

Figure F.66 Illustrations of the use of the Fresnel diffraction in the special lens type known as a Fresnel lens, primarily used in lighthouses and automobile headlights.

Fresnel number ($N_F = a^2/\lambda z$)

[computational, fluid dynamics, general] Dimensionless number that provides the validity of approximation that a SPHERICAL WAVE becomes a PLANE WAVE, which is the asymptotic value, where a is the diameter of the beam at the source (e.g., hole in panel shielding LIGHT SOURCE), λ the wavelength of the light, and z the DISTANCE from the source in the direction of propagation. In a laser cavity a large Fresnel number implies minimal diffraction phenomena.

Fricative consonant

[acoustics, fluid dynamics, mechanics] Turbulent FLOW pattern of AIR in the creation of a SOUND. The TURBULENCE is caused by fluids flowing through a narrowly spaced opening at high velocity. The turbulence creates eddies with inherently random pressure fluctuations that are equivalent to aperiodic sound patterns. In speech, this mechanism references the formation of a consonant that has a sound resulting from forcing air between two tightly spaced objects in the vocal tract, such as the teeth or the tongue and teeth, also referred to as fricatives (*also see* COEFFICIENT OF FRICATION).

Friction

[computational, fluid dynamics, general] Resistive force with respect to two or more media in MOTION with respect to each other. Generally, the friction force is a linear function of the normal force applied by one object with respect to the orthogonal force applied by another object, such as a box sitting on the GROUND. The NORMAL FORCE ($F_N = ma$ [mass m multiplied by acceleration a]) serves as maintaining the equilibrium, provided that there is no motion in the direction of the partition. The proportionality factor is the friction coefficient, separated in static and dynamic conditions as μ_s and μ_d, respectively. The friction force in this situation is $F_f = \mu_i F_N$, in opposite direction of motion. The coefficient of friction is most often a function of temperature, which is why racing cars will warm up both the engine (to minimize the gaps by allowing for optical THERMAL EXPANSION and fit) and the tires for maximum ADHESION to the road surface. Solid-on-solid friction can be modified by introduction of a FLUID; for instance, a FLOW of AIR or a LIQUID/gel lubricant). Teflon has remarkably low surface friction and is used in many designs as a coating or the material choice, also in various other ELASTOMERIC forms (e.g., fluoroplastics: PTFE [polytetrafluoroethylene], FEP [fluorinated ethylene propylene], PFA [perfluoroalkoxy]; Slick50®, etc.). When a body rolls and either/both body deforms (e.g., tire on road surface; rolling-pin on bread dough) there will be additional friction associated with the deformation ENERGY, expressed by the deformation constant μ_r in the same format $F_f = \mu_r F_N$. Coefficients of frictions can be found in tables listed in many ENGINEERING handbooks, for specific conditions and materials. The surface interaction between a fluid and a solid is captured by the VISCOSITY (η_k, the KINEMATIC VISCOSITY; η_s, the static viscosity; stagnant flow) of the fluid, while the SURFACE ROUGHNESS introduces additional restrictive influences. The surface roughness is captured by the DRAG COEFFICIENT D_{drag}, representing the drag force (F_D) over a surface area A for a fluid with density ρ, while traveling at velocity v, expressed as $F_D = D_{drag} A (\rho v^2 / 2)$. The drag coefficient is a function of the MAGNITUDE of the height of the surface "bumps," and the size distribution, next to the REYNOLDS NUMBER, the Froude number, and the MACH NUMBER. In terms of Reynolds number (Re) for flow, the expression for drag becomes $D_{drag} = (Re/24)(1 + 0.15 Re^{0.687})$, for flow $Re = \rho v d_p / \eta$ (d_p the PARTICLE diameter in colloid SOLUTION, respectively the average size of bumps/dips [e.g., GOLF BALL] on surface, and v the flow velocity at the point of contact). Roughness is generally indicated by the roughness parameter, the relative size of the unevenness (δ_s) on the surface to the size $2r_o$ (radius of curvature) of the object in the flow $\delta_s / 2r_o$. Friction under repetitive motion provides a damped OSCILLATION (*also see* **POISEUILLE'S LAW**) (see Figure F.67).

Figure F.67 Friction generating enough heat to provide the initiation of the exothermic oxidation of sulfur in a match tip. The composition of the "safety" match head is actually sulfur, mixed with glass powder, as well as an oxidizing agent (catalyst).

Friction factor ($f = \Delta P / [(L/D)((1/2)\rho v^2)]$)

[mechanics] Dimensionless number pertaining to the influence of WALL SHEAR STRESS in FLUID FLOW calculations. The FRICTION factor depends on two other DIMENSIONLESS NUMBERS: the SURFACE ROUGHNESS (relative ROUGHNESS to be exact) and the REYNOLDS NUMBER (Re). In the expression, the following conventions are used: ΔP, the PRESSURE drop over the entry into a PIPE with diameter D and length L, and ρ, the FLUID DENSITY with v the FLOW VELOCITY. SURFACE ROUGHNESS influence the mathematical approach since there will be a transition to TURBULENT FLOW with increasing roughness, while the friction primarily applies to LAMINAR FLOW. The friction factor can also be included in calculation of the HAGEN–POISEUILLE EQUATION.

Friedmann, Alexander Alexandrovich ("Aleksandr Fridman") [Алекса́ндр Алекса́ндрович Фри́дман] (1888–1925)

[astronomy/astrophysics, computational] Scientist, astrophysicist, and mathematician from the USSR (Soviet Russia). The theoretical efforts of Friedmann showed that the UNIVERSE is expanding and is related to the work of his contemporary GEORGES LEMAÎTRE (1894–1966) and is referred to as the "BIG BANG THEORY," described by the "FRIEDMANN EQUATIONS" (see Figure F.68).

Figure F.68 Alexander Alexandrovich Friedmann ("Aleksandr Fridman") (1888–1925) (Cyrillic, russian: Алекса́ндр Алекса́ндрович Фри́дман).

Friedmann equations

[astronomy/astrophysics, computational] Mathematical expressions with respect to the EXPANSION of the UNIVERSE, within the context of general relativity. The Friedmann equation is based on the hypothesis that the universe is homogeneous and isotropic, the cosmological principle. ALEXANDER FRIEDMANN (1888–1925) published his approximation in 1922, based on the works of ALBERT EINSTEIN (1879–1955) on GRAVITATIONAL FIELD equations. The equation defines the dimensional proportions (r) of the universe with respect to TIME (t) as indication of the galactic evolution based on the assumption $dr^2 = a(t)^2 dr_3^2 - c^2 dt^2$, where the term dr_3^2 defines space in one of three configurations: (1) a hyperbole with negative curvature ($k_F = -1$), (2) two-dimensional flat ($k_F = 0$), or (3) a sphere ($k_F = 1$); $a(t)$ represents a scaling factor (for $k_F = 0$, flat space $a(t) = a_0 t^{2/3(\zeta+1)}$; ζ a boundary condition: $\zeta = 0$ defines "MATTER"-dominated space, and $\zeta = 1/3$ defines RADIATION dominated space) and a_0 is an integration constant, depending on the assumption about the initiation of the universe: $t = 0$ ("BIG BANG THEORY"). The Friedman equations state $(\dot{a}(t)^2 + k_F c^2)/a(t)^2 = (8\pi G\rho + \Lambda c^2)/3$, with G Newton's gravitational constant, density $\rho \sim 0.2$ atoms/m^3 (note that the critical universe density is approximately $\rho_c = 5$ atoms/m^3), Λ the cosmological constant, and c the speed of light. In this definition, the Hubble parameter is incorporated as $H_{\text{Hub}} = (\dot{a}(t))/(a(t))$ $(\dot{a}(t) = \partial a(t)/\partial t)$, the special curvature is defined by $r_{\text{univ}} = k_F/a(t) = (1/6)R_{\text{Ricci}}$, where in the Friedmann model the Ricci curvature scale is defined as $R_{\text{Ricci}} = (6/c^2 a(t)^2)(\ddot{a}(t)a(t) + \dot{a}(t)^2 + k_F c^2)$, and the second equation $\ddot{a}(t)/a(t) = -(4\pi G/3)(\rho + (3P/c^2)) + (\Lambda c^2/3)$ with $P = \varpi \rho c^2$ the pressure (ϖ a NORMALIZATION constant).

FROG

[optics] Acronym for frequency resolved optical gating. Imaging technique used in ultrafast laser pulse interaction, applying gating to perform time-resolved spectral information (see Figure F.69).

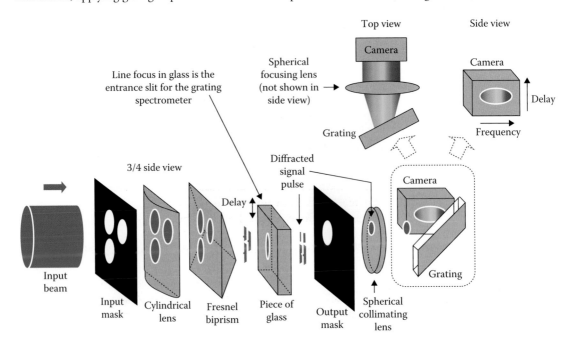

Figure F.69 FROG imaging technique.

Froude, William (1810–1879)

[fluid dynamics] Scientist and architect from Great Britain. William Froude established expressions for the derivation of FLOW resistance in the design of ships. He is most known for his "hull speed equation," relating to the length of the waterline at the hull ($L_{w\ell}$) yielding a VELOCITY (v_{Hull}) for which the bow WAVE (the wavelength of the perturbation at the bow of the ship on the water surface) equals the length of the ship, providing the empirical relation $v_{Hull} \cong 1.34 \times \sqrt{L_{w\ell}}$. His other contributions are in DIMENSIONLESS NUMBERS for the ratio of forces (see Figure F.70).

Figure F.70 William Froude (1810–1879).

Froude number ($Fr = v^2/gL$)

[fluid dynamics, geophysics] Dimensionless number relating the inertial forces in the ratio to gravitational force, where v is the FLOW velocity, g the GRAVITATIONAL ACCELERATION, and L the characteristic length of the phenomenon under investigation. Introduced by WILLIAM FROUDE (1810–1879). Under certain circumstances the Froude number can be used to interpret the ratio of mean flow velocity with respect to the propagation velocity of a SURFACE GRAVITATIONAL WAVE.

Froude's law of similarity

[fluid dynamics] The wave RESISTANCE of an equivalent model of an object in FLOW is equal only when the Froude number is equal from model to reality. Postulated by WILLIAM FROUDE (1810–1879), this applies to the WAVE resistance only, when involving total resistance this would include frictional resistance as well. The LAW OF SIMILARITY would describe the equivalence of a model ship to a real ship.

Fuel cell

[chemical, energy, solid-state, thermodynamics] ENERGY storage and supply mechanism based on the principles of OXIDATION and resulting release of free electrons discovered by the German chemist and scientist Christian FRIEDRICH SCHÖNBEIN (1799–1868) published in 1838. Schönbein was experimenting with ELECTROLYSIS of water, producing oxygen (in fact ozone was part of the product) and hydrogen. This process, when controlled, can be reversed to produce ELECTRICITY, however when uncontrolled it form a HYDROGEN BOMB with considerable explosive energy. An additional accidental discovery of Schönbein was the accidental spill of two liquids: nitric ACID and sulfuric acid, which mixed and were collected with a cotton rag. The soaked cotton spontaneously ignited based on the internal release of oxygen from the nitro group combining with the CELLULOSE generated by the sulfuric acid from the cotton (see Figure F.71).

Figure F.71 Hybrid automobile that relies on hydrogen fuel cell for generation of energy to be converted into translational energy.

Fugacity ($\pi_f(T, P)$)

[thermodynamics] Two-phase equilibrium of a LIQUID substance with its GASEOUS STATE that is not conforming to the IDEAL GAS LAW $\pi_f(T,P) = \pi_f(T,P_0)\exp[(\mu_f(T,P) - \mu_f(T,P_0))/RT]$, a slight deviation from the ideal gas law $PV = nRT$, and the transformation of the CHEMICAL POTENTIAL $\mu_f(T,P)$. For an IDEAL GAS, the chemical potential relates to the VOLUME (V) through the PRESSURE (P) derivative, as a function of TEMPERATURE T as $\left(\partial\mu_f(T,P)/\partial P\right)_T = V = RT/P$ and gas constant $R = 8.3144621(75)$ J/Kmol.

Fugacity coefficient

[thermodynamics] Ratio of FUGACITY ($\pi_f(T,P)$) to PRESSURE (P): $c_F = \pi_f(T,P)/P$, used as indication of MAGNITUDE of deviation from IDEAL GAS behavior.

Fulcrum

[biomedical, engineering, mechanics] Pivot point, as found for a seesaw or draw-bridge (see Figure F.72).

Figure F.72 Fulcrum concept of a drawbridge in the Netherlands.

Fuller, Richard Buckminster (1895–1983)

[computational, general, mechanics] Architect and scientist from the United States. Fuller may be best known for his geometric frame design that uses minimal materials with maximum force, specifically used in the construction of the geodesic dome under his design. Based on his work certain intricate carbon structures were named fullerenes (*see* BUCKMINSTER FULLER, RICHARD).

Functional MRI (fMRI)

[biomedical, biophysics, electromagnetism, energy, engineering, imaging, instrumentation] (Established: 1991) Imaging methodology that uses the hemodynamic and metabolic activities in a biological system for identification of functional disorders and chronic illnesses. It is used to determine the cellular METABOLIC ACTIVITY using a mechanism of radio-frequency modulated MAGNETIC FIELD probing of a biological system and collecting the re-emitted spectrum as a function of location and time. The fMRI distinguishes itself from regular MRI in the fact that the phenomena are time resolved [function], whereas regular MRI is a stationary imaging technique. The IMAGE is reconstructed using the (Fourier-) SLICE theorem, which entails every set of parallel scans at ANGLE θ will produce the values of $F(u, v)$ along one line in the frequency plane (u, v). The Fourier slice theorem can be described in the following two steps: Step 1—create a set of parallel scans at angle θ and produce the projection function $P_\theta(t)$. Step 2—calculate the one-dimensional FOURIER TRANSFORM of the projection function $P_\theta(t)$ to produce the MAGNITUDE of $F(u, v)$ along a line passing though the origin with angle θ with u-axis. See Functional MRI for detailed information (also see magnetic resonance imaging [MRI]). MRI used the SPIN of hydrogen nuclei as the tool for anatomical imaging. The spin has a PRECESSION which can be altered under influence of an externally applied resonant (frequency modulated: RF band) magnetic field. Once the external drive is removed, the hydrogen will return to its original ENERGY state, with the release of the stored energy as RF RADIATION emitted at the Larmor frequency. MRI works without the need for pharmacologic markers, although these are being developed to increase the imaging repertoire; neither radioactive isotopes are required. fMRI is based on MRI technology and is applied specifically for brain function analysis and for the isolation and noninvasive diagnosis of cancer. fMRI uses the variability in BLOOD supply in support of the local METABOLISM while providing a strong RF SIGNAL. By physical stimulation of specific biological functions (VISION: checkerboard), or activities (finger taping), certain segments of the brain will become more active at the time of stimulation, which will appear under fMRI based on the cellular blood supply and interstitial blood content. Based on the higher metabolic activity of cancer than for healthy tissue, the first indication for the presence of cancer is determined. Several types of cancer have additional functional features that are very different from healthy tissues and allow for more specific analysis. In particular, certain cancers use a cyclic mechanism of mimicking impeding CELL death to encourage the surrounding tissue to institute a healing process. The healing process included increased blood vessel formation, providing enhanced PERFUSION with resulting increase in transportation, and delivery of nutrients and oxygen. As a result, the simulating cancer cluster will have created an enhanced environment for growth and will revive and grow with increased vigor. This process may repeat itself and may do so for an extended frequency of revivals. Under long-term (repetitive screening with set intervals) follow-up by fMRI diagnostics, the cancer can be analyzed and the volumetric extend as well as rate of growth can be established for prognosis and evaluation of treatment options. MRI is a tomographic imaging technique that produces images of internal physical and chemical properties of the body by measuring the emitted nuclear magnetic resonance (NMR) signals. In a typical MRI exam, the patient lies inside the machine and a RF-modulated magnetic signal is emitted into the patient body, which responds by emitting another RF signal. The received signal is recorded and processed to yield the MRI image. The NMR phenomenon was reported by FELIX BLOCH (1905–1983) and EDWARD MILLS PURCELL (1912–1997) in 1946 (for which they received the Nobel Prize in Physics) based on the BLOCH EQUATIONS for the temporal evolution of magnetization. In 1973, MRI was made possible by PAUL CHRISTIAN LAUTERBUR (1929–2007) and Sir Peter Mansfield (1933–) (*see* MAGNETIC RESONANCE IMAGING for additional information) (see Figures F.73 and F.74).

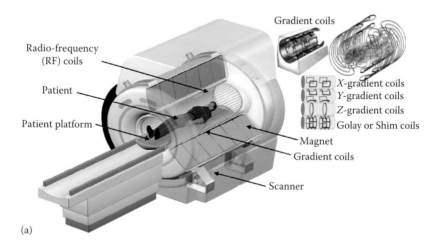

(a)

Figure F.73 (a) MRI machine diagram.

(Continued)

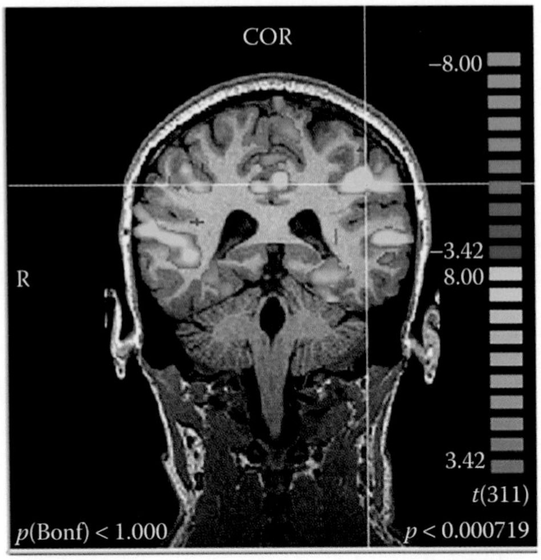

F

COR

−8.00

−3.42
8.00

R

3.42

$t(311)$

$p(\text{Bonf}) < 1.000$ $p < 0.000719$

(b)

Figure F.73 (Continued) (b) False color image of brain activity acquired from MRI scan procedure under the performance of certain tasks. (Courtesy of the University of Missouri, Columbia, MO.)

Figure F.74 Functional magnetic resonance imaging (fMRI) room. (Courtesy of Siemens Healthcare.)

Functional residual capacity (FRC)

[biomedical, fluid dynamics] The amount of GAS that remains in the lungs after normal exhalation during TIDAL BREATHING. Functional residual capacity and other tidal breathing parameters can be measured with a spirometer.

Fundamental frequency

[acoustics, computational, general, imaging, optics, quantum] Baseline frequency. Lowest frequency of a set of harmonics. The fundamental frequency generally conforms to a match in half wavelength or quarter WAVELENGTH (single-ended open system, providing the crest at the opening vs. equilibrium [zero displacement] at the closed end) to the system parameters, either mechanically or energetically as well as QUANTUM mechanically. In FOURIER ANALYSIS, the fundamental frequency is the base frequency that defines the periodic event (whole wavelength).

Fundamental particles

[atomic, high-energy, nuclear, solid-state] The current consensus on nuclear composition is that sub-nuclear particles are composed of clusters containing two or more units called quarks. There are three fundamental PARTICLE types: the electron (which is a LEPTON), the d-quark ("down"-quark), and the u-quark ("up"-quark). The strong force interaction between the quarks creates neutrons, pions, protons, and more ELEMENTARY PARTICLES. These quarks along with the NEUTRINO (which is a lepton) are considered the FIRST GENERATION OF MATTER. With increased ENERGY states there are additional particle configurations in the second and third generation, respectively.

Fundamental wave

[acoustics, mechanics] The frequency in a string with mass m and length ℓ under TENSION (T_s) will reverberate at several harmonic states (n_{harm}), with $n_{harm} = 1$ providing the lowest resonance FREQUENCY (ν):
$\nu_{n_{harm}} = (n_{harm}/2\ell)\sqrt{(T_s/(m/\ell))} = n_{harm}\nu_1$, the n_{harm}th harmonic will be composed of n_{harm} ANTINODES.
On a cello (or other string instrument) the tone will change (increase in PITCH, higher tone) when a string is depressed against the neck of the instrument and hence shortened.

Fusion

[nuclear] Merging of two or more nuclei, short for nuclear fusion. Nuclear fusion is the source of ENERGY for stars, and additionally used in the HYDROGEN BOMB. When two nuclei bond together to form a more massive NUCLEUS, the BINDING ENERGY per nuclide increases. The binding energy can be released based on ALBERT EINSTEIN's (1879–1955) postulate $E = mc^2$. In stars, two protons ($_1^1 H$) (with charge $e = 1.60217657 \times 10^{-19} C$) will overcome their Coulomb repulsion and join (at separation r) with energy $PE = k_e(e^2/r)$, with

Coulomb's constant: $k_C = 1/(4\pi\varepsilon_0)$, where: $\varepsilon_0 = 8.85419 \times 10^{-12}$ C^2/Nm^2 the permittivity of free space. The process is described as $^1_1\text{H} + ^1_1\text{H} \rightarrow ^2_1\text{D} + ^0_{-1}e + \nu_e$, forming deuterium ^2_1D, with emission of a NEUTRINO ν_e, and a positron $^0_{-1}e$. Subsequently, a deuterium PARTICLE will merge with a PROTON to form 3-Helium (^3_2He): $^2_1\text{D} + ^1_1\text{H} \rightarrow ^3_2\text{He} + \gamma$, releasing gamma RADIATION. The final stage of the fusion process is the formation of 4-Helium: $^3_2\text{He} + ^3_2\text{He} \rightarrow ^4_2\text{He} + ^1_1\text{H} + ^1_1\text{H}$; releasing the substantial amount of energy of 26.7 MeV per four protons. These conditions are made possible due to the excessive high temperature of the stellar proton GAS: (using the kinetic energy $KE = (3/2)k_bT$, Boltzmann coefficient: $k_b = 1.3806488 \times 10^{-23}$ m^{2}kg/s^{2}K) $T = 1 \times 10^9$ K. It is also known as THERMONUCLEAR FUSION. The SUN as a whole releases in excess of 3.8×10^{26} J/s, with only a fraction reaching EARTH (see Figure F.75).

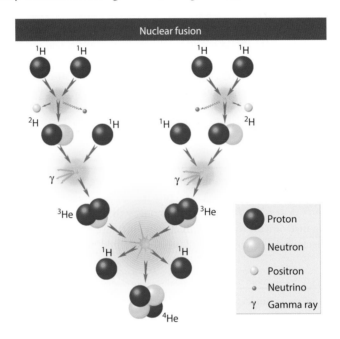

Figure F.75 Solar proton fusion diagram.

Fusion, heat of

[thermodynamics] The ENERGY transfer (i.e., heat Q) in the process of converting one PHASE of a medium into another, measured per unit mass (m): $h_L = Q/m$. The LATENT HEAT of FUSION specifically refers to the SOLIDIFICATION or melting (liquefying) of a FLUID and solid, respectively.

Fusion reactor

[atomic] Based on the ENERGY release for FUSION, artificial fusion between deuterium (^2_1D) and tritium (^3_1T) particles is used to generate energy and from hence ELECTRICITY in fusion power plants. The fusion processes as follows: $^2_1\text{D} + ^2_1\text{D} \rightarrow ^3_2\text{He} + ^1_0 n + 3.3$ MeV, releasing a NEUTRON ($^1_0 n$); alternatively, $^2_1\text{D} + ^2_1\text{D} \rightarrow ^3_1\text{T} + ^1_1\text{H} + 4.0$ MeV, releasing a PROTON (^1_1H). Additional follow-up reactions are $^2_1\text{D} + ^3_2\text{He} \rightarrow ^4_2\text{He} + ^1_1\text{H} + 18.3$ MeV and $^2_1\text{D} + ^3_1\text{T} \rightarrow ^4_2\text{He} + ^1_0 n + 17.6$ MeV; the final process has the largest CROSS SECTION. (*Note*: 1 MeV $= 1.602176565 \times 10^{-13}$ J.) The fusion process needs to be fueled by collision, requiring a PARTICLE ACCELERATOR. No functioning FUSION REACTOR are commercially operational at this time (*also see* FISSION REACTOR).

g

[general] Acceleration due to GRAVITY (*also see* **GRAVITATIONAL ACCELERATION**) (see Figure G.1).

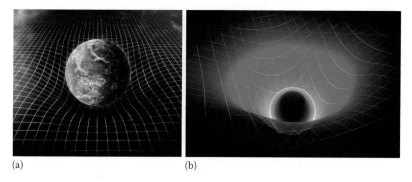

(a)   (b)

Figure G.1 (a) How space may be considered a sheet where the planets form indentations that are a function of their "gravitational acceleration" (i.e., mass/weight), where particles and photons flow through space as if they roll over the surface of the sheet, being drawn closer to the indentation left by the planets and other celestial bodies due to the incline of the "surface of space" in proximity to the planet/star/galaxy, and so on (galactic phenomena are bending space). (b) A black hole in this case would leave a massive through where all photons, passing objects, and stationary object that are coming on the incline as the indentation in space becomes deeper with the ever growing black hole (including planets) are falling into.

Gabor, Dennis (1900–1979)

[computational, general, optics] A physicist and electrical engineer from Hungary. Dennis Gabor is generally considered to be the "inventor" of HOLOGRAPHY, storing three-dimensional data in a two-dimensional matrix, approximately in 1948. He received the Nobel Prize in Physics for his development of the holographic method in 1971 (see Figure G.2).

Figure G.2 Dennis Gabor (1900–1979).

Gadolinium oxysulfide (Gd_2O_2S)

[biomedical] An inorganic chemical molecular assembly. The compound Gd_2O_2S is known for its attenuation spectrum of X-ray RADIATION. The X-RAY attenuation is used for dosimetry purposes in RADIATION THERAPY. The attenuation (μ_{att}) in the keV to MeV photon ENERGY range is expressed as the mass attenuation coefficient μ_{att}/ρ, where ρ is the density, or as the mass-energy attenuation μ_{en}/ρ.

Gait

[biomedical, mechanics] The step rate of a person walking, specifically any deviations from a continuous step sequence. Additional gait-related phenomena are in the difference between the step format of the right and the left LEG, where one foot may be dragging, however not continuously. The gait can be a very discreet phenomenon, which may be difficult to analyze with conventional SIGNAL processing methods. One mechanism of automated computer-based investigation may involve the use of wavelet analysis (see Figure G.3).

Figure G.3 Snapshot of walking gait.

Gait cycle

[biomedical, computational, mechanics] The MOTION of the foot during walking: (1) making contact with floor, (2) foot rests (ankle bends), (3) LIFT, and (4) swing.

Galaxy

[general] Grouping of stars and planets. The affirmation by EDWIN HUBBLE (1889–1953) in 1924 for the cluster of stars at close proximity to our SOLAR SYSTEM (approximately 2×10^6 light-years DISTANCE) formed a scientific description of a neighboring GALAXY, not merely a grouping of stars. The Andromeda nebula discovery by Hubble opened a new dimension to galactic exploration. Further exploration proved that the solar system is part of a galactic formation, named the MILKY WAY. The Milky Way has an approximate CROSS SECTION of 6×10^4 light-years. This size appears average for a galaxy. The Splinter Galaxy (NGC 5907 or "Knife Edge") is a spiral galaxy discovered by WILLIAM HERSCHEL (1738–1822) in 1788 and is at approximately 5×10^7 light-years from our solar system and has a diameter of 2×10^5 light-years. The Splinter Galaxy is located in the constellation Draco. Galaxies also form attractive bonds and make

groupings of their own. One specific example is the Wild's triplet, a grouping of three galaxies that has connecting gravitational bridges, apparently spanning millions of light-years (see Figure G.4).

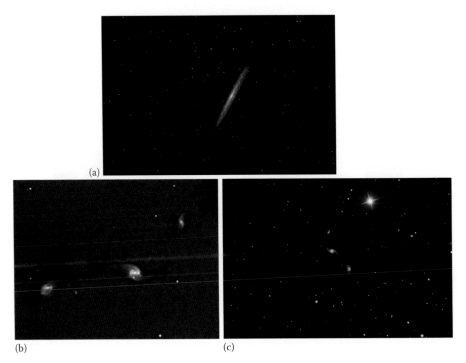

Figure G.4 (a) Splinter Galaxy; NGC 5907, a spiral galaxy located approximately 50 million light-years from Earth. (Courtesy of Stefano Campani.) (b and c) Wild's triplet, three galaxies in three-dimensional arrangement at almost perfect 4π/3 solid angle interval, contorted locking geometry. (Courtesy of Adam Block, Caelum Observatory.)

Gale, Henry Gordon (1874–1942)

[general, optics] A physicist from the United States. In 1908, Henry Gale postulated the concept of accelerating charges (the elusive electrons that still were only the hypothetical building blocks of MATTER) causing the emission of ELECTROMAGNETIC RADIATION, next to the SCATTERING of light by (charged) particles with contemporary ROBERT ANDREWS MILLIKAN (1868–1953). The energetic content of the PHOTON was not fully established until 1915, by Robert Millikan (reluctantly, and still defiant of the postulations made by ALBERT EINSTEIN [1879–1955] in 1905 [*see* PHOTOELECTRIC EFFECT]), after Millikan verified the existence of the electron in 1909 and quantified the charge (see Figure G.5).

Figure G.5   1899 picture of Henry Gordon Gale (1874–1942). (Courtesy of University of Chicago Photographic Archive, [apf digital item number: apf1-06260], Special Collections Research Center, University of Chicago Library, Chicago, IL.)

Galenus, Claudius (c. 130–201)

[biomedical, optics] (*also* Galen of Pergamum or Galen) A Greek anatomist, physiologist, and physician. Galen followed the lead of Plato (427–347 BC) on VISION and EXTRAMISSION; the EYE emits rays to probe the environment. Although misconceived about the vision phenomenon, Galen researched and described the eye in great detail in various publications, also describing the link between the eye and the brain. Galen defined the optic nerve as the "pneuma" in his publications. Galen practiced MEDICINE in The Roman Empire. While treating gladiators, he was offered a glimpse into the HUMAN BODY. He also educated himself with anatomical dissections on pigs and monkeys, due to the fact that vivisection was prohibited under Roman law.

Galerkin method

[computational, fluid dynamics] A method used in finite ELEMENT calculations.

Galilean transformation

[computational, mechanics] The MOTION of a PARTICLE can be observed from several different reference frames simultaneously, each with their respective coordinate systems. Various coordinate transforms are in use; in the Galilean transform, the primary assumption is that the origin of two coordinate systems coincides at time $t = 0$, while the second coordinate system (S') is moving at constant VELOCITY (v^*) with respect to the first reference frame (S). The time frames for observers in both coordinate systems are identical $t = t'$. Assuming motion in the x-direction gives the transform $x' = x + v^* t$, $y = y'$, and $z = z'$ in CARTESIAN COORDINATES. Using the standard differentiation to obtain velocity ($v_i = dx_i/dt$) and acceleration ($a_i = dv_i/dt$), this yields $v_x = v'_x + v^*$, $a_x = a'_x$; $v_y = v'_y$, $a_y = a'_y$; and $v_z = v'_z$, $a_z = a'_z$.

Galilean–Newtonian relativity

[atomic, nuclear] A relativistic theory under the assumption that all observers move with a constant velocity with respect to each other, under "general relativity." All objects have the exact same absolute values and parameters (i.e., Newtonian bodies) for all respective observers. This will make the objects indistinguishable. This reference frame is also referred to as "inertial." Galilean–Newtonian relativity has a strict rule of SIMULTANEITY and absolute objects. The main transition is the introduction of the concept of light CONES versus straight-line propagation (linear for all phenomena and objects under Newtonian principles). The light cone introduces a PROBABILITY aspect, whereas Galilean and Newtonian MOTION and phenomena are well demarcated and linear. The general relativistic approach of gravitational influences still holds, for instance, the existence of GRAVITATIONAL WAVES can be made to adhere, but will generally fall outside the realm. Galileo made it clear that GRAVITATION has the same value for all objects and phenomena. ALBERT EINSTEIN (1879–1955) "unified" the gravitational influences to include that there are no free bodies in the UNIVERSE, all particles experience gravitational attraction, both on small and large scale (*see* DILATION OF TIME).

Galilei, Galileo (1564–1642)

[general, thermodynamics] Galileo Galilei Linceo, an Italian scientist and founding father of the modern PHYSICS concepts. Galilei first studied MEDICINE and followed with physics and mathematics, inspired by his father (Vincenzo Galilei [1533–1591]) who was a musical artist with scientific tendencies. Galilei performed many experimental observations of concepts that were *a priori* postulated under the school of ARISTOTLE (384–322 BC). His first contributions were in the validation of the concept of GRAVITATION acceleration (his works on the laws of free fall: Sermones de Motu Gravium), specifically illustrated by dropping objects from the Tower of Pisa, his place of birth. As an astronomer, he obtained fame in 1609 with his observations made using the TELESCOPE developed by HANS LIPPERHEY (1570–1619) to describe relief on the surface on the MOON with calculated altitude from the shadow cast by solar illumination. Presumably, Galileo made improvements to the telescope by introducing double tubes for improved focusing around 1610. His observations of VENUS made a firm believer out of him of the teachings by NICOLAUS COPERNICUS (1473–1543). Additional work of Galileo was in the determination of the speed of light. The first recorded use of a THERMOMETER can also be attributed to Galilei. In approximately 1595, Galilei provided a mechanism to measure temperature using the THERMAL EXPANSION of AIR, using a GLASS bulb with a narrow (almost CAPILLARY) downward tubular extension to use different levels of VACUUM to draw-in water from an open reservoir, using the level of the water as an indication of temperature. Due to his conflict with the Catholic Church and in particular the inquisition, he fell out of grace and was incarcerated in the dungeons of the ruler of Rome in 1633 and had to denounce his following of Copernicus. The persecution was primarily based on his work "Dialogo sopra i due sistemi del monde" (1632) that was banned and not released until 1822. He was placed on house arrest in 1636. His masterpiece "Discorse e Dimonstrazioni" (1638, Leiden, the Netherlands) described in detail the laws of gravitation and kinematic and momentum as well as the structure and forces on solid media. Most of Galileo's work was not published in Italy but transported to the Netherlands to avoid conflicts with the Catholic Church (see Figure G.6).

Figure G.6 Galileo Galilei (1564–1642). (Courtesy of R. Hart.)

Galilei, Galileo, acceleration experiments

[general] GALILEO GALILEI (1564–1642) postulated that all objects accelerate with the same absolute acceleration but did not have the equipment to measure the duration and velocity during direct vertical fall accurately enough. To compensate for the lack of RESOLUTION in the equipment, he changed the experiment to rolling objects, at different ANGLES of incline and for various masses. The extrapolation to a total vertical drop provided him with proof of a universal absolute GRAVITATIONAL ACCELERATION.

Galilei, Galileo, law of inertia

[general] GALILEO GALILEI (1564–1642), in his experimental design for testing absolute GRAVITATIONAL ACCELERATION, allowed the balls to roll down a curved ramp and subsequently up a curved ramp. The ball never reached the same height as from which it was released, and he concluded that FRICTION must convert a portion of the momentum. He assumed that a frictionless rolling ball would perform a pendulum MOTION on a symmetric ramp for infinity. These observations led

SIR ISAAC NEWTON (1642–1727) to his first law "the LAW OF INERTIA" (see Figure G.7). (Note that Isaac Newton was born within a year after Galileo's death.)

Figure G.7 Galileo Galilei's (1564–1642) law of inertia, the bicycle will not reach higher than the point from which it took off, assuming the person is not paddling.

Galilei, Galileo, pendulum

[general] GALILEO GALILEI (1564–1642) in 1581 observed that a candelabrum in the cathedral in Pisa was pulled and released to reveal a swing MOTION that appeared to have a steady constant rhythm, the period of the swing. He supposedly used his HEART beat as a time-keeping mechanism, because (stop-) watches were not available, only "hour" time pieces. Galilei was studying MEDICINE at the time and the use of his pulse beat may have been an obvious choice (order of SECOND [s]). The period (T) of a PENDULUM is directly proportional to the square root of the length of the pendulum (ℓ) and is independent of the mass attached, for relatively small angular displacement $T \approx 2\pi\sqrt{(\ell/g)}$, where g is the GRAVITATIONAL ACCELERATION (see Figure G.8).

Figure G.8 Pendulum design by Galileo Galilei (1564–1642).

Galilei, Galileo, thermoscope

[general, thermodynamics] A bulb attached to a GLASS pipette invented by GALILEO GALILEI (1564–1642) is held upside down in a reservoir of LIQUID and the bulb on top will expend the GAS or contract in gaseous volume, displacing the liquid in the CAPILLARY tube as function of temperature at the top bulb. The THERMO-SCOPE did not have a scale and was used in relative QUANTIFICATION only. The thermoscope was converted to a THERMOMETER by a French physician J. Rey in 1631 by placing the bulb on the bottom and allowing the liquid to expand up the narrow glass tube and placed marking on the tube for absolute values. Not long after this the water or wine was replaced by mercury or high-concentration ALCOHOL (see Figure G.9).

Figure G.9 Operation of the thermoscope.

Galilei transformation

[mechanics] *See* **GALILEO TRANSFORMATION**.

Galileo number (Ga = $gD3\rho^2/\eta_{kin}{}^2$)

[fluid dynamics] A dimensionless number pertaining to viscous FLUID flow representing the ratio of the REYNOLDS NUMBER multiplied times the gravitational force over the resultant viscous forces, with g the GRAVITATIONAL ACCELERATION, D the characteristic diameter of the FLOW, ρ the density of the fluid, and η_{kin} the KINEMATIC VISCOSITY. The Galileo number is used in viscous flow as well as general flow to scale events in momentum and HEAT TRANSFER (e.g., MIXING/convection/CIRCULATION) and related to THERMAL EXPANSION in particular. The number is dedicated to the efforts by the Italian scientist GALILEO GALILEI (1564–1642).

Galileo transformation

[atomic, mechanics] When two observation reference frames (S, S', respectively) are moving with respect to one and another (with relative velocity $\vec{U} = (u, v, w)$), the observation (can be stationary but changing in place ($\vec{x} = (x, y, z)$) or MOTION) seen from one reference frame can be transformed to the other reference frame by using the relative VELOCITY ($\vec{v} = v_x, v_y, v_z$), using the standard equation for location. This provides the location transformation as $x = x' + ut$; $y = y'$; and $z = z'$, while time is constant $t = t'$. The velocity transformation is now $v_x = v'_x + u$; $v_y = v'_y$; and $v_z = v'_z$, while for the acceleration this yields $a_x = a'_x$; $a_y = a'_y$; and $a_z = a'_z$. Sometimes it is also referenced as Galileo–Newton transformation equations.

Gallium ($^{70}_{31}$Ga)

[general] A semiconductor material ELEMENT used in many transistor and DIODE arrangements to achieve the p–n configuration. Gallium was discovered in 1875 by a French chemist PAUL-ÉMILE LECOQ DE BOISBAUDRAN (1838–1912) and predicted in 1871 by a Russian chemist DMITRI IVANOVICH MENDELEEV (1834–1907).

Galvani, Luigi (1737–1789)

[general] An Italian physicist and scientist, contemporary of and collaborator with Count ALESSANDRO VOLTA (1745–1827). The combined efforts of the invention of the VOLTAIC PILE by Galvani and the theoretical and experimental work of Volta on capacitance led to the definition of the ELECTRIC BATTERY by Volta in 1780 (see Figure G.10).

Figure G.10 Statue erected in honor of Luigi Galvani (1737–1789).

Galvanic mechanism

[electromagnetism, general, thermodynamics] The deflection of a current-carrying conducting ELEMENT (including semiconductor) under the influence of a MAGNETIC FIELD (generally the Lorentz force). The magnetic field results from a current density and generates a Hall effect next to a temperature gradient in perpendicular to the magnetic field direction as well as the direction of the current. Additional phenomena include the induction of magnetoresistance as well as in fourth place the Nernst effect, which is a thermal gradient in the direction of the current (see Figure G.11).

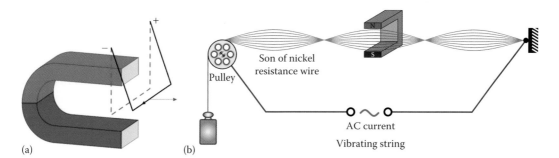

Figure G.11 Force on a current under the Galvanic mechanism: (a) direct current (DC) and (b) alternating current (AC).

Galvanometer

[electromagnetism, general] A device to measure current (I), resulting from a potential difference, using a conducting coil, which is suspended between the NORTH and south POLES of two respective permanent magnets. The MAGNETIC FIELD between the poles ($\vec{B}$) interacting with the current produces a force, generating a torque on the axis that is attached to the coil, that will force the coil to rotate. The force is $\vec{F}_m = \int_{\text{wire}} I d\vec{\ell} \times \vec{B}$ (i.e., Lorentz force), where ℓ is length (and direction; coil) of the Id wire segment containing the charges in MOTION. Placing an (infinitely) high RESISTOR in series with the galvanometer converts the current meter into a VOLTMETER (see Figure G.12).

Figure G.12 Galvanometer.

Gamma absorption

[nuclear] Gamma RADIATION is an ELECTROMAGNETIC RADIATION that has a greater likelihood of being absorbed in heavy ELEMENTS (i.e., high atomic number), mostly dense media. The attenuation follows the Beer's law of attenuation. Since gamma radiation is mass less and free of charge and has a very short wavelength, it is difficult to block but on an atomic level will not cause nuclear transmutations. The penetration is topped by NEUTRON rays; however, alpha and beta rays will be quenched much earlier. Gamma rays can be detected by a GEIGER COUNTER.

Gamma curve

[acoustics, computational, optics] A correction mechanism used in IMAGE processing defined by the POWER LAW correlating the incident radiance/SIGNAL intensity (V_{in}) to the outgoing radiance/signal intensity (V_{out}) expressed as $V_{out} = A V_{in}^{\gamma^*}$, where A represents the "efficiency" and γ^* represents the "encoding gamma."

Gamma decay (γ)

[general, nuclear] Nuclear degradation from an excited state releases high-energy photons with ENERGY ranging from keV to MeV. An example of gamma DECAY is in the degradation of URANIUM (^{238}U) when colliding with a NEUTRON (^1_0n), yielding an excited-state NUCLEUS ^{239}U* further decaying as ^{238}U $+ {}^1_0n \rightarrow {}^{239}U^* \rightarrow {}^{239}U+\gamma$. In positron emission TOMOGRAPHY specifically, a free positron will virtually immediately annihilate with any of the many free electrons in the residing medium. The annihilation process releases a substantial amount of energy: greater than 1 MeV. This energy is released as two gamma QUANTA with 511 keV emitted in perfectly perpendicular direct to each other. The energy balance for the POSITRON ANNIHILATION process and the gamma pair production is described in the following equation (see Figure G.13):

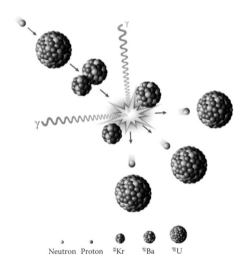

Neutron Proton $^{91}_{36}$Kr $^{142}_{56}$Ba $^{235}_{92}$U

Figure G.13 Gamma decay for the fission process of uranium.

$$E_{pos} + E^{kinetic} + E_e \geq 2E_\gamma$$

The rest energy is determined by the difference in energy of the ISOTOPE and the product. In addition to the rest energy of the positron (E_{pos}) and the electron (E_e), respectively, there may be additional kinetic energy involved from positron or electron MOTION $E^{kinetic}$. The nuclear energy follows from the quantum PHYSICS theorem that mass and energy are equivalent, providing the rest MASS ENERGY equivalent as illustrated by equation $E = mc^2$, where E represents the rest energy of the ELEMENT, m is the atomic mass, and c is the speed of light. Substitution of equation $E = mc^2$ into $E_{pos} + E^{kinetic} + E_e \geq 2E_\gamma$ and using the PHOTON energy $E = h\upsilon$ yields $m_{pos}c^2 + m_e c^2 = 2h\upsilon_\gamma$. Filling in the mass-equivalent energy for each component yields the energies in equation $511\,\mathrm{keV} + E^{kinetic}_{pos} + 511\,\mathrm{keV} \geq 1.022\,\mathrm{MeV}$.

Gamma decay, classification of

[nuclear] Gamma RADIATION is classified based on the QUANTUM principles, in particular the angular momentum of the NUCLEUS (J) prior to emission and post-emission and the angular momentum of the emitted PHOTON (ℓ) need to satisfy $J_{\text{initial}} = J_{\text{final}} + \ell$ (see Figure G.14).

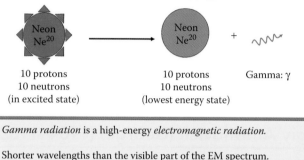

In much the same way that electrons in atoms can be in an *excited state* so can a nucleus

Neon Ne²⁰
10 protons
10 neutrons
(in excited state)

Neon Ne²⁰
10 protons
10 neutrons
(lowest energy state)

Gamma: γ

Gamma radiation is a high-energy *electromagnetic radiation.*

Shorter wavelengths than the visible part of the EM spectrum.

Figure G.14 Gamma (λ) decay process.

Gamma decay, hindrance factor (f_v)

[nuclear] The limitation on the POLARIZATION of gamma emission with respect to restricted nuclear transitions, which appears to be linked with MACROSCOPIC resonance effects and is defined as $f_v = (\tau_{1/2}/\tau_{\text{Weiss}})^{1/v}$, where $\tau_{1/2}$ is the measured half-life of the nuclear transition and τ_{Weiss} is the WEISSKOPF HALF-LIFE.

Gamma decay, Mössbauer effect

[nuclear] *See* MÖSSBAUER EFFECT.

Gamma decay, Weisskopf estimate of decay constant

[nuclear] *See* WEISSKOPF HALF-LIFE, derived by VICTOR WEISSKOPF (1908–2002) (*see also* RADIOACTIVE DECAY).

Gamma particle

[biomedical, electromagnetism, nuclear] A PHOTON, an electromagnetic WAVE, sometimes found in literature based on historic circumstances (*see* GAMMA RAY) (see Figure G.15).

Radiation; effective mass

Particle	Mass* (MeV/c²)	Charge
Gamma (γ)	0	0
Beta (β)	~0.5	−1
Alpha (α)	~3752	+2

$$*m = E/c^2$$

Figure G.15 Comparison of radioactive decay products with respect to effective mass.

Gamma ray

[nuclear] High-frequency ELECTROMAGNETIC RADIATION with wavelength generally shorter than X-RAY (however, an overlap is found depending on the reference), ranging from $\lambda = 10^{-10}\,\mathrm{m} = 0.1\,\mathrm{nm}$ to $\lambda = 10^{-14}\,\mathrm{m} = 10^{-5}\,\mathrm{nm}$, emitted from the NUCLEUS. Gamma rays are emitted from nuclear degradation, whereas x-ray RADIATION generally originates from ELECTRON DECAY. The nomenclature dates back to the early days of RADIOACTIVITY, late 1800s (*also see* GAMMA DECAY).

Gamma ray burst

[astronomy/astrophysics, electromagnetism] Short bursts of gamma radiation with ENERGY ranging from ~20 keV $\leftrightarrow > 10$ GeV with a measured random duration between 10 ms $\leftrightarrow >1000$ s. The RADIATION is associated with unidentified astronomical events and seemingly never originate from the same galactic location. The detection coincides with investigations of global nuclear ACTIVITY (e.g., test discharge of atomic bombs) by the military SATELLITE Vela.

Gamow, Georgiy Antonovich (George) (1904–1968)

[atomic, nuclear] A scientist and physicist from then Russian Empire (also USSR/Soviet Union—now Ukraine). Gamow's main contributions were in the description of alpha-decay principles and the origins to the source of ENERGY in the UNIVERSE. His work in ASTROPHYSICS has contributed to the formulation of the "BIG BANG" THEORY, as initiated by ALEXANDER FRIEDMANN (1888–1925). Gamow also stated in 1930s that the volume of a NUCLEON in an ATOM remains constant with increasing atomic number A. The theoretical work of George Gamow (sometimes found as "Gamov") quantitatively describes the alpha-decay "QUANTUM TUNNELING" process and mechanism of action (see Figure G.16).

Figure G.16 Georgiy Antonovich Gamow (George) (1904–1968) (Cyrillic, Russian: Гео́ргий Анто́нович Га́мов) in 1930.

Gamow–Teller decay

[atomic, general, nuclear, solid-state] Nuclear transition where one unit of angular momentum is relinquished by the emission of a pair of leptons (e.g., electron or NEUTRINO), hence changing the nuclear SPIN and associated angular momentum QUANTUM NUMBER by 0 or 1 $S_\beta = 0,1$. In comparison, a Fermi transition only takes place without a change in angular momentum.

Ganguillet–Kutter equation

[fluid dynamics] Heuristic flow-dynamics equation developed in 1869 by the Swiss engineers Emile-Oscar Ganguillet (1818–1894) and Rudolph Kutter (1818–1888). The equation was based on their observations of the river flow in the Swiss mountains, providing an OPEN CHANNEL in comparison to tube flow. The Ganguillet–Kutter equation approximates "Chézy's C_{chezy}" with respect to the FLOW resistance in a channel, specifically with respect to the bottom SURFACE ROUGHNESS (N) as $\eta_{GK} = 1/C_{chezy} = [1 + [a + (m/s)](N/\sqrt{R})]/[a + (b/N) + (m/s)]$, where a, b, and m are constants, s the slope of the bank, and R is the mean radius of curvature of the flow-bed trough.

Gap junction

[biomedical, chemistry, thermodynamics] In the MEMBRANE of cardiac cells (i.e., HEART MUSCLE) there are ION passages with a diameter of about 1.6 nm, connecting cardiac cells for chemical communication while remaining electrically insulated from the extracellular space. The gap-junction channels have a low RESISTANCE and control the FLOW of ions by means of fixed anions that will regulate the incoming ions based on their size and electro negativity. These channels are complementary to another form of transmembrane ion transport: ION CHANNELS (see Figure G.17).

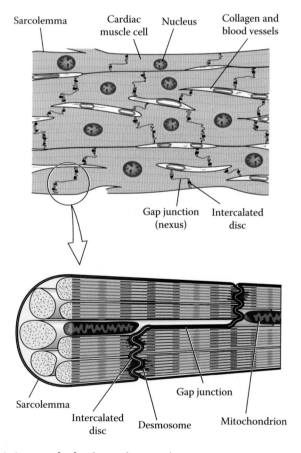

Figure G.17 Gap junction in ion transfer for the cardiac muscle.

Gas

[biomedical, fluid dynamics, general, geophysics] A material PHASE with lower density than solid or LIQUID found to exist above the VAPOR point, or alternatively the TRIPLE POINT. The GAS PHASE of water is water vapor above $100\,^{\circ}\text{C} = 372.16$ K under normal pressure (at sea level), and oxygen is in gas phase above 90.2 K. Definition of "gas" was developed by the alchemist JOHANNES BAPTISTA VAN HELMONT (1579–1644) from the Netherlands in approximately 1620, at the time presumably derived from the Greek word for "gap" or "gaping void": Khaos (χαος). In Greek mythology, Khaos (pronounced "Chaos") is the goddess of AIR. Gas is also considered an AEROSOL. Certain ELEMENTS and molecular structures in gas format obey the IDEAL GAS LAW. A gas generally obey the Gay-Lussac law $PV = nRT$, where P is the pressure of the gas (in Pascal: $\text{Pa} = \text{N/m}^2$), V is the volume (m^3), n is the number of moles of the medium, $R = 8.3144621(75)$ J/Kmol is the gas constant, and T is the temperature (in Kelvin).

Gas, elasticity of

[fluid dynamics] In deviation from an IDEAL GAS, the collisions will no longer be perfectly elastic, resulting in a deformation in the DENSITY (ρ)–PRESSURE (P) relationship with respect to an isothermal situation described by $P/P_0 = \rho/\rho_0$, yielding $\rho/\rho_0 = 1 + (P/k_{\text{elast}})$, with k_{elast} the ELASTICITY of the volume of gas.

Gas, ideal

[fluid dynamics, general] Collisions between the ions, atoms, or molecules are perfectly elastic and the IDEAL GAS LAW applied within a range of pressure and temperature domains. VAPOR is most often unstable and cannot be considered an IDEAL GAS close to the TRIPLE POINT or the vapor point. An ideal gas is a gas that has no imminent condensation potential and the intermolecular collisions are perfectly elastic. The ideal gas obeys the gas law within a certain range of perturbations of the boundary conditions.

Gas, molecule speed in

[general] An IDEAL GAS at a certain TEMPERATURE (T) will have totally elastic interaction in collisions and the absolute value of the (average) VELOCITY ($\bar{v}$) will be maintained based on the kinetic ENERGY $\bar{v} = 0.92 v_{\text{rms}} = \sqrt{3k_b T/m} \xrightarrow{\text{lim mass} \to \text{small}} \sim 3(P/\rho)$, where $k_B = 1.3806488 \times 10^{-23}$ $\text{m}^2\text{kg/s}^2\text{K}$ is the Boltzmann constant, and m is the molecular mass for a single component gas, P is the pressure, and ρ is the density of the gas. The full speed distribution for the constituents of a gas is described by the Maxwell–Boltzmann speed distribution.

Gas, Newton's experiments on

[general, mechanics] ISAAC NEWTON (1642–1727) illustrated the compressibility of gasses and the analogy with the Hook's law and this also related to NEWTON'S THIRD LAW.

Gas, viscosity of

[fluid dynamics] Generally, a GAS is frictionless; however, in order to offer a description for non-ideal gasses, the following equation relates the mean PRESSURE ($\bar{P}$) to the DENSITY (ρ) with respect to an ABSOLUTE TEMPERATURE (Θ) and assign an assumed VISCOSITY (η_{gas}) and a collective of constants ($a + b + c$) that is linked to the internal ENERGY and the rate of deformation ($a + b + c = (1/\rho)(DP/Dt)$) of a volume ELEMENT $\bar{P} = R^*\rho^\Theta - \eta_{\text{gas}}(a + b + c)$, where R^* is a constant that describes the nature of the gas as deviating from ideal. For an ideal gas $\eta_{\text{gas}} = 0$.

Gas chromatograph

[electronics, general, solid-state] An analytical device to assist in the analysis of the composition of media in a process called CHROMATOGRAPHY. The analysis of, for instance, fats and oils can quantify the estimated chain length associated with the fatty-ACID chain, as well as the amount of unsaturation. It is also defined as GAS–liquid CHROMATOGRAPH or liquid–GAS CHROMATOGRAPH (*see* **CHROMATOGRAPH**) (see Figure G.18).

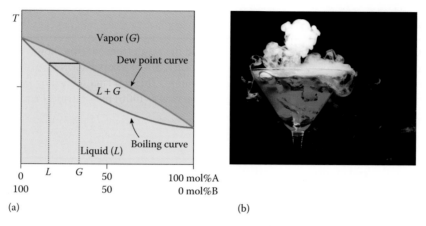

Figure G.18 Gas balance with liquid phase: (a) coexistance of liquid and vapor phase and (b) phase diagram referring to the phase transitions between solid, liquid, and gas, respectively.

Gas constant (*R*)

[thermodynamics] A constant used in the Gay-Lussac law to link the pressure, volume, and temperature to the respective molar content of the constituents of a gaseous substance $R = 8.3144621(75)$ J/Kmol. It is also described as the universal GAS constant. Based on the empirical work by ROBERT BOYLE (1627–1691) and JACQUES CHARLES (1746–1823), where Boyle described the fact that a constant amount (i.e., moles; n with $N = nN_a$ the number of molecules, and Avogadro's number $N_A = 6.022137 \times 10^{-23}$ J/K) of gas at a fixed temperature shows a decrease in pressure that is directly proportional to the increase in volume. Charles, on the other hand, linked the pressure in a fixed volume of gas to the temperature expressed on the Kelvin scale.

Gas diffusion

[general] A GAS and LIQUID with the same molecular structure defines VAPOR. When gas or vapor is exposed to a boundary that is semi-permeable or permeable to the molecules of the gas, the molecules will migrate in the direction of the declining partial pressure, that is, the OSMOTIC PRESSURE or when dissolved as a concentration gradient (see Figure G.19).

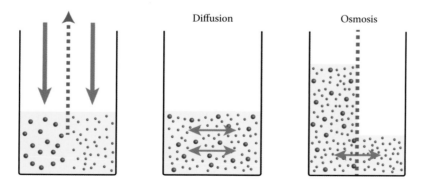

Figure G.19 Diffusion process under partial pressure.

Gas law

[atomic, general] *See* IDEAL GAS LAW.

Gas laws, kinetic theory

[general] PLATO (427–347 BC) already stated that heat and fire result from MOTION; however, the molecular model had not yet been developed. In the sixteenth century, SIR FRANCIS BACON (1561–1626) followed in Plato's footsteps and stated that the essence of heat is motion. This was later refined by JAMES CLERK MAXWELL (1831–1879) and LUDWIG EDUARD BOLTZMANN (1844–1906) in a statistical representation of motion of a collection of "particles."

Gas mixture

[thermodynamics] A blend of several gasses that under ideal conditions behaves as separate gasses, treating each constituent as a component with its own parameters and negligible interaction with other gasses in the same closed container. One example of a GAS mixture is the Gibbs–Dalton mixtures. During MIXING of more than one gas, there will be a change in entropy, which is irreversible, since the gasses cannot be separated without special means and added ENERGY. The conditions of the gas mixture are defined by the Gas laws and can be visually interpreted in PHASE diagrams.

Gas phase

[biomedical, fluid dynamics, general, geophysics] *See* GAS.

Gas pressure (*P*)

[nuclear] The force per unit area exerted by a single molecular quantity or mixture of various quantities of molecules (*n*, the number of moles) in a container of a certain volume (V), maintained at a certain temperature (T), defined by the IDEAL GAS LAW ($PV = nRT$, where $R = 8.3144621(75)$ J/Kmol is the universal gas constant), or more specifically, the Van der Waals equation ($P + (n^2 a / V^2))(V - nb) = nRT$, where *a* and *b* are constants that are specific for the gas in question (*Note: b* represents the volume of the total number of individual molecules, the Avogadro number multiplied by the volume of one single molecule). The pressure is the same on all the walls of the container, when the container is a free-standing device. This is different the LIQUID pressure, which changes with the height (*h*) of the FLUID column with density ρ, described as $P = \rho g h$, where *g* is the GRAVITATIONAL ACCELERATION.

Gas pressure, kinetic theory of

[atomic, nuclear] A theoretical description of the mobility of molecules that are in free movement with respect to each other in random MOTION. The MECHANICS of the relative motion is defined as follows with the 5 postulates: (1) inter-particle collisions are perfectly elastic; (2) the constituents exert no attractive or repulsive forces on each other; (3) the individual molecules are points with no volume; (4) the particles (ions, atoms, molecules) in a GAS traveling in straight-lines obeying Newton's laws; and (5) the temperature is defined by the kinetic ENERGY $KE = 3k_bT/2$, where $k_B = 1.3806488 \times 10^{-23}$ m²kg/s²K is the BOLTZMANN CONSTANT (see Figure G.20).

(a)   (b)

Figure G.20 (a) Gas pressure gauge at natural gas distribution center and (b) gauges on gas tank with compressed gas.

Gas pressure, osmotic (Π_{osm})

[biomedical, chemical, general] The equivalent pressure for the quantity of a mixture of solutes in SOLUTION or suspended in a solid $\Pi_{osm} = (RT/v_{ii})\sum_{j \neq i} n^*_j$, where n^*_j represents the molar fraction of species i $(0 \leq n^*_j \leq 1)$, R = 8.3144621(75) J/Kmol is the universal GAS constant, T is the temperature, and v_{ii} is the specific volume per MOLE of SOLUTE. This definition is known as the VAN'T HOFF EQUATION, sometimes also found described under HENRY'S LAW. The OSMOTIC PRESSURE is sometimes attributed to Ernest Starling (1866–1927), represented by the Starling equation or Frank–Starling equation; however, this generally applies to ION/molecular migration due to a gradient in concentration.

Gastrocnemius

[biomedical, mechanics] LEG muscle below the knee which is split from the lower single connection to the soleus MUSCLE, which in-turn is attached to the Achilles tendon and can hence flex the foot while connected to the calcaneus (heel bone). The gastrocnemius forms the main attribute for the calf of the lower leg.

Gastrointestinal tract

[biomedical] Esophagus, stomach, and intestinal conduit where food is digested and absorbed for consumption by the tissue (also found as GI tract and digestive tract). The gastrointestinal tract has muscles for guided and forced transportation that contract in a PERISTALTIC MOTION (see Figure G.21).

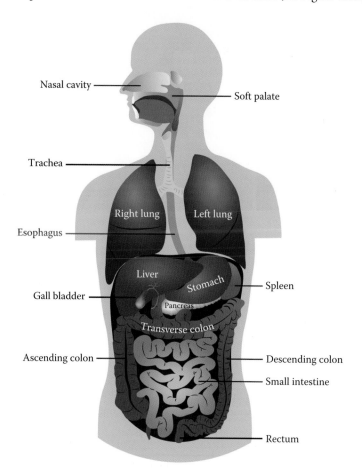

Figure G.21 Diagram of the gastrointestinal tract.

Gate valve

[fluid dynamics] A VALVE mechanism that provides partial to full closure of a FLOW by lowering a wedge or gate in the path of flow (usually a round-tube flow) by means of turning knob attached to a treaded push-rod turning through a screw hole, which drives the wedge. Other advance mechanisms use a worm wheel attached to a slotted shaft or a magnetic SOLENOID (see Figure G.22).

Figure G.22 Gate valve.

Gauge blocks

[general] Verified accurate weight and size blocks used for calibration of instruments, such as a scale. It is usually made of steel to ensure resilience and negligible influences resulting from changing boundary conditions.

Gauge field

[computational, quantum] A mathematical frame of reference in Lagrangian transformations and associated vector fields. Gauge fields are used to describe the elementary forces of nature. Gauge fields are the basis of the Lagrangian. Maxwell's theory of ELECTRODYNAMICS applied the gauge field symmetry implications for the first time. One specific example is in YANG–MILLS FIELD. It is also referred as gauge.

Gauge force field

[general, quantum] A force field between ELEMENTARY PARTICLES that uses a "particulate" connection that is used to transmit the interaction between the fermions, bosons, and quarks. The theory originated in invariant Lagrangian description of field theory is referred as the Lagrangian DEGREES OF FREEDOM. Gauge force field approximation readily found applications in QUANTUM THEORY. The GAUGE THEORY represents the potential for transformations between different potential gauges without influencing the outcome.

Gauge invariant

[computational, electromagnetism, energy, general, thermodynamics] All theories that are proven and accepted are independent of the arbitrary definitions that describe the phenomena and boundary conditions.

Gauge pressure

[fluid dynamics, general] A reference PRESSURE (P) for a system, not necessarily the ATMOSPHERIC pressure. For instance, the gauge pressure for a dam is the pressure at a specific height (z) measured from the bottom of the lake (with total depth h) $P_{\text{gauge}} = \rho g (h - z)$, not accounting for the atmospheric pressure on the surface. In this example, the gauge pressure defines the force exerted on the WALL of the dam.

Gauge symmetry

[computational, electromagnetism, energy, general, thermodynamics] The fact that any physical phenomenon or PHYSICS system that is subjected to a change that results in an observable modification of the reality (e.g., a snow flake is generally unchanged when the flake is rotated 60°, which is repeatable) while the final outcome of the process or ultimate STATE of the system will not be different from previously observed end results. Additional forms of symmetry are explained by the NOETHER'S PRINCIPLE, which states that for every continuous system there is a symmetry that correlates a corresponding alternate conservation law and for every conservation law there is a continuous symmetry. Due to the fact that the electric field, and hence the electric potential, is tied to the CONSERVATION OF CHARGE, it follows that classical ELECTRODYNAMICS is GAUGE INVARIANT.

Gauge theory

[computational, electromagnetism, energy, general, thermodynamics, quantum] The description of the interaction between ELEMENTARY PARTICLES using QUANTUM FIELD THEORY. The theory assumes a symmetry of interchangeability of the fields exerted on the particles by each other. The gauge theory requires that the fields can be described by gauge force fields and should obey the SCHRÖDINGER EQUATION. A part of theoretical PHYSICS that uses quantum ELECTRODYNAMICS and is used in HIGH-ENERGY PHYSICS. The gauge theory hinges on the postulate that all theories are balanced in a GAUGE SYMMETRY. One confirming factor of the gauge symmetry is the GENERAL THEORY OF RELATIVITY, whereas the idea originates in quantum electrodynamics, which is gauge symmetric in its formulation. One of the key ELEMENTS in the gauge symmetry is that all PHYSICS LAWS are identical to all observers. In gauge theory, it is shown how all REFERENCE FRAMES and units can be converted or "transformed" (see Figure G.23).

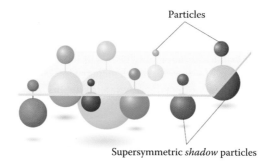

Figure G.23 Supersymmetry as it applies to gauge theory.

Gauss, Johann Carl Friedrich (German: Gauß) (1777–1855)

[biomedical, electronics, general, nuclear] A German scientist, astronomer, physicist, and mathematician. Gauss is most known for his contributions in ELECTROSTATICS and the mechanisms to derive the electric field strength and direction. Gauss contributed significant theoretical descriptions to the field of electrical charge and field theory and electronic interaction. Gauss also introduced the LENS equation relating the object DISTANCE (s_o) to the IMAGE distance (s_i) with respect to the focal length (f_ℓ) of a single lens in 1841 as the GAUSSIAN LENS EQUATION $1/f_\ell = (1/s_o) + (1/s_i)$. The main restriction to the validity of the Gauss' law is the fact that there is a requirement for charge symmetry. Other work of Gauss was in astronomy and general PHYSICS, specifically OPTICS. Gauss defined the focal length (f) of a single thin lens in 1824 as the reciprocal of the sum of the inverse distance for rays crossing the optical axis (s_0, s_1, respectively) that emit parallel to the optical axis for both sides of the lens $1/f = (1/s_0) + (1/s_1)$ (see Figure G.24).

Figure G.24 Johann Carl Friedrich Gauss (German: Gauß) (1777–1855), painted by Christian Albrecht Jensen 1887. (Courtesy of Gauß-Gesellschaft Göttingen e. V. [Foto: A. Wittmann] detail from painting by Gottlieb Biermann, 1887.)

Gauss (unit: G)

[biomedical, electronics, general, nuclear] A unit for MAGNETIC FIELD strength, with respect to the Tesla $1T = 10^4 G$.

Gauss law of electrostatics

[electronics, nuclear, solid-state] The electric FLUX (Φ_e) through an arbitrary closed surface enclosing a net total CHARGE (Q) equates the net charge enclosed inside the surface (S) divided by permittivity of VACUUM (ϵ_0): $\Phi_e = Q/\epsilon_0$. It is also found as Gauss' law.

Gauss's law

[biomedical, electromagnetism, electronics, general, thermodynamics] A description of the electric field resulting from any grouping of charges confined by a chosen virtual surface in free space. The expression of the surface integral over this virtual closed surface (encapsulating a volume filled with net charge) of the outward directed electric field MAGNITUDE normal to the surface is proportional to the enclosed net charge. In case the surface of the medium is chosen to follow a specific contorted outline, the electric field line may exit and reenter the volume to exit at a final point on the egress route. The proportionality factor is defined by the reciprocal DIELECTRIC constant of the medium ($1/\epsilon_0$). Gauss's law is an expression that is included in the set of MAXWELL EQUATIONS related to electric and magnetic fields. Two definitions can be distinguished relating to electric and magnetic fields, respectively. For an electric field in VACUUM Gauss's law states that the electric field created by ELECTRIC CHARGES is directly proportional to the magnitude of the charges contained in an enclosed surface, defined as $\oint_S \vec{E} \cdot \vec{n} dA = Q_{in}/\epsilon_0$, where $\vec{E}$ is the electric field, $\vec{n}$ is the normal to the surface A of the enclosed space, Q_{in} is the enclosed net total charge, and ϵ_0 is the dielectric permittivity of vacuum. Analogues for a MAGNETIC FIELD ($\vec{B}$) Gauss's law is defined as $\oint_S \vec{B} \cdot \vec{n} dA = 0$. Gauss's law is part of ELECTROSTATICS. An expression for the total charge (with charge density $\rho(r)$ and a function of location (r)) enclosed within a surface (S) defined by the electric FLUX emerging (Φ_e) from this surface $\int_S \vec{E} \cdot \vec{n} dS = K_{Gauss} \int_V \rho dV$, where the surface has normal unit vectors to the surface $\vec{n}$, the charges generate an electric field defined by Coulomb's law ($\vec{E} = K_{Gauss}(q/4\pi r^2)\vec{r}'$, in direction $\vec{r}'$), $K_{Gauss} = 1/\varepsilon_0$, ε_0 is the dielectric permittivity of vacuum, V is the volume enclosed by the surface (S), ρ is the net charge density, and q is the net charge. It is also expressed as $\nabla \cdot \vec{E} = K_{\text{Gauss}}\,\rho$ (see Figure G.25).

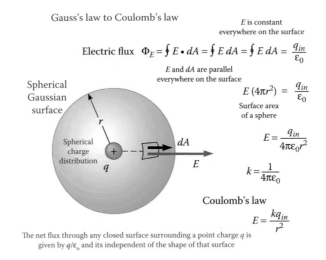

Gauss's law to Coulomb's law

E is constant everywhere on the surface

Electric flux $\Phi_E = \oint E \cdot dA = \oint E\,dA = \oint E\,dA = \dfrac{q_{in}}{\varepsilon_0}$

E and dA are parallel everywhere on the surface

Spherical Gaussian surface

$E\,(4\pi r^2) = \dfrac{q_{in}}{\varepsilon_0}$

Surface area of a sphere

r

Spherical charge distribution

q

dA

E

$E = \dfrac{q_{in}}{4\pi\varepsilon_0 r^2}$

$k = \dfrac{1}{4\pi\varepsilon_0}$

Coulomb's law

$E = \dfrac{k q_{in}}{r^2}$

The net flux through any closed surface surrounding a point charge q is given by q/ε_0 and its independent of the shape of that surface

Figure G.25 Illustration how the Coulomb's law and Gauss's law are inter-related.

Gauss's law of electrodynamics

[nuclear] The Maxwell equation in integral form. This applies to a MACROSCOPIC average for a "non-Abelian" GAUGE THEORY analysis, generally electroweak $\int_S \vec{E} \cdot \vec{n}\,dS = K_c \int_V \rho_e\,dV$, where $\vec{E}$ represents the electric field, $\vec{n}$ is the normal to the surface (S) holding the charge with volumetric charge density ρ_e for the volume V, K_c is a proportionality constant, which depends on the definition of the system ($K_c = 4\pi$ for a system in Gaussian units and $K_c = 1/\epsilon_0$ with ϵ_0 the DIELECTRIC permittivity of VACUUM).

Gaussian distribution

[atomic, computational] A PROBABILITY density function for events and measured values, also known as the NORMAL DISTRIBUTION $f_{\text{Gauss}}(x) = (1/\sigma_{\text{stat}}\sqrt{2\pi})\exp\left[-(21/2\sigma_{\text{stat}}^2)(x - \mu_{\text{stat}})^2\right]$, where μ_{stat} is the mean value of the observation x and σ_{stat} is the statistical variance. Generally, the VARIANCE (σ_{stat}^2) can be derived as follows: $\sigma_{\text{stat}}^2 = [1/(n-1)]\sum_{i=1}^{n}(x_i - \mu_{\text{stat}})^2$, with n values.

Gaussian lens equation

[general, optics] An equation relating the object DISTANCE (s_o) to the IMAGE distance (s_i) with respect to the focal length (f_ℓ) of a single LENS as introduced by CARL FRIEDRICH GAUSS (1777–1855) in 1841 as $1/f_\ell = (1/s_o) + (1/s_i)$. An equation relating the image distance (s_i) to the object placement with respect to a lens (s_o) with focal length f: $1/f = (1/s_i) + (1/s_o)$, as derived by CARL FRIEDRICH GAUSS (1777–1855) in 1841.

Gaussian noise

[biomedical, electronics, thermodynamics] Sensor, detector, or imaging NOISE resulting from randomly generated SIGNAL values due to thermal influences or electronic interactions, which describes the mean value ($\bar{x}_l$) as $\bar{x}_l = (1/n)\sum_{i=1}^{n}x_i$, for a total of n observations of respective values x_i; the standard deviation (σ_{stat}) is defined by $\sigma_{\text{stat}}^2 = [1/(n-1)]\sum_{i=1}^{n}(x_i - \bar{x}_i)^2$, where σ_{stat}^2 represents the variance. This stands in comparison to POISSON NOISE and IMPULSE NOISE.

Gaussian quadrature

[computational] A NUMERICAL algorithm used to obtain the best estimate of an integral $\left(\int_a^b f(x)dx\right)$ by mathematically selecting the optimal the x-coordinates (DISTANCE from a vector point to the y-axis measured parallel to the x-axis) to be used to define the polynomial function $f(x)$ under integration. The fundamental theorem of Gaussian quadrature defines that for an n-point Gaussian quadrature, the optimal abscissas of formulas are exactly the roots of the orthogonal polynomial derived on the same interval and with the same weighting function. Gaussian quadrature accurately fits all polynomials up to the $2n-1$th degree. The extrapolated Lagrange interpolating polynomial used in the Gaussian quadrature through the n-points is derived as $\Phi(x)=\sum_{j=1}^{n}[F(x)/(x-x_j)F'(x_j)]f(x_j)$, where $F(x)=\prod_{j=1}^{n}(x-x_j)$.

Gaussian surface

[biomedical, electronics, general] A symmetric surface of enclosure that confines all the charges in the calculation for the electric field.

Gauss–Jordan elimination

[computational] An analytical mechanism for solving linear equations. The matrix describing the system parameters is systematically reduced as illustrated from the following three equations: $x+y+z=3$; $2x+3y+7z=0$; and $x+3y-2z=17$, which is expressed in matrix format as

$$\begin{bmatrix} 1 & 1 & 1 & 3 \\ 2 & 3 & 7 & 0 \\ 1 & 3 & -2 & 17 \end{bmatrix},$$

which eventually reduces row by row to

$$\begin{bmatrix} 1 & 0 & 0 & 1 \\ 0 & 1 & 0 & 4 \\ 0 & 0 & 1 & -2 \end{bmatrix}$$

with solutions $x=1$, $y=4$, and $z=-2$.

Gauss–Seidel method

[computational] A sequential iteration technique for solving n linear equations ($\vec{y}=A\vec{x}$) by solving each individual equation one-at-a-time and using the prior results in the sequential solutions procedure. This process was introduced by JOHANN CARL FRIEDRICH GAUSS (1777–1855) and PHILIPP LUDWIG VON SEIDEL (1821–1896). The matrix representation has coefficients for the $n\times n$ square matrix.

$$A = \begin{bmatrix} a_{11} & \cdots & a_{1j} \\ \vdots & a_{ij} & \vdots \\ a_{i1} & \cdots & a_{nn} \end{bmatrix}.$$

The parameters are sequentially defined by $x_i^{(k)}=[y_i-\sum_{j<i}(a_{ij}-x_i^{(k)})-\sum_{j>i}(a_{ij}-x_i^{(k-1)})]/a_{ii}$, or in matrix format $\overrightarrow{x^{(k)}}=[1/(D-L)](U\overrightarrow{x^{(k-1)}}+\vec{y})$, where D represents the diagonal of the matrix A, L represents the parameters on the lower half of the diagonal of the matrix A, and U represents the upper triangular part of the matrix A: $A=D-L-U$. The Gauss–Seidel method relies on the fact that there are many zero ELEMENTS in the matrix and requires that the matrix is positive symmetric or diagonally dominant. The Gauss–Seidel method is frequently used to derive the eigenvalues for a complex system.

Gay-Lussac, Joseph Louis (1778–1850)

[atomic, electronics, general] A physicist and chemist from France, contemporary of Jacques Charles. Gay-Lussac is best known for his work in FLUID DYNAMICS and GAS theory, strictly limited to ideal gasses only. It is also known as Charles' law, after JACQUES ALEXANDRE CÉSAR CHARLES (1746–1823) (see Figure G.26).

Figure G.26 Joseph Louis Gay-Lussac (1778–1850), lithograph by François-Séraphin Delpech.

Gay-Lussac's law

[atomic, electronics, general, mechanics] An equilibrium of GAS PHASE and boundary conditions expressed by JOSEPH LOUIS GAY-LUSSAC (1778–1850) in 1802 defining the relationship between PRESSURE (P) and TEMPERATURE (T) as P/T = constant, while the volume and quantity of the gas are kept constant. Also found as Charles' law (see Figure G.27).

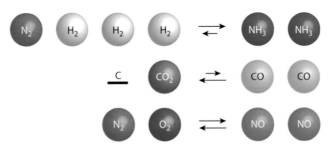

Figure G.27 The implications of the Gay-Lussac law.

Geiger, Johannes "Hans" Wilhelm (1822–1945)

[atomic, nuclear] A physicist and mathematician from Germany. Hans Geiger is world renowned for his device development for the detection of ionizing RADIATION in 1909 (GEIGER COUNTER), in particular emitted from radio-active isotopes, which can prove harmful to human biology. In the joint effort with Ernest Marsden (1889–1970) in 1909, the Geiger–Marsden experiment provided the evidence for the existence of the atomic nucleus (see Figure G.28).

Figure G.28 Johannes "Hans" Wilhelm Geiger (1822–1945) in 1928 with Karl Scheel (left). (Courtesy of Friedrich Hund.)

Geiger and Marsden experiment

[atomic, nuclear] A PARTICLE DISPERSION experiment with gold-foil bombarded by ALPHA PARTICLES designed to define the atomic structure performed by Hans Geiger (1822–1945) and Ernest Marsden (1889–1970) in 1909 under direction of Ernest Rutherford (1871–1937). The experimental data provided conclusive evidence to support the Bohr ATOMIC MODEL, verifying the atomic NUCLEUS as a separate entity of the atomic structure.

Geiger counter

[atomic, biomedical, nuclear, thermodynamics] A device and mechanism to measure ionizing RADIATION based on scintillation designed by Johannes Geiger (1822–1945) in 1909. Gas-filled GLASS bulb with low pressure gas (e.g., argon and AIR) that has a CATHODE and ANODE in the volume. The potential difference applied to the anode–cathode is slightly less than what would be required for IONIZATION. Once a charged PARTICLE enters the tube, it will ionize the GAS and a current will FLOW between the cathode and anode, counting the charge. The current duration is limited by proper configuration of an R–C network (RESISTOR–capacitor) allowing each discharge to be registered individually. Particles are not measured for IONIZATION POTENTIAL; however, the electron count provides an indication per pulse of the ENERGY of the ionizing particle. Ionizing particles measured can be alpha and beta. Geiger recognized that 1 g of radium ISOTOPE generates 3.57×10^{10} ALPHA PARTICLES per SECOND. Using the measured current generated by each discharge can provide the charge of the ionizing particle. Geiger worked closely WITH Ernest Rutherford (1871–1937) on this as a post-doctorate researcher. In literature, one may find this listed as a Geiger–Müller counter, based on the input of Erwin Wilhelm Müller (1911–1977) on later refinements.

Geiger–Müller counter (Geiger–Mueller counter)

[atomic, biomedical, nuclear, thermodynamics] A device designed by Hans Geiger in 1908 and refined and produced in collaboration with Walther Müller (1905–1979) in 1928 (*see* **Geiger counter**).

Geiger–Müller (G–M) tube

[atomic, nuclear] A sealed gas-filled GLASS tube with a special CATHODE–ANODE configuration used in the Geiger counter.

Geiger–Nuttall law

[atomic, nuclear, thermodynamics] The alpha-decay process can be described energetically as if the ALPHA PARTICLE is TUNNELING through a POTENTIAL BARRIER of Coulomb forces, both attractive and repulsive $\log(\tau_{1/2}) \cong aZ/\sqrt{Q_d} + b$, with for heavy nuclei $a = 1.454$ and $b = -46.83$ constants, and the DECAY time depends on the atomic number (Z). The decay ENERGY was derived by JOHN MITCHELL NUTTALL (1890–1958) in collaboration with JOHANNES WILHELM GEIGER (1822–1945) in 1911.

Geiger–Nuttall rule

[atomic, nuclear] *See* GEIGER–NUTTALL LAW.

Geitel, Hans Friedrich Karl (1855–1923)

[nuclear] A physicist and high-school teacher from Germany. Geitel worked on the electron discharge from nonelectrode metals under irradiation by ionizing RADIATION and ULTRAVIOLET light, experimentally verifying the photo-electric effect. Other work by Geitel was related to "atmospheric ELECTRICITY," for example, LIGHTNING and cosmic radiation. The work by Geitel provided elementary revelations about the nature of radioactive isotopes (see Figure G.29).

Figure G.29 Hans Friedrich Karl Geitel (1855–1923) Geitel, with Johann Philipp Ludwig Julius Elster (1854–1920).

Gell-Mann, Murray (1929–)

[energy, general, quantum, relativistic] A physicist from the United States. Murray Gell-Mann along with GEORGE ZWEIG (1937–) provided evidence that BARYONS (as well as the other HADRON group: mesons) are not primary ELEMENTARY PARTICLES (see Figure G.30).

Figure G.30 Murray Gell-Mann (1929–) in 2012. (Courtesy of World Economic Forum, Geneva, Switzerland.)

General nuclear reaction

[nuclear] An equation describing the DECAY of an ISOTOPE under the CONSERVATION LAWS; for example, the decay of uranium-238, producing thorium, is written as $^{238}_{92}\text{U} \rightarrow \, ^{234}_{90}\text{Th} + \, ^{4}_{2}\text{He} + \gamma$, respectively, concerning a collision $^{4}_{2}\text{He} + \, ^{14}_{7}\text{N} \rightarrow \, ^{17}_{8}\text{O} + \, ^{1}_{1}\text{H}$.

General theory of relativity

[computational, general] A theoretical description of the space-time CONTINUUM by ALBERT EINSTEIN (1879–1955) in 1915, linking the PERCEPTION of time to the perception of space. The concept of GRAVITATION is the major component of this theoretical description (1915). In astronomy, Einstein used this theory to predict that stellar binary ORBITS will progressively DECAY through emission of a concept captured by "gravitational radiation." The general theory links gravitation to time distortion, both induced by "neighboring" massive object as well as the rotation of the EARTH with respect to time measured at altitude (moreover moving into outer space). This relates to Einstein's SPECIAL THEORY OF RELATIVITY published in 1905.

Generalized adiabatic availability

[thermodynamics] The transfer of ENERGY when considering two energy states of a complex system with r constituents and ENTROPY (S) that are closely ranked together ("neighbors") has an ADIABATIC AVAILABILITY as the energy difference between the neighboring unstable and stable EQUILIBRIUM STATE $\Psi_{\text{adiab}} = E - E_s(S, n, \vec{\beta})$, where $\vec{\beta} = (\beta_1, \beta_2, \beta_3, \ldots, \beta i_1, \ldots, \beta_r,)$ are the system parameters (e.g., forces, boundary conditions, volume), and E_s is the energy of a stable equilibrium, whereas E is the state that is not necessarily stable and may not have the same number and quantities (n) of constituents. The stable equilibrium will have a stable temperature (T_s) that allows the adiabatic availability to be written in differential form as $d\Psi_{\text{adiab}} = dE - T_s(S, n, \beta)dS$. The "weight" of a system limits the variables to a single independent property only.

Generator

[general] A device that has a mechanical mechanism in place to convert one form of ENERGY into electrical energy. Examples are hydroelectric plant, wind generator, or nuclear power plant. The principle energy conversion comes to a rotating coil interacting with a MAGNETIC FIELD. The magnetic field can be a PERMANENT magnet, a coil or a stator. The magnetic field direction may move or may be fixed in both time and place. A generator can produce either a VOLTAGE (V) and current fluctuating over time with sinusoidal pattern directly related to the ANGULAR VELOCITY (ω) of the rotation of the mechanism $V = V_0 \sin\big((\omega/2\pi)t + \phi\big)$, where ϕ is a phase ANGLE that is device dependent, alternatively may in limited cases be steady state. Linking more than one generator together will illustrate the importance of the phase angle for each device, even when operating at nearly identical revolutions. A cluster of synchronous generators will induce significant risk for electric and mechanical losses and temperature rise with resulting reduced energy generation efficacy (see Figure G.31).

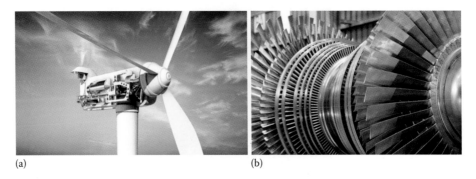

(a) (b)

Figure G.31 (a) Wind generator and (b) rotor for a coal burning steam turbine electric generator.

Genetic algorithm

[computational] An empirical search that emulates a naturally evolving process, incorporating a "learning" mechanism. The heuristic computational mechanism of "natural selection" resembles the biological process, specifically in genetics (Darwin's principle: survival of the fittest). The process originated at the onset of cybernetics in the 1940s.

Genomics

[biomedical, computational, theoretical] A comprehensive breakdown of the inherited information-bearing genetic material residing in the chromosomes (and hence DNA) of an organism's (i.e., genome). DNA is a biopolymer consisting of four nucleotide constituents: A (adenine), C (cytosine), G (guanine), and T (thymine). The study includes the analysis of the full complement of signaling codes that regulate the gene expression. Genomics focuses on species, not on individuals. The name originates from a journal title introduced in 1986 that publishes work related to gene sequencing, mapping and their activities. The information in the chromosomes is digital, consisting of DNA strings whose order provides information in 0s and 1s. Genomics is hence an information science. The study requires the establishment of a "reference genome" for a species of interest to provide a basis for the analysis of the global genomic composition and associated expressions. The most comprehensive DNA sequence-based investigation is the Human Genome Project.

Geocentric

[general, geophysics] A physical model of the UNIVERSE with the EARTH as the center, if not as the center of the SOLAR SYSTEM. This belief was documented by the Greek philosopher PTOLEMY (c. 90–168 AD) around 140 AD and was generally held as the standard until the seventeenth century. This stands opposing the

HELIOCENTRIC model, placing the SUN in the center of our celestial system, the current and verified model (see Figure G.32).

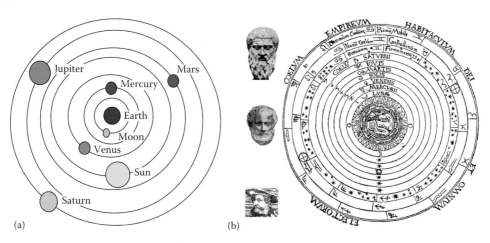

(a)   (b)

Figure G.32 (a) The model of the "universe" as described by Ptolemy (c. 90–168 AD). (b) Dante's (Durante degli Alighieri [Dante Alighieri] [1265–1321], poet from Italy) interpretation of the celestial configuration, portrayed by Ratel; Magasin Pittoresque, Paris, France, in 1850.

Geodesy

[general, geophysics] The study of the shape of the EARTH, divided into two aspects: ellipsoidal shape (in three-dimensional view: spheroid) derived from sea level and the determination of the deviation for the physical sea level (i.e., "geoid") from the perfect spheroid. The Earth deviates from a spheroid in an additional manner that the POLES are "lower" (see Figure G.33).

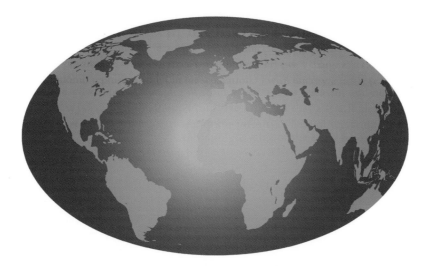

Figure G.33 Geoid shape of the Earth in exaggerated form.

Geographic north

[geophysics, mechanics] A reference point indicating the axis of rotation when referenced against the GEOGRAPHIC SOUTH POLE. Note that the axis of rotation is not the same axis that connects the MAGNETIC NORTH POLE to the MAGNETIC SOUTH POLE. Hence, the magnetic north pole is not in the same location as the geographic north pole (see Figure G.34).

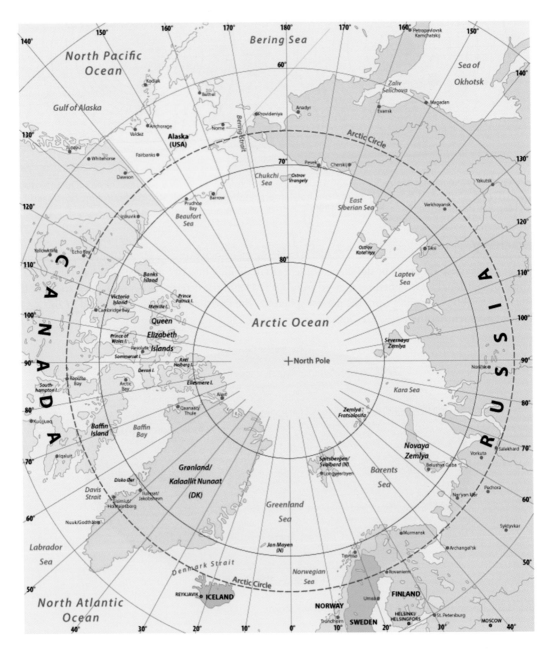

Figure G.34 Geographic location of the north point, the magnetic North and South Pole, respectively, are continuously shifting (drifting) as shown under magnetic pole. The axis of rotation is not indicated on this map, but does not coincide with the location of the North Pole center.

Geographic south

[geophysics, mechanics] A reference point indicating the axis of rotation when referenced against the GEOGRAPHIC NORTH pole. Note that the axis of rotation is not the same axis that connects the MAGNETIC NORTH POLE to the MAGNETIC SOUTH POLE. Hence, the magnetic south pole is not in the same location as the geographic south pole (see Figure G.35).

G

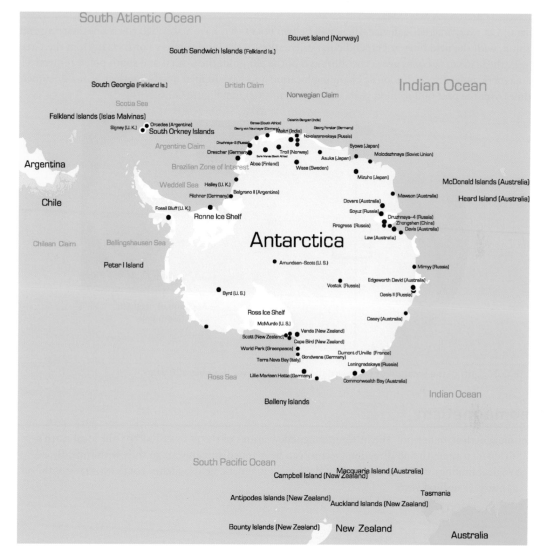

Figure G.35 Geographic location of the south point intersect of the longitudinal lines.

Geography

[astronomy/astrophysics, geophysics] The study of the relief and location of features, phenomena, and inhabitants on the surface of the PLANET EARTH. From Greek γεωγραφία meaning "Earth description." It is also referred to as cartography, the development of charts and maps, illustrating the local statistical values for specific parameters.

Geomagnetic poles

[general] On every magnetic device, there will be two POLES that have MAGNETIC FIELD lines starting and finishing, with the field lines at perpendicular orientation to the local surface. For EARTH (from the Greek γε [ge]), the MAGNETIC POLES are at the short axis of the ellipsoid, the NORTH and south poles, respectively. The magnetic poles are continuously changing location and have flipped north–south at least a documented 65 times over the past 600 million years (*see* MAGNETIC AXIS) (see Figure G.36).

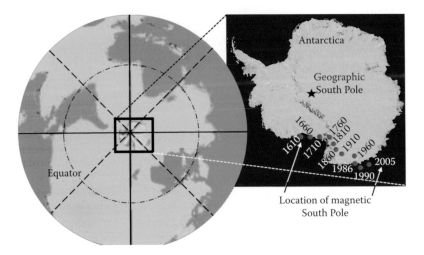

Figure G.36 Drift of the location of the magnetic South Pole over the past 400 years.

Geomagnetism

[electromagnetism, quantum] The inherent magnetic features of the PLANET EARTH. The local earth's MAGNETIC FIELD strength and direction (vector) can be mapped and appears to shift with time. Based on geological findings, it can be shown that the earth's magnetic field switches POLES (NORTH–south) on average every 11 million years.

Geometric center as the center-of-mass

[general] The center of an object found in the intersect of lines drawn from opposite extremes in three or more directions (see Figure G.37).

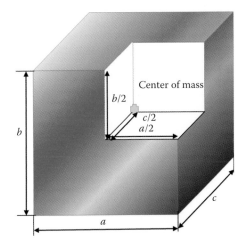

Figure G.37 Geometric center of mass.

Geometrical moment of inertia

[fluid dynamics, general, mechanics] A moment of INERTIA calculated in a simplified manner, using the parallel axis as the reference and calculating the moment of inertia ($I_{\parallel}$) based on global variables such as the sum of the moment of inertia about the axis through the center of mass (I_{cm}) and the total mass (m_{tot}) times the maximum radius of a circular CROSS SECTION (R): $I_{\parallel} = I_{cm} + m_{tot}R^2$ (*see* **MOMENT OF INERTIA**).

Geometrical optics

[general, optics] A geometric description of the tracks of rays of particles, consistent with the WAVE-PARTICLE DUALITY of light. There are five laws that describe geometrical optics: the law of rectilinear propagation, LAW OF REFLECTION, LAW OF REFRACTION, law of diffraction, and the LAW OF CONSERVATION OF ENERGY. Ray tracing is a specific example of geometric optics (see Figure G.38).

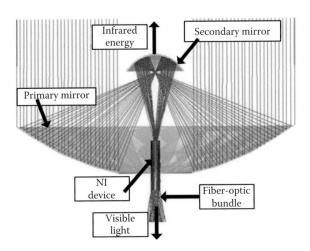

Figure G.38 Ray trace for a lens design.

Geophysics

[acoustics, astronomy/astrophysics, chemical, electromagnetism, fluid dynamics, geophysics, mechanics, solid-state, thermodynamics] The study of the physical phenomena of a PLANET as well as the ATMOSPHERE and LIQUID bodies (i.e., oceans). Geophysics can be subdivided into various subcategories depending on the school of thought, primarily ATMOSPHERIC PHYSICS, CLIMATOLOGY, GEODESY, GEOMAGNETISM and ELECTRICITY, HYDROLOGY, METEOROLOGY, OCEANOGRAPHY, RHEOLOGY, SEISMOLOGY (TECTONICS), and under certain conditions geology. Generally, geophysics entails the mathematical description of the processes and properties observed in the static field of geology (see Figure G.39).

Figure G.39 Geophysics of a geyser.

Geostationary orbit

[astrophysics, general, mechanics] A SATELLITE orbit that completes one revolution around the EARTH in the same time as the Earth completes one full revolution. The placement of the satellite is such that the ORBITAL VELOCITY and DISTANCE to the earth's center of mass creates a CENTRIPETAL FORCE with normal component that cancels out the gravitational attraction. The distance (d) from the center of mass of Earth with mass (M_E) to the satellite is $d = (GM_E/(T/2\pi)^2)^{1/3}$, where G is the gravitational constant and T is the period of the revolution (see Figure G.40).

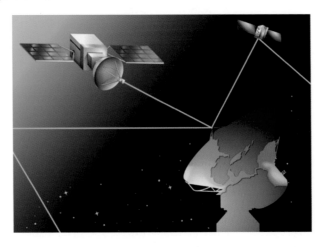

Figure G.40 Geostationary communication satellites.

Geosynchronous orbit

[astrophysics, general] *See* GEOSTATIONARY ORBIT.

Gerlach, Walther (1889–1979)

[atomic, nuclear, quantum] A physicist from Germany. In collaboration with OTTO STERN (1888–1969), Gerlach discovered the atomic ELECTRON SPIN quantization in a MAGNETIC FIELD in 1920, the Stern–Gerlach effect. This observation was made 2 years prior to the official definition of SPIN (see Figure G.41).

Figure G.41 Walther Gerlach (1889–1979).

Germer, Lester Halbert (1896–1971)

[atomic, computational, nuclear] A physicist from the United States. Germer investigated the de Broglie notion that any and every moving object has a WAVE function associated with it. In collaboration with CLINTON JOSEPH DAVISSON (1881–1958), this phenomenon was experimentally verified in 1927, DAVISSON–GERMER EXPERIMENT (see Figure G.42).

Figure G.42 Lester Halbert Germer (1896–1971) on right with Clinton Joseph Davisson (1881–1958); picture taken with a vacuum tube that was used in the electron diffraction experiments performed under the Davisson–Germer experiment, while they were working at Bell Laboratories. Davisson received the Nobel Prize for his efforts in 1937, taken in 1927.

Gerstner, František Josef (1789–1823)

[fluid dynamics] A scientist from Czechoslovakia, now Czech Republic. Gerstner's interests were in fluid-dynamical MOTION, specifically at the surface and related to the rotational nature of SURFACE WAVES, building on the concepts introduced by SIR ISAAC NEWTON (1642–1727) in his 1687 *Principia* (see Figure G.43).

Figure G.43 František Josef Gerstner (1789–1823).

Gerstner's waves

[fluid dynamics] Rotational waves in an infinitely deep LIQUID described by FRANTIŠEK JOSEF GERSTNER (1789–1823) in 1802. Gerstner made the following three assumptions: points at the crest of WAVE travel up and forward; ELEMENTS moving at the trough of the wave travel in downward and backward direction; the combined effects of PARTICLE motions outline a circle. Gerstner defined the location of the MOTION of a particle in a surface wave with horizontal location (x) and vertical direction (y) as $x = a + (1/k)e^{kb} \sin\left(k\{a + vt\}\right)$, where $k = 2\pi/\lambda$ is the wave number for wavelength λ, v is the propagation velocity of the wave, a and b are parameters used to identify a specific particle in the wave. The rotational motion was verified by both ERNST WEBER (1795–1878) and WILHELM WEBER (1804–1891); the latter built a wave tank for experimental verification (first of its kind) (see Figure G.44).

Figure G.44 Gerstner wave.

g-factor, electron orbital (g_L)

[atomic, energy, nuclear, optics] A correction factor on the total magnetic moment (μ_L) resulting from the orbital angular momentum ($\vec{L}$), expressed as $\vec{\mu_L} = -g_L(\mu_B/\hbar)\vec{L}$, where $\hbar = h/2\pi$ with $h = 6.62606957\times10^{-34}$ m^{2}kg/s the Planck's constant, and $\mu_B = e\hbar/2m_e$ the BOHR MAGNETON. For the components that align with the precision axis, this reduces to $\mu_L = g_L\mu_B m_\ell$, where m_ℓ is the MAGNETIC QUANTUM number and $g_L = 1$.

g-factor, electron spin (g_e)

[atomic, energy, nuclear, optics] The MUON has a SPIN similar to the electron with magnetic moment $\vec{\mu_m} = g_e(\mu_B/\hbar)\vec{S}$, where $\vec{S}$ is the spin angular momentum of the electron, $\hbar = h/2\pi$ with $h = 6.62606957\times10^{-34}$ m^{2}kg/s the Planck's constant, $\mu_B = e\hbar/2m_e$ the BOHR MAGNETON, m_e the electron mass, and $e = 1.60217657\times10^{-19}$ C the electron charge. Note also used $g_s = |g_e|$. The value for g_s can be derived from the DIRAC EQUATION.

g-factor, muon (g_μ)

[atomic, energy, nuclear, optics] The MUON has a SPIN similar to the electron with magnetic moment $\vec{\mu_m} = g_\mu(e/2m_\mu)\vec{S}$, where $\vec{S}$ is the spin angular momentum of the muon, e the electron charge, and m_μ the muon mass.

g-factor, nuclear (g_n)

[atomic, energy, nuclear, optics] The magnetic moment as a result of the nuclear SPIN with respect to protons, NEUTRON is linked to the nuclear spin angular momentum (I_{spin}) as $\mu_n = g_n(\mu_N/\hbar)I_{\text{spin}}$, where $\mu_N = e\hbar/2m_p$ the nuclear MAGNETON, and $\hbar = h/2\pi$ with $h = 6.62606957\times10^{-34}$ m^{2}kg/s the Planck's constant, m_p the PROTON mass, and $e = 1.60217657\times10^{-19}$ C the electron charge. It is also referred to as NUCLEON g-factor.

g-factor, total angular momentum (g_J)

[atomic, energy, nuclear, optics] A correction factor for the total magnetic moment resulting from the TOTAL ANGULAR MOMENTUM ($\vec{J} = \vec{L} + \vec{S}$, $\vec{L}$ the orbital angular momentum of the electron and $\vec{S}$ the SPIN angular momentum) $\vec{\mu_J} = -g_J(\mu_B/\hbar)\vec{J}$, where $\hbar = h/2\pi$ with $h = 6.62606957\times10^{-34}$ m^{2}kg/s the Planck's constant, $\mu_B = e\hbar/2m_e$ the BOHR MAGNETON. It is also referred to as the Landé factor. (*Note:* the total angular momentum g-factor is correlated with the electron-orbital g-factor (g_L) and the ELECTRON SPIN g-factor ($g_s = |g_e|$, the absolute value of the electron g-value) by QUANTUM mechanical principles defined through $\vec{\mu_J} = \vec{\mu_L} + \vec{\mu_S} = \vec{L}g_L\mu_B + \vec{S}g_S\mu_B$, yielding $g_J = g_L\left[|\vec{J}(\vec{J}+1) - \vec{S}(\vec{S}+1) + \vec{L}(\vec{L}+1)|/[2\vec{J}(\vec{J}+1)]\right] + g_S\left[|\vec{J}(\vec{J}+1) + \vec{S}(\vec{S}+1) - \vec{L}(\vec{L}+1)|/[2\vec{J}(\vec{J}+1)]\right]$, where $\vec{J}$ is the total electronic angular momentum ($|\vec{J}| = j(j+1)\hbar$, j the angular QUANTUM NUMBER, expressed as operator $\vec{J}^2\Psi = \hbar j(j+1)\Psi$), $\vec{L}$ the orbital angular momentum (also referred to as rotational angular momentum) derived from the WAVE function Ψ as an operator ($\vec{L}^2$): $\vec{L}^2\Psi = \hbar\ell(\ell+1)\Psi$, g_L the electron-orbital g-factor, where the wave function is a complex function analog to a harmonic OSCILLATOR although in QUANTUM MECHANICS describing the PROBABILITY function of the position.) The value $|\Psi|^2$ is real and provides a probability distribution for the location of the PARTICLE in MOTION at a given time when the location is derived from observation, g_S the electron-spin g-factor, and $\vec{S} = \epsilon_0\int(\vec{E}\times\vec{A})d^3r$ the spin angular momentum, with ϵ_0 the permittivity of VACUUM, $\vec{E}$ the electrical field, and $\vec{A}$ the area or vector potential.

G

Gibbon, John Heysham, Jr. (1903–1973)

[biomedical] A cardiovascular surgeon from the United States who invented the first working lung–HEART machine in 1937. The work of Gibbon was based on the pioneering efforts of ex vivo oxygen exchange performed by another vascular surgeon ALEXIS CARREL (1873–1944) (see Figure G.45).

Figure G.45 John Heysham Gibbon, Jr. (1903–1973): photograph by Fabian Bachrach. (Courtesy of National Academy of Sciences, Washington, DC.)

Gibbs, Josiah Willard (1839–1903)

[biomedical, general, mechanics, thermodynamics] A physicist from the United States. Josiah Gibbs introduced and verified the Gibbs free ENERGY concept. Additional work of Gibbs was in vector analysis and ELECTROMAGNETIC THEORY (see Figure G.46).

Figure G.46 Josiah Willard Gibbs (1839–1903).

Gibbs free energy (*G*)

[biomedical, general, solid-state, thermodynamics] ENERGY that determines whether a reaction will be spontaneous or forced, introduced by JOSIAH WILLARD GIBBS (1839–1903). The Gibbs free energy is defined as the change in enthalpy minus the temperature times the entropy $G = E_{\text{Helmholtz}} + PV = U + PV - TS$, where $E_{\text{Helmholtz}}$ is the Helmholtz free energy, P is the pressure, V is the volume, T is the temperature, and S is the entropy of the system. The Gibbs free energy for forced or unfavorable (not spontaneous) reactions is positive ($G > 0$) and are identified as endergonic. Reactions will be spontaneous if the free energy is negative ($G < 0$) and are classified as EXERGONIC. During redox reactions, the Gibbs free energy is equal to the maximum electric work $\Delta G^0 = -nF_{\text{Helmholtz}}\Delta V^0$, where the superscript "o" indicates standard conditions (25°C temperature and 1 atm pressure) and ΔV^0 is the electric potential. It can be shown that at any temperature, the energy balance is represented as $\Delta G = \Delta G^0 + RT \ln Q_r$, where $R = 8.314$ J/molK is the GAS constant and Q_r is the reaction quotient. In biological cells, this can be used to illustrate the transmembrane electrical POTENTIAL (V or ε_{emf}) that occurs as a result of the difference in ionic concentrations $\left[C_{\text{ion}}\right]$ in the intracellular versus the extracellular space. The result is expressed as the NERNST EQUATION, also referred to as the NERNST POTENTIAL $\Delta\varepsilon_{\text{emf}} = \Delta\varepsilon_{\text{emf}}^{\ 0} - (RT/nF_{\text{Helmholtz}})\ln Q_r$, with ε_{emf} the electromotive transmembrane potential. This equation also provides the potential difference generated in a galvanic CELL as a function of the ION concentrations in both compartments. The Nernst potential is defining the RESTING POTENTIAL during equilibrium. In this equation, Q_r states the ratio of the extracellular to intracellular ion concentrations. The Nernst equation only represents the most simple case in which the CELL MEMBRANE is permeable to one type of ions (for instance, only POTASSIUM (K^+) ions) (see Figure G.47).

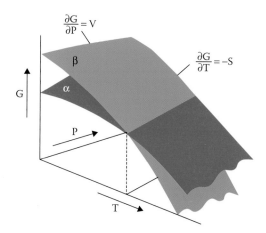

Figure G.47 Graphical representation of the Gibbs free energy concept.

Gibbs free energy of van der Waals gas

[thermodynamics] The van der Waals GAS has the characteristics of a liquid–GAS PHASE TRANSITION at constant pressure. The Gibbs free ENERGY is adjusted as follows: $G(T,V,N) = [NTV/(V - Nb_w)] - (2N^2a_w/V) - NT\left\{\log\left(n_Q[(V - Nb_w)/N]\right) + 1\right\}$, where T is the temperature, N is the number of atoms, V is the volume, the definition of the QUANTUM CONCENTRATION (*Note:* the "van der Waals gas" is a QUANTUM GAS) $n_Q = \left(mT/2\pi\hbar^2\right)^{3/2}$ with $\hbar$ the Planck constant (*h*) divided by 2π and m the mass of the gas, a_w and b_w constants tying the pressure ($P_c = a_w/27b_w^{\ 2}$), temperature ($T_c = 3a_w/27b_w$), and volume ($V_c = 3Nb_w$) to a reference point, primarily the critical point, as derived from the PV-curve (*also see* LAW OF CORRESPONDING STATES). The $\hat{P}\hat{V}$-curve (*Note:* $\hat{X} = X/X_c$) allows for the determination of a_w and b_w at the point (critical point) that satisfies the following two conditions (minimum or maximum): $\left(\partial \hat{P}/\partial \hat{V}\right)_{\hat{T}} = 0$ and $\left(\partial^2 \hat{P}/\partial \hat{V}^2\right)_{\hat{T}} = 0$.

Gibbs paradox

[thermodynamics] QUANTUM thermodynamics concept introduced by JOSIAH WILLARD GIBBS (1839–1903) describing the discontinuity in entropy for a two-component system, each with its own entropy, energy, and number of molecules that changes under external influences. When both constituents suddenly become alike (for reasons not to be discussed), the difference in entropy between the two constituents instantaneously reduces to zero.

Gibbs relation

[thermodynamics] The differential internal ENERGY (dU) between two states that are neighbors and at equilibrium but differ in entropy by amount dS, expressed as $dU = TdS - PdV + \mu_1 dn_1 + \mu_2 dn_2 + \mu_3 dn_3 + \cdots + \mu_r dn_r$, where P is the pressure, V is the volume, T is the temperature, S is the entropy of the system, μ_i is the chemical potential of the respective constituents, and n_i is the respective quantities of constituents.

Gibbs–Dalton mixture

[thermodynamics] Description of ideal gases that are composed of multiple constituents (r) of respective quantity n_i, each with a pressure P_{ii} assembled in a volume V at temperature T. The mixture has negligible interaction between the respective constituents. In case all except one of the components are removed there will be no change in the volume occupied by that constituent. This can be defined theoretically as a collective that satisfies the following three conditions: volume $n_i V_{ii}(T, P_{ii}) = V$ for each respective component, where V_{ii} is the volume of one ELEMENT of the component; internal ENERGY $U(T, P, \vec{n}) = \sum_{i=1}^r n_i U_{ii}(T, P_{ii})$; and entropy $S(T, P, \vec{n}) = \sum_{i=1}^r n_i S_{ii}(T, P_{ii})$.

Gibbs–Duhem relation

[thermodynamics] When combining the GIBBS RELATION and the EULER RELATION for enthalpy ($H = U + PV$) to give $H = TS + \mu_1 n_1 + \mu_2 n_2 + \cdots + \mu_r dn_r$, and the HELMHOLTZ FREE ENERGY ($E_{\text{Helmholtz}} = U - TS$) as $E_{\text{Helmholtz}} = -PV + \mu_1 n_1 + \mu_2 n_2 + \cdots + \mu_r dn_r$ and the Gibbs free ENERGY as defined under the Euler relation: $G = \mu_1 n_1 + \mu_2 n_2 + \cdots + \mu_r dn_r$ to yield another condition for the change in temperature and pressure under constant entropy as: $SdT - VdP + \mu_1 n_1 + \mu_2 n_2 + \cdots + \mu_r dn_r = 0$. Introduced by the French physicist and mathematician PIERRE DUHEM (1861–1916).

Gilbert, William (1540–1603)

[general] Physician, astronomer, and physicist from Great Britain who proposed a new concept of the earth's MAGNETIC FIELD. He stated that the earth's magnetic field originates at the POLES of a spherical LODESTONE-shaped EARTH in 1600. This was an advancement over the spheroid model introduced by NICOLAUS COPERNICUS (1473–1543). Initially, it was thought that a compass needle pointed to the "heavens." Gilbert outlined the path of the magnetic field lines of a BAR MAGNET with a compass to illustrate the concept. Descartes followed through on this with IRON filing on a sheet of paper placed on top of a bar MAGNET in around 1645. William Gilbert also described the influence of the MOON on the tides. He was

also an advocate of a sun-centered SOLAR SYSTEM. Gilbert also erroneously assumed that MAGNETISM was the source for driving the planetary attraction, not GRAVITATION (see Figure G.48).

Figure G.48 William Gilbert (1540–1603).

Gillet, Joseph Anthony (1837–1908)

[general] Scientist from the United States who proposed the existence of an AETHER for the transmission of ELECTROMAGNETIC RADIATION in 1882. Other work by Gillet together with WILLIAM JAMES ROLFE (1827–1910) describes the chemical interaction and the interchanges in forces between chemistry and ELECTRICITY (see Figure G.49).

Figure G.49 Joseph Anthony Gillet (1837–1908).

Glancing collision

[general] PARTICLE collision where the paths do not align. In this situation, the momentum is split in linear and orthogonal components in order to satisfy the CONSERVATION LAWS (see Figure G.50).

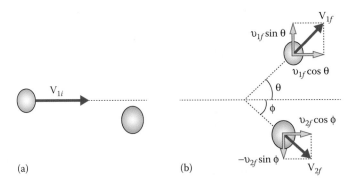

Figure G.50 Glancing collision: (a) before the collision and (b) after the collision.

Glashow, Sheldon Lee (1932–)

[atomic, general] Scientist from the United States who introduced the formal assumption of an idea that was discussed regarding the existence of a total of four quarks, similar to the known existence of four leptons. Nobel Prize winner for his work on the ELECTROWEAK force theory (see Figure G.51).

Figure G.51 Sheldon Lee Glashow (1932–) on the right during the Origins Symposium at Arizona State University in Tempe, Arizona, in 2009.

Glass

[solid-state] Amorphous material that has no distinguishable order or periodicity (not a crystal) that is transparent to most visible and INFRARED light and some ULTRAVIOLET as well as long wavelength ELECTROMAGNETIC RADIATION (e.g., RADIO, television). Most glass substances (depending on the chemical structure) are not a solid but display several LIQUID phase characteristics. Window panes use glass. Old window panes will show sagging over long periods of time, where the structure becomes less defined, making the IMAGE formation through the glass subject to deformities and the bottom side will increasingly become thicker, while the top decreases in thickness. Materials used for the fabrication of the general garden variety glasses are silicon oxide ["vitreous silica"] (SiO_2), arsenic trisulfide (As_2S_3), and the incorporation of other oxides such as Na_2O or B_2O_3 with different transmission spectra. Generally, the use of the word glass for a transparent medium is classified based on the production process (see Figure G.52).

Figure G.52 Examples of the use and appearance of glass products: (a) house with glass windowpanes and (b) heating glass in furnace for glassblower applications.

Global positioning system (GPS)

[general, geophysics] Coordination system relying on the earth's MAGNETIC FIELD lines to determine the longitude and LATITUDE of the recording device, and as such define the location of the user. The GPS relies on the fixed (geostationary) position of a number of satellites (24 total), each in a 12 h orbit. The satellites are referenced with respect to each other and with respect to the earth's magnetic field. The beacon signals emitted by the respective satellites are captured by receiver stations (e.g., mobile phones) that will define the location of the receiver station with respect to the global geometry. The satellites continuously send out their position information and this is used to cross-correlate with respect to the remaining satellites in orbit for global referencing. A minimum of four satellites are used to define the position of a device. Additional linking with existing maps a path which can be outlined from point "A" to point "B," all with respect to the observer's location. The GPS navigation system will need to include the SPECIAL THEORY OF RELATIVITY to maintain accuracy. Without special relativistic calculations the determination of the location would be accurate only within several kilometers, whereas state of the art is accurate within meters. Land-based reference signals can enhance the positioning accuracy further (see Figure G.53).

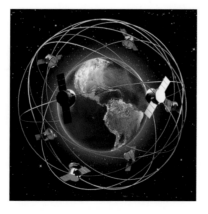

Figure G.53 Representation of the satellite configuration for the global positioning system.

Global symmetry

[computational, general] A system that is invariant under transforms, such as rotation of the reference frame or linear translation (i.e., GLOBAL TRANSFORMATION). The length and respective ANGLES to the edges remain constant. A car driving on the highway will not change configuration, while retaining its own reference frame, or within the reference frame of a passing car.

Global transformation

[computational, general] Linear transformations of a system with associated parameters (e.g., mass, length) and described by functions (e.g., OSCILLATION), such as rotation or translation.

Globe, oscillations of a liquid

[fluid dynamics] Finite oscillations of a LIQUID in an ellipsoidal configuration. The ellipsoid has three principal axes: a, b, and c in a Cartesian coordinate system (x, y, z). The axes will have a rate of change associated with the OSCILLATION that is described as: $\dot{a} = \partial x/\partial t$, $\dot{b}$, and $\dot{c}$. The VELOCITY POTENTIAL (Φ_v) profile will satisfy $\Phi_v = -(1/2)\left((\dot{a}/a)x^2 + (\dot{b}/b)y^2 + (\dot{c}/c)z^2\right)$, with boundary conditions: $(\dot{a}/a) + (\dot{b}/b) + (\dot{c}/c) = 0$, since this is for traveling RESONANCE, not EXPANSION.

Globe valves

[fluid dynamics] Tortuous FLOW and gating mechanism with high flow RESISTANCE as one of several VALVE configuration (see Figure G.54).

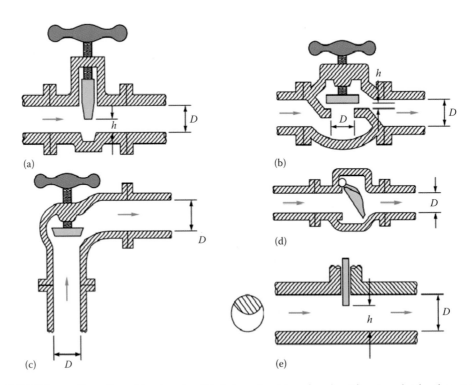

Figure G.54 Valve configurations: (a) gate valve, (b) globe valve, (c) angle valve, (d) swing-check valve, and (e) disk-type gate valve.

Glow discharge

[atomic, nuclear] Corona electron discharge between two charged (opposite) "electrodes" when the pressure (and hence the potential for IONIZATION of the GAS at low density) drops below a critical pressure. This luminous discharge is diffuse. This is in contrast to SAINT ELMO'S FIRE (see Figure G.55).

Figure G.55 Glow discharge on roof top. (Courtesy of Joe Thomissen.)

Glucose

[biomedical, chemical] CARBOHYDRATE chemical as one of the primary sources of cellular nutrition: $C_6H_{12}O_6$. Glucose oxides as follows: $C_6H_{12}O_6 + 6O_2 \rightarrow 6CO_2 + 6H_2O + energy$. The breakdown of glucose is referenced as the KREBS CYCLE. The Krebs cycle generates a total of 30 molecules of ATP per molecule of glucose. The conversion efficiency is very high, perhaps the highest for biologically viable ENERGY production. A specific variety of glucose is used in positron emission TOMOGRAPHY, FLUORODEOXYGLUCOSE.

Gluon

[general, high-energy, nuclear] STRONG NUCLEAR FORCE between quarks is represented by the exchange of a BOSON messenger designated as a gluon to form hadrons. The force represented by the gluon is called COLOR force. In QUANTUM mechanical theory, the gluon is a massless PARTICLE also called a VECTOR BOSON with "four-dimensional" Schrödinger WAVE functions with SPIN 1 and is magnetically neutral. Emission of a gluon can change the "color" of a QUARK, however it cannot change the "FLAVOR" OF THE QUARK. There are eight different gluons associated with the eight different potential quark couplings. The gluon color phenomenon is described by QUANTUM CHROMODYNAMICS.

Glycogen

[biomedical, chemical] Biopolymer form of GLUCOSE that supports storage in a polysaccharide form (i.e., starch) (see Figure G.56).

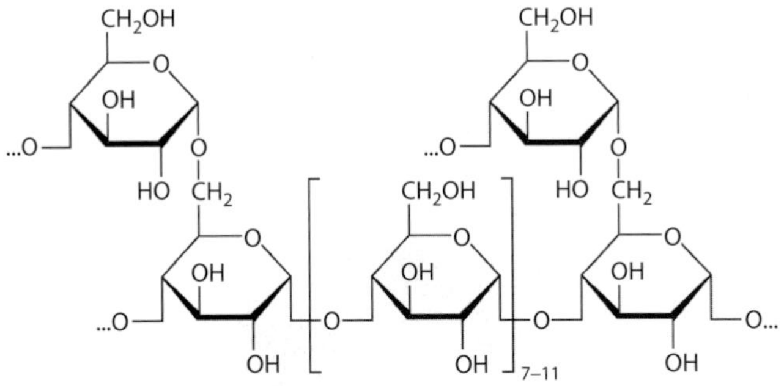

Figure G.56 Glycogen.

Glycosaminoglycan (GAG)

[biomedical, chemical] Building block for regenerative scaffolding. Mucopolysaccharides, long unbranched polysaccharide. GAG consists of a repeating disaccharide unit. Disaccharides are linked saccharides such as GLUCOSE, galactose, and fructose. For instance, sucrose has a linked fructose and glucose and maltose has two glucose molecules in a specific molecular configuration.

Glycosidic bond

[chemical] Type of covalent bond. The glycosidic bond specifically refers to bonding CARBOHYDRATE groups to other chemical groups or MACROMOLECULES. One specific example of chemical bonds of this nature are found in saccharides. More specifically, the bond between amino groups and carbohydrates or similar nitrogen-based chemical groups covalently bonding with carbohydrate molecules.

Goddard, Robert Hutchings (1882–1945)

[general, mechanics] Scientist from the United States, best known for his pioneering efforts on rocket design and rocket launch. In 1935, Goddard performed feasibility with rockets having $889.6\,N$ trust force. Goddard faced much RESISTANCE against his attempted space flight program based on the common misconception that the rocket would "push" against an AETHER for propulsion (and space is a VACUUM), instead of the correct philosophy that the combustion in a JET is providing an impulse that obeys the CONSERVATION LAWS, granting a forward MOTION with respect to the open-ended exhaust, in the opposite direction. Goddard is considered the father of modern space travel. Goddard applied a combination of mechanisms to control the flight of the rockets, consisting of gyroscopes, three-axis control, and thruster jets at strategically placed location (see Figure G.57).

Figure G.57 Robert Hutchings Goddard (1882–1945) on US postal stamp.

Goeppert-Mayer, Maria (Göppert) (1906–1972)

[electromagnetism, general, nuclear] Scientist from Germany (location is currently Poland). Nobel laureate (1963) for her contribution to the description of the electron SHELL MODEL of the ATOM. Maria shared her Nobel Prize with JOHANNES HANS DANIEL JENSEN (1907–1973). Dr. Goeppert-Mayer pioneered the investigations of the double QUANTUM emission phenomenon. She also experimentally verified the phenomenon of double beta DECAY. Maria Goeppert-Mayer was also the first to study the influence of magnetic susceptibility on the index of REFRACTION (also known as refractive index) of specific gasses, with her collaborator KARL FERDINAND HERZFELD (1892–1978) (see Figure G.58).

Figure G.58 Maria Goeppert-Mayer (1906–1972). (Courtesy of University of Chicago Photographic Archive, [apf1-04167], Special Collections Research Center, University of Chicago Library, Chicago, IL.)

Goldman, David E. (1910–1998)

[biomedical, chemical, energy] Scientist from the United States. Goldman developed the description of the electrochemical potential of the cellular MEMBRANE, based on the work of WALTHER HERMANN NERNST (1864–1941) (see Figure G.59).

Figure G.59 David E. Goldman (1910–1998). (Reproduced with permission from *Journal of the Acoustical Society of America*, 106[3], 1999 Obituaries. Copyright 1999, Acoustical Society of America.)

Goldman (voltage) equation

[biomedical, chemical, energy] Electrical potential across biological CELL MEMBRANE (i.e., MEMBRANE POTENTIAL: V_m) based on differences in ionic concentrations on either side. For example, assume that only the following ions are allowed to FLOW through the MEMBRANE of a biological cell: POTASSIUM (K^+), SODIUM (Na^+), and chlorine (Cl^-), with respective concentrations $[K^+]$, $[Na^+]$, and $[Cl^-]$. The transmembrane RESTING POTENTIAL is based on the NERNST EQUATION. The derived Goldman potential includes multiple ions and the respective ability of migration through the ION CHANNELS and GAP JUNCTIONS, indicated by the respective permeability for each ION: P_{ion}, expressed as: $V_m = (RT/F_{Far}) \ln \left\{ (P_{Na}[Na^+]_{out} + P_K[K^+]_{out} + P_{Cl}[Cl^-]_{in})/P_{Na}[Na^+]_{in} + P_K[K^+]_{in} + P_{Cl}[Cl^-]_{out} \right\}$, where $F_{Far} = 96{,}485$ C/mol is the FARADAY CONSTANT, $R = 8.314462175$ J/Kmol the universal GAS constant, and T the temperature during the process. The Goldman potential describes a static PHASE, where the dynamic configuration describes the ACTION POTENTIAL formulated by SIR ALAN LLOYD HODGKIN (1914–1998) and SIR ANDREW FIELDING HUXLEY (1917–2012). In this equation is the permeability constant for the Na^+ ion. A typical value for the resting potential is −90 mV. It is negative because cells are more negative relative to the surrounding medium. Cells that are excitable have the ability to rapidly reverse the potential, causing it to be slightly positive. The potential that is generated in this process is known as the action potential.

Goldman–Hodgkin–Katz equation

[biomedical] *See* GOLDMAN EQUATION.

Goldstein, Eugen (1850–1930)

[atomic, nuclear] Physicist from Poland/Germany, who worked on the first model of GAS discharge tubes in the investigation of CATHODE and ANODE rays. Goldsteins's work lead to the discovery of the PROTON. His interpretation of the rays emitted from the anode was "Kanalstrahlen," specifically based on the fact that his investigational device used a perforated disk as anode, as well as for the cathode. The gas in the sealed tube is ionized on contact with the surface of the holes of the perforations of the anode and hence the PHASE "canal ray" was coined (see Figure G.60).

Figure G.60 Eugene Goldstein (1850–1930). (Courtesy of Bettmann/CORBIS.

Golf ball

[computational, fluid dynamics, mechanics] Aerodynamically designed ball that can be steered with great accuracy resulting from a single stroke. The TURBULENCE induced by the dimples on the surface of the ball provide an increase in DRAG. Combining the rotation of the ball with the turbulent BOUNDARY LAYER provides a LIFT that more than compensates for the increase in FRICTION. The resulting increase in upward force will promote a greater DISTANCE of flight. The dimples thus enhance the MAGNUS LIFT. The wedge shape of the golf club generally ensures that the rotation of the ball is in the direction of MOTION on the bottom, thus creating a greater pressure on the bottom (according to Bernoulli's law; DANIEL BERNOULLI [1700–1782]) and lower pressure on the top with the surface moving in the same direction. When the velocity of the golf ball reaches a threshold (i.e., REYNOLDS NUMBER $Re < 2300$) and when slowing down the FLOW suddenly becomes laminar. The LAMINAR FLOW will detach from the surface that moves in the direction of flow (which goes laminar prior to the opposite side that move against the flow), which will be the top. The break-away laminar part of the flow on the top will result in a negative Magnus lift force, resulting in a rapidly declining trajectory. The concept of the impact of the dimples on the golf ball was realized in the early nineteenth century, by a professor at St. Andrews University in Scotland. Note that for an airplane the laminar flow is more desirable (see Figure G.61).

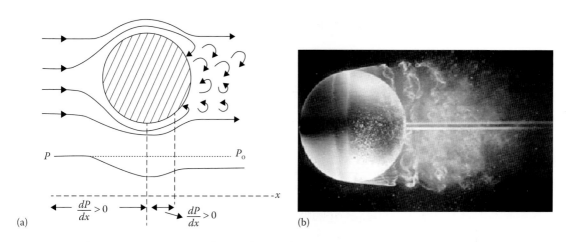

Figure G.61 (a,b) Fluid dynamic aspect of the golf ball in flight.

Goudsmit, Samuel Abraham (1902–1978)

[atomic, nuclear] Scientist from the Netherlands who introduced the concept of ELECTRON SPIN in 1925, simultaneously but independently from GEORGE UHLENBECK (1900–1988) while working under PAUL EHRENFEST (1880–1933), based on the description of the four QUANTUM numbers associated with the electron introduced by WOLFGANG PAULI (1900–1958) (see Figure G.62).

Figure G.62 Samuel Abraham Goudsmit (1902–1978). (Courtesy of Nederlands Tijdschrift voor Natuurkunde.)

Gould, Gordon (1920–2005)

[optics] American physicist attributed to the invention of the laser (acronym for light amplification by STIMULATED EMISSION of RADIATION), however not universally accepted, since there still remains controversy with the work by Theodore Maiman (1927–2007) (see Figure G.63).

Figure G.63 Gordon Gould (1920–2005) picture taken in 1940 by his brother Geoffrey Gould.

G-parity

[atomic, nuclear, quantum] Quantum number indicative of the rotation of μ-meson in isospin space around the two axis which culminates in charge conjugation. The G-parity follows from the generalization of C-parity applied to multiple clusters of particles, resulting in a concise quantum number resulting from multiplication of associated quantum numbers. The value of G-parity is defined as $G_{par} = Ce^{i\pi I_2}$, where C is a constant related to the C-parity (i.e., C-parity operator), and I_2 is the second component of the isospin of the nucleon. The elementary particle mesons have a specific parity referred to as G-parity.

Graded index of refraction—lens

[optics] Lens that relies on the gradual or discrete changes in index of refraction to provide focusing action, also known as GRIN lens. The GRIN lens does not necessarily require curvature to focus. GRIN lens design is often applied to fiber-optics to center the optical information for maximum efficiency in ray transfer.

Graf's addition theorem

[acoustics, computational] Bessel function summation algorithm, expressing the Bessel function: $C_v(\eta r_j)e^{iv(\theta_j - \vartheta_{j\ell})}$, in polar coordinates (r_j, θ_j), as a series of two functions in respective coordinate systems with respect to each other. Each coordinate system has an origin: O_ℓ and O_j, where O_ℓ can be placed with respect to O_j in coordinates $(R_{j\ell}, \vartheta_{j\ell})$. The function C_v can in this case be represented by any of the Bessel functions (zeroth-order, first-order and second-order): J_v, I_v, Y_v, K_v, $H^{(1)}_v$, and $H^{(2)}_v$. The representation is as the following recursive relations: $C_v(\eta r_j)e^{iv(\theta_j - \vartheta_{j\ell})} = \sum_{\mu=-\infty}^{\infty} C_{v+\mu}(\eta R_{j\ell})J_\mu(\eta r_\ell)e^{i\mu(\pi - \theta_\ell - \vartheta_{j\ell})}$ under the condition that $j \neq \ell$; as well as $H^{(1)}_v(\eta r_j)e^{iv(\theta_j - \vartheta_{j\ell})} = \sum_{\mu=-\infty}^{\infty} H^{(1)}_{v+\mu}(\eta R_{j\ell})J_\mu(\eta r_\ell)e^{i\mu(\pi - \theta_\ell - \vartheta_{j\ell})}$, and $K_v(\eta r_j)e^{iv(\theta_j - \vartheta_{j\ell})} = \sum_{\mu=-\infty}^{\infty} K_{v+\mu}(\eta R_{j\ell})I_\mu(\eta r_\ell)e^{i\mu(\pi - \theta_\ell - \vartheta_{j\ell})}$ under the same boundary condition.

Graham, Thomas (1805–1861)

[biomedical, chemical] Chemist from Great Britain (Scotland) and physical scientist. Graham was a pioneer in dialysis and made significant contributions to the theoretical description of gas diffusion. In dialysis, colloids (microscopic solutes, or suspension) are separated from dissolved molecules or ions (crystalloids) by means of chemical gradients and the inherent differences in diffusion rates across a semipermeable membrane. Graham is also called the father of colloid chemistry (see Figure G.64).

Figure G.64 Thomas Graham (1805–1861) by Maull & Polyblank, 1856.

Graham's law of diffusion

[biomedical, chemical] Gas DIFFUSION expression defined by THOMAS GRAHAM (1805–1861) in 1833. The rate of gas diffusion is inversely proportional to the square root of the MOLECULAR WEIGHT (W) of the substance (e.g., SOLUTE): $D_{\text{diff}} \sim MW^{-(1/2)}$.

Grain boundary

[acoustics, optics, solid-state] The mismatch in surface junction mating two separate crystal structures. Crystals with different direction can be joined as a result of ANNEALING or they can be grown as such. In vertical direction this type of discontinuities are classified as "staking faults."

Gram (g)

[general] Unit of mass, whereas weight is a function of GRAVITATIONAL ACCELERATION. Mass is a QUANTIFICATION of the amount of MATTER in an object. Even though the gram is elementary, the use of the KILOGRAM (kg) is the international standard. The mass of a platinum–iridium cylinder (height = 39 mm; diameter = 39 m), held at the INTERNATIONAL BUREAU OF WEIGHTS AND MEASURES is the recognized standard for 1 kg. A set of 40 virtually identical cylinders (slight errors in reproduction, even though the dimensions can be verified with extremely high degree of accuracy, on a wavelength basis) were produced and distributed to regional calibration stations (in the United States the National Institute of Standards and Technology in Boulder, Colorado). Due to the inherent difficulties associated with measuring the mass of a single ATOM this has still not evolved into a easily reproducible measure and remains the only artificial standard.

Gram atomic weight

[atomic, nuclear] The relative atomic weight of an ELEMENT, expressed in grams.

Gram mole

[general, thermodynamics] The quantity of a substance containing the equivalent number of components (atoms/molecules) to the 12 g mass of carbon 12 (^{12}C); a mol of ELEMENT, containing Avogadro's number in molecules: $Na = 6.022 \times 10^{23}$ molecules/mol.

Gram molecular mass (gram formula)

[atomic, nuclear] The mass of a volume of GAS at 22.4 L = 0.0224 m^3 will measure a mass in grams that equals the molecular mass for the ELEMENT, all at standard temperature (0°C = 273.15 K) and standard pressure (1 atm = 101.325 kPa).

Gram molecular weight (gram mole)

[nuclear] The relative molecular weight of a compound, expressed in grams.

Gramme, Zénobe-Théophile (1826–1901)

[energy, general] Engineer from Belgium, inventor of the electric GENERATOR (see Figure G.66).

Figure G.66 Zénobe-Théophile Gramme (1826–1901).

Gramme dynamo

[energy, general] Early concept electrical current GENERATOR designed by the French engineer ZÉNOBE-THÉOPHILE GRAMME (1826–1901) in late 1860s. The electric motor (invented by WILLIAM STURGEON [1783–1850] in 1832) and the dynamo were developed independent from each other and in 1873 at an exposition in Vienna, Austria, both devices were simultaneously displayed, while the dynamo was accidentally wired wrong and became a MOTOR. The GRAMME DYNAMO was originally driven by a steam ENGINE to create ELECTRICITY (see Figure G.65).

Figure G.65 Gramme Dynamo.

Grand unification theories (GUTs)

[computational, general, quantum] Also referred to as GRAND UNIFIED THEORY, (*see* SCHRIEFER UNIFIED THEORY).

Grand unified theory

[computational, general, quantum] *See* SCHRIEFER UNIFIED THEORY.

Grashof, Franz (1826–1893)

[fluid dynamics, thermodynamics] German mechanical engineer, introduced the concept of the Grashof number for scaling free convection. Additional work by Grashof was in free convection and the FLOW of steam in relation to steam engines and steam generators as well as mechanical quality control for MACHINES (see Figure G.67).

Figure G.67 Franz Grashof (1826–1893), around 1870.

Grashof number ($Gr = (g\beta\rho^2(T_s - T_\infty)L^3)/\eta_{kin}^2$)

[biomedical, fluid dynamics, geophysics, thermodynamics] Dimensionless number showing the proportional factor between buoyant force and the net viscous force, where β is the linear EXPANSION coefficient, g the GRAVITATIONAL ACCELERATION, ρ density, T_s the surface temperature, T_∞ the main body of LIQUID temperature, L the characteristic length of the object, and η_{kin} is the KINEMATIC VISCOSITY. The Grashof number is particularly important in FLUID flow containing NATURAL CONVECTION. The Grashof number describes the conditions for turbulent versus LAMINAR FLOW, where at low Grashof number the FLOW is laminar, the transition from laminar to turbulent flow occurs from $10^8 < Gr < 10^9$ beyond which the flow is fully turbulent, specifically for natural convection involving vertical flat plates. The multiplication of the Grashof number by the PRANDTL NUMBER yields the RAYLEIGH NUMBER. For mass transfer there is an equivalent Grashof number describing the natural convection, the mass transfer Grashof number (Gr_m), *see* **GRASHOF NUMBER, MASS TRANSFER**.

Grashof number, mass transfer ($Gr_m = -(1/\rho)(\partial\rho/\partial[a])_{T,P}$ $(g([a]_s - [a]_a L^3/\eta_{kin}^2))$

[fluid dynamics, thermodynamics] Dimensionless number describing the quotient of the mass transfer over the viscous force, where $[a]$ is the concentration of constituent "a," $[a]_s$ the concentration of "a" at the surface, $[a]_a$ the ambient environment concentration of species "a," g the GRAVITATIONAL ACCELERATION, ρ density, L the characteristic length of the object, η_{kin} is the KINEMATIC VISCOSITY, and $_T$ and $_P$ represent that the derivative is made under constant temperature and constant pressure, respectively.

Grating

[atomic, general, nuclear] *See* DIFFRACTION GRATING.

Grating, flow of a liquid through a

[fluid dynamics] Viscous FLOW through a mechanical GRATING or LATTICE (e.g., square array or lattice of cylinders) that produces a specific flow pattern of wave INTERFERENCE. Viscous flow of a LIQUID perpendicular to a line (or "grating") of identical cylinders that are evenly spaced can provide a scenario that can be described by "lubrication theory" when the spacing between the cylinders is significantly smaller than the CROSS SECTION of the respective cylinders. The pressure drop and the associated force of the respective cylinders can be derived for various configurations.

Grating, reflection and transmission of sound waves by a

[fluid dynamics] In ACOUSTICS just as much as in OPTICS a "RESONANCE" pattern can be defined based on the width (b_{gap}) and spacing (d_{gap}) between openings in a plate where SOUND with WAVE number $k = 2\pi/\lambda$, with wavelength λ for a system in one, two, or three dimensions can be defined to have maximum INTERFERENCE at spacings λ_{space} to be $\lambda_{space} = \{(4s_i^2\pi^2/(b_{gap} + d_{gap})) - k^2\}^{1/2}$, where s_i is different for the three dimensional configurations; in one dimension $s_1 = \tau/2v_s\{(t - (x/v_s))^2 + \tau^2\}^{-1}$ with location x; in two dimensions: $s_2 = \sqrt{([v_s r/r]\{\sin((\pi/4) - (3/2)\xi)\cos^{3/2}\xi\})(4\sqrt{2v_s^2 r^2})^{-1}}$ where ξ follows indirectly from $t = (r/v_s) + r\tan\xi$; and in three dimensions: $s_3 = (\tau/2\pi v_s^2)((r/v_s) - t)\left[2\{(t - (x/v_s))^2 + \tau^2\}^2\right]^{-1}$, location r, where τ_{period} is the time constant over which the phenomenon is observed, and v_s the speed of sound for the medium.

Grating equation

[general] The formation of INTERFERENCE lines as the result of light projected on a DIFFRACTION GRATING with line separation d_{line}, operating under illumination at wavelength λ, creating several orders of maxima, starting directly on the optical axis with $m = 0$ and subsequent maxima ($m = 0, \pm1, \pm2, \ldots$), where the ANGLE of the order of the maximum is θ_m defined as $d_{line} \sin\theta_m = m\lambda$. The first documented use of a diffraction GRATING was by JOSEPH VON FRAUNHOFER (1787–1826) in 1823. A diffraction grit with n slits, separated by DISTANCE d and width b (diffraction constant) produces a radiant intensity profile (i.e., radiance measured electronically I) as a function of the angle with the normal (θ) with the diffraction plate expressed as $I = I_0 (\sin n\alpha'/\sin\alpha')(\sin\beta'/\beta')$, where I_0 is the incident MAGNITUDE of radiance, $\alpha' = (\pi d_{line} \sin\theta)/\lambda$, and $\beta' = (\pi b \sin\theta)/\lambda$. For nonnormal incidence (at angle θ_i) the respective ($m = 0, 1, 2, \ldots$) principal maxima will fall in the following directions: $m\lambda = d_{line}(\sin\theta_i + \sin\theta_m)$.

Grating space of crystals

[atomic, electromagnetism, nuclear] Crystal structure diffraction based on REFLECTION, in comparison to optical transmission diffraction from a plate with lines (DIFFRACTION GRATING). In X-RAY diffraction, the GRATING used is a crystal with a highly regular crystal spacing, identified by the DISTANCE between crystal planes (d_{plane}) yielding the angular distribution of the respective maxima as $2d_{plane} \sin\theta_{Bragg} = m\lambda$, where θ_{Bragg} is the Bragg ANGLE (i.e., angle of incidence) for the crystal, X-ray wavelength λ, and order of maximum $m = 1, 2, \ldots$. The distribution of the maxima is known as Bragg's law. This mechanism is used to analyze the structure of crystals.

Grätz, Leo (1825–1941; "Graetz")

[fluid dynamics, thermodynamics] Physicist and mathematician from the Prussian Empire (now Germany) involved in electromagnetic WAVE propagation, contemporary of HEINRICH HERTZ (1847–1894) and

WILLIAM RÖNTGEN (1845–1923). In addition to electromagnetic dispersal his work included FRICTION and ELASTICITY as well as heat conduction (see Figure G.68).

Figure G.68 Leo Grätz (1825–1941) painted by Franz von Stuck (1906).

Grätz number ($Gz = \dot{m}c_p/kL = (L/d)(\kappa/(d \times v\rho c_p)) = (L/d)(\alpha_{th}/(d \times v))$)

[biomedical, fluid dynamics, geophysics, thermodynamics] Dimensionless number representing the ratio of thermal capacity with respect to convective HEAT TRANSFER, where c_p is the heat capacity, $\dot{m}$ the mass flow rate of the FLUID, κ the THERMAL CONDUCTIVITY, L the characteristic length of the object, d the FLOW diameter, ρ density, v velocity, and α_{th} the thermal DIFFUSIVITY. The Grätz number applies in particular to heat transfer under general flow conditions and in LAMINAR FLOW the convection phenomenon in particular. In biomedical applications, there are several configurations of the flow diagram involving many odd and contorted bi- and trifurcation splits reducing or increasing in vessel diameter. In addition, the pulsatile character will generate flow patterns that will have opposing direction from WALL to center of flow. In the CAPILLARY beds (arterioles, capillaries, and venules), the Grätz number is generally less than 0.4, whereas in the arterial and venous system Gz > 10. With these Grätz numbers as a reference the capillary bed will have an intense heat exchange, while in the tissue surrounding the veins and arteries negligible heat exchange will take place, with congruent insignificant drop in BLOOD temperature over the course of the flow. In GEOPHYSICS applications, the Grätz number can be used, for instance, to determine the extent of lava flow, where a high Grätz number (e.g., Gz > 230) pertains to a fast, wide, short, and thick flow of lava which is seemingly unstoppable with little intention to cool down due to heat exchange with its environment, whereas slow, long, and thin lava flow with low Grätz number will be expected to stop and solidify.

Gravimeter

[general, geophysics] Instrument used to measure the gravitational force as a function of location. The gravitational attraction is a function of the local density and material structure of the EARTH to extensive depths. The gravimeter can provide indications of mineral formations and fossil fuel pockets due to their specific material properties and volumetric presence.

Gravitation

[general, geophysics] Physical attraction force (F_G) between two object based on their respective masses; large mass (M) versus small mass (m), separated at DISTANCE from respective centers of mass (r): $F_G = G(mM/r^2)$, where G is the gravitational constant. It is also known as Newton's LAW OF UNIVERSAL GRAVITATION (see Figure G.69).

Figure G.69 Influence of gravity.

Gravitational acceleration

[general, mechanics, thermodynamics] The continuous increase in velocity of an object moving under the influence of attraction from another object with mass, where GRAVITY is the only force involved. Phenomenon first described by Strato of Lampsacus (c. 335–289 BC) from his observations of streaming rainwater running down a roof, starting out in LAMINAR FLOW and gradually developing an erratic flow pattern, TURBULENCE. The gravitational acceleration is a function of LATITUDE and altitude. The acceleration ($g = d^2r/dt^2$) of any object (irrespective of mass or shape and size) under the influence of gravitational force as a function of the mass of the object (M) exerting the gravitational pull at a DISTANCE from the CENTER OF GRAVITY (r) multiplied by the gravitational constant expressed as: $d^2r/dt^2 = G(M/r^2)$. In reality, the true acceleration (a) will be subject to other external forces, such as FLUID DRAG ($\sum F = ma$) (*also see* GRAVITY).

Gravitational constant (G)

[general] In Newtonian PHYSICS, the acceleration ($a = d^2r/dt^2$) of an object under the influence of GRAVITY is a function of the mass of the object (M) exerting the gravitational pull at a DISTANCE from the CENTER OF GRAVITY (r) multiplied by the gravitational constant expressed as: $d^2r/dt^2 = G(M/r^2)$.

Gravitational energy

[general] *See* GRAVITATIONAL POTENTIAL ENERGY.

Gravitational field

[general] Theoretical field of influence exerted by an object with mass. The definition assumes the exchange of massless particles across a DISTANCE that exchange the gravitational force in similarity with electroforce resulting from free and "bound" charges (*see* **COULOMB LAW**). The gravitational attraction force has a

MAGNITUDE and direction that can be measured for instance with the GRAVIMETER. The units for the magnitude of the gravitational field are 10^{-2} m/s^2. For the NORTH American landmass, the gravitational field is weakest in the peaks of the Rocky Mountains and strongest in the Atlantic basin, specifically off the continental shelf (*see* G) (see Figure G.70).

General relativity (gravity)

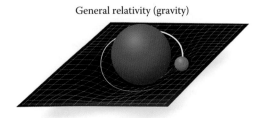

Figure G.70 Gravity in the two-dimensional fabric of space, where space acts as a sheet on which celestial objects are spread out and make an indentation representative of their gravitational attraction.

Gravitational potential energy (PEG: U)

[general] ENERGY of a MASS (*m*) resulting from the gravitational attraction due to the DISTANCE (*h*) from the center of mass of the "attracting" object with exerted GRAVITATIONAL ACCELERATION (*g*): $U = mgh$.

Gravitational waves

[astrophysics, general, geophysics, mechanics, relativistic] Propagation of gravitational disturbance from a point of origin. This phenomenon is predicted under EINSTEIN's THEORY OF GENERAL RELATIVITY. Examples are the GRAVITATIONAL FIELD of a double STAR that rotates and generates a stable OSCILLATION, whereas an exploding star (e.g., SUPERNOVA) will have an instantaneous and nonreproducible fluctuation in the gravitational attraction influence span (*also see* TIME PERTURBATION). Ripples in the space–time CONTINUUM. GRAVITY, according to Einstein's GENERAL THEORY OF RELATIVITY, is how mass deforms the fabric of space. In the proximity of any body with mass, the shape of space turns into curved array. There is still no conclusive statement about the extent and "mechanism of action" with regard to the extended reach (in DISTANCE to the mass) of GRAVITATION at this time (*also see* STRING THEORY). Apparently, this spatial curving does not always remain only in the proximity of the massive body. Einstein hypothesized that this deformation can propagate throughout the GALAXY, similar to the propagation of SEISMIC waves in the EARTH'S CRUST; captured partially by the EINSTEIN EQUATIONS. Gravitational waves will travel through the "empty" UNIVERSE, unlike seismic waves, propagating at the speed of light. Next to the gravitational waves that are linked to the theoretical formation model for the creation of the universe described by "inflation" process following the "Big Bang," additional gravitational waves are observed in association with black holes. The rotation of the incursion disk surrounding a BLACK HOLE, and the potential PRECESSION of the helical MAGNETIC FIELD emerging from the core of the incursion disk in perpendicular direction are theoretically showing the potential to generate gravitational WAVE twists, which propagate similar to the movement of giant whip. The helical magnetic field will interact with charged particles (i.e., ions) to become coiled into a helix-shaped pattern. The helix of charged particles is ejected in the direction perpendicular to the incursion disk at speeds approaching the speed of light. Note that the MOTION of the charged particles also generates an electric field. The electric and magnetic field from both the black hole and the ION JET generate an additional force on the moving particles, influencing the trajectory, bending away from the central axis, forming a whip-action pattern. The trajectory of the PARTICLE jet can change direction, even change back-and-forth. This changing motion generally takes place in a period scale ranging from thousands to millions of years, but can be detected with a RESOLUTION of months or weeks. The gravitational waves are in the direction perpendicular to the incursion disk. These gravitational waves are also known as Aflvén waves. The Aflvén waves appear to migrate at speeds just slower than the speed of light. Gravitational waves were predicted by ALBERT EINSTEIN (1878–1955) in 1916.

Gravitational waves are verified based on observations made under the Background Imaging of Cosmic Extragalactic POLARIZATION experiment, such as performed in locations as the Advanced Laser Interferometer Gravitational-Wave Observatory (LIGO, founded 1992), Livingston, Louisiana and more pronounced, at the South Pole. The gravitational waves are derived from observation made by radiotelescopes, detecting a broad spectrum of ELECTROMAGNETIC RADIATION and associated DOPPLER shifts in specific spectral signatures resolved by interferometric techniques. Inflation theory, which was first proposed in 1979, states that the universe, originally chaotic quantum NOISE made of unstable particles and space–time TURBULENCE, expanded at an unimaginable rate, creating these gravitational waves, smoothing out the chaos, and leaving the orderly cosmos we see today. Additional gravitation waves can be linked to the collision of black holes, and the potential FUSION of the respective black holes into one. The periodic changing with respect to the gravitational WAVE concept are linked to the scaling of the systems involved, and can range from microseconds to millions of years. In theoretical analyses, the gravitational wave concept is subdivided into four ranges: (1) high-frequency regime (microsecond scale) potentially associated with COLLIDING BLACK HOLES, creating ripples in time; (2) the low-frequency regime, has not yet been confirmed, not associated with physical phenomena and is derived from PHASE shifts detected in the MICROWAVE RADIATION from galactic background emissions.; (3) the very-low frequency regime (period in the order of months and several years) is measured and detected based on PULSAR stars and black hole mechanism of action principles; (4) the ultra-low frequency regime (millions of years) is associated with the ENERGY release of a theoretical phenomenon, such as described by, for instance, the theoretical "Big Bang" concept. The effect of the "Big Bang" with respect to gravitational waves are captured by a concept proposed in 1979, "inflation theory." Inflation theory states that the universe expanded from a chaotic QUANTUM noise, consisting of unstable particles and space–time turbulence at an unimaginable rate, generating gravitational waves in the process. This process could be thought of as resembling a MACH WAVE in the ACOUSTICS domain. This chaotic energy balance subsequently was prone to a damped OSCILLATION process, rendering an orderly cosmos after extinction of the wave (with presumed oscillatory period of millions of years). The ultra-low frequency regime so far has not been confirmed since it requires several decades and centuries worth of data collection. The perturbation process can be described in various ways, based on the assumption made regarding the boundary conditions. The four prevailing mechanisms are, according to their presumed relevance, (1) Isaacson shortwave approximation; (2) small perturbation in flat space; (3) linearization to the Isaacson shortwave approximation with respect to the Einstein equations; (4) gauge condition, referred to as Donder gauge or Lorentz gauge, also found as "harmonic gauge" and Hilbert gauge. A theoretical SOLUTION for Einstein's equations with respect to a nonrotating black hole was derived under severely limiting boundary conditions by KARL SCHWARZSCHILD (1873–1916) in 1915, however generally a computer is required to perform the required NUMERICAL analyses (see Figure G.71).

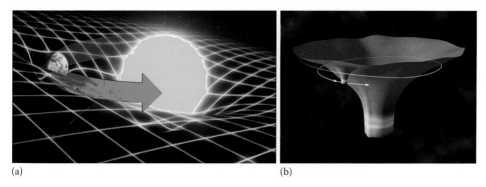

(a) (b)

Figure G.71 (a) Graphical representation of the curvature of space in the gravitational model. (b) An artist's impression of the space–time "geometry" in the event of an extreme mass ratio in spiral, representing a smaller black hole orbiting around a supermassive black hole. (Courtesy of NASA.) *(Continued)*

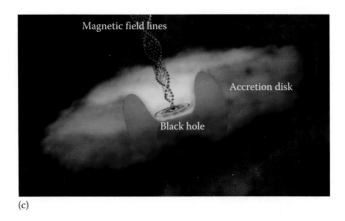

(c)

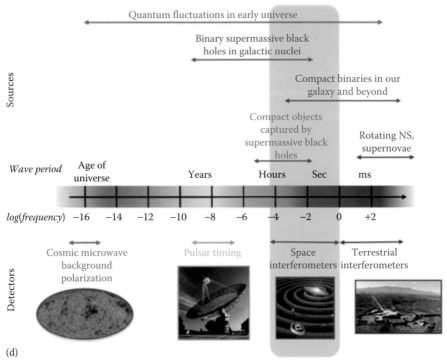

(d)

Figure G.71 (Continued) (c) Artist's impression of RGG 118 in which a black hole is surrounded by an accretion disk consisting of hot gas, with a torus consisting of dust and cooler gas. The light blue ring on the rear side of the torus is the predicted fluorescence resulting from iron atoms, which are excited by X-rays emitted from the hot gas torus. (Courtesy of NASA/CXC/M.Weiss.) Graphical representation of the accretion disk surrounding a black hole and perpendicularly emerging "magnetic field." (d) The gravitational wave spectrum, sources, and detectors. (Courtesy of NASA.) *(Continued)*

(e)

Figure G.71 (Continued) (e) LISA Pathfinder, launched 2015. Laser Interferometer Gravitational wave. BICEP2 (Background Imaging of Cosmic Extragalactic Polarization). (Courtesy of European Space Agency [ESA]/Astrium/ IABG, Paris, France.)

Graviton

[computational, energy, general, nuclear, quantum, relativistic, thermodynamics] Virtual PARTICLE that has been theoretically deduced in QUANTUM FIELD THEORY that represent the QUANTUM of the GRAVITATIONAL FIELD. The graviton mediates the GRAVITATIONAL FORCE, as part of the ELEMENTARY PARTICLES and FOUR BASIC FORCES. The graviton is a hypothetical MASSLESS, UNCHARGED particle that travels with the SPEED OF LIGHT and has presumably infinite range. Based on ALBERT EINSTEIN's (1879–1955) SPECIAL THEORY OF RELATIVITY, masses trapped in a GRAVITATIONAL FIELD that are accelerating should radiate GRAVITATIONAL WAVES, similar to accelerated charges emitting ELECTROMAGNETIC RADIATION. The graviton has spin 2 expressed in Plank's constant for SPIN: h, as $\hbar = h/2\pi$. In this configuration the particle obeys the BOSE–EINSTEIN STATISTICS.

Gravity

[geophysics, mechanics] (syn.:) {use: planetary objects ORBITS, PLANET, comet} The earth's terrestrial gravitational pull is influenced by the fact that the EARTH is rotating and as such expanding at the equator and compressed between the two POLES. The attraction between objects was described by ISAAC NEWTON in 1685. In addition, the first documented observation of gravitational attraction between ordinary objects was made in 1798 by HENRY CAVENDISH (1731–1810). The ABSOLUTE GRAVITATIONAL ACCELERATION g_o at any location on the planet, and for that MATTER of any object, is a linear function of the mass of the object M and inversely proportional to the local radial DISTANCE from the center point of the core to a point of relevance outside the mass (r) squared: $g_o = G(M/r^2)$, where $G = 6.67 \times 10^{-11}$ Nm^2/kg^2 is the universal gravitational constant. Compensating for the earth's rotational velocity at the equator (v_E) and distance to the core (R_E) gives the corrected weight as a function of location with respect to the equator as: $W_{corr} = m\left(g - (v_E^2/R_E)\right)$. *See* G (*also see* GRAVIATION).

Gravity, acceleration due to (g)

[general] Gravitational acceleration at various latitudes at sea level.

LATITUDE	GRAVITATIONAL ACCELERATION: G (M/S^2)
0°	9.7804
15°	9.7838
30°	9.7933
45°	9.8062
60°	9.8192
75°	9.8287
90°	9.8322

Gravity, Kepler's laws

[general] *See* KEPLER'S LAW.

Gray, Stephen (1666–1736)

[electronics, general] Scientist and astronomer from the United Kingdom/Great Britain who discovered the difference between conductors and nonconductors with the newly discovered availability of electric current in 1729 (*also see* RESISTANCE) (see Figure G.72).

Figure G.72 Stephen Gray (1666–1736).

Green, George (1793–1841)

[computational] Mathematician from Great Britain. Green is known for his methods of solving differential equations with complex number basis. Green also introduced the concept of POTENTIAL FUNCTION.

The theoretical work by George Green also included a mathematical description of ELECTRICITY and MAGNETISM, influencing the work of JAMES CLERK MAXWELL (1831–1879) and CARL FRIEDRICH GAUSS (1777–1855) (see Figure G.73).

Figure G.73 George Green (1793–1841).

Green's function

[acoustics, computational] Mathematical description of the field established from a POINT SOURCE. Note that the field can be electric or magnetic, but also VELOCITY POTENTIAL. SOLUTION method for the determination of electronic ENERGY levels (E) and WAVE FUNCTIONS (Ψ) of a system with respect to the HAMILTONIAN (H) energy expression: $H\Psi = E\Psi$, where the Hamiltonian can often be periodic. The Green's function $G(x,t|x_i,\tau)$ is a family of solutions to linear differential equations. One specific example of a Green's function is the solution to one-dimensional thermal DIFFUSION expressed by the heat equation as: $\alpha(\partial^2 T/\partial x^2) + (\alpha_T/k_T)g(x,t) = \partial T/\partial t$, where $T(x,t_0) = U(x)$, with $t_0 = 0$, and $k_i(\partial T/\partial n_i)_{x_i} + h_i T(x_i,t) = f_i(t)$; $i = 1,2$. The solution yields:

$$T(x,t) = \int_{x'} G(x,t|x',0)U(x')dx' + \frac{\alpha}{k}\int_{\tau=0}^{t}\int_{x'} g(x',\tau)G(x,t|x',\tau)dx'd\tau$$

$$+ \alpha\int_{\tau=0}^{t}\sum_{i=1}^{2}\left\{\frac{u_i(\tau)}{k_i}G(x,t|x_i,\tau)\right\}d\tau - \alpha\int_{\tau=0}^{t}\sum_{i=1}^{2}\left\{u_i(\tau)\frac{\partial G(x,t|x_i,\tau)}{\partial n_i}\bigg|_{x'=x_i}\right\}d\tau,$$

where the first term represents the initial conditions and the second term provides the generation of ENERGY. Computational technique applied to an inhomogeneous differential equation describing the approach to solve for an IMPULSE RESPONSE: $f(t) = \delta(t - t')$ (i.e., delta function). The impulse will be limited in time and place within a spatial domain and well-defined initial conditions. The Green function solution for a point source in $x_i = \xi$ is given by $G(x,t|\xi,\tau) = -\left[a^2/(4\pi(t-\tau))\right]^{3/2}(1/a^2)\exp\left(-(a^2|\vec{r}-\vec{\rho}|/(t-\tau))\right)$ (see Figure G.74).

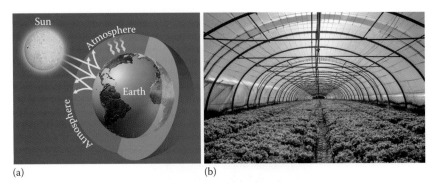

(a)　　　　　(b)

Figure G.74 (a) The atmospheric conditions providing a shell of gasses that reflect the infrared emission from the Earth as a result from the exposure to sunlight, thus retaining a larger quantity of heat close to the earth's surface in comparison to the absence of the reflective gaseous shell and (b) glass greenhouse, confining the heat from the Sun to promote plant growth in geographic locations that may not be ideally suited for the type of plant.

Green's law

[fluid dynamics] Definition used in the description of the WAVE MOTION in a canal with variable outline. Using the EQUATION OF CONTINUITY with respect to the CROSS SECTION of the tidal wave motion ($\mathbb{S}$), this provides for the ordinate of the upward displacement (ψ) of the free surface of the LIQUID medium as a function of location in the direction of FLOW (x): $\psi = -(\mathbb{S}/b)(\partial\xi/\partial x)$, where $\xi = \int u\,dt$, the time-dependent integral of the surface displacement within the location range past point x with u the average velocity and b the local width of the canal. The equation of continuity for a variable CROSS SECTION is now defined as $\psi = -(1/b)(\partial(\xi\mathbb{S})/\partial x)$. Using the ENERGY content of the wave RIPPLE in a riverbed, the AMPLITUDE can be shown to adhere to Green's law: $\psi \propto -(1/\sqrt{b})(1/\sqrt{b})$ (see Figure G.75).

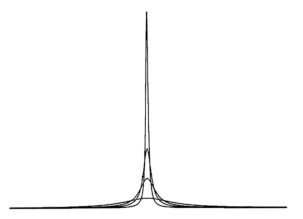

Figure G.75 Greens function limit.

Green's theorem

[fluid dynamics] Calculus theorem relating the integration of a permutation of derivatives (e.g., complex-valued functions from G and U, with both the first and second partial derivatives continuous and single-valued) within a closed surface S (enclosing a volume V) to an integration over a volume. This is formulated with respect to the partial derivative to the normal ($\vec{n}$) in the outward direction ($\partial/\partial n$), expressed as

$\iiint_V \left(U\nabla^2 G - G\nabla^2 U \right) dV = -\iint_S \left(U(\partial G/\partial n) - G(\partial U/\partial n) \right) dS$, where $\nabla^2 = (d^2/dx^2) + (d^2/dy^2) + (d^2/dz^2)$ is the LaPlace operator. This theorem applies, for instance, to diffraction theory, irrotational MOTION, and ELECTROSTATICS.

Greenhouse effect

[geophysics, meteorology, thermodynamics] Gradual global increase in average temperature attributed to the increase in gasses such as CO, CO_2, CH_4, N_2O, and CFC's the latter is used as a propellant in AIR canister-based delivery of cosmetic and industrial spraying aerosols as well as consumer paints. The changes in atmospheric composition are increasing the transmission for broadband ULTRAVIOLET through visible RADIATION, inherently heating the EARTH's CRUST and water surface. One notable difference is the depletion of the ozone layer, in the lower portion of the STRATOSPHERE, or between the TROPOSPHERE $(0-10 \text{ km})$ and the main stratosphere $(10-50 \text{ km})$. Additionally, the formation of specific air mass compositions apparently acts as a "blanket," insulating the natural cooling process of black-body radiation, thus encapsulating the black-body radiation to remain close to the earth's surface (see Figure G.76).

Figure G.76 Surface wave phenomenon.

Greenwich Meridian

[astronomy, general, geophysics] Starting point for the longitude demarcation running through the city of Greenwich in Great Britain: 0°. It is also referred to as the Prime Meridian (see Figure G.77).

Figure G.77 Demarcation of the Greenwich Meridian in London near the Greenwich Observatory.

Gregorian telescope

[astronomy/astrophysics, general, optics] REFLECTION telescope concept invented by JAMES GREGORY (1638–1675) in 1661. Gregory was not satisfied with the design, based on the theoretical PROBABILITY of aberrations in shape and COLOR reproduction, and did not finish constructing his invention.

Gregory, James (1638–1675)

[astronomy/astrophysics, general, optics] Scientist and mathematician from Great Britain (Scotland) who invented the REFLECTION telescope in 1661 (i.e., GREGORIAN TELESCOPE). The physical construction of the reflection TELESCOPE was performed after several iterations for aberrations by SIR ISAAC NEWTON (1642–1727) in 1668 (see Figure G.78).

Figure G.78 James Gregory (1638–1675).

Grimaldi, Francesco Maria (1618–1663)

[optics] Jesuit priest from Italy who described the concept of edge diffraction. Grimaldi documented the formation of an IMAGE of a pinhole on a distant WALL to be larger than the pinhole with undefined edge.

He attributed this phenomenon to be akin to the Christiaan Huygens (1629–1695) water-wave phenomenon, in this case for light (see Figure G.79).

Figure G.79 Francesco Grimaldi (1618–1663).

GRIN

[optics] *See* **GRADED INDEX OF REFRACTION LENS.**

Ground

[energy, general] Volume of conduction material with virtually unlimited amount of ELECTRIC CHARGE that can be added to or retrieved from without a significant change in the total net electric charge. As a result of the preservation of "equilibrium," the ENERGY requirements for the successive electron exchanges will be identical and minimal. Specifically, since more than one component will be connected to this "ground" the depletion and replenishing of free electrons will be interchanging from location to location, providing an average state of equilibrium (see Figure G.80).

Figure G.80 "Ground" in electronic conditions.

Ground state

[atomic, general, solid-state, thermodynamics] Stable ENERGY configuration for atomic or molecular configuration with respect to the lowest level of energy that does not allow for perturbations. Any energy levels above ground-state will involve energy exchange or the increasing PROBABILITY to liberate free electrons. During interaction of ELECTROMAGNETIC RADIATION with MATTER, the process will be a resonant process for the electrons, in contrast to excitation which is a metastable condition that will eventually revert back to the ground state by single or multiphoton emission. The ground state satisfies the minimum requirements for the Bohr model with respect to the electron orbit filling. Ground state ELEMENTS are also nonradioactive. The ground state is usually attributed with entropy to equal zero: $S_g = 0$, which entails that different constituents all in equilibrium are at zero entropy, each with the lowest possible energy.

Ground state mass

G

[thermodynamics] The mass (m_g^{free}) of a free constituent at the lowest ENERGY (E_g^{free}), defined by $E_g^{\text{free}} = m_g^{\text{free}} c^2$, where $c = 2.99792458 \times 10^8$ m/s is the speed of light.

Ground state of hydrogen

[atomic, thermodynamics] Hydrogen (H_2) is often referred to as the standard in ENERGY configuration due to it relatively simple structure as the smallest ELEMENT.

Ground wave

[general] Electromagnetic WAVE that travels in the surface layer of the EARTH or close to the surface in the ATMOSPHERE, primarily for AM RADIO wavelengths. These waves follow the curvature of the globe (*also see* SEISMIC WAVE).

Ground-energy state

[thermodynamics] *See* GROUND STATE, ENERGY.

Ground-energy state, temperature of a

[thermodynamics] Based on the THIRD LAW OF THERMODYNAMICS the temperature of all GROUND states of a system will equal zero.

Ground-energy value, degeneracy of

[thermodynamics] The ENTROPY (S) of any GROUND STATE is defined as $S_g(n,\beta) = k_b \ln(D_g(n,\beta))$, where $D_g(n,\beta)$ is the DEGENERACY of the ground state, for a system with n components and β the parameters describing the conditions of the constituents and $k_b = 1.3806488 \times 10^{-23}$ m^{2}kg/s^{2}K the Boltzmann constant. For a nondegenerate state $D_g(n,\beta) = 1$, while when $D_g(n,\beta) \neq 1$ the system is degenerate.

Ground-state configurations

[atomic, general] Electronic configuration for the GROUND STATE of an ELEMENT. Examples:

ELEMENT	SYMBOL	ATOMIC NUMBER (Z)	ELECTRONIC CONFIGURATION
Hydrogen	H	1	$1s$
Helium	He	2	$1s^2$
Lithium	Li	3	$1s^2 2s$
Beryllium	Be	4	$1s^2 2s^2$
Boron	Be	5	$1s^2 2s^2 2p$
Carbon	C	6	$1s^2 2s^2 2p^2$
Nitrogen	N	7	$1s^2 2s^2 2p^3$
Oxygen	O	8	$1s^2 2s^2 2p^4$
Fluorine	F	9	$1s^2 2s^2 2p^5$
Neon	Ne	10	$1s^2 2s^2 2p^6$
Sodium	Na	11	$1s^2 2s^2 2p^6 3s$
Magnesium	Mg	12	$1s^2 2s^2 2p^6 3s^2$

Group velocity of waves

[atomic, general, nuclear] Propagation speed for the AMPLITUDE of a modulated waveform constructed from a range of wavelengths defined as $v_g = d\omega/dk = v - \lambda(dv_p(\lambda)/d\lambda)$, where $\omega = 2\pi v$ is the ANGULAR VELOCITY, for the respective wave FREQUENCY (v), $k = \lambda/2\pi$ the WAVE number for the respective WAVELENGTH (λ), $v_p = \omega/k$ is the wavelength-dependent PHASE VELOCITY, respectively, across the bandwidth of waves, and v the propagation velocity linked to the wave frequency as. $v\lambda = v$; in this case, the propagation speed for the average wavelength of the group of wavelengths traveling through a dispersive medium. When the phase velocity is identical for all waves the group velocity equals the phase velocity. This applies to both mechanical and electromagnetic waves. An example of a compound wave is the electrocardiogram, which can be approximated by a FOURIER SERIES truncated at eight harmonics, with minimal error. The constituent waves are combined through the superposition principle (see Figure G.81).

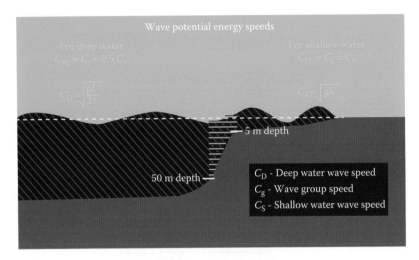

Figure G.81 Group velocity for water waves, the phase velocity of the higher frequencies becomes faster for shallow water, generating the conditions for the "surf" where the higher frequency waves "walk-away" from the base frequency.

Grüneisen coefficient (γ_{Gr})

[thermodynamics] Parameter used to indicate the conversion efficiency of THERMAL ENERGY into mechanical stress expressed as $\gamma_{Gr} = -(V/\Theta_{Debye})(d\Theta_{Debye}/dV)$, where V is the volume of the medium under stress and Θ_{Debye} represents the Debye temperature. This accounts for LATTICE vibrations.

Grüneisen equation

[thermodynamics] Expression for thermal volumetric EXPANSION for a crystalline structure, where generally the volume expansion is the superposition of the linear expansion (α_x) in three dimensions: $\beta_{ex} = (1/V)(\partial V/\partial t)_p = \alpha_{ex,1} + \alpha_{ex,2} + \alpha_{ex,3} = a + bT + cT^2$ as a function of TEMPERATURE (T) with empirical coefficients (a, b, c), which can be converted using the Grüneisen equation to read as $\beta_{ex} = \gamma_{Gr}(c_p/VB_s) = (1/VB_s)(\gamma_{Gr,electron}\alpha_{electron}T + \gamma_{Gr,lattice}\alpha_{lattice}T^3)$, where c_p is the heat capacity at constant PRESSURE (P), V the initial volume, B_s the adiabatic bulk modulus, $\gamma_{Gr,electron}$ and $\gamma_{Gr,lattice}$ are the Grüneisen coefficients for electron and LATTICE contribution, respectively, which accounts for the atomic vibrations as a function of temperature.

Guericke, Otto von (1602–1682)

[astronomy, general] *See* VON GUERICKE, OTTO (1602–1682).

Guggenheim, Edward Armand (1901–1970)

[thermodynamics] Physicist from Great Britain, seen as the founder of chemical THERMODYNAMICS. The work of Guggenheim provides the definition of PHASE transitions and ENERGY balances involved in the process of PHASE TRANSITION. His work was largely inspired by JOSIAH WILLARD GIBBS (1839–1903) (see Figure G.82).

Figure G.82 Edward Armand Guggenheim (1901–1970).

Guided wave

[acoustics, general, optics] WAVE bound by closely confined boundaries, where one boundary may be the bulk medium, but represented by a discontinuity or gradient in the density (mechanical or optical) or in elastic properties. Surface wave are in ACOUSTICS and OPTICS, also described as Rayleigh wave, EVANESCENT WAVE, or surface acoustic wave. Additionally, guided waves are part of FIBER-OPTIC transmission. One specific example in found under LAMB WAVE, describing an atmospheric wave phenomenon. Guided waves also define SEISMIC phenomena (see Figure G.83).

Figure G.83 Impression of guided wave on the surface of sand. (Courtesy of The Royal Society.)

Guitar

[acoustics, computational] Musical instrument using strings to generate a WAVE pattern that is amplified by a case or is detected by a magnetic ELEMENT that converts the local changes in MAGNETIC FIELD strength into electric signals based on the modulation FREQUENCY SPECTRUM of the strings.

Guldberg, Cato Maximilian (1836–1902)

[computational, thermodynamics] Chemist and mathematician from Norway. Guldberg is known for his definition of the temperature (T) of chemical reaction equilibrium as "the LAW OF MASS ACTION of Guldberg, Waage, van't Hoff and Horstmann" (*see* LAW OF MASS ACTION).

Gurney, Ronald Wilfred (1898–1953)

[computational, nuclear] Scientist from Great Britain. Gurney worked on the theoretical description of ionic crystals and discovered the concept of QUANTUM TUNNELING (with respect to ALPHA DECAY). The full quantitative description of the TUNNELING phenomenon was provided by GEORGIY ANTONOVICH GAMOW (1904–1968).

Gutenberg–Richter law

[computational, geophysics] Empirical correlation between the MAGNITUDE (M_{seis}) of earthquakes with respect to the total number of SEISMIC events (N_{seis}) in a specific region as well as over a certain time period, where the magnitudes are in the same range (equal to or greater than M_{seis}), expressed as $N_{seis} = 10^{a-bM_{seis}}$, where a and b are constants. The values of b depend on the following conditions: the depth of the WAVE, the locally applied stress, the mechanical mechanism that is the focus of the origin, the heterogeneity of the material, as well as the DISTANCE between the source and any of MACROSCOPIC failure (which creates a discontinuity, which may give rise to REFLECTION and diffraction of the wave). This relationship was defined by BENO GUTENBERG (1889–1960) and CHARLES FRANCIS RICHTER (1900–1985).

Gyro

[general] Generally known as a flywheel or GYROSCOPE, also found under gyrodyne.

Gyromagnetic ratio

[atomic, biomedical] The ratio of the atomic or molecular MAGNETIC DIPOLE moment to the angular momentum of the PARTICLE system. Specific examples are in MRI, more precisely with respect to the LARMOR PRECESSION and for a classical rotation body as well as for a NUCLEUS or isolated electron next to a special relativistic case. For the Larmor precession the gyromagnetic ratio indicates the link between the induced gyroscopic orbital action (with the resulting ELECTROMAGNETIC RADIATION) and the applied MAGNETIC FIELD. For Larmor precession the expression reads $\nu = (\gamma/2\pi)B$, where ν is the PRECESSION frequency, $\gamma_{Larmor} = -(eg_a/2m_e)$ the gyromagnetic ratio with g_L the electron orbital g-factor, e = the electron charge, m_e the mass of the mass of the precessing electron, and B the applied magnetic field strength. For a classical rotation of a body, the expression is as follows: $\gamma_{classic} = q/2m$, q the total charge of the rotating body with mass m; applied as $\mu = \gamma_{classic}L$, where $\mu = IA$, I the current represented by the moving charge and $A = \pi r^2$ the cross-sectional area of the MOTION with radius r and $L = mvr$ the angular momentum at velocity v. For a nucleus, the gyromagnetic ratio is $\gamma_N = eg_N/2m_p = g_N(\mu_N/\hbar)$, where g_N is the nuclear g-factor, m_p the PROTON mass, $\hbar = h/2\pi$ PLANCK'S CONSTANT $h = 6.626070040(81) \times 10^{-34}$ m^{2}kg/s (Js), and $\mu_N = e\hbar/2m_p$ the nuclear MAGNETON. The relativistic approach uses the DIRAC EQUATION, providing the gyromagnetic ratio of 2.

Gyroscope

[general] Instrument that is persistent in maintaining its spatial orientation due to an applied angular momentum, first constructed by JEAN BERNARD LÉON FOUCAULT (1819–1868). A gyroscope has a mechanism of high angular velocity rotation while suspended in two or more axis. The main axis of the gyroscope will precess at a fixed angular velocity gives as $\omega_{\text{precess}} = mgr/I\omega$, where m is the gyroscope mass, I the moment of INERTIA for the gyroscope, g the GRAVITATIONAL ACCELERATION, r the DISTANCE from the axis for the center of mass, ω the angular velocity of the spinning mass, and $L = I\omega$ the angular momentum of the rotor inside the gyroscope. A slight error may result from the fact that the vector describing the angular momentum does not exactly line up with the axis of symmetry. This misalignment may cause the gyroscope to wobble up and down or bob. Gyroscopes form a critical component of many guidance systems due to their inherent stability in orientation resilient to displacement, primarily due to the virtually frictionless operation (no external torque applied) (see Figure G.84).

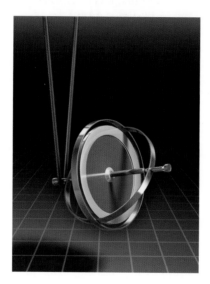

Figure G.84 Stability of a gyroscopic design.

Gyroscopic stability

[general] The angular momentum of the GYROSCOPE $L = I\omega$, where I is the moment of INERTIA of the rotating device and ω the angular velocity, makes any change in direction experience a force (i.e., torque) die to the reorientation of the angular momentum. As a result, a gyroscope will tend to remain in the same position when no other forces are acting, which is illustrated when a bicycle is pushed out while the driver jumps off and the bicycle will remain straight and erect until the angular velocity of the wheels is reduced below a critical value and any slight tilt is acted on by GRAVITY.

Gyrostat

[fluid dynamics] Rotating and moving solid within a FLUID environment, generally closed system. Examples can be found in the following: bicycles, flywheels and spherical dampers (SATELLITE; *Note:* has dual SPIN), helicopters (main rotor tied to tail rotor), toy gyroscopes, and unicycles (symmetric wheel) (see Figure G.85).

Figure G.85 A gyrostat, helicopter.

Index of Names

Note: Names are arranged in chronological order by date of birth. Page numbers followed by f refer to images.

Index of Subjects